RSC Drug Discovery Series

Titles in the Series:
1: Metabolism, Pharmacokinetics and Toxicity of Functional Groups
2: Emerging Drugs and Targets for Alzheimer's Disease; Volume 1
3: Emerging Drugs and Targets for Alzheimer's Disease; Volume 2
4: Accounts in Drug Discovery
5: New Frontiers in Chemical Biology
6: Animal Models for Neurodegenerative Disease
7: Neurodegeneration
8: G Protein-Coupled Receptors
9: Pharmaceutical Process Development
10: Extracellular and Intracellular Signaling
11: New Synthetic Technologies in Medicinal Chemistry
12: New Horizons in Predictive Toxicology
13: Drug Design Strategies: Quantitative Approaches
14: Neglected Diseases and Drug Discovery
15: Biomedical Imaging
16: Pharmaceutical Salts and Cocrystals
17: Polyamine Drug Discovery
18: Proteinases as Drug Targets
19: Kinase Drug Discovery
20: Drug Design Strategies: Computational Techniques and Applications
21: Designing Multi-Target Drugs
22: Nanostructured Biomaterials for Overcoming Biological Barriers
23: Physico-Chemical and Computational Approaches to Drug Discovery
24: Biomarkers for Traumatic Brain Injury
25: Drug Discovery from Natural Products
26: Anti-Inflammatory Drug Discovery
27: New Therapeutic Strategies for Type 2 Diabetes: Small Molecules
28: Drug Discovery for Psychiatric Disorders
29: Organic Chemistry of Drug Degradation
30: Computational Approaches to Nuclear Receptors
31: Traditional Chinese Medicine

32: Successful Strategies for the Discovery of Antiviral Drugs
33: Comprehensive Biomarker Discovery and Validation for Clinical Application
34: Emerging Drugs and Targets for Parkinson's Disease
35: Pain Therapeutics; Current and Future Treatment Paradigms
36: Biotherapeutics: Recent Developments using Chemical and Molecular Biology
37: Inhibitors of Molecular Chaperones as Therapeutic Agents
38: Orphan Drugs and Rare Diseases

How to obtain future titles on publication:
A standing order plan is available for this series. A standing order will bring delivery of each new volume immediately on publication.

For further information please contact:
Book Sales Department, Royal Society of Chemistry, Thomas Graham House, Science Park, Milton Road, Cambridge, CB4 0WF, UK
Telephone: +44 (0)1223 420066, Fax: +44 (0)1223 420247,
Email: booksales@rsc.org
Visit our website at www.rsc.org/books

Orphan Drugs and Rare Diseases

Edited by

David C Pryde
Pfizer Global Research and Development, Cambridge, UK
Email: David.Pryde@pfizer.com

Michael J Palmer
Medicinal Chemistry Consultant, UK
Email: mikepalmer@live.co.uk

RSC Drug Discovery Series No. 38

Print ISBN: 978-1-84973-806-4
PDF eISBN: 978-1-78262-420-2
ISSN: 2041-3203

A catalogue record for this book is available from the British Library

Published by The Royal Society of Chemistry,
Thomas Graham House, Science Park, Milton Road,
Cambridge CB4 0WF, UK

Registered Charity Number 207890

For further information see our web site at www.rsc.org

Foreword

In the decade building up to the adoption of the Orphan Medicinal Products Regulation[1] by the EU in 2000 there was almost no interest in developing interventions for small populations of patients with rare conditions.

Rare diseases were generally considered as being unusual, generally incurable and not very important – the subject for academic curiosity for a small group of research active clinicians, but not a significant issue for healthcare planners or providers. 'Rare' equated to 'small numbers', which were not significant and probably too difficult to do much about.

Since the adoption of the Orphan Medicinal Products Regulation there has been a sea change in thinking about rare diseases. The Committee for Orphan Medicinal Products (COMP) and the European Medicines Agency has recommended (June 2013) the grant of 1123 orphan designations, and 66 [2] drugs for rare diseases have been given a marketing authorisation (MA). New therapies are being made available for hitherto intractable and incurable rare conditions at the rate of 8–10 new MA per year and there is now strong interest from academics, clinicians and the commercial sector in this issue. Developments in Europe followed a similar pathway to those in the USA, where the adoption of the Orphan Drugs Act in 1983 led to a sustained interest in developing therapies for rare conditions in what had previously been a desert for investment opportunities, with little publically funded investment in basic research, and even less private sector money going into creating commercially viable drugs from promising research.

Why has this happened? What has changed that has brought rare diseases from the margins of medicine to centre stage? It's tempting to see this as a planned development, underpinned by logic and a rational decision-making process, but the truth is that serendipity has played a significant role

RSC Drug Discovery Series No. 38
Orphan Drugs and Rare Diseases
Edited by David C Pryde and Michael J Palmer

Published by the Royal Society of Chemistry, www.rsc.org

in creating the opportunity to make serious progress in the care and support available for patients and families with rare disease.

Crucial to successfully placing rare diseases on the health and policy agenda were the advances in understanding the molecular basis of disease made possible by the Human Genome Project. This created the framework for systematic biomedical research, and the opportunity to try to hypothesise systematically about logical interventions in basic genotypic mechanisms that resulted in the phenotypic manifestation of rare diseases.

Simultaneously there was an emerging confidence amongst patients and patient organisations that they had a right to have a say in the development of services and support for those affected. This resulted in a more assertive and unified approach to engage with other stakeholders, in politics, industry, academia and medicine, based on partnership rather than gratitude for crumbs that fall from the table that was the pursuit of blockbuster cures for common diseases.

Coincidentally, this blockbuster strategy adopted by the large pharmaceutical companies was faltering and pharma pipelines were emptying. Science lead development was taking industry away from a 'small molecules: big populations' model to a 'big molecules: small populations' approach, often with start-ups and small or medium sized companies (SMEs) providing the entrepreneurial leadership and the innovative thinking. Today, rare disease research and development is a key plank in the planning of most large pharmaceutical companies. This would have been unthinkable in the years leading up to the millennium. The success of a small number of pioneering biotechnology companies such as Genzyme and Amgen demonstrated that it was possible to create feasible alternative business models that would deliver health gain and a return on investment.

A fourth element in this mix was the granting of competence in the field of public health to the European Commission by the Maastricht Treaty in 1992. While the delivery of health care remains a jealously guarded national prerogative, rare diseases were seen as being one issue where there was a clear potential for demonstrating European Added Value and in so doing giving a purpose to DG Sanco (the Directorate for Health and Consumer Affairs) to involve itself in health policy without alienating Member States. This led to the creation of the Rare Disease Task Force in 2003, and ultimately to the publication of the Communication of the Commission[3] in 2008 and the subsequent unanimous adoption by the Council of Health Ministers of the EU of a recommendation on rare diseases in 2009.[4] This called on Member States to produce national plans or strategies for improving services and support for rare diseases by the end of 2013. These documents contained calls to action in respect of coding and classification, research, centres of expertise, the clinical added value of orphan drugs and patient empowerment. These are the key elements of any effective strategy, and the adoption of the Recommendations opened a window that allowed for the creation of concerted campaigns to respond to the unmet medical needs of rare disease patients across the whole of Europe.

The insights from the Human Genome Project and other international collaborations which aimed to address the fundamental biology of health and disease had an impact on the research community too. Traditional priority areas were adjusted to make space for rare diseases, both because they were seen as important in themselves, but also because of the insight they could provide into common complex diseases in the context of a growing awareness of the importance of personalised (or stratified) medicine and the development of targeted therapies for genotypically distinct but phenotypically similar common diseases. Ultimately, this resulted in the formation of a strategic partnership between DG Research in the European Commission, the National Institutes of Health in the USA and a growing number of research funding agencies such as the UK's National Institute for Health Research, known as the International Rare Disease Research Consortium (IRDiRC).[5] This has set itself the ambitious target of a diagnostic for every rare disease and two funded new orphan therapies by 2020.

Advances in biology, increased opportunities for treating rare diseases, a growing sense of empowerment and engagement by patients at patient advocacy groups and realisation amongst politicians and policy makers that 'rare' did not equate with uncommon, simply because of the large number of different rare diseases that have been identified, has resulted in rare diseases being recognised as a significant public health issue. Paradoxically this has coincided with a downward trend in the economies of much of the developed world, putting healthcare decision makers between a rock and a hard place. The opportunity to do more, coupled with increasing awareness of finite resources necessitated the creation of new systems for licensing novel therapies and determining policy with regard to clinical utility and reimbursement. This brings me to the final element in my thesis, namely the role of the regulations and Health Technology Assessment (HTA) bodies. The transitional gold standard of the large, multicentre randomised double-blind trial does not work for drugs developed for tiny populations and while patients have no interest in drugs that do not work, new methods for proving quality, safety and efficacy need to be developed if orphan drugs are to make it onto the market. Financial constraints are increasingly important, and the role of HTA bodies is rapidly growing in significance. Again traditional methods do not work, simply because its data is lacking in most instances to form a robust assessment of clinical effectiveness and a rational policy for determining patient access, pricing and reimbursement. The challenge facing many healthcare providers and payment agencies today is to develop systems which will provide a framework that carries the trust of patients and families, or which will enable fair decision making in healthcare and resource allocation now and sustainably into the future. Patients and other key stakeholders are actively engaging to try and bring this about, but the shape of a sustainable healthcare future for rare disease patients is not yet clear anywhere across the globe.

So, we have a potent mix of elements, all of which have come together to raise the profile of rare diseases and the patients of families affected by them.

Rising to the challenge is a critical priority for all involved if patients are to see their expectations for effective therapies realised, scientists to generate the insights that will create clinical service improvements for doctors and the possibility of a return on investment for industry. Regulators, policy makers and politicians are intimately involved as well if we are to see the promise realised and a sustainable, affordable healthcare system that incorporates current scientific understanding and good clinical practice intertwined in systems that provide timely, appropriate, user-friendly care for rare disease patients wherever they live across the globe.

The years since the adoption of the Orphan Drugs Act in the USA, and the Orphan Medicinal Product Regulations in the EU have seen rapid developments that have brought hope to patients and their families living with the daily consequences of unmet medical needs. This book is a timely contribution to the literature in this fast-changing field. It will serve as a useful primer to those new to the subject, and for those already engaged it provides a reminder of the breadth of the field today, reinforcing the importance of rare diseases as a subject for pioneering research, as a commercial opportunity for innovative pharma and biotech companies, but most of all as a call to action on behalf of the patients and families with rare diseases for whom there is as yet no effective therapy. This book demonstrates clearly that there is no disease that is too rare, too difficult or too expensive not to warrant the attention of the scientific and clinical community, industry, policy makers and politicians alike in the search for a response to the needs of patients, and the importance of sustainable progress towards the continued production of novel therapies for currently unmet needs.

Alastair Kent *Director, Genetic Alliance UK and Chair, Rare Disease UK*

1. Regulation (EC) no. 1141/2000.
2. http://www.ema.europa.eu/ema/, accessed 1 September 2013.
3. Rare Diseases: Europe's Challenges, *European Commission, Luxembourg,* 2008.
4. Council Recommendations on action in the field of rare diseases, *European Commission, Luxembourg,* 2009.
5. http://www.irdirc.org, accessed 1 September 2013.

Preface

As a term, 'rare diseases' covers an enormous and hugely diverse range of diseases, disorders and conditions. The 'rare' label can be deceptive in that many of the diseases bracketed within this class can affect a substantial number of people and when taken in totality, rare diseases affect a significant proportion of the world's population. The unmet medical need within this population is vast. In a similar way, the term 'orphan drug' is also subject to some confusion and misconception within the drug discovery community.

When we decided to undertake the editing of this book, we had a number of aims in mind. First and foremost, we wished to produce a broadly accessible book that would set out clearly what is meant by the terms rare diseases and orphan drugs. In so doing, we wanted to highlight the critical role that disease advocacy has played and continues to play in building drug discovery efforts in this area of biomedical science, and discuss some of the unique challenges that this field presents. Secondly, we wished to present the range of innovative science taking place to create therapies directed at rare diseases through a combination of review and case studies, highlighting the breadth of drug modalities that research in the field has produced. Research and clinical development in this area has often been both path-finding and innovative, and in many cases this has been pioneered by small biotechnology companies, or in some cases small parts of much larger companies. As such, undertaking to write a book chapter from within a small group is a significant commitment, and we are most grateful to the chapter authors for contributing their time to the writing of this book. Thirdly, we wanted to give the reader a sense of which rare diseases are currently being tackled by the drug discovery community; while to be anything like comprehensive would be impossible in a single volume, we have selected case studies from different disease classes and different drug modalities as exemplars of

RSC Drug Discovery Series No. 38
Orphan Drugs and Rare Diseases
Edited by David C Pryde and Michael J Palmer

Published by the Royal Society of Chemistry, www.rsc.org

successful drug discovery efforts. Finally, in what is an expanding and evolving area of drug discovery research, we wanted to provide some perspective on where the field may evolve to in the near future.

We found the planning and editing of this book hugely informative and enjoyable, and armed with the knowledge that this book provides, we hope the reader will also share our enthusiasm for this important area of drug discovery.

David C Pryde and Michael J Palmer

Contents

What are Rare Diseases and Orphan Drugs?

RSC Drug Discovery Series No. 38
Orphan Drugs and Rare Diseases
Edited by David C Pryde and Michael J Palmer

Published by the Royal Society of Chemistry, www.rsc.org

Rare Muscle Disorders

Rare Cancers

Respiratory Disorders

Blood Disorders

Outlook

WHAT ARE RARE DISEASES AND ORPHAN DRUGS?

CHAPTER 1

Definitions, History and Regulatory Framework for Rare Diseases and Orphan Drugs

DAVID C. PRYDE*[a] AND STEPHEN C. GROFT[b]

[a]Worldwide Medicinal Chemistry, Pfizer Neusentis, Portway Building, Granta Park, Great Abington, Cambridge CB21 6GS, UK; [b]Office of Rare Diseases Research, National Center for Advancing Translational Sciences, National Institutes of Health, Bethesda, Maryland 20892-4874, USA
*E-mail: David.Pryde@pfizer.com

1.1 Orphan Drugs

An orphan drug or orphan medicine is a formal regulatory term used to describe a drug product that has been granted orphan status by a regulatory agency. Orphan designation is reserved for medicines that will treat diseases with prevalence below the threshold set for rare diseases, and may have additional factors such as the lack of availability of alternative treatments. The word 'orphan', from the Greek word *orphanus* for a child that has lost a parent,[1] is taken from the ground-breaking legislation in the USA enshrined in the Orphan Drug Act of 1983,[2–6] designed to stimulate the development of pharmaceutical products that target rare diseases, which were at the time largely neglected as they affected relatively few people.

RSC Drug Discovery Series No. 38
Orphan Drugs and Rare Diseases
Edited by David C Pryde and Michael J Palmer

Published by the Royal Society of Chemistry, www.rsc.org

1.2 Rare Diseases

There is considerable diversity among conditions that are defined as rare diseases and include neurological conditions, infectious diseases, rare cancers, autoimmune disorders, respiratory disorders, muscle disorders, blood disorders and a wide range of inherited genetic disorders. It has been estimated that there are more than 7000 rare diseases known,[7] but only around 5% of these have therapies available[8,9] and the unmet medical need across the breadth of rare diseases remains high. Over 80% of rare diseases are genetic in origin.[10] Most of these are caused by defects in a single gene (that may be dominant or recessive), but some rare diseases are caused by multiple gene defects or a multitude of factors. Fifty percent of all rare diseases affect children and 85% are classified as serious or life-threatening. Some rare diseases may only affect literally a handful of individuals around the world, while others may affect hundreds of thousands of patients. In the developed world alone, rare diseases are thought to affect some 6% of the population, with estimates of more than 25 million North Americans and more than 30 million Europeans affected by a rare disease. Across the thousands of highly heterogeneous rare diseases that are known, there is no unifying classification that links them all, with the exception that they affect a relatively small number of people.

There is no single, widely accepted definition for rare diseases. In the USA, rare diseases are defined as any disease or condition affecting fewer than 200 000 people.[11] In Europe, a condition is considered rare if it affects fewer than 1 in 2000 people[12] and in Japan 1 in 50 000.[13] There are a few diseases that affect more than 200 000 people where certain subpopulations that carry a particular disease fall below the prevalence threshold for a rare disease.

1.3 Developing an Orphan Drug

Developing drugs to treat rare diseases poses many unique challenges. Designing and conducting clinical trials is constrained, as there is usually little understanding or information about the natural progression of the disease to inform end point selection.[8] Many rare diseases do not have clearly identifiable symptoms and investigators often have difficulty identifying and enrolling a large number of patients. Basic tools, such as validated animal models, may not exist. Small sample sizes pose statistical hurdles. These challenges increase the uncertainty that a research programme will lead to a new therapy, resulting in historically less investment into these therapies. An interesting example was raised by Tambuyzer,[8] who highlighted that for Gaucher disease patients in Germany, only around 5% of all possible patients are being treated despite treatments being available for more than 15 years. This example also highlights the difficulties of obtaining accurate prevalence data for rare diseases, and how variable different sources of these data are. Certain rare diseases are also known to have very different prevalence rates in different populations and geographical regions, for example the glycogen storage disease Pompe disease, which can range in prevalence from 1 in 200 000 in Caucasians to as much as 1 in 14 000 in African Americans.[14]

In recognition of these specific issues facing drug development for rare diseases, many governments around the world have developed orphan drug regulations to support those working to develop new products intended for the diagnosis, prevention or treatment of rare conditions. While provisions vary from country to country, the key incentives created under various orphan drug regulations generally include marketing exclusivity, which prevents similars from competing with the original approved product during the exclusive period but is in no way intended to create a monopoly if clinical differentiation can be demonstrated. For example, several small molecule treatments (imatinib, dasatinib and nilotinib) have been approved in parallel for chronic myeloid leukaemia. There is also support for sponsors taking their orphan drug through the regulatory approval process in the form of fee waivers, additional scientific advice and expedited review. Some regulations also include research grants or R&D tax credits.

These incentives have successfully increased drug development activities within the orphan drug space. Orphan drugs can offer faster development timelines, lower R&D costs, lower marketing costs and lower risk of generic competition. An analysis has suggested that orphan drug approval rates were greater than those of mainstream drugs, and the proportion of overall new drug approvals in recent years that are orphan drugs has steadily grown.[15]

1.4 The Orphan Drug Act

The USA passed the first legislation of this type when the Orphan Drug Act of 1983 was signed into law.[5] Similar legislation has been created in Australia, Europe, Japan and Singapore, with Canada and Russia set to introduce their own regulatory frameworks in the near future. The Orphan Drug Act sought to encourage development of drugs, diagnostics and vaccines intended to improve the treatment options for rare diseases by designating them as an orphan drug.

Orphan drug designation does not imply that a medicine is safe, effective or legal to develop and manufacture, but simply that the sponsor qualifies for certain benefits in the course of the drug development process.[16]

In the USA, the Office of Orphan Products Development (OOPD) within the Food and Drug Administration (FDA) grants an orphan designation to any product that is indicated for a rare disease as per the above definition.[17] Orphan designation may be granted at any point through the drug development process. An orphan-designated product may subsequently gain market approval only if data derived from clinical trials demonstrate the safety and efficacy of the product. Orphan designation confers certain benefits to a sponsor; 50% tax credits for clinical development costs, exemption from application user fees, subsidies for conducting clinical trials and market exclusivity for 7 years. These incentives have clearly made a significant impact on rare disease drug development. In the decade leading up to the Orphan Drug Act being passed, only 10 products for rare diseases received marketing approved while in the period since, more than 10 products have received marketing approval every year, and to date some 430 orphan products for rare diseases have been approved.[18] The OOPD also administers a related programme that is intended to stimulate the development of medical devices that are intended for use in the treatment or diagnosis of a rare disease.[19]

The Orphan Drug Act is widely accepted as having been hugely successful in driving R&D into rare diseases.[20] In Figure 1.1, the number of orphan drug designations made by the OOPD since the adoption of the Orphan Drug Act in 1983 in the USA is illustrated, along with the number of orphan drugs that received market authorisation. The number of designations has increased markedly in the last decade to an average of well over 100 per year, reflective of generally increased interest from R&D companies in rare diseases. However, one can also see that the number of market authorisation approvals in the same period has remained relatively constant, and in fact relatively constant going back to the previous decade also, which at first glance may look like diminished, or at best flat, productivity.

Using the data set charted in Figure 1.1, over the entire period 1984–2013, 15% of all designations resulted in an approved product. Year on year, overall approval rates as a proportion of designations was plotted in Figure 1.2 and is relatively flat, with a peak towards the end of the last decade.

This picture is of course atypical in a period where overall drug approval rates have fallen, and therefore the proportion of orphan drugs being approved as a percentage of overall drug approvals is actually rising and appear to have higher approval rates than more mainstream drug applications in recent years.[15]

1.5 Outside of the USA

In the European Union, the Committee for Orphan Medicinal Products (COMP) is responsible for reviewing requests for drug products being given an orphan medicinal product (OMP) designation that are being developed for the purpose of treating a rare disease.[21] Compounds that are given this designation are then assessed by the Committee for Medicinal Products for Human Use (CHMP) to receive formal market authorisation should they have demonstrable efficacy and safety. In Europe, the incentives for drugs that have been designated as having orphan status include 10 years of market exclusivity, grants for conducting clinical trials and fee reductions for requests made to the European Medicines Agency (EMA). The OMP legislation came into force in 2000, the same year the COMP was established.[22,23]

Several other countries also now have dedicated legislation, development incentives and approvals procedures for rare disease treatments, including Japan, Australia and Singapore.

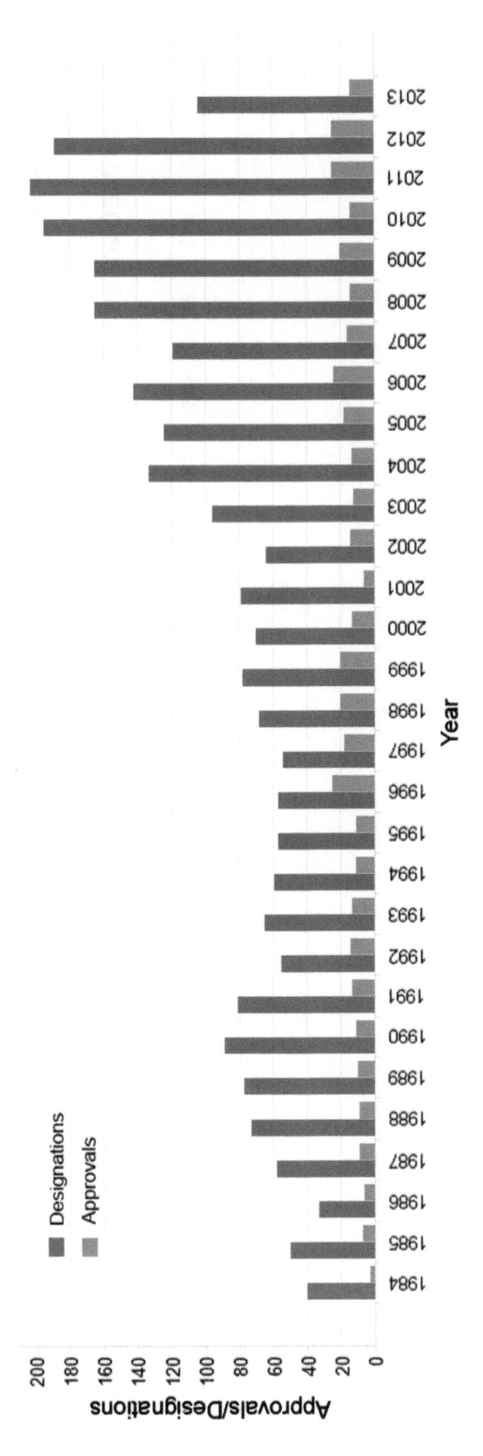

Figure 1.1 Orphan drug designations and market authorizations granted by year in the USA since 1983. Data obtained from the US FDA OOPD website.[18] Data from 2013 is only partial data, correct up to and including May 2013.

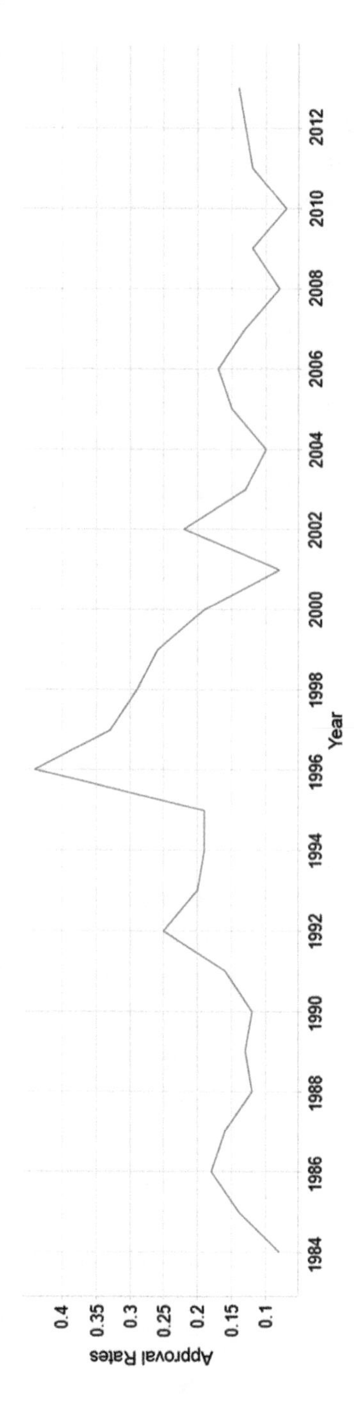

Figure 1.2 Approval rates in the USA as a proportion of orphan drug designations. Data obtained from the US FDA OOPD website.[18] Data from 2013 is only partial data, correct up to and including May 2013.

1.6 Thirty Years on from the Orphan Drug Act

Since these regulations were established, the number of licensed therapies for rare diseases has increased markedly. Since 1983, there have been some 400 orphan product approvals in the USA now available for use by patients around the world, compared to just 10 orphan drugs approved in the previous decade.[24] Focusing on the last decade only, the number of drug approvals has been both high (in total more than 180) and consistent and appears set to continue this trend into the future (Figure 1.3).

In Europe, within the same period of time, more than 65 orphan drug products have been granted market authorisation since the EU orphan drug legislation was enacted (Figure 1.4).

These regulatory guidelines can therefore be described as very successfully stimulating orphan drug development. However, the Orphan Drug Act and its sister regulations in other regions have sparked some controversy, not least through the advent of blockbuster orphans.[24,25] These are drugs that generate annual revenues of at least several hundred million dollars through

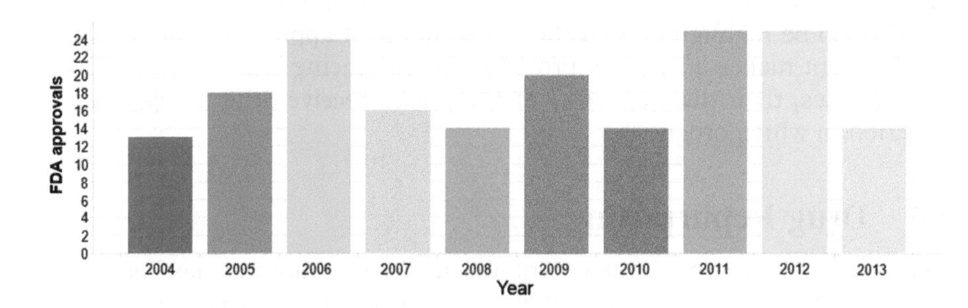

Figure 1.3 FDA orphan drug approvals by year in the last decade. Data obtained from the US FDA OOPD website.[18] Data from 2013 is only partial data, correct up to and including May 2013.

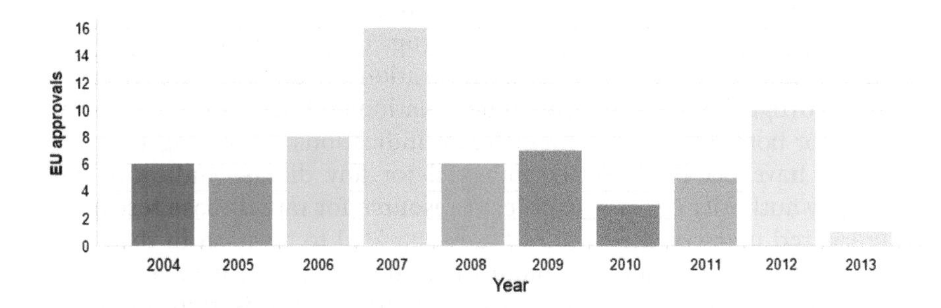

Figure 1.4 EU orphan drug approvals by year in the last decade. Data obtained from the US FDA OOPD website.[18] Data from 2013 is only partial data, correct up to and including May 2013.

high per unit cost, sometimes in excess of $100 000 per patient per year or by widespread use of the drug outside of its primary orphan indication. Some studies have identified orphan drugs that generate significantly more revenue through off-label use than for any orphan indication.[26]

Most orphan drugs are approved for a single indication only, but now almost 50 drugs have been approved for multiple rare diseases. One of the most commercially successful orphan drugs is imatinib (Gleevec), with sales in excess of $4.5 billion in 2011 and seven separate orphan drug approvals. Indeed, in 2006–2008, no less than 16 orphan drugs made the top 200 list of best-selling drugs in the USA with annual sales in the range $200 million to $2 billion,[27] and fuels the perception that orphan drugs can be economically viable and offer attractive business opportunities for biomedical drug development organisations.[28] It should be borne in mind, however, that each market authorisation requires separate clinical trials for each additional indication added to a product, which needs to be paid for by the sponsor. It is also clear that for small market sizes and first-in-class medicines, a sponsor needs to embark on a R&D programme in the knowledge that their investment can be recouped, which does imply higher drug pricing, without which many of the products invented to date may never have come to market. It should also be highlighted that the legislation as it applies to orphan drug development makes no explicit provision for enhancing basic research into rare diseases, their diagnosis or which diseases receive drug development attention in which order.

1.7 Drug Repurposing

A crucial aspect of drug development activity for rare diseases has been the repurposing of existing drugs that had previously received marketing approval for a more common disease. This is particularly important as previously approved compounds will already have completed pre-clinical toxicity testing and been deemed to have demonstrated pharmacological activity in another disease indication. The OOPD has recently published data on the FDA website that details all 97 drugs that have an orphan designation and have previously been approved for a more mainstream indication.[29] In addition, the same resource details the 71 drugs that are orphan-designated and have received previous market authorisation for another rare disease, and the 36 drugs that have an orphan designation and have been previously approved for both orphan and mainstream indications. Taken together, all drugs that have been previously approved for any disease indication by a regulatory authority offers a significant resource for rare disease research, having cleared many of the hurdles that often lead to attrition in the drug development process. There are more than 200 drugs that have a current orphan drug designation and benefit from market authorisation for some disease indication, but of course this is but a small fraction of the totality of approved drugs that could have some utility against a rare disease.

An example of a drug that was approved for a mainstream indication and subsequently approved for a rare disease is sildenafil from Pfizer (as Viagra, approved for the treatment of male erectile dysfunction in 1998), which was approved for the treatment of pulmonary arterial hypertension in 2005 as Revatio. Examples of drugs that were initially launched as orphan drugs and then were repurposed for broader indications include rituximab from Genentech (as Rituxan, initially approved for the treatment of non-Hodgkin lymphomas in 1997) and epoetin alpha from Amgen (as Epogen initially approved for the treatment of anaemia in 1989).[27]

An interesting example of a drug that was never intended for use as a human therapeutic is nitisinone, which was developed originally as a herbicide. Nitisinone is a 4-hydroxyphenyl pyruvate oxidase inhibitor that interrupts the formation of excess tyrosine in the blood and helps to prevent liver damage in children with hereditary tyrosinemia.[30]

1.8 Treatments and Modalities

The breadth and diversity of rare diseases makes all therapeutic modalities potentially applicable. Of the currently approved orphan drugs, many are small molecules, but others include, for example, monoclonal antibodies or enzyme replacement therapy (ERT) that at the time of approval were considered to be highly innovative.[31,32] A wide range of experimental therapies based on the next generation of novel and innovative technologies that are especially well suited to gene defects include anti-sense oligonucleotides, RNA interference and stem cell-based therapies. Applications of all of these technologies to the treatment of rare diseases are illustrated below.

1.8.1 Small Molecules

The vast majority of rare disease treatments available today are small molecules, and mirrors the research focus of most mainstream disease indications across the industry. It seems likely that this trend will continue, particularly where a specific distribution feature of a rare disease (a neurological disease, for example) or an intolerance to other modalities (for example the use of the small molecule miglustat for the treatment of metabolic disorders where ERTs are poorly tolerated) dictates that small molecule options are best suited. The origins of the small molecule agents that are currently approved as a rare disease treatment again mirrors those of more mainstream small molecule drugs, and include phenotypic screens, high-throughput single target screens and natural product semi-synthesis as well as drug repurposing.

An interesting example of how small molecule therapies (and their delivery methods) have evolved through the years comes from the portfolio of approved products for the treatment of pulmonary arterial hypertension. The first agent approved was the vasodilating prostaglandin derivative epoprostenol, which had to be administered by IV injection. This was followed by the small molecule endothelin receptor antagonists, for example bosentan, which are taken orally. More recently, synthetic derivatives of prostaglandins have been developed using advances in formulation and drug delivery, for example the inhaled iloprost.

1.8.2 Antibody Therapies

Therapeutic monoclonal antibodies (mAbs) are large heterodimeric molecules composed of a heavy and a light chain that offer exquisite selectivity for their intended biological target. mAbs do not readily cross the blood–brain barrier or cell membranes, and as such are suitable for extracellular, non-central targets. Initially, murine mAbs were manufactured using hybridoma technology, but due to toxicity and variable immunologic response have since been replaced by other, more human versions.

Chimeric mAbs are murine-based in which the mAb constant region is replaced by a human equivalent. Chimeric mAb drugs include infliximab, a mAb that targets tumour necrosis factor and decreases intestinal inflammation in Crohn's disease.

Humanised mAbs are human antibody-based, in which murine hypervariable regions are grafted on. Example products of this type that have been developed for rare diseases include Soliris, for the treatment of paroxysmal nocturnal haemoglobinuria.

Human mAbs are produced by vaccinating transgenic mice, which contain human genes, with the antigen of choice, leading to the production of fully human mAbs. An example is Ilaris®, which is approved for the treatment of cryopyrin-associated periodic syndrome.

1.8.3 Therapeutic Proteins

Protein therapies aim to deliver a protein that is either absent or depleted in a disease state. Most often this is an endogenous protein, for example human growth hormone (marketed as Somatropin) that stimulates cell production and growth in conditions such as growth hormone disorders and paediatric growth disorders. More recent examples include Amgen's Neupogen, a granulocyte colony-stimulating factor analogue that is used to stimulate neutrophil production in patients with neutropenia.

1.8.4 Enzyme Replacement Therapies

Enzyme replacement therapies (ERTs) are the regular injection of a native or a recombinant enzyme to patients throughout their lives to mitigate for a lack or dysfunction of an endogenous enzyme. Lysosomal storage disorders are especially suitable for this type of therapy as they are caused by a deficiency in a single enzyme and respond well to ERT. Initially, the replacement enzymes were isolated from human organs, but enzyme yields were often low and ultimately recombinant versions were developed. Manufacture of the recombinant enzymes is often expensive and this is reflected in the high annual treatment cost of ERT which commonly exceeds $100 000 per patient per year. Successful ERT applications include Fabry disease, Gaucher disease and Hunter syndrome.

The Medical Research Council is funding research at the University of London into mitochondrial neurogastrointestinal encephalomyopathy (MNGIE),[33] which is caused by a defect in the gene that produces the enzyme thymidine phosphorylase, and leads to mitochondrial dysfunction, muscle defects and a malfunctioning central nervous system. Only 200 cases of MNGIE have been identified worldwide. A new ERT is being developed using erythrocyte encapsulated thymidine phosphorylase (EE-TP) in which TP is introduced directly into a patient's red blood cells. The University of London, in partnership with Orphan Technologies, is continuing to develop the approach following the demonstration of efficacy in a small pilot study.

1.8.5 Gene Therapies

Diseases that are caused by single or multiple gene changes make them technically attractive to genetic or cell-based therapies to correct the malfunctioning gene(s). A number of innovative gene therapies have been developed for rare diseases which at a top level involves the use of a viral vector to deliver either a DNA cassette containing the missing gene to be expressed or an anti-sense or interfering RNA molecule to silence an over-expressing gene. In many ways the viral vector delivery system is critical to the success of the approach. The viruses of choice are adeno-associated viruses (AAVs), which are small human parvoviruses. They are single-stranded DNA viruses that can be delivered in high titres to both dividing and non-dividing cells and very effectively integrate into the host genome. Most importantly, they are safe and non-pathogenic, and can produce an effect that lasts for years. This is the vector used in uniQure's Glybera, the first ever gene therapy product to be approved in the EU[34] as a treatment for lipoprotein lipase deficiency (LPL) through intramuscular expression of lipoprotein lipase.[35] Contrast this with some of the false starts for gene therapy in which adenoviruses caused massive immune reaction or lentiviruses induced leukaemias in patients. Now, several gene therapy trials are under way for rare diseases, including alpha-1 antitrypsin deficiency, age-related macular degeneration, glioblastoma, Leber congenital amaurosis and Duchenne muscular dystrophy (DMD), and the results of these trials are eagerly anticipated.[36]

Recently, Sangamo Biosciences has reported the successful application of their *in vivo* protein replacement platform to modulate levels of the coagulation protein, Factor VIII, to offer a potential therapeutic approach to haemophilia.[37] This approach is based on zinc-finger DNA-binding protein genome editing technology, which enables the precise insertion of a replacement Factor VIII gene into the albumin gene, thereby allowing high and stable expression of Factor VIII in the liver following a single systemic treatment in a mouse model. Other monogenic diseases that are being targeted by the Sangamo technology include sickle cell anaemia, Gaucher disease and beta thalassaemia.

Sarepta Therapeutics has reported on a potential new treatment for DMD using an anti-sense oligonucleotide AVI-4658 (Eteplirsen) in a small patient study. AVI-4658 is a phosphoramidate morpholino-oligomer that interacts with and then silences the long dystrophin exon 51 gene and enables expression of a shorter but functional dystrophin protein in DMD patients. AVI-4658 is well tolerated in dystrophic mice, normal mice and in non-human primates and has shown efficacy in Phase 2 clinical trials.[38]

1.8.6 Cell Therapies

Pluripotent stem cells are derived from adult somatic cells and can be induced (induced pluripotent stem cells, iPSCs) to take on the properties of human embryonic stem cells (hESCs); potentially this therapeutic approach could have a massive impact on rare disease research. Stem cells have the unique ability to renew themselves continuously and could be applied to the supply of native-like cell types for screening purposes, used to repair mutated systems caused by a rare disease before being transplanted back into the patient or directly targeting disease-producing cell types (*e.g.* the so-called cancer stem cells). Several reports have been made of the manufacture of iPSCs from rare disease patients including those for Gaucher disease, DMD, Huntington's disease and Hurler syndrome, but to date no iPSC clinical trial has been initiated. Several stem cell trials are, however, under way for a number of rare diseases using stem cells derived from bone marrow, for example retinitis pigmentosa, age-related macular degeneration and sickle cell disease. The biotechnology company Bluebird Bio has clinical stage assets based on genetically altered haematopoietic stem cells for the treatment of adrenoleukodystrophy and beta thalassaemia.[39]

Another viable approach involves the application of gene-altered stem cells as a combination of both gene and cell therapies that allows components of both technologies to operate. For example, altered stem cells have been used to express Factor VIII in a mouse model of haemophilia.

Finally, micro RNAs (miRNAs) are short, non-coding RNAs that function as gene regulators and can control the switching on and off of genes within stem cells. This is being applied in experimental approaches to regulating gene switching and elucidating the role of miRNAs in primarily monogenic rare diseases. miRNAs can also be detected in blood and could offer a simple biomarker for certain rare diseases. Santaris Pharma A/S is a drug development company that specialises in the discovery and development of RNA-based therapies using its locked nucleic acid (LNA) platform and reportedly has been looking into the application of its technology to rare genetic disorders.[40]

1.9 Orphan Drugs and the Diseases they are Used to Treat

In the above sections, the numbers of orphan drugs that have been approved and/or designated as orphan drugs have been presented along with the modalities that have been or are being researched within the rare disease field. In this section, examples of what the orphan drugs actually are, when they were approved, which modality they concern and which rare disease they are used to treat is now detailed. In Table 1.1, a range of orphan drug examples are shown that have been approved in the EU and/or the USA along with their target indication.

One can see from the table that through the 30 years of orphan drug approvals, a range of therapeutic modalities are represented and a large cross-section of the industry are represented as sponsors of orphan drug development programmes. A large range of rare diseases have been served by the drug approvals shown in the table, but when one considers the breadth of total rare disease space (>7000), the products shown in the table only cover a tiny percentage of all rare diseases.

1.10 Rare Diseases and their Prevalence

In Table 1.2, we have focused this time on rare diseases, and attempted to classify them using similar criteria to those applied by the PhRMA resource.[31] Once again, we have not attempted to be exhaustive but highlighted several rare diseases of differing origins and causative links, where known, for illustrative purposes. It is important to point out that the prevalence data compiled in the table was obtained from several sources, including Orphanet,[41] Eurordis[42] and is quite variable, most likely because accurate figures in many cases are lacking.

One can see that some categories, for example lysosomal, genetic and respiratory disorders, are reasonably well served by the drugs displayed in Table 1.2, while in other categories such as gastrointestinal and movement disorders, these are much less well served.

It is also notable that while for many of the diseases listed in the table, a definite causative link has been elucidated, in many more cases there is no definitive molecular target for the disease. In some cases, even where a molecular target is implicated it is not always known in detail exactly how this creates the disease state. This does imply that there is a lack of basic research into many rare diseases.

In looking more closely at the range of rare diseases targeted by approved orphan drugs and how this picture has changed over the years, the chart in Figure 1.5 shows all the orphan drugs approved in the USA in the first 5 year period since the introduction of the Orphan Drug Act arranged by class.

The largest proportion of drugs target blood disorders, with approximately half of all disease classes showing no drug approval. The chart in Figure 1.6

Table 1.1 Example orphan drugs and the rare diseases they are intended to treat.[a]

Active principle	*Trade name*	*Year of approval*	*Indication*	*Product class*	*Market authorization holder*
Somatropin	Nutropin	1985 (USA)	Growth failure due to lack of growth hormone	Protein	Genentech
Clofazimine	Lamprene	1986 (USA)	Lepromatous leprosy	Small molecule	Novartis
Etidronate	Didronel	1987 (USA)	Hypercalcaemia of malignancy	Small molecule	MGI Pharma
Epoetin alpha	Epogen	1989 (USA) 2007 (EU)	Anaemia	Protein	Amgen
Eflornithine	Ornidyl	1990 (USA)	Treatment of *Trypanosoma brucei*	Small molecule	Hoechst Marion Roussel
Baclofen	Lioresal	1992 (USA)	Treatment of intractable spasticity	Small molecule	Medtronic
Felbamate	Felbatol	1993 (USA)	Lennox–Gastaut syndrome	Small molecule	Wallace Laboratories
Imiglucerase	Cerezyme	1994 (USA)	Gaucher disease	ERT	Genzyme
Filgrastim	Neupogen	1994 (USA)	Severe chronic neutropenia	Protein	Amgen
Riluzole	Rilutek	1995 (USA)	Amyotrophic lateral sclerosis	Small molecule	Rhone-Poulenc Rorer
Clonidine	Duraclon	1996 (USA)	Cancer pain	Small molecule	Roxane Laboratories
Tobramycin	Tobi	1997 (USA)	Bronchopulmonary infections	Small molecule	Novartis
Infliximab	Remicade	1998 (USA)	Crohn's disease	mAb	Centocor
Temozolomide	Temodar	1999 (USA)	Malignant glioma	Small molecule	Schering Plough
Arsenic trioxide	Trisenox	2000 (USA)	Acute promyelocytic leukaemia	Small molecule	Cephalon
Agalsidase beta	Fabrazyme	2001 (EU) 2003 (USA)	Fabry disease	ERT	Genzyme
Agalsidase alpha	Replagal	2001 (EU)	Fabry disease	ERT	Shire HGT
Imatinib	Glivec	2001 (EU) 2001 (USA)	Chronic myeloid leukaemia	Small molecule	Novartis
Bosentan	Tracleer	2002 (EU) 2001 (USA)	Pulmonary arterial hypertension	Small molecule	Actelion
Miglustat	Zavesca	2002 (EU) 2003 (USA)	Gaucher disease	Small molecule	Actelion
Laronidase	Aldurazyme	2003 (EU) 2003 (USA)	Mucopolysaccharidosis I	ERT	Genzyme
Cladribine	Litak	2004 (EU)	Hairy cell leukaemia	Small molecule	Lipomed

Nitisinone	Orfadin	2005 (EU) 2002 (USA)	Hereditary tyrosinaemia type 1	Small molecule	Swedish Orphan International AB
Ziconotide	Prialt	2005 (EU) 2004 (USA)	Chronic pain	Peptide	Elan Pharma
Sildenafil	Revatio	2005 (EU) 2005 (USA)	Pulmonary arterial hypertension	Small molecule	Pfizer
Galsulfase	Naglazyme	2006 (EU) 2005 (USA)	Mucopolysaccharidosis VI	ERT	BioMarin
Sunitinib	Sutent	2006 (EU) 2006 (USA)	Metastatic renal cell carcinoma	Small molecule	Pfizer
Sitaxentan	Thelin	2006 (EU)	Idiopathic pulmonary arterial hypertension	Small molecule	Encysive
Dasatinib	Sprycel	2006 (EU) 2006 (USA)	Chronic myeloid leukaemia	Small molecule	Bristol-Myers Squibb
Stiripentol	Diacomit	2007 (EU)	Severe myoclonic epilepsy in infancy	Small molecule	Biocodex
Idursulfase	Elaprase	2007 (EU) 2006 (USA)	Hunter's syndrome (Mucopolysaccharidosis II)	ERT	Shire HGT
Betaine	Cystadane	2006 (EU)	Homocystinuria	Small molecule	Orphan Europe
Lenalidomide	Revlimid	2007 (EU) 2006 (USA)	Multiple myeloma	Small molecule	Celgene
Eculizumab	Soliris	2007 (EU) 2007 (USA)	Paroxysmal nocturnal haemoglobinuria	mAb	Alexion
Trabectedin	Yondelis	2007 (EU)	Soft tissue sarcoma	Small molecule	PharmaMar
Temsimorilus	Torisel	2007 (EU) 2007 (USA)	Renal cell carcinoma	Small molecule	Pfizer
Icatibant	Firazyr	2008 (EU) 2011 (USA)	Angioedema	Small molecule	Jerini
Sapropterin	Kuvan	2008 (EU) 2007 (USA)	Hyperphenyl-alaninaemia	Small molecule	BioMarin
Azacytidine	Vidaza	2008 (EU) 2004 (USA)	Myelodysplastic syndromes	Small molecule	Celgene
Romiplostim	Nplate	2008 (EU) 2008 (USA)	Chronic immune thrombocytopenia	Protein	Amgen

(continued)

Table 1.1 (*continued*)

Active principle	*Trade name*	*Year of approval*	*Indication*	*Product class*	*Market authorization holder*
Everolimus	Afinitor	2009 (EU) 2009 (USA)	Renal cell carcinoma	Small molecule	Novartis
Plerixafor	Mozobil	2009 (EU) 2008 (USA)	Mobilize progenitor stem cells prior to stem cell transplantation	Small molecule	Genzyme
Rilonacept	Arcalyst	2009 (EU) 2008 (USA)	CAPS	Protein	Regeneron
Canakinumab	Ilaris	2009 (EU) 2009 (USA)	CAPS	mAb	Novartis
Ofatumumab	Arzerra	2010 (EU) 2009 (USA)	Chronic lymphocytic leukaemia	mAb	GSK
Eltrombopag	Revolade	2010 (EU) 2008 (USA)	Idiopathic thrombocytopenic purpura	Small molecule	GSK
Velaglucerase alpha	Vpriv	2010 (EU) 2010 (USA)	Gaucher disease Type 1	ERT	Shire HGT
Cinacalcet	Sensipar	2011 (USA)	Severe hypercalcaemia	Small molecule	Amgen
Crizotinib	Xalkori	2011 (USA)	Metastatic non-small cell lung cancer	Small molecule	Pfizer
Clobazam	Onfi	2011 (USA)	Lennox–Gastaut syndrome	Small molecule	Lundbeck
Ruxolitinib	Jakafi	2012 (EU) 2011 (USA)	Myelofibrosis	Small molecule	Incyte
Ivacaftor	Kalydeco	2012 (EU) 2011 (USA)	Cystic fibrosis	Small molecule	Vertex
Mifepristone	Korlym	2011 (USA)	Cushing's syndrome	Small molecule	Corcept Therapeutics
Lomitapide	Juxtapid	2012 (USA)	Familial hypercholesterolemia	Small molecule	Aegerion Pharmaceuticals
Alipogene	Glybera	2012 (EU)	Lipoprotein lipase deficiency	Gene therapy	uniQure
Pomalidomide	Pomalyst	2013 (USA)	Multiple myeloma	Small molecule	Celgene
Tocilizumab	Actemra	2013 (USA)	Juvenile idiopathic arthritis	mAb	Genentech
Nimodipine	Nymalize	2013 (USA)	Subarachnoid haemorrhage	Small molecule	Intercell

[a]Data obtained from the US FDA OOPD website,[18] the Orphanet website[12] and individual company websites.

Table 1.2 Selected rare diseases, their prevalence and causative link where known.

Disease	*Estimated prevalence per 100 000*	*Causative link*
Lysosomal storage disorders		
Fabry disease	1.8	Alpha-galactosidase A
Pompe disease	1	Alpha-1,4-glucosidase
Hunter's syndrome	1	Iduronate-2-sulfatase deficiency
Gaucher disease	1	Glucocerebrosidase
Niemann-Pick disease	2.5	NPC
Tay–Sachs disease	5	Hexosaminidase A
Growth disorders		
Growth hormone deficiency	10	Growth hormone releasing hormone + other causes
Acromegaly	5	Pituitary gland tumour
Autoimmune diseases		
Kawasaki syndrome	1.4	Unknown
Muckle–Wells syndrome	0.1	Cryopyrin
Sarcoidosis	15	Unknown
Stevens–Johnson syndrome	0.5	Severe adverse drug reaction
Scleroderma	42	Unknown
Blood disorders		
Thalassaemia	0.5	Haemoglobin
Paroxysmal nocturnal haemoglobinuria	0.8	Haematopoietic stem cells
Myelofibrosis	2.7	Bone marrow stem cells, JAK2
Cyclic thrombocytopenia	—	Growth factors, platelet autoantibodies
Eye disorders		
Retinitis pigmentosa	30	Unknown
Chronic uveitis	38	Various primary diseases
Refractory glaucoma	—	Unknown
GI disorders		
Paediatric ulcerative colitis	2	Unknown
Paediatric Crohn's disease	2	Innate immune system
Pouchitis	—	Unknown
Infectious diseases		
Anthrax	—	*Bacillus anthracis*
Pulmonary infection in cystic fibrosis patients	—	Various infectious pathogens
Leishmaniasis	—	Protozoan parasites
Malaria	—	*Plasmodium* parasites
Chagas disease	—	*Trypanosoma cruzi*
Rare cancers		
Acute myeloid leukaemia	16	Various genetic and environmental factors
Ewing's sarcoma	0.1	EWS protein

(*continued*)

Table 1.2 (*continued*)

Disease	*Estimated prevalence per 100 000*	*Causative link*
Renal cell carcinoma	36	Various lifestyle and genetic factors
Kaposi's sarcoma	2.1	Human herpes virus 8
Movement disorders		
Amyotrophic lateral sclerosis	5	Superoxide dismutase and various other genetic and environmental factors
Huntington's disease	5	Huntingtin
Freidreich's ataxia	3	Frataxin
Duchenne muscular dystrophy	3.7	Dystrophin
Neurological disorders		
Narcolepsy	26	HLA complex
Post-herpetic neuralgia	—	Varicella zoster virus
Trigeminal neuralgia	5	Demyelination of ganglia axons
Lennox–Gastaut syndrome	15	Various genetic factors and inflammatory brain diseases
Charcot–Marie–Tooth disease	24	Myelin and axonal proteins
Cardiovascular diseases		
Dilated cardiomyopathy	40	Myocardial damage by various pathogens and cytoskeletal genetic factors
Genetic disorders		
Cystic fibrosis	10	Cystic fibrosis transmembrane conductance regulator
α1-Antitrypsin deficiency	25	α1-Antitrypsin
Sickle cell disease	15	β-Globin chain of haemoglobin
Cryopyrin-associated periodic syndrome	0.1	Cryopyrin
Hyperphenylalaninaemia	0.4	Phenylalanine hydroxylase
Epidermolysis bullosa	2.5	Collagen and laminin
Respiratory disorders		
Acute respiratory distress syndrome	—	Various primary diseases and traumas
Idiopathic pulmonary fibrosis	26	Various viral infections and environmental factors
Pulmonary arterial hypertension	20	Glucose-6-phosphatase
Transplantation		
Graft-*versus*-host disease	2.8	Graft immune cells
Ischemia-reperfusion injury	—	Various inflammatory mediators
Other		
Primary focal segmental glomerulosclerosis	0.3	Viral and toxin agents, and renal haemodynamic changes
Acute radiation syndrome	—	Widespread cellular damage
Hepatic veno-occlusive disease	11	Unknown

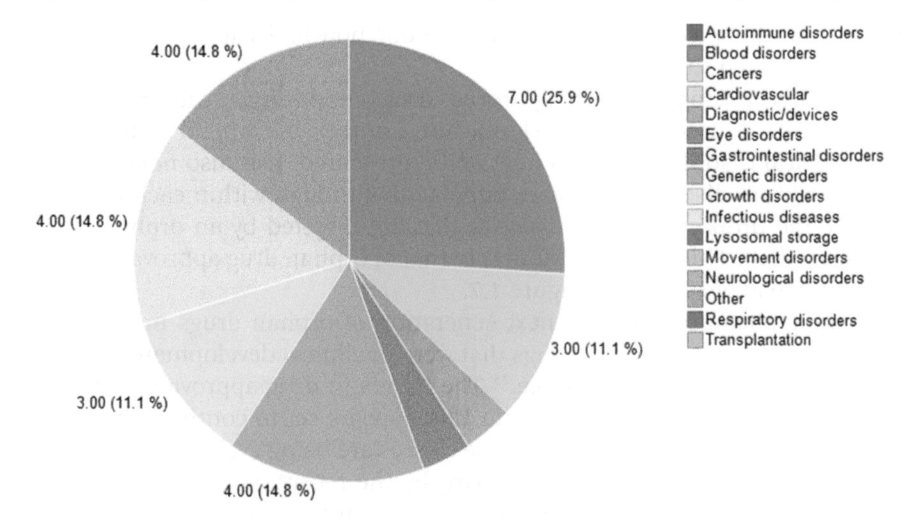

Figure 1.5 FDA approvals in the first 5 years of the Orphan Drug Act. Data obtained from the US FDA OOPD website.[18] The classification used follows a similar pattern to that used by the PhRMA resource.[31]

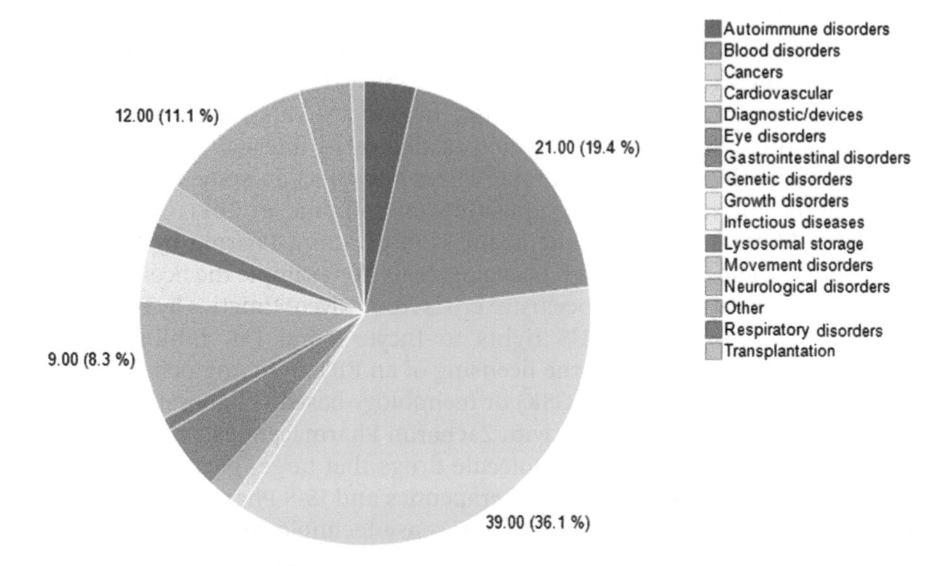

Figure 1.6 FDA approvals in the last 5 years. The classification used follows a similar pattern to that used by the PhRMA resource.[31]

shows how this picture has changed when one now looks at approved drugs in the USA from the last 5 years.

The largest proportion of approved drugs now target cancers. Blood disorders still account for a significant number of orphan drugs, but neurological disorders are also now well represented. It is also notable that almost all rare disease classes are populated, although within each class of disease the proportion of all diseases that are targeted by an orphan drug remains small. This picture is similar to that of orphan drug approvals in the EU since 2000, as shown in Figure 1.7.

Looking ahead to what the next generation of orphan drugs may target, Figure 1.8 shows the orphan drugs that were in clinical development in 2011 according to the PhRMA resource.[31] The trends in drug approvals from the last 5 years seen in both the EU and USA appears set to continue, but it is striking that a greater variety of disease classes are being evaluated clinically. This is probably driven, at least in part, by the recent advances in genetic screening and analysis technologies, and a significant increase in understanding of the genetic basis for some diseases. This is an encouraging sign that basic science advances of the last decade are fuelling the clinical advances of the next.

1.11 The Changing Face of Orphan Drug Developers

In the early years after the Orphan Drug Act was signed into law, there were relatively few industry drug developers focused on rare disease treatments. Companies such as Genzyme, Genentech, Shire Human Genetic Therapies, Amgen and Actelion were most closely associated with rare disease drug discovery.

In recent years, the companies involved in rare disease R&D have become much more diverse, as was highlighted in Table 1.1, and it is notable how many large pharma companies are now involved.[43] This involvement has come in some cases from the creation of dedicated internal research and business units (including for example Pfizer and GSK), in some cases from acquisitions (for example the acquisition of Genzyme in 2011 by Sanofi, FoldRx by Pfizer in 2010 and the acquisition of Amira Pharma by BMS in 2011) and in some cases through licensing deals (for example the licensing of a selectin antagonist for vaso-occlusive crisis from GlycoMimetics by Pfizer in 2011, the licensing of non-US rights to Incyte's oral JAK inhibitor for myelofibrosis by Novartis and the licensing of an ERT from Angiochem for a lysosomal storage disorder by GSK) or technology-based collaborations (for example the Pfizer collaboration with Zacharon Pharmaceuticals to find rare disease treatments using small molecule drugs that target glycans and the GSK collaborations with Prosensa Therapeutics and ISIS Pharmaceuticals to seek rare disease therapies based on anti-sense technology).[44]

It is also notable that several smaller companies that have specialised exclusively in rare disease research have been hugely commercially successful. Alexion Pharmaceuticals is a small biotechnology company that

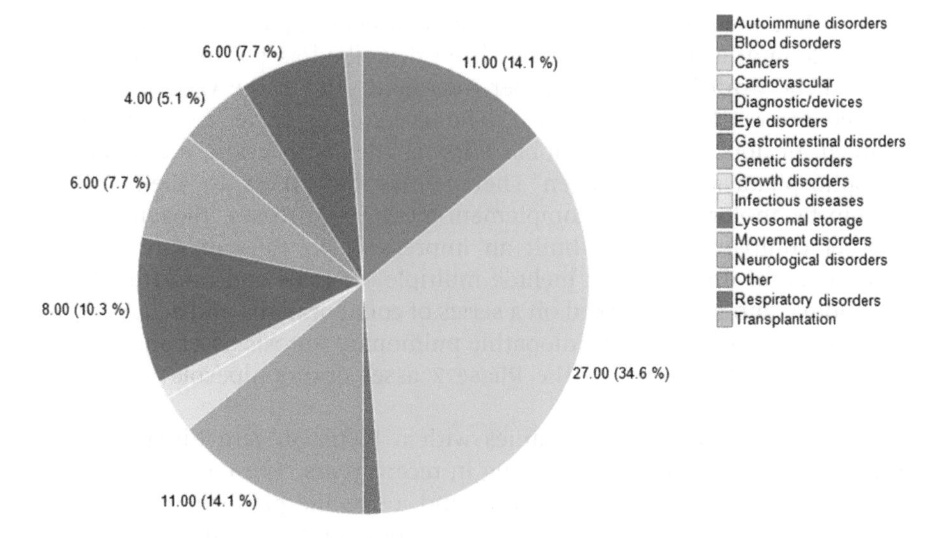

Figure 1.7 EU approvals since 2000. The classification used follows a similar pattern to that used by the PhRMA resource.[31]

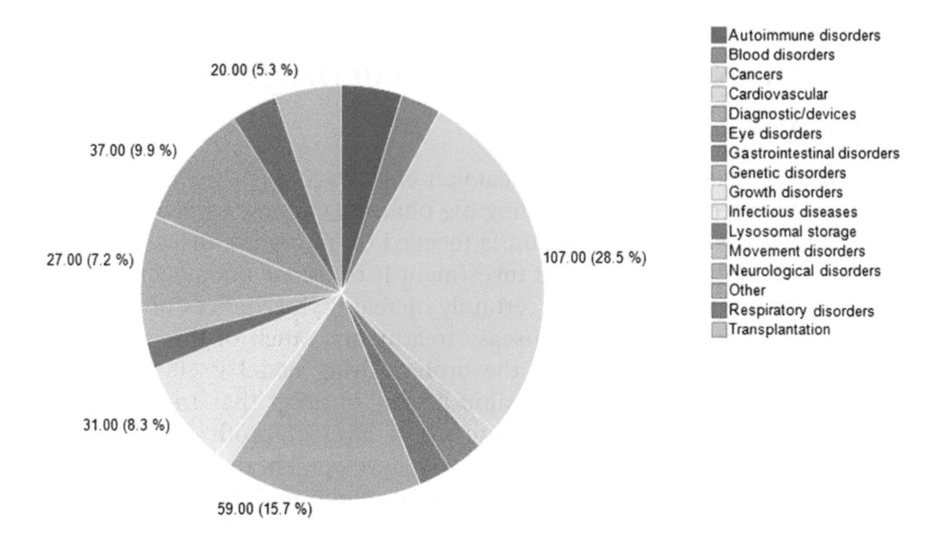

Figure 1.8 Orphan drugs in development in 2011. Data obtained from the PhRMA website.[31]

markets but one product – Soliris for the treatment of paroxysmal nocturnal haemoglobinuria (PNH), which has been described as the most expensive drug in the world[45] at $400 000 per treatment year in the USA, and sales projected to reach >$1 bn in 2012.[46] The revenues generated by Soliris have enabled Alexion to acquire Enobia Pharma Inc., with a Phase 2 asset for hypophosphatasia and Taligen Therapeutics for access to their fusion protein technology to treat complement-mediated diseases. Biogen Idec is another company that has built an impressive portfolio of rare disease treatments for diseases that include multiple sclerosis and non-Hodgkin's lymphoma, and has embarked on a series of collaborations and acquisitions (Stromedix in 2012 for their idiopathic pulmonary fibrosis asset and Knopp Neurosciences for access to the Phase 2 asset dexpramipexole for amyotrophic lateral sclerosis).

Start-up biotechnology companies with a focus on rare diseases have attracted significant investor funding in recent years. These companies are often, but not always, established around a specific platform technology. Rare disease start-ups include Ultragenyx[47] (developing an extended-release formulation of sialic acid for the treatment of hereditary inclusion body myopathy), Synageva[48] (developing sebelipase, an ERT for the treatment of lysosomal acid lipase deficiency), Retrophin[49] (developing an asset for the treatment of focal segmental glomerulosclerosis) and Sarepta[38] (developing eteplirsen for the treatment of DMD).

The pricing debate will undoubtedly be reignited when the pricing of uniQure's gene therapy product Glybera is announced. The pricing of a one-time series of intramuscular injections of Glybera is likely to exceed that of all existing orphan drugs that are dosed chronically, and could exceed the $1 m per patient level.

1.12 The Next 30 Years of Orphan Drug Development

1.12.1 Value for Money

The analyses presented above indicate an encouraging era for rare disease research. More companies, including big pharma, are now involved in drug R&D than ever, and their attention is focused on more rare disease classes than ever. This increased level of investment is of course not a guarantee of successful drug products, but it certainly increases the chances of realising more new and innovative rare disease treatments. Much of this increased level of investment, and indeed the orphan drug model itself, has been predicated on the promise of premium pricing of drugs that are eventually brought to market, thereby guaranteeing a level of profit for the drug sponsor, and it is this aspect of orphan drug development that does seem set to evolve in the near to medium term. While the overall budget spend by healthcare systems around the world on orphan drugs is small compared to more mainstream products such as cardiovascular or anti-inflammatory treatments, and the rare diseases that those drugs treat are often serious and life-threatening, there does appear to be increasing scrutiny of orphan drug pricing.[50] This can only intensify as more treatments come online and the cluster of rare disease treatments as a class starts to grow.

The key evidence for successful orphan drugs in the future will more than ever be safety and above all efficacy. This will be especially true in disease classes where multiple products exist, and one could envisage a system of 'risk sharing' in which a sponsor will be required to lower the cost of a drug treatment if it is shown to have less than expected efficacy.[51] The corollary to this could be the advent of gene therapy, which in all likelihood will command ultra-premium prices but overall will provide greater value for money if the therapy offers a cure.

1.12.2 The Expanding Role of Government Translational Research Programmes with Industry Partnerships

Science will continue to advance in the coming decades on a number of fronts, and the cost of genetic diagnosis will continue to decrease. More accurate data of rare disease prevalence and genetic causal links will become available. Translational data sets to refine the targeting of small patient populations and measurement of meaningful clinical biomarkers to assess outcome measures as reliable indicators of drug efficacy will evolve. In the USA, the National Institutes of Health (NIH) provided more than $823 million for 1600 orphan product development activities in the fiscal year 2012. The special research emphasis on rare diseases at NIH has resulted in the allocation of $3.623 billion for 9400 research projects. Several research institutes and centres of the NIH have made available personnel and financial resources for translational research initiatives to respond to scientific opportunities and the need for interventions for the prevention, diagnosis or treatment of rare and common diseases. These resources bridge existing data gaps to complete the necessary studies to provide pre-clinical and clinical data required for regulatory purposes. Of particular interest are the translational research programmes offered by the National Center for Advancing Translational Sciences, the National Cancer Institute, The National Institute of Allergy and Infectious Diseases, The National Heart, Lung, and Blood Institute, and the National Institute of Neurological Disorders and Stroke. Although similar in many respects, there are differences in each of the translational research programmes, including the application and review processes. Some institutes make the traditional grants and contracts available for clinical trial planning and implementation. To avoid confusion with the different processes, it is advisable to identify those institutes with a research portfolio that includes a specific disease interest. These activities within the translational research programmes are expected to complement or supplement the existing biopharmaceutical industry efforts and not to replace the extensive activities related to rare disease research and orphan product development activities.

1.12.3 Expanding Role of Academic Centres and Industry to Identify and Develop Orphan Products for Rare Diseases

It is likely that with an increase in the role of big pharma in rare disease research and the opening of their large chemical files, drug repurposing will increase, to uncover new drugs for rare diseases, and potentially this could impact in the relatively near term (within the next 5 years). It will also be interesting to see if big pharma, and indeed smaller biotech companies, can be incentivised to work on rare diseases for which there is very little known and take the lead role in driving the basic science behind such diseases. Rare diseases can be staggering if you consider the need for sufficient resources to discover and develop products to diagnose, treat or prevent rare diseases experienced by approximately 6–8% of the population who have one of the more than 6000 rare diseases. Partnering and collaborating with the academic research community is an essential component of R&D efforts for the biopharmaceutical and medical device industries to develop a portfolio of potential interventions and diagnostics. The pharmaceutical industry, with its unique product R&D infrastructure and expertise, provides the academic research community with the capability of moving a discovery to the marketplace. Rare diseases do not respect geographical or national borders and offer numerous research and regulatory challenges requiring global efforts as we observe expansion of activities that include the academic research communities from around the world. Numerous academic and government technology transfer programmes are now available to industry. Many of these programmes are formal partnerships between the industry and the academic partners. Both initiatives can lead to products for rare diseases but require a keen understanding of these programmes and the responsible programme staff who provide the links to the existing resources. Collaborative public–private partnership efforts required for rare diseases use disease-specific steering committees often led by patient advocacy groups (PAGs) and their scientific or medical advisory board members, most of whom are research investigators and medical specialists from the academic research and medical care communities. It is equally important for the academic community to have a clear path to the biopharmaceutical and medical device industries by having knowledge of appropriate contacts and available programmes from the potential industry partners. Many of these arrangements require considerable time for resolution of legal considerations between two or more parties, involving intellectual property and the estimated value of this property, and the milestones associated with drug development.

1.12.4 Recognition and Acceptance of PAGs and Patients/ Families as Research Partners

As mentioned earlier, there are now several successful models of PAGs directing partnerships to reach their organisational goals of providing treatments for their patients. Many of these organisations, such as the Cystic

Fibrosis Foundation, Pulmonary Hypertension Association, the National Urea Cycle Diseases Foundation, the Parent Project for Duchenne Muscular Dystrophy, and the Progeria Research Foundation, continue to extend their traditional advocacy roles of emphasising rare diseases research and orphan products development and have led to the identification of potentially useful products for their diseases. This new coordinating role has relied upon guidance from the pharmaceutical, biotechnology and medical device industries and contract research organisations. Active relationships with PAGs has helped reduce some of the barriers to research of rare diseases and orphan products including gaining access to a sufficient number of patients to participate in clinical trials. Lack of access to a sufficient number of patients to open a clinical trial through numerous recruitment procedures at a particular research site requires a collaborative global effort including research investigators and PAGs around the world. The PAGs utilise their unique resources to develop patient registries, encourage patients and their families to participate in natural history studies, facilitate the development of acceptable informed consent documents, and encourage research investigators to develop and adhere to common research protocols. The leadership of the PAGs has been able to establish global communication and social media links. Many of these working relationships advancing to global research investigations are the result of sponsorship and attendance at patient or family and scientific conferences. Many members of the biopharmaceutical industry now include active staff liaison and outreach activities between the industry and the patient communities. These activities facilitate the transfer of valuable information about the disease and possible interventions to patients, families, physicians and other healthcare providers, and the public. The development of strong and knowledgeable leadership in the PAG community with their scientific and medical advisory boards is a key to the successful development of interventions for rare diseases. The leadership of many of the PAGs continues to be recognised as the knowledgeable voice of specific rare diseases to the scientific community, the pharmaceutical industry and the media.

1.12.5 Role and Value of Research Networks and Consortia for Rare Diseases Research and Orphan Products Development

For most rare diseases and conditions, local clinics and academic medical centres do not have access to a sufficient number of patients and clinical researchers to complete clinical trials in a timely fashion. The Rare Diseases Clinical Research Network (RDCRN) was created to respond to these needs and provide the infrastructure to conduct different types of studies with multiple investigators available at numerous research sites and ready access to a sufficient number of patients to initiate a clinical study and to complete the trial without unreasonable delays. The majority of studies in the RDCRN are longitudinal or natural history studies. Other types of study include pilot studies, and Phase I, II and III studies. Industry sponsorship of studies is encouraged. Individual consortia in the RDCRN support collaborative clinical research studies, provide training programmes for new clinical investigators, and enable the partnership role of PAGs in rare disease research programmes. The RDCRN utilises a Data Management and Coordinating Center (DMCC) as a facilitating centre to assist in the design of clinical protocols, data management and analyses from multiple diseases and multiple clinical research sites. The DMCC maintains a web-based patient contact registry as a recruitment and referral tool for all consortia to use.

The 17 consortia in the RDCRN are supported by the Office of Rare Diseases Research (ORDR) and eight research institutes at NIH. Each consortia is required to focus on a group of at least three related disorders and receives 5 years of support. The data from ongoing and completed RDCRN clinical studies are shared openly with the scientific community. ORDR-developed data is placed in the National Library of Medicine's dbGaP database. The RDCRN studies more than 200 diseases at 225 research sites around the world, with 86 active protocols and an additional 37 protocols in different stages of development. There have been more than 130 trainees in the second 5 year cycle of RDCRN, with approximately 175 trainees in both 5 year periods. More than 15 000 patients have enrolled in studies in the second 5 year period for a total of 22 000 people. A total of 119 studies have been activated since inception and 76 studies activated during the current grant period. There are 97 PAGs and collectively they have formed the Coalition of Patient Advocacy Groups (CPAG) to support the numerous activities of the individual RDCRN consortia. Several consortia have established international collaborations in 14 countries other than the USA, including Australia, Austria, Belgium, Canada, England, France, Germany, Iceland, India, Italy, the Netherlands, Scotland, Spain and Switzerland. The RDCRN Contact Registry, which enhances participation in clinical trials and disseminates information, has approximately 11 000 registrants from 90 countries for more than 200 rare diseases.

1.12.6 Collaborative Partnerships are Required for Advances

Most rare diseases affect multiple organ systems and each have their own clinical and research disciplines involved in research through the multidisciplinary research teams. With limited resources available, newer models have evolved that utilise the resources available from public–private partnerships. These models include major participatory efforts by the academic research community, federal research and regulatory agencies, private foundations, PAGs, and members of the pharmaceutical, biotechnology and medical device industries. The rare diseases community recognises and encourages the different multi-organisational approaches to drug discovery and development, especially if there is limited or no commercial interest in developing a product for rare diseases. These models also require resources and commitments be made from many private and public organisations to facilitate the development of products. A global approach is required to coordinate research efforts at multiple research sites working under a common protocol and utilising the skills and knowledge from multidisciplinary research teams. Coordinated and systematic efforts to research and product development require numerous highly motivated global partners utilising the strengths of the individual organisations towards a common goal of developing treatments or diagnostics for rare diseases.

References

1. J. K. Aronson, *Br. J. Clin. Pharmacol.,* 2006, **61**, 243–245.
2. M. E. Haffner, *Retina,* 2005, **25**, S89–S90.
3. U.S. Department of Health and Human Services, U.S. Food and Drug Administration, http://www.fda.gov/ForIndustry/DevelopingProductsforRareDiseasesConditions/HowtoapplyforOrphanProductDesignation/ucm135122.htm, accessed 5 July, 2013.
4. U.S. Food and Drug Administration, *Fed. Regist.,* 1992, **57**, 62076.
5. United States Congress, Orphan Drug Act, 1983, Public Law Number 97-414, 96, Stat. 2049.
6. M. E. Haffner, *N. Engl. J. Med.,* 2006, **354**, 445–447.
7. Global Genes, Rare List, http://globalgenes.org/rarelist/, accessed 5 July, 2013.
8. E. Tambuyzer, *Nat. Rev. Drug Discovery,* 2010, **9**, 921–929.
9. *Rare Diseases and Orphan Products: Accelerating Research and Development*, ed. M. J. Field and T. F. Boat, The National Academies Press, Washington DC, USA, 2011.
10. Eurordis, European organization for rare diseases, http://www.eurordis.org/, accessed 10 June 2013.
11. U.S. National Institute of Health, Office of rare disease research, http://rarediseases.info.nih.gov, accessed 10 June 2013.
12. Orphanet, The portal for rare diseases and orphan drugs, http://www.orphanet.net/consor/cgibin/Education AboutRareDiseases.php?Ing=EN, accessed 10 June 2013.
13. J. Llinares, in *Rare Diseases Epidemiology*, ed. M. Posada de la Paz and S. C. Groft, Springer, Netherlands, 2010, vol. 686, pp. 193–207.
14. R. Hirschhorn and A. J. J. Reuser, in *The Online Metabolic and Molecular Bases of Inherited Disease*, ed. C. R. Scriver, McGraw Hill, New York, USA, 8th edn, 2001.
15. M. Orfali, L. Feldman, V. Bhattacharjee, P. Harkins, S. Kadam, C. Lo, M. Ravi, D. T. Shringarpure, J. Mardekian, C. Cassino and T. Cote, *Clin. Pharmacol. Ther.,* 2012, **92**, 262–264.
16. W. Yin, *J. Health Econ.,* 2008, **27**, 1060–1077.

17. U.S. Department of Health and Human Services, U.S. Food and Drug Administration, http://www.fda.gov/ForIndustry/DevelopingProductsforRareDiseasesConditions/default.htm, accessed 12 June 2013.
18. U.S. Department of Health and Human Services, Food and Drug Administration, http://www.accessdata.fda.gov/scripts/opdlisting/oopd/index.cfm, accessed 12 June 2013.
19. U.S. Department of Health and Human Services, Food and Drug Administration, http://www.fda.gov/ForIndustry/DevelopingProductsforRareDiseasesConditions/DesignatingHumanitarianUseDevicesHUDS/default.htm, accessed 12 June 2013.
20. M. E. Haffner, J. Whitley and M. Moses, *Nat. Rev. Drug Discovery,* 2002, **1**, 821–825.
21. The Committee for Orphan Medicinal Products of the European Medicines Agency Scientific Secretariat, *Nat. Rev. Drug Discovery,* 2011, **10**, 341–349.
22. R. Joppi, V. Bertele' and S. Garattini, *Eur. J. Clin. Pharmacol.,* 2013, **69**, 1009–1024.
23. European Parliament, *Off. J. Eur. Communities: Legis.,* 2000, **141**(L18), 1–5.
24. O. Wellman-Labadie and Y. Zhou, *Health Pol.,* 2010, **95**, 216–228.
25. D. Rohde, *Food Drug Law J.,* 2000, **55**, 125–143.
26. A. S. Kesselheim, J. A. Myers, D. H. Solomon, W. C. Winkelmayer, R. Levin and J. Avorn, *PLoS One,* 2012, 7, e31894.
27. I. Melnikova, *Nat. Rev. Drug Discovery,* 2012, **11**, 267–268.
28. K. N. Meekings, C. S. M. Williams and J. E. Arrowsmith, *Drug Discovery Today,* 2012, **17**, 660–664.
29. U.S. Department of Health and Human Services, Food and Drug Administration, http://www.fda.gov/ForIndustry/DevelopingProductsforRareDiseasesConditions/HowtoapplyforOrphanProductDesignation/ucm216147.htm, accessed 6 July 2013.
30. P. McKiernan, *Drugs,* 2006, **66**, 743–750.
31. PhRMA, Orphan Drugs in Development for the Treatment of Rare Diseases, http://www.phrma.org/sites/default/files/pdf/rarediseases2011.pdf, accessed 6 July 2011.
32. PhRMA, A Decade of Innovation: Advances in the Treatment of Rare Diseases, http://phrma.org/sites/default/files/pdf/phrma_rare_diseases_06.pdf, accessed 6 July 2006.
33. University of London, Mitochondrial Neurogastrointestinal Encephalomyopathy treatment programme, http://www.sgul.ac.uk/media/latest-news/mrc-awards-a33.3million-to-st-george2019s-for-rare-disease-treatment-development, accessed 20 June 2013.
34. S. Yla-Herttuala, *Mol. Ther.,* 2012, **20**, 1831–1832.
35. uniQure, http://www.uniqure.com/, accessed 18 June, 2013.
36. M. I. Phillips, *Clin. Pharmacol. Ther.,* 2012, **92**, 182–192.
37. Sangamo Biosciences, http://www.sangamo.com/pipeline/rare-diseases.html, accessed 18 June 2013.

38. Sarepta Therapeutics, http://www.sareptatherapeutics.com/our-programs/, accessed 18 June 2013.
39. BluebirdBio, http://www.bluebirdbio.com/product-overview.php, accessed 20 June 2013.
40. Santaris, http://www.santaris.com/product-pipeline, accessed 18 June 2013.
41. Orphanet, The portal for rare diseases and orphan drugs, http://www.orpha.net/orphacom/cahiers/docs/GB/Prevalence_of_rare_diseases_by_alphabetical_list.pdf, accessed 24 June 2013.
42. Eurordis, European organization for rare diseases, http://ec.europa.eu/health/archive/ph_threats/non_com/docs/rdnumbers.pdf, accessed 24 June 2013.
43. A. Philippidis, *Hum. Gene Ther.*, 2011, **22**, 1037–1040.
44. Elsevier Business Intelligence, The Orphan Drug Boom: gold rush or flash in the pan, http://www.elsevierbi.com/publications/in-vivo/30/10/the-orphan-drug-boom-gold-rush-or-flash-in-the-pan, accessed 14 May 2013.
45. Fierce Pharma, The most expensive drugs in the world, http://www.fiercepharmamanufacturing.com/pages/chart-worlds-most-expensive-drugs, accessed 24 May 2013.
46. Nature Blogs, The most expensive drugs in the world, http://blogs.nature.com/spoonful/2011/09/soliris.html, accessed 21 June 2013.
47. Ultragenyx, http://www.ultragenyx.com/, accessed 15 May 2013.
48. Synageva, http://www.synageva.com/programs-pipeline.htm, accessed 15 May 2013.
49. Retrophin, http://www.retrophin.com/, accessed 17 May 2013.
50. E. Picavet, M. Dooms, D. Cassiman and S. Simoens, *Drug Dev. Res.*, 2012, **73**, 115–119.
51. J. Cook, J. Vernon and R. Manning, *PharmacoEconomics,* 2008, **26**, 551–556.

SCREENING, DIAGNOSIS AND BASIC SCIENCE

CHAPTER 2

Diagnosis of Rare Inherited Diseases

WILLIAM G. NEWMAN* AND GRAEME C. BLACK

Manchester Centre for Genomic Medicine, Institute of Human Development, Faculty of Medical and Human Sciences, University of Manchester, Manchester Academic Health Science Centre (MAHSC), Manchester M13 9WL, UK
*E-mail: William.newman@manchester.ac.uk

2.1 Introduction

One in 17 individuals is affected by a rare disease,[1] defined as a condition that affects fewer than 5 in 10 000 of the population. Even though individually each of these 6000 conditions is rare, they are estimated to affect 30 million people in the USA and 3.5 million in the UK,[2,3] and consequently place an enormous burden on individuals, families and healthcare systems.[1] A large number of rare disorders are inherited and therefore have a disproportionate effect on families. Approximately one-quarter of all children that are inpatients in hospital have an inherited disorder.[4] The discipline of clinical genetics has for the past 50 years provided services to these families by providing diagnosis and clinical management. Over the past decade the practice of clinical genetics has been transformed by technological advances that have enhanced our ability to define different genetic conditions, refined clinical management and provided insights to allow the development of specific treatments (Table 2.1). This chapter details the approaches to disease gene identification and how this knowledge is changing the practice of genomic medicine in the 21st century.

RSC Drug Discovery Series No. 38
Orphan Drugs and Rare Diseases
Edited by David C Pryde and Michael J Palmer

Published by the Royal Society of Chemistry, www.rsc.org

Table 2.1 Examples of inherited disorders where accurate molecular diagnosis has led to significant improvements in outcome through precise clinical management and treatment.

Condition [MIM reference]	*Causative gene(s)*	*Frequency*	*Treatment option(s)*	*Effect of treatment*
Familial adenomatous polyposis [175100]	*APC* or *MYH*	1 : 7000	Surveillance colonoscopy, prophylactic colectomy	Reduction in colorectal cancer from 43% to 3% [5]
Long QT syndrome [192500]	Multiple genes, *e.g. KCNQ1*, *KCNH2*	1 : 3000	Beta-blockade, implantable cardio-defibrillator; avoid QT-prolonging drugs	Avoidance of sudden cardiac death
Fabry disease [301500]	*GLA*	1 : 50 000 males	Enzyme replacement therapy	Prevention of end-stage renal failure and cardiomyopathy and other complications
Maturity onset diabetes of the young (MODY) [125850]	Many including *HNF4A*, *GCK*	1 : 10 000	Different oral hypoglycaemics	Effective glycaemic control, avoidance of insulin and diabetic complications

2.2 Phenotypic Characterisation

The study of rare disease has at its foundations precise and accurate phenotypic characterisation. Such detailed clinical assessments of individuals with a range of rare disorders have defined patterns of unifying signs and symptoms with additional haematological, biochemical or radiological features which have defined a catalogue of disorders. In the early 1960s, Victor McKusick created the Mendelian Inheritance in Man resource,[6] which defines approximately 4000 inherited disorders and their causative basis and a further 3000 conditions where the cause is unknown. In the 1980s, Albert Schintzel created a parallel database of chromosomal disorders.[7]

Increasingly it is being recognised across all specialties that detailed clinical phenotyping can describe individuals with rare subtypes of common disorders. An illustrative example is maturity onset diabetes of the young (MODY), where genetically defined subgroups can respond to different oral hypoglycaemics (Table 2.1).[8]

For familial conditions delineation of the inheritance pattern is key both to appropriate advice as well as to relevant genetic and genomic analysis. Precise diagnosis allows accurate risk assessment, provision of advice regarding the natural history of a disorder, the development of multidisciplinary approaches to patient care and a starting point to define these rare disorders at a molecular level. Definitive diagnosis of a rare disorder removes uncertainty for individuals and families and facilitates accurate clinical management and the development of personalised approaches to treatment. It can routinely take over 5 years of investigations and medical assessments for a patient with a rare disease to receive an accurate diagnosis.[9] Prior to this the average patient receives two to three misdiagnoses.[9] Therefore access to specialised services and powerful genetic testing approaches is vital to address this suboptimal situation. Broadly speaking rare inherited disorders can be categorised as chromosomal (genomic) or genetic disorders.

2.3 Chromosomal Disorders

For many years cytogenetic analysis was reliant on Giesma stain karyotyping of dividing cells, either lymphocytes or amniocytes. Standard karyotyping was able to define chromosomal copy number changes at a resolution of approximately 4 Mb (approximately 4 million base pairs). Essentially only large chromosomal changes could be diagnosed – autosomal aneuploidies, for example trisomy 21 (Down syndrome), 47, XXY (Klinefelter syndrome) and 45, X (Turner syndrome) and cytogenetically visible deletions, many of which were initially crucial in localising a number of disease genes, including *PAX6* (11p13, aniridia),[10] *RB1* and *WT1* (13q; retinoblastoma and Wilms' tumour).[11] New techniques at the end of the last century, including fluorescent *in situ* hybridisation (FISH) allowed the detection of sub-microscopic deletions and duplications that

Table 2.2 Microdeletion and microduplication disorders.

Condition	*Causative locus*	*Frequency*	*Clinical features*
Williams syndrome	Microdeletion of chromosome 7q11	1 : 7500	Characteristic appearance, developmental delay, cardiovascular anomalies, transient hypercalcaemia
Velocardiofacial (DiGeorge) syndrome	Microdeletion of chromosome 22q11	1 : 3000	Characteristic appearance, congenital heart defect, hypocalcaemia, reduced immunity, cleft palate
Smith–Magenis syndrome	Microdeletion of chromosome 17q11	1 : 25 000	Characteristic appearance, behavioural difficulties
Potocki–Lupski syndrome	Microduplication of chromosome 17q11	1 : 20 000	Learning disability, autism, hypotonia

were beyond the resolution of karyotypic analysis (Table 2.2). Generally these allowed diagnosis of genetic conditions that are suspected clinically and confirmed by molecular testing.

The true complexity of the genomic architecture and the variation in terms of copy number variants was not apparent until the emergence of significant advances in microarray technology. These arrays have taken many forms but essentially are based on the hybridisation of genomic DNA labelled with a fluorescent marker taken from an individual to glass slides containing thousands of fragments of DNA, either as oligonucleotides or in bacterial artificial chromosomes (BACs). These arrays are able, at a much higher level of resolution, to define microdeletions or microduplications (copy number variants, CNVs) in terms of kilobases. Very high-resolution arrays are even able to detect exonic deletions/duplications within specific genes. Application of this technology has defined the degree to which the genome can vary in terms of gene deletion, duplication or even amplification. A hitherto unsuspected level of copy number variation, within the normal population, has been uncovered using these technologies and in this manner, thousands of different such chromosomal changes which cause a clinical phenotype have now been identified. Such variants can lead to a very broad range of clinical features, which often include learning and developmental disorders, growth disturbance and congenital birth defects. Some of these chromosomal changes are surprisingly common; for example deletions at chromosome 1p36 have a frequency of 1 in 5–10 000 live births and have been shown to underlie discrete clinical entities, with predictable phenotypic features.[12] Demonstrating the important link between rare and more common disorders, CNVs have also been identified in individuals with schizophrenia, bipolar disorder and autism.[13]

2.4 Single Gene Disorders

In the 1980s, the discovery of the polymerase chain reaction (PCR) coupled to DNA sequencing approaches facilitated the seminal identifications of the genes underlying some of the more common inherited disorders, including cystic fibrosis, Huntington's disease and thalassaemia.

2.4.1 Autosomal Dominant and X-Linked Disorders

Positional clues from rare children with chromosomal aberrations detectable by standard karyotyping defined loci for some single gene disorders, including balanced X:autosome translocations in females manifesting Duchenne muscular dystrophy, or X chromosome deletions in patients with Norrie disease and X-linked retinitis pigmentosa. Autosomal genes identified in this way include *PAX6* underlying aniridia.[14] Linkage studies relied on the collection of multiple affected individuals, usually from single families or from multiple families with truly monogenic conditions. Such analyses were then followed by laborious and time-consuming positional cloning strategies—such protocols defined disease-causing genes for numerous disorders, including adult polycystic kidney disease, Huntington's disease and myotonic dystrophy throughout the 1990s. Although painstaking, these processes increased the numbers of genes identified for autosomal dominant and X-linked disorders over a number of years.

2.4.2 Autosomal Recessive Disorders and Autozygosity Mapping

The sharing of ancestral chromosomal regions (identity by descent) by individuals in families where relatives marry (consanguinity) facilitated the identification of the causative genes underlying a number of recessive disorders.[15] Autozygosity mapping is the technique where genome-wide polymorphic markers (microsatellites or SNPs) are genotyped and chromosomal regions shared by affected individuals (and not in unaffected family members) are prioritised for further study. Such regions are hypothesised to contain a rare homozygous pathogenic mutation inherited by descent from a common ancestor. This technique continues to be used by research groups to identify the genes responsible for autosomal recessive disorders, in particular those that are more common in geographical isolates, *e.g.* Northern Finland, Newfoundland and in communities where consanguinity is common. In Manchester, our group has worked with many families from the British Pakistani community to define the genes that cause many rare recessive disorders, including brittle cornea syndrome,[16] urofacial syndrome[17] and dihydrofolate reductase deficiency.[18]

The completion of the Human Genome Project in 2003 and some modest technological advances, together with the substantial efforts of the genetics research community, led to the molecular definition of additional rare

Table 2.3 Autosomal dominant retinitis pigmentosa is a genetically heterogeneous condition. Clinically indistinguishable phenotypes are caused by different genes that can be inherited in different patterns.

Locus	*Gene*	*Chromosome*
RP18	*PRPF3*	1q13–q23
RP4	*RHO*	3q21–q24
RP7	*PRPH2*	6p21.2–cen
RP9	*PAP1*	7p13–p15
RP10	*IMPDH1*	7q31.3
RP1	*ORP1*	8q11–q13
RP27	*NRL*	14q11.2
RP13	*PRPF8*	17p13.3
RP17	*CA4*	17q22
RP11	*PRPF31*	19q13.4
RP30	*FSCN2*	17q25.3
RP48	*GUCA1B*	6p21.1
RP42	*KLHL7*	7p15.3
RP37	*N2RE3*	15q23
RP53	*RDH12*	14q24.1
RP35	*SEMA4A*	1q22

disorders. However, difficulties in identification of individuals with the same rare conditions, genetic heterogeneity (different genes causing the same clinical disorders; see Table 2.3) and the lack of multi-generational families limited these successes, and the majority of inherited disorders remained undefined.

2.5 Next Generation Sequencing Technology

In 2009, the first proof of principle studies were published on the application of a novel technology called massively paralleled or next generation sequencing (NGS) to define the causes of rare diseases,[19,20] previously not amenable to standard approaches. This technology has led to the molecular characterisation of numerous rare disorders, at an accelerating pace, over the past 4 years. The technology has rapidly positioned itself to revolutionise medical practice.

NGS is a technology that is able to generate large sets of sequence data from individual DNA strands and allows interrogation of the entire genome of an individual in days or weeks at a cost of thousands of pounds/dollars, which is comparable to the costs of standard diagnostic genetic testing. The huge data sets require large computing data storage capacity and analysis undertaken by dedicated bioinformaticians as well as detailed interpretation at the biological level by scientists and clinicians.

The primary aim of the technology was to undertake whole genome sequencing, and this has been achieved for pathogens, lower organisms as well as plants and mammals, including humans. However these applications have been refined to accelerate rare disease gene identification. The majority of these disease gene identification studies to date using NGS have undertaken exome sequencing. The exome encompasses all coding and non-coding exons, some intronic and untranslated regions and promoters often produced as off-the-shelf reagents that allow hybridisation or 'capture' of the relevant sequences. This approach has primarily been championed as an effective method of identifying disease-causing mutations underlying rare disorders, which are predicted to be in protein-coding sequence; the significantly smaller data sets (when compared to complete genomes) mean that the computing challenges are more easily surmountable.

The majority of successes using NGS in rare disease gene identification have been focused within three areas (Figure 2.1); (i) Truly monogenic disorders – the comparison of sequence data from unrelated individuals with identical clinical phenotypes who share novel genetic changes in the same gene has been especially successful in defining the causes of rare autosomal recessive disorders.[20–22] Although challenging, this has been achieved in autosomal dominant conditions, as shown through our work on multiple spinal meningiomas[23] and Ohdo syndrome;[24] (ii) *De novo* disorders – disorders that affect a child due to a novel mutation not present in either parent include disorders with significantly reduced reproductive fitness often associated with severe learning disability, *e.g.* Bohring–Opitz[25] and Coffin–Siris syndromes.[26,27] For such disorders it is unusual to have more than one affected family member, which historically made disease gene identification resistant to traditional approaches. The DNA sequence of the affected child is compared to that of the parents (trio analysis) and only novel genetic variants not present in the parents are considered. Usually only one or two novel *de novo* loss of function, nonsense or frameshift mutations are present in an individual. Consequently, their identification can rapidly define a causative or mutated gene, especially if more than one trio with the same underlying disorder is sequenced; (iii) Accelerated positional cloning – NGS has been

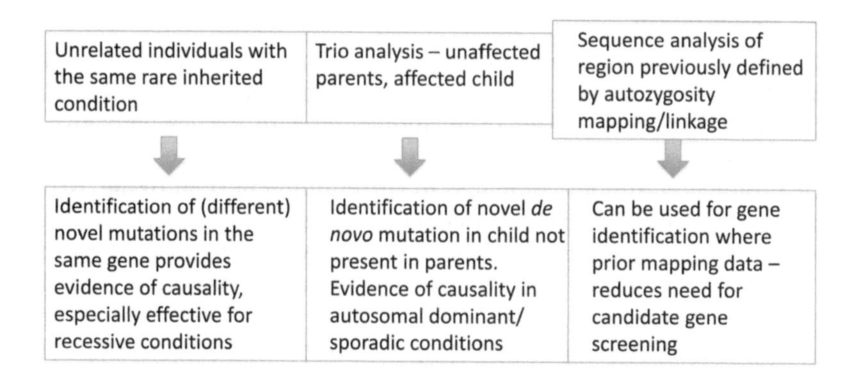

Unrelated individuals with the same rare inherited condition
Trio analysis – unaffected parents, affected child
Sequence analysis of region previously defined by autozygosity mapping/linkage
Identification of (different) novel mutations in the same gene provides evidence of causality, especially effective for recessive conditions
Identification of novel de novo mutation in child not present in parents. Evidence of causality in autosomal dominant/ sporadic conditions
Can be used for gene identification where prior mapping data – reduces need for candidate gene screening

Figure 2.1 Uses of next generation sequencing (exome analysis) in rare disease gene identification.

used to define the causative gene for rare disorders where there is some prior evidence about the likely chromosomal location of the responsible gene. Such prior information is generated through linkage studies by, for example, genotyping distantly related individuals who are both affected by the same condition and defining shared chromosomal regions or by autozygosity mapping in consanguineous families.[28] Once the shared chromosomal regions are defined this information can be used to refine the sequence data that needs to be screened, limiting the bioinformatic challenges.

2.6 Complexities in Disease Gene Identification

2.6.1 Heterogeneity

A major challenge of disease gene identification arises with clinical and genetic heterogeneity. Clinical heterogeneity – multiple conditions with similar, but not identical, clinical features – creates complexity as the conditions are unlikely to be caused by changes in the same gene. Therefore precise phenotypic characterisation is key to successful disease gene identification. Genetic heterogeneity arises where changes in more than one gene can lead to indistinguishable clinical conditions (Table 2.3). Many common disorders with a genetic basis, including sensorineural deafness, non-syndromal learning disability and retinitis pigmentosa, demonstrate high genetic heterogeneity. Novel gene identification is hindered by the low frequency of mutations among the remaining 'undiscovered' genes. The subsequent, and important, processes for delivery of clinical diagnostic services are substantially hindered by the requirement to screen effectively such a large number of genes. The bioinformatic and clinical interpretation in such situations remains complex. Many rare inherited disorders exhibit more limited heterogeneity, including those defined by our group such as brittle cornea[16] and urofacial syndromes.[17,22]

2.6.2 Mosaicism

Mosaicism arises secondary to post-zygotic (changes that arise after fertilisation of the ovum by the spermatozoa) mutations or chromosomal aberrations. Such alterations result in an individual with tissues with distinct genetic profiles. A classic example of this is the difference between the genetic profile of a tumour compared to surrounding normal tissue. NGS is being used to define somatic (acquired) changes in genes that lead to a range of rare disorders associated with tissue overgrowth and cancer. Examples include Proteus syndrome[29] and macrocephaly-cutis marmorata telangiectasia (MCMT).[30] Activating mutations in members of the *AKT-PIK3* pathway have been identified in the affected overgrown tissues, which are not present in the surrounding tissues. A number of the genes related to the overgrowth disorders have been targets for a number of cancer treatments and therefore immediate exciting therapeutic opportunities have arisen.[29]

2.6.3 Non-Coding Causal Variants

To date most uses of NGS in rare disease gene identification have focused on exome sequencing that target coding sequences around expressed sequences. As studies rapidly move towards whole genome sequencing they have been able to complement the identification of sequence variants with a simultaneous ability to identify CNVs. However, such an approach also identifies mutations in introns, regulatory promoters and enhancers or in non-genetic sequences that regulate genes already known to cause rare disorders. The challenges of whole genome analysis, particularly the analysis of larger data sets – containing up to 6000 novel sequence variants in each individual – and the interpretation of the consequences of the sequence alterations require consideration to determine how this approach will be used to maximally exploit the data produced.

2.7 Bioinformatic Challenges

The confirmation that a genetic variant is truly causal for a rare disease remains challenging when the analysis of a whole exome or genome can identify hundreds of variants of unknown effect. There are a number of recognisable approaches that can help to filter such extensive lists of genetic changes: segregation of the putative causal variant with a given phenotype in affected family members and its absence in unaffected family members can be helpful. However, for conditions and families where there is only limited family history information this may be impossible, while non-penetrance and variable expression of the phenotype can make interpretation difficult.

(i) When examining genomic data, the specific type of mutation can be ranked according to the likelihood that it is pathogenic. Thus, loss of function mutations, such as nonsense or frameshift mutations, are more likely to be pathogenic compared to splicing, missense or synonymous changes.

(ii) Absence of the genetic variant in large cohorts of ethnically matched controls also provides supporting evidence of pathogenicity, in particular for variants that may underlie autosomal dominant and X-linked conditions. Large publically available databases, including the Exome Variant Server (EVS, http://evs.gs.washington.edu/EVS/), provide resources of sequence data from healthy individuals, although there is currently a limited range of ethnic groups represented by such resources.

(iii) Different mutations in the same gene in unrelated individuals affected by the same condition provide definitive genetic evidence regarding causality. Rarely is the same mutation in a gene responsible for a specific condition, *e.g.* Myhre syndrome due to activating mutations in *SMAD4*.[31]

(iv) Ultimately functional evidence is required to be certain that a variation substantially diminishes or alters protein function. Again there are strategies that may give a supporting indication of this.

Comparison of sequences across species and evidence of conservation of amino acid residues indicates a higher likelihood that any change would result in a deleterious effect on the protein.

Modelling the potential effects on the resultant protein of an amino acid substitution or the functional effects through disruption of a specific motif can be informative.

In silico algorithms, including PolyPhen-2 (http://genetics.bwh.harvard.edu/pph2/) and SIFT (http://sift.jcvi.org/) have been developed to provide likelihood scores for pathogenicity of a predicted amino acid change.

For a minority of variants, in particular those hypothesised to underlie novel genetic causes of human disease, functional studies using cell culture systems can be employed to examine the effects of specific variants. However these are often time-consuming and require specialist skills. Such approaches can be further complemented by animal models, including in *Drosophila*, zebrafish and mice with defined genetic alterations. Patient-derived induced pluripotent stem cell (iPSC) models are also being used to generate the relevant tissues for functional studies to study disorders affecting tissues previously not amenable to interrogation, *e.g.* neural tissues. Currently most functional and/or animal studies do not have the throughput to be practical to inform routine diagnosis, but where available are useful in providing evidence to support the role of the causative gene.

2.8 Clinical Diagnostic Services

Of the approximately 7000 rare genetic disorders that have been defined, 3500 have so far been characterised at a molecular level.[2] For clinicians it is often difficult to know how to access genetic testing for their patients. The Genetic Testing Registry (http://www.ncbi.nlm.nih.gov/gtr/) has collated the details on 10 300 tests for 3350 conditions analysing 2318 different genes. The majority of these tests are still undertaken on a research basis in a range of laboratories.

The traditional testing model has been for a clinician to define, through detailed clinical investigation, a specific phenotype and to develop a clinical hypothesis. This would result in the ordering of a specific genetic test on a single gene (or at most a very small number of potentially relevant genes) to test that hypothesis. The pick-up rate of such a testing approach varies considerably, from approximately 0.5% for Fragile X [32] to over 40% for CHARGE syndrome (personal communication). In general this has been an inefficient approach which is by its very nature limited to patients, and their relatives, with phenotypes consistent with a genetic disease. For the majority of conditions, or for population testing (*e.g.* widespread *BRCA1/2* testing for breast cancer), clinical testing has not been possible. Testing has been especially challenging in heterogeneous conditions, including developmental delay, deafness, retinal dystrophies and glycogen storage disorders. Here, NGS has started to have a major impact. The development of panel testing, where a selected array of genes can be analysed in a single assay, has been successfully introduced. Our own experience with testing of a panel of 105 retinal dystrophy genes has seen an increase in detection of the causal variant from 14 to 60% over the past 2 years of providing this service.[33] Such diagnostic information has clarified recurrence risks, facilitated carrier testing and allowed identification of individuals for whom specific treatment strategies may be relevant.

Clinical exome sequencing has been launched at a number of centres in the USA and is being developed by clinical laboratories across Europe. This approach has the potential to transform the practice of medicine. At present clinical reports are generated providing feedback on specific phenotypes relevant to the presentation of the tested individual. Reports may also provide information about carrier status for a range of recessive disorders, so informing future reproductive risks, and of unexpected dominant disorders for which preventive screening may be appropriate. Initial clinical exome testing has focused on the testing of children with learning disabilities, developmental disorders and neurological phenotypes. In approximately 25% of cases the causative genetic change is being identified. Studies have assessed the utility of exome testing in a number of settings including improving diagnosis of children on intensive care units or affected by likely recessive disorders when born to consanguineous parents. The next challenge is to introduce this testing into other areas of mainstream medicine including cardiology, renal and gastrointestinal medicine.

A number of studies have started to consider how this extra information generated from exome or genome analysis should be fed back to tested individuals. Information about increased risks of coronary artery disease, cancer and rare inherited disorders like Marfan syndrome lend themselves to targeted interventions. However, concerns have been raised about individual autonomy, inappropriate use of this information to discriminate in terms of employment and insurance and the burden placed upon health professionals to feed back accurate information that can have a benefit rather than indicating increased risk with no potential to alter natural history, for example in providing information about neurodegenerative disorders. The improved technology, reduction in costs and advances in bioinformatics mean that exome sequencing and in time whole genome sequencing will become routine in clinical diagnosis over the next decade. Many challenges exist to ensure that the potential is harnessed to improve health care but the opportunities are too great for this not to happen.

2.9 Concluding Remarks

The last few years has seen a transformation in disease gene identification for rare diseases, with the introduction of microarrays and NGS. These advances have been rapidly adopted into clinical practice. Exome/whole genome sequencing will become a routine part of the diagnostic armamentarium. Identification of the genetic cause of individual disorders through exome sequencing offers enormous opportunity for personalised medicine. Gene therapy based approaches based on knowledge of the specific altered gene or alternative strategies, *e.g.* stop mutation, read through or exon skipping based on the specific mutational type, for example in DMD, will be increasingly available.

Already examples have emerged of individuals who have had successful diagnosis and treatment due to exome sequencing.[34,35] Such examples of NGS aiding the diagnosis and treatment of previously intractable medical cases provide real-life stimuli to overcome the challenges and ensure the broadest access to genomic diagnostics and the related benefits.

Acknowledgements

GCB is an NIHR Senior Clinical Investigator. The authors thank the Manchester Biomedical Research Centre for support.

References

1. Rare is Common, http://www.sthc.co.uk/Documents/CMO_Report_2009.pdf, accessed 1 October 2013.
2. National Institutes of Health website, "Office of Rare Disease Research", http://rarediseases.info.nih.gov/AboutUs.aspx, accessed 23 September 2013.
3. Rare Disease UK website, http://www.raredisease.org.uk/, accessed 23 September 2013.
4. S. E. McCandless, J. W. Brunger and S. B. Cassidy, *Am. J. Hum. Genet.,* 2004, **74**, 121–127.
5. E. K. Mallinson, K. F. Newton, J. Bowen, F. Lalloo, T. Clancy, J. Hill and D. G. Evans, *Gut,* 2010, **59**, 1378–1382.
6. Online Mendelian Inheritance in Man, OMIM®, McKusick-Nathans Institute of Genetic Medicine, Johns Hopkins University (Baltimore, MD), http://omim.org/, accessed 1 October 2013.
7. *Catalogue of Unbalanced Chromosome Aberrations in Man*, ed. A. Schinzel, Pub. Walter de Gruyter, Germany, 2001.
8. A. Hattersley, J. Bruining, J. Shield, P. Njolstad and K. C. Donaghue, *Pediatr. Diabetes,* 2009, **10**(suppl. 12), 33–42.
9. Rare Disease Impact Report: Insights from patients and the medical community, April 2013, http://www.raredisease.org.uk/documents/Website%20Documents%20/rare-disease-impact-report.pdf.
10. J. K. Cowell, R. B. Wadey, B. B. Buckle and J. Pritchard, *Hum. Genet.,* 1989, **82**, 123–126.
11. J. M. Dunn, R. A. Phillips, A. J. Becker and B. L. Gallie, *Science,* 1988, **241**, 1797–1800.

12. M. Gajecka, K. L. Mackay and L. G. Shaffer, *Am. J. Med. Genet., Part C,* 2007, **145**, 346–356.
13. D. Malhotra and J. Sebat, *Cell,* 2012, **148**, 1223–1241.
14. T. Jordan, I. Hanson, D. Zaletayev, S. Hodgson, J. Prosser, A. Seawright, N. Hastie and V. van Heyningen, *Nat. Genet.,* 1992, **1**, 328–332.
15. E. S. Lander and D. Botstein, *Science,* 1987, **236**, 1567–1570.
16. E. M. Burkitt Wright, H. L. Spencer, S. B. Daly, F. D. Manson, L. A. Zeef, J. Urquhart, N. Zoppi, R. Bonshek, I. Tosounidis, M. Mohan, C. Madden, A. Dodds, K. E. Chandler, S. Banka, L. Au, J. Clayton-Smith, N. Khan, L. G. Biesecker, M. Wilson, M. Rohrbach, M. Colombi, C. Giunta and G. C. Black, *Am. J. Hum. Genet.,* 2011, **88**, 767–777.
17. S. B. Daly, J. E. Urquhart, E. Hilton, E. A. McKenzie, R. A. Kammerer, M. Lewis, B. Kerr, H. Stuart, D. Donnai, D. A. Long, B. Burgu, O. Aydogdu, M. Derbent, S. Garcia-Minaur, W. Reardon, B. Gener, S. Shalev, R. Smith, A. S. Woolf, G. C. Black and W. G. Newman, *Am. J. Hum. Genet.,* 2010, **86**, 963–969.
18. S. Banka, H. J. Blom, J. Walter, M. Aziz, J. Urquhart, C. M. Clouthier, G. I. Rice, A. P. M. de Brouwer, E. Hilton, G. Vassallo, A. Will, D. E. C. Smith, Y. M. Smulders, R. A. Wevers, R. Steinfeld, S. Heales, Y. J. Crow, J. N. Pelletier, S. Jones and W. G. Newman, *Am. J. Hum. Genet.,* 2011, **88**, 216–225.
19. S. B. Ng, E. H. Turner, P. D. Robertson, S. D. Flygare, A. W. Bigham, C. Lee, T. Shaffer, M. Wong, A. Bhattacharjee, E. E. Eichler, M. Bamshad, D. A. Nickerson and T. Shendure, *Nature,* 2009, **461**, 272–276.
20. S. B. Ng, K. J. Buckingham, C. Lee, A. W. Bigham, H. K. Tabor, K. M. Dent, C. D. Huff, P. T. Shannon, E. W. Jabs, D. A. Nickerson, J. Shendure and M. J. Bamshad, *Nat. Genet.,* 2010, **42**, 30–35.
21. E. M. Jenkinson, A. U. Rehman, T. Walsh, J. Clayton-Smith, K. Lee, R. J. Morell, M. C. Drummond, S. N. Khan, M. A. Naeem, B. Rauf, N. Billington, J. M. Schultz, J. E. Urquhart, M. K. Lee, A. Berry, N. A. Hanley, S. G. Mehta, D. Cilliers, P. E. Clayton, H. Kingston, M. J. Smith, T. T. Warner, G. C. Black, D. Trump, J. R. E. Davis, W. Ahmad, S. M. Leal, S. Riazuddin, M. C. King, T. B. Friedman and W. G. Newman, *Am. J. Hum. Genet.,* 2013, **92**, 605–613.
22. H. M. Stuart, N. A. Roberts, B. Burgu, S. B. Daly, J. E. Urquhart, S. Bhaskar, J. E. Dickerson, M. Mermerkaya, M. S. Silay, M. A. Lewis, B. O. Olondriz, B. Gener, C. Beetz, R. E. Varga, O. Gulpınar, E. Suer, T. Soygur, Z. B. Özçakar, F. Yalçınkaya, A. Kavaz, B. Bulum, A. Gucuk, W. W. Yue, F. Erdogan, A. Berry, N. A. Hanley, E. A. McKenzie, E. N. Hilton, A. S. Woolf and W. G. Newman, *Am. J. Hum. Genet.,* 2013, **92**, 259–264.
23. M. J. Smith, J. O'Sullivan, S. S. Bhaskar, K. D. Hadfield, G. Poke, J. Caird, S. Sharif, D. Eccles, D. Fitzpatrick, D. Rawluk, D. du Plessis, W. G. Newman and D. G. Evans, *Nat. Genet.,* 2013, **45**, 295–298.
24. J. Clayton-Smith, J. O'Sullivan, S. Daly, S. Bhaskar, R. Day, B. Anderson, A. K. Voss, T. Thomas, L. G. Biesecker, P. Smith, A. Fryer, K. E. Chandler, B. Kerr, M. Tassabehji, S. A. Lynch, M. Krajewska-Walasek, S. McKee,

J. Smith, E. Sweeney, S. Mansour, S. Mohammed, D. Donnai and G. Black, *Am. J. Hum. Genet.*, 2011, **89**, 675–681.
25. A. Hoischen, B. W. van Bon, B. Rodríguez-Santiago, C. Gilissen, L. E. Vissers, P. de Vries, I. Janssen, B. van Lier, R. Hastings, S. F. Smithson, R. Newbury-Ecob, S. Kjaergaard, J. Goodship, R. McGowan, D. Bartholdi, A. Rauch, M. Peippo, J. M. Cobben, D. Wieczorek, G. Gillessen-Kaesbach, J. A. Veltman, H. G. Brunner and B. B. de Vries, *Nat. Genet.*, 2011, **43**, 729–731.
26. Y. Tsurusaki, N. Okamoto, H. Ohashi, T. Kosho, Y. Imai, Y. Hibi-Ko, T. Kaname, K. Naritomi, H. Kawame, K. Wakui, Y. Fukushima, T. Homma, M. Kato, Y. Hiraki, T. Yamagata, S. Yano, S. Mizuno, S. Sakazume, T. Ishii, T. Nagai, M. Shiina, K. Ogata, T. Ohta, N. Niikawa, S. Miyatake, I. Okada, T. Mizuguchi, H. Doi, H. Saitsu, N. Miyake and N. Matsumoto, *Nat. Genet.*, 2012, **44**, 376–378.
27. G. W. Santen, E. Aten, Y. Sun, R. Almomani, C. Gilissen, M. Nielsen, S. G. Kant, I. N. Snoeck, E. A. Peeters, Y. Hilhorst-Hofstee, M. W. Wessels, N. S. den Hollander, C. A. Ruivenkamp, G. J. van Ommen, M. H. Breuning, J. T. den Dunnen, A. van Haeringen and M. Kriek, *Nat. Genet.*, 2012, **44**, 379–380.
28. D. Hanson, P. G. Murray, J. O'Sullivan, J. Urquhart, S. Daly, S. S. Bhaskar, L. G. Biesecker, M. Skae, C. Smith, T. Cole, J. Kirk, K. Chandler, H. Kingston, D. Donnai, P. E. Clayton and G. C. Black, *Am. J. Hum. Genet.*, 2011, **89**, 148–153.
29. M. J. Lindhurst, J. C. Sapp, J. K. Teer, J. J. Johnston, E. M. Finn, K. Peters, J. Turner, J. L. Cannons, D. Bick, L. Blakemore, C. Blumhorst, K. Brockmann, P. Calder, N. Cherman, M. A. Deardorff, D. B. Everman, G. Golas, R. M. Greenstein, B. M. Kato, K. M. Keppler-Noreuil, S. A. Kuznetsov, R. T. Miyamoto, K. Newman, D. Ng, K. O'Brien, S. Rothenberg, D. J. Schwartzentruber, V. Singhal, R. Tirabosco, J. Upton, S. Wientroub, E. H. Zackai, K. Hoag, T. Whitewood-Neal, P. G. Robey, P. L. Schwartzberg, T. N. Darling, L. L. Tosi, J. C. Mullikin and L. G. Biesecker, *N. Engl. J. Med.*, 2011, **365**, 611–619.
30. J. B. Rivière, G. M. Mirzaa, B. J. O'Roak, M. Beddaoui, D. Alcantara, R. L. Conway, J. St-Onge, J. A. Schwartzentruber, K. W. Gripp, S. M. Nikkel, T. Worthylake, C. T. Sullivan, T. R. Ward, H. E. Butler, N. A. Kramer, B. Albrecht, C. M. Armour, L. Armstrong, O. Caluseriu, C. Cytrynbaum, B. A. Drolet, A. M. Innes, J. L. Lauzon, A. E. Lin, G. M. Mancini, W. S. Meschino, J. D. Reggin, A. K. Saggar, T. Lerman-Sagie, G. Uyanik, R. Weksberg, B. Zirn and C. L. Beaulieu, *Nat. Genet.*, 2012, **44**, 934–940.
31. C. Le Goff, C. Mahaut, A. Abhyankar, W. Le Goff, V. Serre, A. Afenjar, A. Destrée, M. di Rocco, D. Héron, S. Jacquemont, S. Marlin, M. Simon, J. Tolmie, A. Verloes, J. L. Casanova, A. Munnich and V. Cormier-Daire, *Nat. Genet.*, 2011, **44**, 85–88.

32. J. Macpherson and H. Sawyer, *Best practice guidelines for molecular diagnosis of Fragile X Syndrome*, http://www.cmgs.org/BPGs/pdfs%20current%20bpgs/Fragile%20X.pdf, accessed 1 October 2013.
33. J. O'Sullivan, B. G. Mullaney, S. S. Bhaskar, J. E. Dickerson, G. Hall, A. O'Grady, A. Webster, S. C. Ramsden and G. C. Black, *J. Med. Genet.*, 2012, **49**, 322–326.
34. E. A. Worthey, A. N. Mayer, G. D. Syverson, D. Helbling, B. B. Bonacci, B. Decker, J. M. Serpe, T. Dasu, M. R. Tschannen, R. L. Veith, M. J. Basehore, U. Broeckel, A. Tomita-Mitchell, M. J. Arca, J. T. Casper, D. A. Margolis, D. P. Bick, M. J. Hessner, J. M. Routes, J. W. Verbsky, H. J. Jacob and D. P. Dimmock, *Genet. Med.*, 2011, **13**, 255–262.
35. M. N. Bainbridge, W. Wiszniewski, D. R. Murdock, J. Friedman, C. Gonzaga-Jauregui, I. Newsham, J. G. Reid, J. K. Fink, M. B. Morgan, M. C. Gingras, D. M. Muzny, L. D. Hoang, S. Yousaf, J. R. Lupski and R. A. Gibbs, *Sci. Transl. Med.*, 2011, **3**, 87re3.

DRUG DEVELOPMENT FACTORS

CHAPTER 3

The Challenges of Conducting Clinical Trials in Diseases with Small Target Populations

STEVEN ARKIN

BioTherapeutics Clinical Research, Pfizer Worldwide R&D, 200 Cambridge Park Drive, Cambridge, MA 02140, USA
E-mail: steven.arkin@pfizer.com

3.1 Regulatory Considerations

Rare diseases comprise a substantial spectrum of human disease pathology. The National Institutes of Health (NIH) Office of Rare Diseases reports almost 7000 rare diseases in its searchable database, and EURORDIS Rare Diseases Europe reports there are between 6000 and 8000 rare diseases affecting 30 million individuals in the EU, with 80% of genetic origin.[1,2] By their nature, rare diseases are described as affecting a small portion of the population. While there is no universally accepted definition of a rare disease, the US Food and Drug Administration (FDA) and the European Medicines Agency (EMA) have provided regulatory criteria for orphan drugs. These criteria, a prevalence of <200 000 and an incidence no greater than 5 in 10 000 persons respectively, provide acceptable operational definitions for what constitutes a rare disease.[3]

In recent years, drug development for treatment of rare diseases has been increasingly prioritised by pharmaceutical sponsors, resulting in an increased number of orphan drug designations and an increased number of novel molecular entities receiving regulatory approval in orphan indications.[4,5] Factors that have created enhanced opportunities for drug development in

RSC Drug Discovery Series No. 38
Orphan Drugs and Rare Diseases
Edited by David C Pryde and Michael J Palmer

Published by the Royal Society of Chemistry, www.rsc.org

rare diseases include recent advances in characterisation of the human genome, molecular profiling and target and biomarker selection.[6] Favourable regulatory incentives have also contributed to a climate that is more supportive of drug development for orphan diseases. The US Orphan Drug Act confers 7 years of market exclusivity from date of approval, tax incentives, fee exemptions and priority review vouchers for companies that develop drugs for orphan diseases.[7] EU orphan drug designation provides 10 year marketing exclusivity, protocol assistance and fee reductions.[3] These incentives provide substantial rationale for prioritisation of rare diseases when pharmaceutical sponsors make decisions regarding allocation of their R&D resources. There are also a number of regulatory enablers that can help to facilitate clinical development of drugs for rare diseases. For the US FDA these include Fast Track, Breakthrough Therapy, Accelerated Approval and Priority Review designations.[8] For the EMA these include conditional marketing authorisation and market authorisation under exceptional circumstances.[9,10]

Despite recent scientific advances and available regulatory incentives and enablers, drug development for rare diseases poses substantial challenges, some unique to clinical investigation in rare patient populations and some germane to drug development in general. By their nature, clinical trials for rare diseases are conducted in small patient populations. However drugs developed for rare disease populations are subject to the same rigour in the assessment of safety and efficacy as drugs developed for more common diseases. The US FDA requires that reports of adequate and well-controlled investigations provide the primary basis for determining whether there is substantial evidence to support the claims of effectiveness of new drugs.[11] The EMA requires that the particulars and documents which accompany an application for market authorisation for a medicinal product demonstrate that the potential risks are outweighed by the therapeutic efficacy of the product and the EMA Guideline on Clinical Trials in Small Populations states that market authorisation of orphan drugs will be judged according to the same standards as for other products.[12,13] In a 2012 analysis, Orfali and colleagues compared the clinical trial experience described in the approved US product label for a sampling of 37 orphan drugs and 58 non-orphan drugs across the pulmonary, neuroscience, haematology and endocrinology therapeutic areas. While the number of clinical trials described in the label was smaller for orphan than for non-orphan drugs (2.81 *vs.* 3.51), and clinical trials for orphan drugs had fewer participants than did trials for orphan drugs, there was no difference between the proportion of orphan and non-orphan drug labels that gave evidence of randomisation, blinding and clinically relevant end points.[14]

3.2 Challenges in Rare Disease Populations – General Overview

To succeed in meeting regulatory requirements for market authorisation, clinical investigations that are conducted in small patient populations, often in diseases without precedented treatments, must surmount

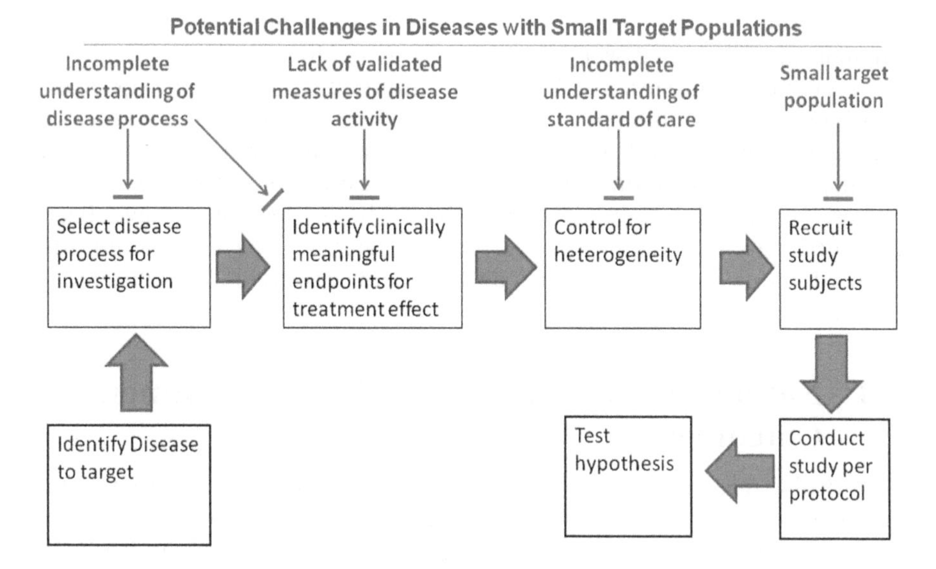

Figure 3.1 Potential challenges in diseases with small target populations. Development of therapeutics for rare diseases requires identification of the disease to study, the disease process/stage of disease progression for investigation, clinically meaningful end points for treatment effect, controlling for heterogeneity, recruitment, conduct of the study and hypothesis testing. In rare diseases with small target patient populations, incomplete understanding of the disease process, lack of validated measures for disease activity, incomplete understanding of the standard of care and a small target population from which to recruit may present unique challenges that must be mastered if the study hypothesis is to be properly tested.

multiple challenges (Figure 3.1). These challenges may include (i) incomplete understanding of the disease clinical course, with a resulting adverse impact on ability to anticipate outcomes for placebo or active comparator treatment groups; (ii) lack of validated measures of disease activity/progression resulting in limitations on end point(s) capable of supporting regulatory approval in a reasonable time frame; (iii) incomplete understanding of the standard of care for disease management, potentially increasing the heterogeneity that may be encountered among subjects in small data sets; (iv) small numbers of potential subjects, which creates difficulty accessing sufficient sample size to support hypothesis testing and/or characterisation of benefit risk. Engaging a sufficient number of study sites to support subject recruitment plus heightened competition with other sponsors/investigators for the limited quantum of patients available for investigation may create logistical challenges.[15] As well, the challenges that are inherent in drug development for large populations, selection of appropriate and clinically relevant end points for efficacy (and/or safety) that are accessible in a sufficiently compressed time frame to permit drug development, management of heterogeneity in the patient population and proper oversight of the investigations to assure quality of the registration dossier, are frequently amplified when conducting investigations in small patient populations. While not intended to be an exhaustive survey of all issues encountered for development of drugs in small patient populations, this chapter will explore some of these challenges, provide relevant vignettes where these challenges have been managed, often in an innovative manner, and set the context for the detailed case studies of drug development provided in subsequent chapters.

3.3 Natural History and Likely Placebo Treatment Outcomes

Historically, rare diseases have been defined by clinical features, frequently on the basis of experience in a very limited number of patients. For disorders of genetic origin, this often results in aggregation of groups of (i) individuals affected by the same mutation against a background of differing background genetic modifiers, (ii) individuals affected by different mutations in the same gene, or (iii) by mutations in different genes that contribute to a common physiologic pathway. For acquired disorders, individuals with similar pathologic outcomes are often aggregated despite incompletely understood differences in underlying aetiology. The resulting limitations in clinical experience and poorly understood heterogeneities in patient populations contribute to an incomplete understanding of the disease clinical course and adversely impact the ability to anticipate outcomes for placebo or active comparator treatment groups. Natural history study data, collected prospectively or retrospectively, can be invaluable in modelling anticipated outcomes in such populations (Table 3.1). The rare genetic disorders

Table 3.1 Understanding disease history and anticipating control arm outcomes.

Disease	*Data source*	*Type of data collection*
Autosomal dominant polycystic kidney disease	Consortium of Radiologic Imaging Study of Polycystic Kidney Disease study (CRISP study, CRISP 2 study)	Prospective natural history study
Sickle cell anaemia	Cooperative Study of Sickle Cell Disease (CSSCD)	Prospective natural history study
Intracerebral haemorrhage	Virtual International Stroke Archive (VISTA)	Archive of clinical trial data sets
Congenital factor XIII deficiency	Sponsor initiated survey of patient data	Retrospective registry type data collection

autosomal dominant polycystic kidney disease and sickle cell anaemia provide examples where data from prospective natural history studies was successfully exploited to inform the design of pivotal therapeutic clinical studies. In contrast, intracerebral haemorrhage (ICH) provides an example where natural history data, had it been available, could have impacted favourably on design elements of a failed pivotal study. For ultra-rare diseases, prospectively collected natural history data is often not available but the experience with hereditary deficiency of coagulation factor XIII (FXIII deficiency) demonstrates that retrospectively collected natural history data can also effectively inform design of a pivotal study.

3.3.1 Autosomal Dominant Polycystic Kidney Disease

Autosomal dominant polycystic kidney disease (ADPKD) is a hereditary disorder affecting approximately 1 in every 1000 individuals and is most frequently associated with inheritance of a single mutated allele of either PKD1 (polycystin-1) or PKD2 (polycystin-2).[16,17] The disease is characterised by development of multiple fluid-filled cysts in both kidneys; other symptoms during the course of the disease may include hypertension, episodes of renal pain, haematuria, liver cysts and brain aneurysms. Progression of ADPKD is characterised by growth of numerous renal cysts leading gradually to the loss of renal function and eventually end-stage renal disease (ESRD) in approximately 50% of patients, typically in the fourth to sixth decade of life.[18] The disease course for ADPKD has marked variability and evaluating pharmacological treatments for this disorder is difficult due to the associated variability and typically long time course for progression.[16] The Consortium of Radiologic Imaging Studies of Polycystic Kidney Disease (CRISP) study represents an NIH sponsored observational study of ADPKD that has measured rates of change in total kidney volume (TKV), total kidney cyst volume, glomerular filtration rate (GFR) and blood pressure.[19] Results from the CRISP study have been published and demonstrated that kidney volume and cyst volume increase exponentially, and that baseline kidney volume predicted the subsequent rate of increase in volume independent of age. Individuals with a baseline TKV $\geq$ 750 mL experienced more rapid enlargement of their kidneys than did individuals with TKV $<$ 750 mL and relative to other subjects, individuals with baseline TKV $>$ 1500 mL experienced a significantly greater rate of decline in GFR.[20] Data sets from the CRISP publicly funded study are available to support approved analyses. Findings from that study are reflected in design of the TEMPO 3:4 Phase 3 study that was sponsored by Otsuka Pharmaceuticals to evaluate the vasopressin V_2 antagonist tolvaptan for treatment of ADPKD. Key eligibility criteria in that study, a diagnosis of ADPKD, TKV $\geq$ 750 mL and a creatinine clearance $\geq$ 60 mL per minute, assured a study population of ADPKD subjects at the appropriate stage of disease progression to test the primary study end point (annual change in TKV) and the key composite secondary end point of time to investigator-assessed disease progression. Over a 3 year treatment period the rate of increase in TKV was lower and fewer composite secondary end point events were observed per 100 person-years of follow-up in the tolvaptan treated group than in the placebo group. Thus the TEMPO 3:4 study achieved its primary and key secondary efficacy end points.[21]

3.3.2 Sickle Cell Anaemia

Sickle cell anaemia is a hereditary haemolytic anaemia, affecting 1 in 600 African Americans in the USA, caused by a T to A substitution in the sixth codon of the gene coding for the β globin chain resulting in a glutamic acid to valine substitution. The resulting sickle haemoglobin (haemoglobin S, $\alpha_2 B_2{}^S$) has the property of forming polymers when deoxygenated in the tissues with the extent and rapidity of this polymerisation dependent on the concentration of haemoglobin S.[22] Individuals with sickle cell disease (homozygotes for haemoglobin S [HbSS] or compound heterozygotes for haemoglobin S and other haemoglobinopathies [Hb SB^0 thalassaemia, HbSC, $HbSB^+$ thalassaemia]) experience the pathology of sickle cell disease. In these individuals polymerisation of haemoglobin SS results in occlusion of small and sometimes large blood vessels causing vascular injury. Contributors to this process include haemolysis, adhesive interactions between blood cells and endothelial cells and disruption of the balance between vasodilators and vasoconstrictors. Affected individuals incur a chronic disease characterisation that can include acute episodes of intense vaso-occlusive crisis pain affecting the chest, back, abdomen or extremities, vaso-occlusive events affecting the lung (acute chest syndrome) or brain (ischemic stroke), priapism, episodes of splenic sequestration of blood, bacterial sepsis as well as chronic end organ damage that includes progressive retinopathy, renal insufficiency, skin ulcers, osteonecrosis and chronic respiratory insufficiency.[22] Individuals with sickle cell disease have had a shortened life expectancy with median age at death of 42 years for males and 48 years for females.[23] Despite the substantial morbidity and mortality associated with sickle cell disease, frequency of painful crisis is one of the few end points readily accessible on a timeline that permits drug development. The NIH Cooperative Study of Sickle Cell Disease (CSSCD) was a multicentre prospective natural history study in patients from the newborn period through adulthood.[24] Results from that study demonstrated, over a 5.13 year mean observation period, that the rate of vaso-occlusive crisis varied widely among patients with sickle cell disease. Thirty-nine percent of study subjects had no episodes of pain and 1% had more than six episodes of pain per year. The 5.2% of study subjects that had 3–10 episodes of pain per year accounted for 32.9% of all painful episodes. Among study subjects, lower levels of foetal haemoglobin, higher haematocrit and the genotypes of Hb SS and Hb SB^0 thalassaemia were associated with more frequent episodes of painful crisis.[25] Hydroxyurea is currently the only disease-modifying treatment that is approved for treatment of sickle cell disease. Results from the CSSCD were reflected in the design for the pivotal study that supported regulatory approval of hydroxyurea for treatment of sickle cell disease.[26] Inclusion criteria restricted enrolment to individuals with the HBSS variant of the disease who experienced three or more episodes of painful crisis in the prior year, a population clearly characterised in the CSSCD as having a severe disease phenotype. The observed reduction in frequency of painful crisis, median 2.5 episodes of crisis per year in hydroxyurea-treated patients *versus* 4.5 episodes of crisis per year in placebo-treated patients, was sufficient to stop the clinical trial early and to support approval of hydroxyurea for treatment within the proposed indication. More recently multiple pharmaceutical sponsors have exhibited interest in sickle cell disease, initiating clinical programmes to increase expression of foetal haemoglobin, antagonise cell adhesion, modulate inflammatory responses or modulate deoxygenation of haemoglobin SS.[27–31] These clinical investigations will probably be conducted in the more severely affected patients, against a background of hydroxyurea treatment, the current standard of care for such patients. A thorough understanding of the clinical course for sickle cell disease in hydroxyurea-treated patients will become critical for anticipating outcomes for placebo-treated subjects in these future safety and efficacy investigations. Whereas the CSSCD was conducted in a pre-hydroxyurea era, the Registry and Surveillance System for Hemoglobinopathies (RuSH) pilot project is a US Centers for Disease Control registry, conducted in collaboration with the NIH, that is collecting healthcare statistics surveillance data including hospitalisations, emergency room care, other sources of medical care and information on complications and death rates.[32] Prospectively collected data sets from this registry will be reflective of the current standard of care and can augment published clinical trial data for planning of clinical investigations of new therapeutic agents in patients with severe sickle cell disease.

3.3.3 Spontaneous Intracerebral Haemorrhage

Spontaneous intracerebral haemorrhage (ICH) is an acquired haemorrhagic event where bleeding occurs into the parenchyma of the brain and may extend into the ventricles of the brain. ICH accounts for at least 10% of stroke cases worldwide and estimates of the number of individuals affected annually in the USA range from 37 000–52 400 to 79 500.[33,34] ICH is associated with dismal clinical outcomes; approximately 40–50% of patients die within 1 month and a large proportion of individuals fail to regain functional independence by 6 months.[35] It is a major cause of permanent partial or complete disability, there are no approved therapies for its treatment and current management is largely supportive.[36] Volume of ICH has been shown to be an independent predictor of mortality and morbidity in patients with ICH.[37] Volume of ventricular blood has also been described as an important determinant of outcome in the subset of patients with supratentorial ICH.[38] Substantial growth in volume of ICH occurs in approximately 26% of patients within 1 hour of the baseline diagnostic CT, and several studies indicate that growth in volume of ICH is also an independent predictor of mortality and poor outcome.[39–41] Based on these observations, interventions targeted at reducing growth in ICH volume have been proposed as a mechanism to reduce the morbidity and mortality associated with this condition. A major challenge with this strategy is demonstrating that reduction in growth of ICH volume results in improved functional neurological outcomes and/or mortality. The potential impact of multiple baseline haemorrhage variables on outcomes, including ICH volume, location of ICH within the brain, presence and volume of intraventricular haemorrhage (IVH), as well as the potential impact of other variables such as age of study subjects and background morbidities such as hypertension, creates uncertainty regarding the outcomes that can be anticipated for placebo control groups in clinical studies of ICH. The Virtual International Stroke Archive (VISTA) database is a collaborative archive that contains data sets from 21 clinical trials, representing 13 029 subjects with ischemic stroke and 1202 subjects with ICH. Data available within the VISTA archive include results for commonly used neurological functional end points such as the Barthel Index, NIH Stroke Scale and the modified Rankin Scale as well as medical history data, onset of event to treatment time and, for selected studies, computed tomography lesion data.[42] Analyses of the VISTA data can be restricted to placebo groups as well as type of stroke and can therefore be used to examine the relationship between baseline prognostic factors and outcome measures in order to inform effective design of clinical trials. The utility of such natural history data in modelling outcomes and in designing appropriate stratifications to protect against unbalanced randomisation between treatment groups is underscored by the unexpected experience of the FAST trial in patients with ICH. Following a successful Phase 2 study in which treatment with activated coagulation factor VII (eptacog alfa) reduced growth in volume of ICH and improved neurological outcomes the Phase 3 FAST study was initiated to confirm these promising initial findings.[43] Study subjects were randomly assigned to one of three treatment groups: low dose eptacog alfa, high dose eptacog alfa or placebo. While the FAST trial confirmed a reduction in growth of ICH within the initial 24 hours for the high dose eptacog alfa treatment arm (relative to placebo), there was no significant difference among the study treatment groups for the primary efficacy end point, frequency of poor neurological outcomes.[44] Review of the FAST study results revealed an unbalanced randomisation occurred wherein mean baseline IVH volume in the high dose eptacog alfa treatment group was 5.3 mL, a value greater than the mean placebo group IVH volumes of 2.7 mL at baseline and 4.6 mL at the 24 hour post-treatment time point. Had VISTA natural history data for ICH been available at the time the FAST study was planned, correlation of IVH volume with neurological outcomes might have prompted the sponsor to include stratification for baseline IVH volume in the randomisation assignments for study treatments, thereby avoiding a bias against successful study outcome.

3.3.4 Severe Congenital Factor XIII Deficiency

Severe congenital factor XIII deficiency is an ultra-rare hereditary bleeding disorder characterised by umbilical bleeding during the neonatal period, delayed/poor wound healing, delayed soft tissue bruising, mucosal bleeding, recurrent miscarriages and increased risk for life-threatening intracranial haemorrhages. Factor XIII deficiency is generally caused by mutations in the gene coding for the FXIII-A subunit or, much less frequently, in the gene coding for the factor XIII-B subunit. Overall incidence of the disorder is 1 in 3–5 million people with an autosomal recessive mode of transmission. Treatment of factor XIII deficiency is based on factor replacement therapy using cryoprecipitate or more commonly factor XIII concentrates, with prophylactic treatment generally recommended for patients with a history of intracranial haemorrhage.[45] Comprehensive natural history data for this ultra-rare disorder are lacking and current knowledge is largely derived from registry and survey data. The lack of natural history data posed a substantial challenge for the study sponsor in modelling efficacy outcomes as it designed a pivotal efficacy and safety study for its recombinant factor XIIIA-subunit product catridecacog. To support a primary end point of number of bleeding episodes per year requiring treatment with factor XIII replacement therapy, the sponsor conducted a comprehensive global survey of physicians. Data were collected regarding how many bleeds per year patients had experienced based on 92 patients with congenital factor XIII deficiency. Of the 92 patients, 23 were receiving exclusively episodic (on-demand) treatment in response to bleeding episodes and data on frequency of bleeding episodes was available for 16. Mean frequency of bleeding episodes per year among these 16 patients was 2.91, with a range of 0–12 episodes per year and a 95% confidence interval calculated as 0.95–4.86.[46] Based on these results the sponsor designed an open-label, single-arm, Phase 3 study evaluating prophylactic treatment with catridecacog in subjects with genetically proven deficiency of factor XIII-A subunit using a primary efficacy end point of frequency of bleeding episodes. Study subjects on prophylactic treatment with catridecacog experienced a mean of 0.048 bleeding episodes per year (95% CI, 0.0094–0.2501), a bleeding frequency significantly less than that observed in the above historic group. These treatment results, when compared with the historic control group receiving episodic treatment, supported approval of catridecacog by the US FDA.

3.4 Development of Clinically Meaningful End Points

For rare diseases, data are often lacking to support identification of clinically meaningful end points for disease activity/progression that are accessible within a time frame that can support regulatory approval of new and novel treatments. When there are no precedented treatments to inform clinical development of new agents, this challenge can become particularly daunting. Validation/qualification of surrogate end points predictive of beneficial effect

Table 3.2 Development of clinically meaningful end points to support drug approval.

Disease	*End point*	*Party responsible for development*
Autosomal dominant polycystic kidney disease	Clinical composite of disease severity	Sponsor initiated
Duchenne muscular dystrophy	6 minute walk distance	Sponsor-academic collaboration
Chronic myeloid leukaemia (chronic phase)	Freedom from disease progression	Sponsor and investigators
	Complete cytogenetic response (surrogate end point for recently diagnosed chronic phase disease)	Sponsor and investigators (based on 5 year long-term study results)

from treatment can also be challenging. Respective approaches that can be employed to identify end points or surrogate end points for disease activity/disease progression include analysis of data from natural history studies and analysis of existing data from natural history and interventional studies to qualify/validate end points or surrogate end points. Development of a clinical composite of disease severity as an end point for the study of ADPKD, the 6 minute walk distance (6MWD) as an end point for the study of DMD and complete cytogenetic response as a surrogate end point for efficacy in treatment of recently diagnosed chronic phase chronic myeloid leukaemia provide instructive examples for these respective approaches to identifying efficacy end points for investigation in rare patient populations (Table 3.2).

3.4.1 Autosomal Dominant Polycystic Kidney Disease

ADPKD, the CRISP natural history study and the TEMPO 3:4 study of tolvaptan for treatment of ADPKD have been described above. In developing the TEMPO 3:4 study, the sponsor encountered the challenge that the primary efficacy end point, rate of change in TKV, a direct mechanistic end point for tolvaptan treatment effect, had not been previously demonstrated to be predictive for favourable effect on ADPKD disease progression.[47] To address this issue in the TEMPO 3:4 study, the sponsor developed a key secondary composite end point of time to investigator-assessed disease progression, defined as worsening kidney function, clinically significant kidney pain, worsening of hypertension and worsening albuminuria. This composite end point of disease severity, reflective of elements identified in the CRISP study as contributing to disease, was evaluated using a hierarchical model of hypothesis testing immediately after the primary end point. Over a 3 year period tolvaptan treatment conferred a favourable effect on both the primary and key secondary end points and, supported by TEMPO 3:4 study results, regulatory submissions were made seeking approval of tolvaptan for treatment of ADPKD.[21,48]

3.4.2 Duchenne Muscular Dystrophy

Duchenne muscular dystrophy (DMD) is an X-linked inherited neuromuscular disease affecting 1 in 3500 male births that is caused by a mutation in the gene for the protein dystrophin, resulting in its complete absence from muscle.[49] Boys experience onset of symptoms at around 2–3 years of age, followed by loss of the ability to walk between the ages of 8 and 10 years. The disease continues with progressive cardiac problems manifesting in the second decade of life and death from pneumonia or cardiac involvement in the late teens or early 20s is commonly observed in affected patients.[50] The progressive loss of muscle function, first in the lower extremities, then in the upper extremities, then in cardiac muscle and in the axial muscles that control respiration, creates discrete stages in disease progression. Natural history studies have helped to define the temporal chronology of this disease progression. In a longitudinal natural history study in 18 ambulatory boys with Duchenne/Becker muscular dystrophy and in 22 healthy boys aged 4–12 years, McDonald and colleagues characterised changes in the 6MWD over 58 and 69 weeks, respectively.[51] That investigation demonstrated an age and stride length dependent decrement in 6MWD in boys with DMD relative to their healthy counterparts and identified the stage of disease progression at which this clinically meaningful end point can be applied. Building on these observations, in a recent analysis, McDonald and colleagues evaluated baseline clinical data for 174 subjects from an international multicentre study, PTC124-GD-007-DMD, evaluating the agent ataluren in boys aged ≥5 years with phenotypic evidence of a dystrophinopathy, a non-sense mutation in the dystrophin gene and a baseline unassisted 6MWD ≥ 75 metres.[52] In their analysis the authors were able to provide evidence for reliability and concurrent validity of the 6MWD and were able to define minimal clinically important differences of the 6MWD that support its use as primary end point in DMD interventional studies that are focused on preserving ambulation.

3.4.3 Chronic Myeloid Leukaemia

Chronic myeloid leukaemia (CML) is a malignant clonal disorder of the haematopoietic stem cells, affecting 1.6 cases per 100 000 adults (approximately 5000 new cases per year in the USA). In 95% of affected patients CML is caused by a balanced chromosomal translocation that fuses the BCR gene on chromosome 22 to the ABL gene on chromosome 9, a cytogenetic abnormality termed the Ph chromosome. The protein product of the resulting BCR-ABL fusion gene (BCR-ABL) results in continuous cell growth and replication and is responsible for the disease.[53] Cytogenetic testing of the cells from the bone marrow can readily detect presence of the Ph chromosome and the burden of disease is reflected in the percentage of bone marrow metaphases containing the Ph chromosome and in the level of BCR-ABL gene transcripts in the blood as measured by quantitative PCR. The natural history of CML is progression from a benign chronic phase to a rapidly fatal blast crisis within 3–5 years.[54] Imatinib, an ABL kinase inhibitor, was developed as a targeted therapy to inhibit activity of the BCR-ABL fusion protein. Imatinib received an initial accelerated regulatory approval from the US FDA for treatment of patients with chronic phase CML following failure of interferon-α therapy, for treatment of accelerated phase CML and for CML in blast crisis. That accelerated approval was based on the high observed frequency of haematological remissions and cytogenetic response rates and the high likelihood that these results would lead to a real benefit.[55] In 2000, to secure regulatory approval of imatinib for initial treatment of recently diagnosed chronic phase CML, the sponsor initiated the IRIS study. The IRIS study evaluated imatinib against the then standard of care for CP CML, combination therapy with interferon-α plus cytosine arabinoside with provisions for treatment arm crossover in the event of treatment failure or treatment toxicity.[56] Between June 2000 and January 2001, 553 subjects were enrolled in each treatment arm. By 12 months median follow-up, the imatinib treatment arm had demonstrated superior results with 96.6% of the imatinib treated patients free of disease progression (death from any cause, development of the accelerated phase of disease, development of the blast phase of disease, loss of a previously attained haematological remission or loss of a previously attained major cytogenetic response [<36% of metaphases containing the Ph chromosome] with an absolute increase of ≥30% Ph positive cells) *versus* 79.9% of the combination therapy subjects.

Regulatory approval for imatinib within this indication was sought and a large proportion of the combination therapy subjects subsequently switched to imatinib treatment. Initial publication of study results after median follow-up of 19 months also described a clear benefit in the imatinib treatment arm for the primary end point of freedom from disease progression. Observation of the imatinib treated study population was then continued on an extended basis to provide long-term data for treatment of newly diagnosed chronic phase CML. In an analysis of 5 year follow-up data from the imatinib treatment arm, 97% of the subjects with complete cytogenetic responses (undetectable Ph chromosome on bone marrow cytogenetic analysis) within 12 months of beginning treatment and 100% of the subjects with a reduction of at least 3 log in levels of BCR-ABL fusion gene transcripts within 12 months of beginning treatment had remained free of progression to accelerated or blast phase disease.[57] On the basis of these clinical study results, complete cytogenetic response in the bone marrow and major (≥3 log) reduction in level of BCR-ABL gene transcripts were qualified as surrogate end points for efficacy. These surrogate end points of complete cytogenetic response and major molecular response were subsequently used as primary efficacy end points in the Phase 3 clinical studies that supported registration of the second-generation ABL kinase inhibitors dasatinib and nilotinib, respectively, for initial treatment of chronic phase CML.[58,59]

3.5 Understanding the Standard of Care

In rare diseases, the body of evidence to guide clinical management of patients can be very limited and treatment practices can vary substantially. Incomplete understanding of the resulting standard of care may introduce excessive heterogeneity into clinical studies, confound sponsor efforts to control for heterogeneity *via* eligibility criteria, supportive care guidelines or randomisation stratifications and compromise the ability to detect treatment effect from the therapeutic intervention. A number of strategies can be employed to better understand the standard of care in rare diseases and thereby inform design of clinical studies. These include accessing supportive care guidelines from clinical experts, review of clinical study databases for information on frequently used concomitant medications and non-pharmacological supportive care and access to disease registries of individual patient data. Respective examples where each of these types of background information may be accessed include published guidelines for the management of CML, the VISTA database for stroke and the Diamond Blackfan Anemia Registry of North America, a comprehensive database of individuals with this rare bone marrow failure syndrome (Table 3.3).

3.5.1 Chronic Myeloid Leukaemia

Chronic myeloid leukaemia, a malignancy of the haematopoietic system, is described above. While ABL kinase inhibitors such as imatinib, nilotinib and dasatinib are the standard of care for initial treatment of chronic phase CML, there is substantial complexity in management of the disease including initial diagnostic studies, timing and scope follow-up assessments, criteria for optimal/suboptimal clinical responses and for treatment failure at specified times in the disease course, management of disease associated leucocytosis and/or thrombocytosis, management of treatment-associated

Table 3.3 Understanding the standard of care.

Disease	*Data source*	*Category of source*
Chronic myeloid leukaemia	European LeukemiaNet NCCN guidelines for CML	Clinical expert recommendations based on review of current data
Intracerebral haemorrhage	Virtual International Stroke Archive (VISTA)	Archive of clinical trial data sets
Diamond Blackfan anemia	North American Diamond Blackfan Anemia Registry	Disease registry

toxicities, as well as standardised definitions for stage of disease and for clinical response. Based on reviews of published literature and available data, panels of clinical experts for the European LeukemiaNet (ELN) and for the National Comprehensive Cancer Network (NCCN) in the USA have published a series of largely concordant guidelines for the clinical management of CML, with the most recent updates issued in 2013.[60,61] Essential elements of the published ELN and NCCN guidelines for CML were reflected in the design, assessments and supportive care guidelines for the pivotal Phase 3 investigations of nilotinib and dasatinib for treatment of newly diagnosed chronic phase CML.[58,59]

3.5.2 Spontaneous Intracerebral Haemorrhage

Spontaneous intracerebral haemorrhage (ICH) or haemorrhagic stroke has been described earlier. Given the dismal outcomes for this condition and the limited avenues for pharmacological intervention, substantial efforts have been devoted to improving outcomes by optimising supportive care. Among the interventions being investigated by clinical study consortia is intensive blood pressure control as a means to limit growth in ICH haemorrhage and thereby improve neurological outcomes. Results from these studies, whether positive or not for the primary end point, have the potential to influence the standard of care used by practitioners based on results for secondary end points. The INTERACT2 study evaluated intensive lowering of blood pressure as compared with the more conservative blood pressure control currently reflected in treatment guidelines. Recently reported results for that study did not demonstrate a significant reduction in the rate of the primary outcome, mortality or major disability 90 days post-event. However, in an ordinal analysis of the primary outcome event, to enhance statistical power for assessing physical functional outcomes, there were significantly better functional outcomes in patients who received intensive blood pressure control.[62] These results have the potential to influence the standard of care. A second clinical consortium is evaluating the intervention of intensive blood pressure control in the ongoing ATTACH II study. Treatment regimens being evaluated by that study have the potential to become integrated into subsequent practices for treatment of patients with ICH even in advance of analysis and publication of study results. Sponsors for subsequent interventional studies in ICH will need to allow for the potential impact of supportive care practices such as blood pressure control. The VISTA database, described previously, contains the data sets for both placebo and actively treated subjects from selected studies in ischemic stroke and in ICH.[42] To the extent that data sets for the INTERACT2 and ATTACH II studies become integrated into this data repository, sponsors will be able to benefit from accessing VISTA repository data to inform randomisation stratification strategies and supportive care guidelines based on results from these recent studies, in order to manage potential heterogeneity in the standard of care.

3.5.3 Diamond Blackfan Anaemia

Diamond Blackfan anaemia (DBA) is a rare genetic disorder affecting about 7 individuals per million live births. The disease is characterised by red cell aplasia that classically presents with severe anaemia in early infancy, often in association with physical anomalies and short stature. DBA is also associated with an increased risk of malignancy with a higher than predicted incidence of acute myeloid leukaemia and osteogenic sarcoma.[63] Emerging data indicate that the disease is a disorder of ribosomal biogenesis or function with 50% of patients having mutations affecting a single allele of a ribosomal protein gene. A substantial subset of patients with DBA, up to 80%, respond to glucocorticoid treatment with a rise in haemoglobin and can then be tapered to determine the minimum dosage required to maintain transfusion independence. Across affected individuals the maintenance dose is highly variable; in over 20% of patients glucocorticoids can be completely stopped with maintenance of adequate haemoglobin levels, whereas some patients become refractory to glucocorticoid therapy and require ongoing transfusion support.[63] As in many ultra-rare diseases there is limited evidence to guide patient management and the standard of care can be variable with practices guided by judgement of clinicians. The limitations in epidemiological knowledge, the variability in clinical responses to treatment and a lack of evidence-based guidance for supportive care creates challenges in anticipating the standard of care for subjects with this disorder and can compromise the outcome of clinical studies. To address these limitations in knowledge, investigators established the Diamond Blackfan Anemia Registry of North America. With informed consent, the registry collects demographic, laboratory, clinical and survival information and has generated analyses of disease epidemiology, genetics, congenital anomalies, treatment practices, treatment responses and treatment-related toxicities.[64] Based on an understanding of the disease epidemiology and standard of care, investigators have been able to define a study population and eligibility criteria, including red blood cell transfusion dependence, and commence an investigator-initiated Phase 1/2 investigation of the drug sotatercept (ACE-011), a soluble activin receptor type 2A IgG-Fc fusion protein, for the treatment of DBA.

3.6 Creating Efficiencies in Sample Size to Support Hypothesis Testing

Available clinical substrate, by definition, is limited for conduct of clinical investigations in small patient populations. However, regulatory approval of pharmaceutical agents to treat rare diseases requires adequate and well-controlled investigations as the primary basis for determining whether there is substantial evidence to support claims of effectiveness and that particulars and documents in an application for market authorisation for a medicinal product that demonstrate the potential risks are outweighed by the therapeutic efficacy of the product.[11,12] Thus, even for rare diseases, sponsors must

Table 3.4 Reducing sample size required to support hypothesis testing.

Disease	*Clinical programme*	*Strategy*
Chronic myeloid leukaemia	Imatinib	Enrichment. All subjects have the fusion oncogene responsible for disease
Haemophilia A	Kogenate, Recombinate	Enrichment. Monogenic basis for protein deficiency
Autosomal dominant polycystic kidney disease	mTOR pathway inhibition	Enrichment. Population best suited to test treatment effect selected *via* clinical study design
Haemophilia A	Moroctocog alfa (AF-CC)	Efficient statistical design. Bayesian statistical model to evaluate immunological safety
Haemophilia B	RIXUBIS	Historic control group. Reduced overall sample size requirement

access sufficient sample size to support hypothesis testing with regard to claims of efficacy and to support conclusions of benefit/risk. Several approaches are available to study sponsors that may be used alone or in combination to manage this challenge. These approaches include strategies to reduce the sample size required to test the respective study hypothesis (Table 3.4) and measures to more efficiently recruit study subjects from the available substrate.

3.6.1 Enrichment Strategies

Clinical investigations in rare diseases lend themselves to enrichment strategies, whether driven by a common genetic basis for the disease or based on investigation of subjects at a common stage in disease progression. In either scenario the objective is to more consistently observe responses to therapeutic agents across a greater proportion of the study subjects, sometimes resulting in a greater magnitude of treatment effect than in other settings, and permitting corresponding reductions in the number of subjects required for hypothesis testing. Innovative statistical models can also be used to support hypothesis testing in small clinical studies. Finally, under certain circumstances historic control groups can be utilised, permitting allocation of the limited clinical substrate to the investigational treatment arm.

Examples where enrichment strategies have been used to support clinical development of new drugs include investigation of the ABL kinase inhibitor imatinib in CML, the clinical investigation of recombinant factor VIII products in the monogenic disorder haemophilia A and an elegant enrichment study design that was used to enhance ability to detect treatment effect in an investigator-initiated investigation of sirolimus for treatment of ADPKD.

3.6.1.1 *Chronic Myeloid Leukaemia*

CML, a malignant clonal disorder of the haematopoietic stem cells, is described above.[54] The underlying basis for this disease is the gene product of the BCR-ABL fusion oncogene that results from the balanced translocation responsible for formation of the Ph chromosome. Imatinib, a targeted therapy, very effectively exploits the pathophysiology of this monogenic disease by inhibiting activity of the BCR-ABL gene product. As a consequence of this common mechanism the initial investigations of imatinib were able to demonstrate haematological improvements and reductions in disease burden, as evidenced by cytogenetic response, in a substantial proportion of poor prognosis patients with chronic phase CML who had failed prior therapy. The same improvements were also seen in CML patients with accelerated disease and even in a subset of patients in blast crisis.[65] These findings supported initial US FDA accelerated approval in 2001. An even more dramatic enrichment of response to treatment with imatinib was observed in recently diagnosed patients with chronic phase CML, where efficacy results included a dramatic reduction in the frequency of disease progression and, following 18 months of treatment, 96.8% of subjects were in haematological remission and 39.6% of subjects had no cytogenetic evidence of disease.[56]

3.6.1.2 *Haemophilia*

Haemophilia is a hereditary X chromosomal recessive bleeding disorder, caused by factor VIII deficiency (haemophilia A) or factor IX deficiency (haemophilia B), which is characterised by bleeding, either spontaneous or following recognised or unrecognised trauma. Common anatomical locations for bleeding are joints and muscles, although bleeding can also occur in other locations such as the central nervous system, with the potential for life-threatening consequences.[66] Clinical severity of haemophilia is defined based on the residual level of clotting factor activity: factor activity <1% defines severe disease, factor activity 2–5% defines moderate disease and factor activity >5% to <40% defines mild disease.[67] The hallmark of treatment for haemophilia is clotting factor replacement therapy administered 'on demand' to establish haemostatic levels of the missing clotting factor in response to bleeding episodes and an emerging standard of care to administer clotting factor prophylactically, to prevent onset of bleeding episodes. The monogenic nature of haemophilia A ensures a very consistent disease phenotype and great uniformity in clinical responses to factor VIII replacement therapy. Clinical responses are integrally tied to the level of factor VIII activity established by replacement therapy, therefore pharmacokinetics are considered a surrogate for clinical efficacy when factor VIII products are evaluated clinically. These properties of the disease have been consistently exploited in the clinical investigations of factor VIII concentrates. The clinical studies supporting approval of the initial recombinant factor VIII products,

Kogenate and Recombinate, made efficient use of small pivotal studies with limited sample size, characterising pharmacokinetics (PK), a surrogate end point of efficacy, by evaluating PK in 17 and 69 subjects, respectively, and by characterising safety and efficacy in 76 and 69 subjects, respectively.[68,69] Evidence for enrichment and consistency of clinical response is provided by the comparably high rates of efficacy observed in these two studies, with 94% and 91% of bleeding episodes responding to two or fewer infusions of the respective clotting factor concentrate.[69,70]

3.6.1.3 Autosomal Dominant Polycystic Kidney Disease

Enrichment of rare disease patient populations can also be achieved on the basis of study design. The entity autosomal dominant polycystic kidney disease (ADPKD) has been described above. In an investigator-initiated study Serra and colleagues evaluated treatment effect of the mTOR inhibitor sirolimus on growth in kidney volume in individuals with ADPKD. To reduce the sample size required to detect a treatment effect, the investigators used an enrichment strategy to select for subjects with rapidly growing kidneys. Subjects with ADPKD and adequate renal function (GFR > 70 mL min^{-1}) were recruited and underwent a baseline MRI to assess TKV. Following a run-in period of 6 months these subjects had a repeat MRI, those individuals demonstrating a rapid (≥2%) increase in TKV during the run-in period were randomly assigned to treatment with the investigational agent or to standard care. Using this approach and, by assuming a 6% annual rate of kidney enlargement, the investigators were able to enrol a population with rapidly enlarging kidneys and to power the study to detect a 50% difference in annual kidney enlargement at 80% power with a two-sided alpha of 0.05, and allow for 20% attrition, using only 50 subjects in each study arm. The study failed to demonstrate a treatment effect on kidney enlargement in the population studied. However the observed rates of kidney enlargement in subjects selected for study treatment were 9.7% in the group receiving active treatment and 9.5% in the control group, underscoring the operational success of this strategy for enriching the study with individuals sustaining rapid kidney enlargement.[71]

3.6.2 Innovative Statistical Models

Innovative statistical designs can also be used to achieve the statistical power necessary for hypothesis testing in small studies. The disease entity haemophilia has been described above. Occurrence of inhibitors (neutralising antibodies) to factor VIII replacement therapy is a known complication of haemophilia A treatment.[72] Previously treated patients (PTPs) who have been extensively treated with factor VIII (>150 prior exposure days) without occurrence of inhibitors are considered at minimal risk for this complication with a very low, but finite, rate of inhibitor development. A potential safety concern with new factor VIII products is the possibility that they have been rendered immunogenic and will cause an increased frequency of inhibitors. Therefore current guidance is that new products be evaluated for risk of immunogenicity in PTPs who are considered at minimal risk for new onset of inhibitor. In a 2005 publication, Lee and Roth analysed the challenges of demonstrating immunological safety of new factor VIII products in small studies in the setting where the test population has a low but finite rate of inhibitor formation that can be as high as 3%. Based on guidance from the US FDA, for a clinical study to successfully demonstrate inhibitor safety the upper bound of the two-sided 95% confidence interval for the product inhibitor incidence rate had to be <6.8% using an intention to treat (ITT) paradigm. Under these conditions only the occurrence of ≤1 inhibitors in a study population of 80 subjects would meet criteria for immunological safety. When the authors applied this paradigm to completed pivotal studies for then approved recombinant factor VIII products, each of which was considered safe, they found that several would have been unduly prohibited from licensure.

As an alternative approach, Lee and Roth proposed use of a Bayesian statistical model. In their model, when incorporating prior knowledge into the Bayesian polynomial describing the distribution of factor VIII inhibitor rate, all then approved recombinant factor VIII products would meet the stringency of the above FDA criteria. Furthermore this conclusion of safety held up true for most approved products, even when no prior knowledge was incorporated into the Bayesian polynomial. In contrast the factor VIII products known to have been associated with an epidemic of inhibitors would fail the above criteria.[73] Subsequently, this Bayesian statistical approach was successfully used to test the primary safety end point, frequency of inhibitors, in a pivotal study that supported US FDA approval of the B-domain deleted factor VIII product moroctocog alfa (AF-CC). In that study the sponsor reviewed the data for development of inhibitors from four licensed full-length recombinant factor VIII products and using a Bayesian analysis with a non-informative prior distribution, generated an upper threshold for acceptable inhibitor frequency of 4.4%. For moroctocog alfa (AF-CC) an informative Bayesian prior distribution was established using data from two previous studies evaluating the same active substance and, after discounting the prior data by 50% to align with sample size from the new study, the Bayesian distribution for the inhibitor rate was updated with results from the new pivotal study. The observed incidence of inhibitors in the new study was 2 in 94 subjects by an ITT analysis, and when the Bayesian distribution was updated with data from the pivotal study, results met the primary safety end point, to demonstrate an inhibitor rate below the 4.4% threshold with 95% probability.[74]

3.6.3 Historical Control Groups

Use of a historical or external control group is another strategy that has been used to mitigate the challenge of assembling sufficient sample size for clinical studies in small patient populations. While this approach has major disadvantages, including loss of randomisation as a tool to minimise bias and risk that the external control group and the study population are dissimilar with respect to a wide range of factors, under certain conditions this approach can be entertained. ICH E10 provides a description of circumstances where an external control group may be entertained.[75] The inability to control bias restricts use of external control designs to situations in which the effect of treatment is dramatic and the usual course of disease is highly predictable. The study end points should be objective, impact of baseline and treatment variables on the end point should be well characterised, there should be detailed information on the control group, the control group should be as similar as possible to the population expected to receive the test drug in the study and should be selected before performing any comparative analyses.

3.6.3.1 Recombinant Coagulation Factor IX

Recent approval of a recombinant coagulation factor IX product (RIXUBIS) by the US FDA provides an example where an external, historic control group was used to support regulatory approval. For the entity haemophilia B, described above, the hallmark of treatment is replacement therapy to establish haemostatic levels of the missing clotting factor and the emerging standard of care is to administer clotting factor prophylactically to prevent onset of bleeding episodes and thereby avoid the morbidity of their sequelae. At the time of US FDA approval of RIXUBIS there were no marketed factor IX products approved for prophylaxis to prevent or reduce the frequency of bleeding episodes, but the clinical practice of using marketed products for prophylaxis treatment was widely prevalent and supported by results from randomised prospective clinical trials.[72] To support regulatory approval of RIXUBIS for the indication of routine prophylaxis to prevent or reduce the frequency of bleeding episodes in adults with haemophilia B, the respective sponsor utilised an external, historic control group. Haemophilia B patients aged 12–65 years at time of screening, with factor IX activity $\leq$2% and with at least 12 documented bleeding episodes requiring treatment within the year prior to study enrolment were enrolled into a clinical study.[76] Subjects were allocated to episodic (on-demand treatment) or to 26 weeks of prophylactic treatment by the respective investigator. Results from the prophylactic treatment group (N = 56) were compared to results from a historical control group treated on demand instead of to the 14 subjects enrolled in the on-demand arm of the study. The mean annualised bleed rate (ABR) during prophylaxis treatment was 4.20, which represented a 75% reduction compared to the mean ABR of 20 for on-demand treatment in the historical control group.[77] Based upon this comparison to a historic control group, RIXUBIS received US FDA approval for the indication of prophylaxis to prevent or reduce the frequency of bleeding episodes.

3.7 Augmentation of Study Enrolment

Augmentation of study enrolment through effective planning and recruitment strategies represents a complementary approach to attaining sufficient study sample size to support hypothesis testing (Table 3.5). Numerous advocacy groups have been founded to advocate for patients suffering from rare diseases, particularly those affecting children.[6] These advocacy groups, described in detail in Chapter 5, are often very effective in supporting outreach in order to direct subjects affected by the respective disease to interventional study sites where new potential treatments are being evaluated. Established clinical treatment and/or clinical trial networks, frequently sponsored by advocacy groups, provide a means for study sponsors to engage experienced investigators at sites caring for patients with the respective rare diseases. Examples include networks sponsored by groups such as the Cystic Fibrosis Foundation Therapeutics Development Network and the Muscular Dystrophy Association network of specialised clinics in the USA and the neuromuscular care and trial centres that are reflected in the TREAT-NMD registry in Europe. Disease-specific treatment centres sponsored by government and public agencies can also be a powerful tool in gaining access to experienced investigators caring for patients with rare diseases. The US network of hospital-based haemophilia diagnostic and treatment centres that were established in the 1970s with support from advocacy groups and funding from several states and from the US Congress is a prime example. This network currently comprises more than 130 treatment centres and provides comprehensive services within a single treatment facility for over 10 000 patients with bleeding disorders and their families.[78] Similar networks of specialised treatment centres for persons with bleeding

Table 3.5 Augmentation of enrolment to meet sample size requirements.

Disease area	*Potential collaborators*	*Organisation type*
Cystic fibrosis	Cystic Fibrosis Foundation Therapeutics Development Network	Patient advocacy group sponsored network
Muscular dystrophy	Muscular Dystrophy Association network of clinics	Patient advocacy group sponsored network
	Treat NMD	Advocacy group sponsored registry of clinical care and trial centres
Haemophilia	Comprehensive haemophilia treatment centres	Government/publically sponsored network of treatment facilities
Paediatric cancer	Children's Oncology Group (COG)	Cooperative study group
	Innovative Therapies for Children with Cancer Consortium (ITCC)	Cooperative study group

disorders exist across the developed world and collectively constitute a highly effective means to recruit patients with bleeding disorders to clinical trials. For disease areas such as paediatric oncology, highly structured clinical trial networks such as the Children's Oncology Group in North America and the Innovative Therapies for Children with Cancer consortium in Europe can be accessed by sponsors for varying degrees of collaboration in evaluating new investigational agents to treat childhood cancer.

3.8 Summary and Future Prospects

Recent advances in characterising the human genome, in molecular characterisation of physiological pathways and in identifying therapeutic targets and selection of biomarkers, combined with regulatory incentives and enablers, are creating unprecedented opportunities for drug development in rare diseases. While there are substantial challenges entailed with conduct of clinical studies in small patient populations, there are a host of strategies available to sponsors to facilitate the conduct of rigorous, hypothesis-driven investigations that can support regulatory approval. Natural history studies, disease registry data and repositories of clinical trial data can provide insights into the underlying disease process, disease heterogeneity and progression and standards of care. These tools can also be used to identify appropriate patient populations for clinical investigation and to identify or validate clinically meaningful end points that are accessible in a timeline permissive for drug development. The underlying monogenetic nature of many rare diseases creates the opportunity for enrichment strategies to increase the proportion of subjects likely to respond to effective treatments, permitting robust hypothesis testing while reducing sample size requirements. The genetic nature of many rare diseases can also ensure a predictable disease course and an accurate characterisation of patient populations that may be permissive for use of historical controls in order to reduce study sample size requirements. Challenges in gaining access to affected patient populations can be mitigated through collaborations with patient advocacy groups and by engagement with clinical trial networks and consortia dedicated to improving treatment outcomes among patients affected by the respective rare diseases.

Ongoing evolution in the landscape of drug development promises to provide continued opportunity for development of new therapies to treated individuals affected by rare diseases. Several major pharmaceutical companies have established new research divisions dedicated to orphan diseases. The scope of gene corrective therapies, many uniquely poised to correct underlying genetic causes of rare diseases, continues to expand, creating exciting possibilities for response enrichment strategies in small patient populations. In the EU, a draft policy under review at the EMA would make patient level clinical study data sets publically available, with appropriate privacy protections. This policy, if implemented, could provide enhanced data to inform clinical study designs, facilitate identification of end points that can support drug development, facilitate validation of surrogate end points to support accelerated or conditional regulatory approval and when justified, facilitate use of historic control groups, all to the potential benefit of investigators conducting research in small patient populations.[79] While the challenges are substantial, the climate for clinical research in rare diseases remains promising.

The views presented in the chapter reflect those of the author and do not necessarily reflect those of Pfizer.

References

1. Rare diseases resources, http://rarediseases.info.nih.gov/resources/2/rare-diseases-resources#category4, accessed 4 November 2013.
2. Eurordis Rare Diseases Europe, http://www.eurordis.org/about-rare-diseases, accessed 4 November 2013.
3. I. Hudson and A. Breckenridge, *Clin. Pharmacol. Ther.*, 2012, **92**, 151–153.
4. The Committee for Orphan Medicinal Products and the European Medicines Agency Scientific Secretariat, *Nat. Rev. Drug Discovery,* 2011, **10**, 341–349.
5. J. Woodcock, *Clin. Pharmacol. Ther.,* 2012, **92**, 146–148.
6. N. Azie and J. Vincent, *Clin. Pharmacol. Ther.,* 2012, **92**, 135–139.
7. A. S. Kesselheim, *Clin. Pharmacol. Ther.,* 2012, **92**, 153–155.
8. R. E. Sherman, J. Li, S. Shapley, M. Robb and J. Woodcock, *N. Engl. J. Med.,* 2013, **369**, 1877–1880.
9. Guideline on the scientific application of the practical arrangements necessary to implement commission regulation (EC) No. 507/2006 on the conditional marketing authorization for medicinal products for human use falling within the scope of regulation (EC) No. 726/2004, http://www.ema.europa.eu/docs/en_GB/document_library/Scientific_guideline/2009/10/WC500004908.pdf, accessed 4 November 2013.
10. Guideline on procedures for the granting of a marketing authorization under exceptional circumstances pursuant to article 14(8) of regulation (EC) No. 7262004, http://www.ema.europa.eu/docs/en_GB/document_library/Regulatory_and_procedural_guideline/2009/10/WC500004883.pdf, accessed 4 November 2013.
11. Part 314 Applications for FDA approval to market a new drug 21 CFR ch 1 (4-1-13 edition) 314.a26, http://www.ecfr.gov/cgi-bin/searchECFR, accessed 16 November 2013.
12. Directive 2001/83/EC of the European Parliament and of the council of 6 November 2001 on the community code relating to medicinal products for human use, http://www.edctp.org/fileadmin/documents/ethics/DIRECTIVE_200183EC_OF_THE_EUROPEAN_PARLIAMENT.pdf, accessed 16 November 2013.
13. Guideline on clinical trials in small populations, CHMP/EWP/83561/2005, http://www.ema.europa.eu/docs/en_GB/document_library/Scientific_guideline/2009/09/WC500003615.pdf, accessed 16 November 2013.

14. M. Orfali, L. Feldman, V. Bhattacharjee, P. Harkins, S. Kadam, C. Lo, M. Ravi, D. T. Shringarpure, J. Mardekian, C. Cassino and T. Coté, *Clin. Pharmacol. Ther.*, 2012, **92**, 262–264.
15. K. Hull, *Study designs for rare diseases*, http://rarediseasesnetwork.epi.usf.edu/conference/documents/slides/HULL-StudyDesigninRareDiseases.pdf, accessed 16 November 2013.
16. V. E. Torres, P. C. Harris and Y. Pirson, *Lancet,* 2007, **369**, 1287–1301.
17. P. C. Harris and V. E. Torres, *Annu. Rev. Med.,* 2009, **60**, 321–337.
18. J. J. Grantham, *N. Engl. J. Med.,* 2008, **359**, 1477–1485.
19. A. B. Chapman, L. M. Guay-Woodford, J. J. Grantham, V. E. Torres, K. T. Bae, D. A. Baumgarten, P. J. Kenney, B. F. King Jr, J. F. Glockner, L. H. Wetzel, M. E. Brummer, W. C. O'Neill, M. L. Robbin, W. M. Bennett, S. Klahr, G. H. Hirschman, P. L. Kimmel, P. A. Thompson and J. P. Miller, *Kidney Int.,* 2003, **64**, 1035–1045.
20. J. J. Grantham, V. E. Torres, A. B. Chapman, L. M. Guay-Woodford, K. T. Bae, B. F. King Jr, L. H. Wetzel, D. A. Baumgarten, P. J. Kenney, P. C. Harris, S. Klahr, W. M. Bennett, G. N. Hirschman, C. M. Meyers, X. Zhang, F. Zhu and J. P. Miller, *N. Engl. J. Med.,* 2006, **354**, 2122–2130.
21. V. E. Torres, A. B. Chapman, O. Devuyst, R. T. Gansevoort, J. J. Grantham, E. Higashihara, R. D. Perrone, H. B. Krasa, J. Ouyang and F. S. Czerwiec, *N. Engl. J. Med.,* 2012, **367**, 2407–2418.
22. M. H. Steinberg, *N. Engl. J. Med.,* 1999, **340**, 1021–1030.
23. O. S. Platt, D. J. Brambilla, W. F. Rosse, P. F. Milner, O. Castro, M. H. Steinberg and P. P. Klug, *N. Engl. J. Med.,* 1994, **330**, 1639–1644.
24. Cooperative study of sickle cell disease (CSSCD), https://biolincc.nhlbi.nih.gov/studies/csscd/, accessed 18 December 2013.
25. O. S. Platt, B. D. Thorington, D. J. Brambilla, P. F. Milner, W. F. Rosse, E. Vichinsky and T. R. Kinney, *N. Engl. J. Med.,* 1991, **325**, 11–16.
26. S. Charache, M. L. Terrin, R. D. Moore, G. J. Dover, F. B. Barton, S. V. Eckert, R. P. McMahon, D. R. Bonds and The Investigators of the Multicenter Study of Hydroxyurea in Sickle Cell Anemia, *N. Engl. J. Med.,* 1995, **332**, 1317–1322.
27. K. Ataga, *Novel therapies in sickle cell disease*, http://asheducationbook.hematologylibrary.org/content/2009/1.toc, accessed 18 December 2013.
28. A. Kutlar, P. S. Swerdlow, S. E. Meller, K. Natrajan, L. G. Wells, B. Clair, S. Shah and R. Knight, *Blood,* 2013, **122**, 777.
29. J. J. Field, K. I. Ataga, E. Majerus, C. A. Eaton, R. Mashal and D. G. Nathan, *Blood,* 2013, **122**, 977.
30. J. Howard, S. L. Thein, F. Galacteros, A. Inati, M. Reid, P. E. Keipert, T. N. Small and F. Booth, *Blood,* 2013, **122**, 2205.
31. G. J. Kato, M. P. Lawrence, L. G. Mendelsohn, R. Saiyed, X. Wang, A. K. Conrey, J. M. Starling, G. Grimes, J. G. Taylor, J. McKew, C. P. Minniti and W. Stern, *Blood,* 2013, **122**, 1009.
32. Centers for Diseases Control and Prevention RuSH overview, http://www.cdc.gov/ncbddd/hemoglobinopathies/, accessed 14 Nov 2013.

33. A. I. Qureshi, S. Tuhrim, J. P. Broderick, H. H. Batjer, H. Hondo and D. F. Hanley, *N. Engl. J. Med.,* 2001, **344**, 1450–1460.
34. V. L. Roger, A. S. Go, D. M. Lloyd-Jones, E. J. Benjamin, J. D. Berry, W. B. Borden, D. M. Bravata, S. Dai, E. S. Ford, C. S. Fox, H. J. Fullerton, C. Gillespie, S. M. Hailpern, J. A. Heit, V. J. Howard, B. M. Kissela, S. J. Kittner, D. T. Lackland, J. H. Lichtman, L. D. Lisabeth, D. M. Makuc, G. M. Marcus, A. Marelli, D. B. Matchar, C. S. Moy, D. Mozaffarian, M. E. Mussolino, G. Nichol, N. P. Paynter, E. Z. Soliman, P. D. Sorlie, N. Sotoodehnia, T. N. Turan, S. S. Virani, N. D. Wong, D. Woo, M. B. Turner and American Heart Association Statistics Committee and Stroke Statistics Subcommittee, *Circulation,* 2012, **125**, e2–e220.
35. J. Broderick, S. Connolly, E. Feldmann, D. Hanley, C. Kase, D. Krieger, M. Mayberg, L. Morgenstern, C. S. Ogilvy, P. Vespa and M. Zuccarello, *Stroke,* 2007, **38**, 2001–2023.
36. F. Rincon and S. A. Mayer, *Crit. Care,* 2008, **12**, 237–250.
37. J. P. Broderick, T. G. Brott, J. E. Duldner, T. Tomsick and G. Huster, *Stroke,* 1993, **24**, 987–993.
38. S. Tuhrim, D. R. Horowitz and M. Sacher, *Crit. Care Med.,* 1999, **27**, 617–621.
39. T. Brott, J. Broderick, R. Kothari, W. Barsan, T. Tomsick, L. Sauerbeck, J. Spilker, J. Duldner and J. Khoury, *Stroke,* 1997, **28**, 1–5.
40. S. M. Davis, J. Broderick, M. Hennerici, N. C. Brun, M. N. Diringer, S. A. Mayer, K. Begtrup and T. Steiner, *Neurology,* 2006, **66**, 1175–1181.
41. C. Delcourt, Y. Huang and H. Arima, *Neurology,* 2012, **79**, 314–319.
42. M. Ali, P. M. W. Bath, J. Curram, S. M. Davis, H.-C. Diener, G. A. Donnan, M. Fisher, B. A. Gregson, J. Grotta, W. Hacke, M. G. Hennerici, M. Hommel, M. Kaste, J. R. Marler, R. L. Sacco, P. Teal, N.-G. Wahlgren, S. Warach, C. J. Weir and K. R. Lees, *Stroke,* 2007, **38**, 1905–1910.
43. S. A. Mayer, N. C. Brun, K. Begtrup, J. Broderick, S. Davis, M. N. Diringer, B. E. Skolnick and T. Steiner, *N. Engl. J. Med.,* 2005, **352**, 777–785.
44. S. A. Mayer, N. C. Brun, K. Begtrup, J. Broderick, S. Davis, M. N. Diringer, B. E. Skolnick and T. Steiner, *N. Engl. J. Med.,* 2008, **358**, 2127–2137.
45. L. Hsieh and D. Nugent, *Haemophilia,* 2008, **14**, 1190–1200.
46. A. Inbal, J. Oldenburg, M. Carcao, A. Rosholm, R. Tehranchi and D. Nugent, *Blood,* 2012, **119**, 5111–5117.
47. T. Watnick and G. G. Germin, *N. Engl. J. Med.,* 2010, **363**, 879–881.
48. European Medicines Agency (EMA) accepts Otsuka's marketing authorization application (MAA) for tolvaptan, an investigational compound for autosomal dominant polycystic kidney disease (ADPKD), http://www.otsuka.com/en/hd_release/index.php?year=2013, accessed 30 December 2013.
49. A. E. Emery, *Neuromuscular Disord.,* 1991, **1**, 19–29.
50. K. Bushby, R. Finkel, D. J. Birnkrant, L. E. Case, P. R. Clemens, L. Cripe, A. Kaul, K. Kinnett, C. McDonald, S. Pandya, J. Poysky, F. Shapiro, J. Tomezsko and C. Constantin, *Lancet Neurol.,* 2010, **9**, 77–93.

51. C. M. McDonald, E. K. Henricson, J. J. Han, R. T. Abresch, A. Nicorici, L. Atkinson, G. L. Elfring, A. Reha and L. L. Miller, *Muscle Nerve,* 2010, **42**, 966–974.
52. C. M. McDonald, E. K. Henricson, R. T. Abresch, J. Florence, M. Eagle, E. Gappmaier, A. M. Glanzman, R. Spiegel, J. Barth, G. Elfring, A. Reha and S. W. Peltz, *Muscle Nerve,* 2013, **42**, 357–368.
53. C. A. Schiffer, *N. Engl. J. Med.,* 2007, **357**, 258–265.
54. C. L. Sawyers, *N. Engl. J. Med.,* 1999, **340**, 1330–1340.
55. M. H. Cohen, J. R. Johnson and R. Pazdur, *Clin. Cancer Res.,* 2005, **11**, 12–19.
56. S. G. O'Brien, F. Guilhot and R. A. Larson, *N. Engl. J. Med.,* 2003, **348**, 994–1004.
57. B. J. Druker, F. Guilhot, S. G. O'Brien, I. Gathmann, H. Kantarijian, N. Gattermann, M. W. N. Deininger, R. T. Silver, J. M. Goldman, R. M. Stone, F. Cervantes, A. Hochhaus, B. L. Powell, J. L. Gabrilove, P. Rousselot, J. Reiffers, J. J. Cornelisson, T. Hughes, H. Agis, T. Fischer, G. Verhoef, J. Shepherd, G. Saglio, A. Gratwohl, J. L. Neilsen, J. P. Radich, B. Simonsson, K. Taylor, M. Baccarani, C. So, L. Letvak and R. A. Larson, *N. Engl. J. Med.,* 2006, **355**, 2408–2417.
58. H. Kantarjian, N. P. Shah, A. Hochhaus, J. Cortes, S. Shah, M. Ayala, B. Moiraghi, Z. Shen, J. Mayer, R. Pasquini, H. Nakamae, F. Huguet, C. Boque, C. Chuah, E. Bleickardt, M. B. Bradley-Garelik, C. Zhu, T. Szatrowski, D. Shapiro and M. Baccarani, *N. Engl. J. Med.,* 2010, **362**, 2260–2270.
59. G. Saglio, D. W. Kim, S. Issaragrisil, P. le Coutre, G. Etienne, C. Lobo, R. Pasquini, R. E. Clark, A. Hochhaus, T. P. Hughes, N. Gallagher, A. Hoenekopp, M. Dong, A. Haque, R. A. Larson and H. M. Kantarjian, *N. Engl. J. Med.,* 2010, **362**, 2251–2259.
60. M. Baccarani, M. W. Deininger, G. Rosti, A. Hochhaus, S. Soverini, J. F. Apperley, F. Cervantes, R. E. Clark, J. E. Cortes, F. Guilhot, H. Hjorth-Hansen, T. P. Hughes, H. M. Kantarjian, D.-W. Kim, R. A. Larson, J. H. Lipton, F.-X. Mahon, G. Martinelli, J. Mayer, M. C. Muller, D. Niederwieser, F. Pane, J. P. Radich, P. Rousselot, G. Saglio, S. Saubele, C. Schiffer, R. Silver, B. Simonsson, J.-L. Steegmann, J. M. Goldman and R. Hehlmann, *Blood,* 2013, **122**, 872–884.
61. The National Comprehensive Cancer Network, NCCN Guidelines Version 2.2014 Chronic Myelogenous Leukemia, http://www.nccn.orgprofessionals physician_glspdfcml.pdf, accessed 30 December 2013.
62. C. S. Anderson, E. Heeley, Y. Huang, J. Wang, C. Stapf, C. Delcourt, R. Lindley, T. Robinson, P. Lavados, B. Neal, J. Hata, H. Arima, M. Parsons, Y. Li, J. Wang, S. Heritier, Q. Li, M. Woodward, R. J. Simes, S. M. Davis and J. Chalmers, *N. Engl. J. Med.,* 2013, **368**, 2355–2365.
63. A. Vlachos, S. Ball, N. Dahl, B. P. Alter, S. Sheth, U. Ramenghi, J. Meerpohl, S. Karlsson, J. M. Liu, T. Leblanc, C. Paley, E. M. Kang,

E. J. Leder, E. Atsidaftos, A. Shimamura, M. Bessler, B. Glader and J. M. Lipton, *Br. J. Haematol.*, 2008, **142**, 859–876.
64. A. Vlachos, G. W. Klein and J. M. Lipton, *J. Pediatr. Hematol./Oncol.*, 2001, **23**, 377–382.
65. J. R. Johnson, P. Bross and M. Cohen, *Clin. Cancer Res.*, 2003, **9**, 1972–1979.
66. P. M. Mannucci and E. G. Tuddenham, *N. Engl. J. Med.*, 2001, **344**, 1773–1779.
67. G. C. White, F. Rosendaal, L. M. Aledort, J. M. Lusher, C. Rothschild and J. Ingerslev, *Thromb. Haemostasis*, 2001, **85**, 560.
68. R. S. Schwartz, C. F. Abildgaard, L. M. Aledort, S. Arkin, A. L. Bloom, H. H. Brackmann, D. B. Brettler, H. Fukui, M. W. Hilgartner, M. J. Inwood, C. K. Kasper, P. B. A. Kernoff, P. H. Levine, J. M. Lusher, P. M. Mannucci, I. Scharrer, M. A. MacKenzie, N. Pancham, H. S. Kuo and R. U. Allred, *N. Engl. J. Med.*, 1990, **323**, 1800–1805.
69. G. C. White, S. Courter, G. L. Bray, M. Lee and E. D. Gomperts, *Thromb. Haemostasis*, 1997, 77, 660–667.
70. S. Seremetis, J. M. Lusher, C. F. Abildgaard, C. K. Kasper, R. Allred and D. Hurst, *Haemophilia*, 1999, **5**, 9–16.
71. A. L. Serra, D. Poster and A. D. Kistler, *N. Engl. J. Med.*, 2010, **363**, 820–829.
72. E. Berntorp and A. D. Shapiro, *Lancet*, 2012, **379**, 1447–1456.
73. M. L. Lee and D. A. Roth, *Haemophilia*, 2005, **11**, 5–12.
74. M. Recht, L. Nemes, M. Matysiak, M. Manco-Johnson, J. Lusher, M. Smith, P. Mannucci, C. Hay, T. Abshire, A. O'Brien, B. Hayward, C. Udata, D. A. Roth and S. Arkin, *Haemophilia*, 2009, **15**, 869–880.
75. International Conference on Harmonisation of Technical Requirements for Registration of Pharmaceuticals For Registration of Pharmaceuticals for Human Use, Choice of control group and related issues in clinical trials E10, http://www.fda.gov/downloads/Drugs/GuidanceCompliance RegulatoryInformation/Guidances/ucm073139.pdf, accessed 30 December 2013.
76. Pivotal study (pharmacokinetics, efficacy, safety) of BAX 326 (rFIX) in hemophilia B patients, ClinicalTrials.Gov, http://www.clinicaltrial s.gov/ct2/show/NCT01174446?term=hemophilia&recr=Completed&intr=rixubis &spons=baxter &phase=2&fund=2&rank=1, accessed 30 December 2013.
77. Summary Basis for Regulatory Action – RIXUBIS, http://www.fda.gov/BiologicsBloodVaccines/BloodBloodProducts/ApprovedProducts/Licensed ProductsBLAs/FractionatedPlasmaProducts/ucm358781.html, accessed 30 December 2013.
78. S. D. Grosse, M. S. Schechter, R. Kulkarni, M. A. Lloyd-Puryear, B. Strickland and E. Trevathan, *Pediatrics*, 2009, **123**, 407–412.
79. H.-G. Eichler, F. Petavy, F. Pignatti and G. Rasi, *N. Engl. J. Med.*, 2013, **369**, 1577–1579.

COMMERCIAL CONSIDERATIONS

CHAPTER 4

Treating Rare Diseases: Business Model for Orphan Drug Development

CORY WILLIAMS

Pfizer China R&D Center (CRDC), 5th Floor, German Centre, Tower One, 88 Keyuan Road, Pudong Zhangjiang Hi-Tech Park, Shanghai 201203, China
E-mail: cory.williams@pfizer.com

4.1 Orphan Drug Development: Background and History

The 1983 Orphan Drug Act (ODA) in the USA, enacted on 4 January 1983, was the first major milestone to address the significant unmet need for, and the inadequate attention of pharmaceutical R&D activity focus on, small-population or rare diseases.[1] The awareness of this gap in global pharmaceutical development was raised, in the period leading up to the 1983 ODA, by a number of individuals and interest groups close to patients and families suffering from rare diseases, which were not considered profitable to invest in by the pharmaceutical industry.[2,3] Given that the original 1983 ODA provisions were imprecise and impractical (*i.e.* 'to provide incentives in the development of drugs for the treatment of rare diseases that would normally be unprofitable or unpatentable'), there were two subsequent ODA amendments in 1984 (*i.e.* numeric prevalence threshold implemented to define disease scope, supporting drug therapies for diseases which either affect fewer than 200 000 people in the USA, or diseases for which sales in the USA are unlikely to recoup

RSC Drug Discovery Series No. 38
Orphan Drugs and Rare Diseases
Edited by David C Pryde and Michael J Palmer

Published by the Royal Society of Chemistry, www.rsc.org

R&D costs) and 1985 (*i.e.* marketing exclusivity extended to 7 years for both patentable and unpatentable products), respectively (Table 4.1).[2,3] These two amendments brought greater definition, recognition and focus on orphan or rare diseases, as well as the therapies to treat these diseases (*i.e.* orphan drugs), where 'orphan' indicated that only few pharmaceutical companies were willing to 'adopt' products to treat these diseases.[2,3] Reviews of the evolution of orphan drug designations and approvals in the context of specific ODA provisions (*e.g.* market exclusivity) have indicated that the ODA has been associated with increased orphan drug development.[4,5] For example, there were only 10 orphan drug approvals in the decade before the 1983 ODA, compared to more than 90 in the decade after, underscoring the impact of this legislation in the USA.[5] Since 1983, the US Food and Drug Administration (FDA) has granted more than 2700 designations and more than 400 approvals for orphan drug indications.[6]

Other countries and regions have introduced similar policies (*e.g.* Singapore, Japan, Australia and the EU), as illustrated in Table 4.2, with differences in the prevalence definition and incentive structure.[2] The cumulative orphan drug designations in the EU and Japan, the two major markets for orphan drug development outside of the USA, are 898 and 293, respectively.[7] The majority of designations are for lower prevalence diseases (*i.e.* <10 000 patients). A review of US orphan drug approvals by Braun *et al.* showed an inverse correlation between orphan disease prevalence and orphan drug designations and approvals, for the overall orphan disease universe as well as for lower prevalence diseases (*i.e.* <10 000 patients).[8] In addition, this study demonstrated a higher 'orphan drug approval proportion' for lower

Table 4.1 Orphan drug policy by key country.

	USA	*Japan*	*EU*
Key administrative details			
Nature of policy	1983 – Legislation	1999 – Regulation	2000 – Regulation
Prevalence definition[a]	<200 000	<50 000	<250 000
Prevalence per 10 000 population[a]	<6.37 in 10 000	<4 in 10 000	<5 in 10 000
Authority[b]	FDA OOPD	MHLW OPSR	EMA COMP
Key incentives			
R&D tax credits[c]	Yes (50%)	Yes (6%)	No
R&D grants	Yes	Yes	Yes
Application fee waivers	Yes	No	Fee reductions
Trial design support	Yes	Yes	Yes
Market exclusivity	7 years	10 years	10 years

[a]USA and Japan each use an absolute prevalence for defining orphan diseases, while the EU uses prevalence per 10 000 population. Prevalence per 10 000 population estimates derived for comparison based on current population sizes: USA (314 m), Japan (128 m) and EU (506 m).
[b]Administrative offices include the Food & Drug Administration's Office of Orphan Product Development (OOPD) in the USA, the Ministry of Health, Labor & Welfare's Organization for Pharmaceutical Safety & Research (OPSR) in Japan and the European Medicines Agency's Committee for Orphan Medicinal Products (COMP) in the EU.
[c]R&D tax credits apply to clinical costs in the USA, and to both clinical and non-clinical costs in Japan. There may be tax credits within EU member states.

Table 4.2 Orphan drug designations by key country.

	Annual results													
	1999 total[a]	*2000*	*2001*	*2002*	*2003*	*2004*	*2005*	*2006*	*2007*	*2008*	*2009*	*2010*	*2011*	*2012*
Orphan designations														
USA	994	70	78	64	95	131	123	142	117	165	165	194	203	188
Japan	136	12	7	5	7	8	5	14	10	16	7	10	24	32
EU	0	8	31	31	36	56	69	63	79	66	91	119	102	147

[a]'1999 total' represents the cumulative total orphan designations up until 1999. The EU regulation came into effect the year after in 2000. However, the EU has had a higher growth rate in annual orphan designations, surpassing Japan's cumulative total by 2005.

prevalence *vs.* higher prevalence orphan diseases (*i.e.* 20% *vs.* 15% for cohorts above and below the median prevalence of 39 000; $p = 0.01$). A review of European orphan drugs showed a similar trend of an inverse correlation between orphan disease prevalence and orphan drug approvals by the EMA.[9]

US orphan drug development has gone through a three-stage market evolution based on output (*e.g.* the volume of designations and approvals; productivity), and the nature of drug development approaches and therapeutic area focus.[10] This three-stage evolution, first described by Wellman-Labadie *et al.* and adapted here based on consistent trends to an updated timeline, roughly corresponds to three 10 year periods since the 1983 ODA (ending 1990, 1999 and 2011, respectively). Each period outlines tremendous growth in output to a peak at the end of the period. In addition, the second and third periods are characterised by a nadir at the beginning, resulting from a fall-off in output from the preceding period's peak due to macro-level market factors, after which the growth uptick restarts. This paradigm and associated trends, confirmed based on updates with more recent output data, are illustrated in Table 4.3. Key points to highlight from this three-stage distribution of orphan drug development market output include the importance of market shocks resulting from a fall-off in output from the previous peak (*i.e.* proposed ODA amendments that threatened to strip orphan drug pharmaceutical companies of their status in the early 1990s; the global economic recession of 2000–01, which curtailed investment), the importance of oncology to orphan drug development (*i.e.* the growth in annual orphan drug designations for oncology surpasses all other therapeutic areas, over the three periods), and peak output in 2011, which is a landmark year for orphan drug development (*i.e.* orphan drug designations and approvals numbered 203 and 26, respectively, making 2011 the year with the highest number ever).[6,7] Indeed, the compound annual growth rate (CAGR) for FDA orphan designations for 2001–10, the decade leading up to this landmark year and which coincides with the third evolutionary period, approximated 10%, in contrast to a negative CAGR for FDA new molecular entity (NME) approvals for the same period.[11]

The core aetiological paradigm underlying the orphan diseases being addressed by drug development has expanded from enzyme disorders (often

Table 4.3 US orphan drug R&D output – three distinct periods.

Period definition	*Period 1*	*Period 2*	*Period 3*
Timeline	Through 1990	Through 1999	Through 2011
Key milestones in the growth trajectory for orphan drug designations	1 (1983) to peak of 89 (1990)	56 (1992) to peak of 73 (1999)	64 (2002) to peak of 203 (2011)
Key market challenges to sustained growth in orphan drug designations[a]	Proposals to reverse orphan status (1990)	Global economic recession (2001)	N/A
Orphan drug R&D output			
Designations – average number per year (all)	47	65	129
Designations – average number per year (oncology)	10	17	43
Approvals – average number per year	8	14	16
Approvals – annual growth or CAGR[b]	29%	6%	7%

[a]There were proposed ODA amendments in 1990–1992 to reverse orphan drug status for products where the population increased above 200 000 and/or the sales surpassed $200 million. The annual designation peaks in 1990 and 1999 were each followed by decreases in annual designations (probably due to related market challenges), to a trough in the following 2 or 3 years, after which the growth trajectory restarted.
[b]CAGR is compound annual growth rate.

monogenic diseases), where repurposed and novel enzyme replacement therapies flourish, to include more complex, polygenic diseases, especially oncology disorders. Indeed, among therapeutic areas, oncology represents a majority share of new orphan disease therapies that come to market (*i.e.* up to 33% in the USA and 40% in the EU, in recent reports), as well as the majority share of the pharmaceutical industry's orphan drug pipeline (*i.e.* more than 30%), driven by a higher growth in annual orphan drug designations for oncology compared to other therapeutic areas, as mentioned earlier.[8–12] Moreover, this trend has contributed to orphan drug development playing a key role in the evolution of overall pharmaceutical R&D to more targeted and personalised approaches.[3,9,11,13] This has also led to the expansion of the orphan disease taxonomy to now include 'sub-diseases', uniquely characterised and defined based on important genetic and clinical features (*e.g.* subsets of late-stage melanoma and non-small cell lung cancer).[8,9,11,14,15] This interconnection between orphan drug development and overall pharmaceutical drug development should continue to provide insights and approaches to the continued evolution to personalised and stratified drug development.

Current estimates indicate that there are 5000–8000 rare diseases in the world for which Orphanet, a European organisation, has done systematic identification and classification.[16–21] The universe of rare diseases corresponds to ~30 to 35 million people across the EU and ~25 to 30 million in North America, many of whom are children, and includes the ultra-rare diseases (*i.e.* those with less than 20 patients per million of population or less than 5000 patients worldwide).[19,22,23] Despite the tremendous growth in orphan drug development, and accomplishments to date, only ~300 of the 6000–8000 rare diseases have approved therapies and less than 10% of patients affected with orphan diseases are treated currently.[3] Taken together, there is still significant unmet need to be addressed across the universe of orphan diseases.

Many orphan diseases are characterised by a tight-knit community of patients, care-givers and treating healthcare professionals, which shares information among members on symptoms, disease characteristics, treatment options and new therapies being investigated, including potential clinical trials to participate in. Quite often, these communities are built on the backbone of formal organisations (*i.e.* patient advocacy groups or PAGs), dedicated to the betterment of patients with the particular disease.[3] These PAGs essentially function as a first and key resource for patients and their families affected by the disease (*e.g.* general disease awareness and education; links to informational resources and community members; patient referrals; testimonials; investigational therapies and clinical trials), which is important for many orphan diseases, where not much is widely known. Also important is the key role PAGs play in advocacy and facilitation of policy making to support a variety of causes related to improving the lives of patients and families suffering from rare diseases (*e.g.* disease definition, recognition, awareness, funding, research, drug development, distribution and access). Indeed, NORD (National Organization for Rare Disorders) was established in 1983 in the USA, comprising many of the individuals and groups who were influential in passing of the 1983 ODA.[3] Orphan disease PAGs may exist at global, regional and national levels, with a network of affiliations and partners, as they provide critical support to the community (*e.g.* for adrenoleukodystrophy: 'ALD-AMN' is global, 'ALD LIFE' is one of many in Europe, 'The Stop ALD Foundation' is one of many in the USA, and the 'Australian Leukodystrophy Support Group' operates in Australia). The degree of community 'stickiness' for many orphan diseases will influence many of the key factors that underpin overall product development approaches for new orphan drugs.

Orphan drugs, with current global revenues of $83 billion, have become an increasingly large and important part of the global pharmaceutical market, for which global sales in 2012 amounted to $645 billion.[7] The first decade of the 21st century has been the most productive period in the history of orphan drug development, when considering orphan drug designations and approvals.[7] Orphan drugs comprise an increasing share of FDA NME approvals – one-third on average over the 2007–12 period, with a peak of 44% in 2009 (Table 4.4).[7] The 2001–12 CAGR in FDA orphan drug approvals, essentially mapping to the third evolutionary period of orphan drug development as previously discussed, is significant at approximately 14%, compared to less than 3% for FDA NME approvals for the same period, as shown in Figure 4.1.[7] Moreover, a recent analysis of 'strategically important orphan drugs within the pharmaceutical industry' (*i.e.* those orphan drugs for which revenues are comprised predominantly from orphan disease indications), demonstrated that orphan drugs currently represent 22% of total pharmaceutical sales with a 2001–10 CAGR of ~26% (compared to 20% for a set of matched controls of non-orphan drugs).[11] Additionally, EvaluatePharma market estimates suggest that the growth in orphan drug sales will have a disproportionately favourable impact on pharmaceutical revenue growth, compared to non-orphan drugs.[7] Key metrics describing this impact include:

- The CAGR of orphan drugs at 7.4% will double that of the overall pharmaceutical market (excluding generics) for the period 2012–18, as shown in Figure 4.2.[7]
- Orphan drug sales will total $127 billion, representing 15.9% of the overall prescription drug market in 2018, which is a significant increase from the 5.1% share in 1998.

Table 4.4 US FDA drug approvals – overall and orphan.

	2005	*2006*	*2007*	*2008*	*2009*	*2010*	*2011*	*2012*
US FDA approvals								
All orphans	19	24	16	14	20	14	26	26
NME orphans	10	8	8	9	15	6	14	15
All NMEs	28	29	26	31	34	26	35	43
NME orphan share[a]	36%	28%	31%	29%	44%	23%	40%	35%

[a]'NME orphan share' represents the number of NME or BLA orphan drug approvals expressed as a share of total NME or BLA approvals for each year.

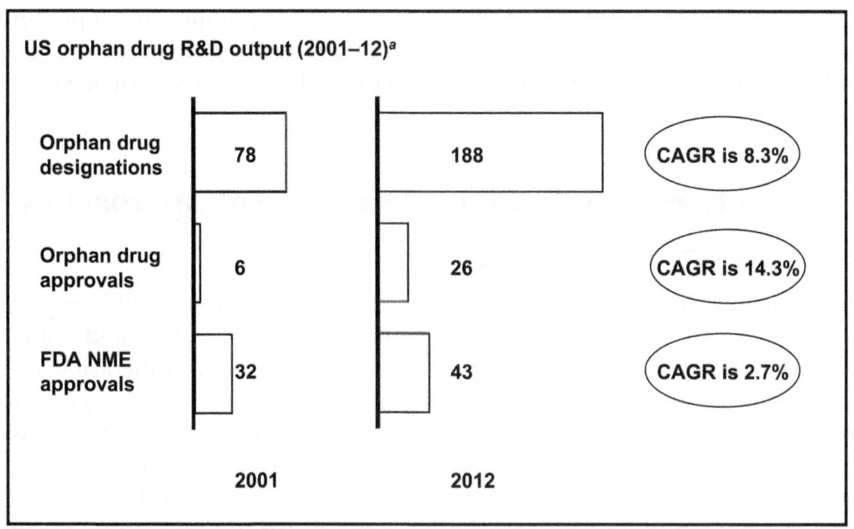

[a]There has been significant growth in orphan drug R&D output in the USA over the last decade, compared with FDA NME approvals, based on data reported by EvaluatePharma. CAGR is compound annual growth rate.

Figure 4.1 Recent orphan drug designation and approval trajectory.

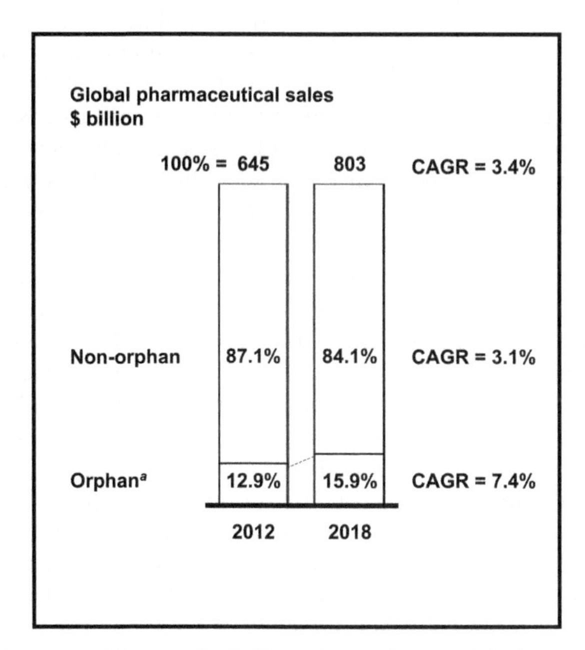

[a]Orphan drugs, which comprised 5.1% of the overall prescription drug market in 1998, will increase to a 15.9% share by 2018, corresponding to total orphan drug sales of $127 billion, based on an EvaluatePharma report.

Figure 4.2 Orphan drug market growth outlook.

The following sections will explore orphan drug product development, commercialisation, investment and economics, as well as the future outlook for this rapidly developing and exciting sector within the biopharmaceutical industry.

4.2 Orphan Drug Product Development Approaches

4.2.1 Background Challenges

Although clinical development for orphan drugs may 'theoretically' appear to be straightforward and easy (*e.g.* small number of patients and sometimes only one Phase 3 trial needed), there can be several difficulties and challenges to be addressed.[3] The product development challenges for orphan drugs fall into three categories: (i) disease understanding and knowledge, (ii) clinical programme strategy and execution, and (iii) interpretation of clinical trial results and regulatory negotiation.[3,9,12,24] First, for many orphan diseases, there is suboptimal understanding of the natural histories of disease progression, as well as significant heterogeneity in pathophysiology, both of which pose difficulties for drug development. Second, orphan drug clinical programme strategy and execution is further challenged by uncertainties in the selection and characterisation of appropriate trial end points and treatment durations, paucity of robust biomarkers, recruiting the right patient populations and identifying qualified investigators.[12] Indeed, the FDA often encourages sponsors to perform observational trials to understand the natural history of the relevant orphan disease. Special attention must be paid to ensure that the small patient populations for orphan diseases are characterised appropriately, given biological and pharmacological heterogeneity/variability, as well as geographical distribution and scarcity.[25] There is a need for robust statistical approaches to address these population-related challenges, especially with the increasing regulatory requirement for clinical trial control arms.[3,25] Third, companies working to get orphan drugs approved must demonstrate favourable benefit–risk profiles for their products, while addressing variabilities in treatment effect, as well as the lack of harmonisation across regulatory and national agencies. Additionally, for many orphan diseases, the lack of regulatory 'precedent' presents a challenge in itself.[12] Although frequently applied in oncology, the applicability of the FDA's Accelerated Approval pathway to non-oncology orphan drugs (*i.e.* for rare genetic diseases) has been variable, given the difficulty/inability to validate surrogate end points and challenges with completing Phase 4 studies. Despite language on Subpart H in the Food and Drug Administration Safety and Innovation Act (FDASIA), such as the use of biomarkers and consideration of rare diseases, some orphan drug players believe there is need for more clarity. The scientific basis, level of evidence required, and context for biomarker use are areas to clarify.[24] The FDA is expected to issue a guidance document that recognises the special contextual features of rare diseases (*e.g.* low prevalence; serious, life-threatening nature of many rare diseases).

Accordingly, regulatory success often requires robust and frequent interactions with regulatory agencies to gain alignment on key programme design elements including trial design, patient population requirements, clinical end points (*e.g.* focused on prevention, reversal and/or stabilisation of disease), and the selection of biomarkers (the strength of which may facilitate accelerated approval). Best practice would suggest that this engagement starts very early (*e.g.* before or during clinical programme design stages) and continues throughout programme execution. The next few sections will discuss traditional and emerging orphan drug product development platforms.

4.2.2 Enzyme Replacement Therapy (ERT) and the Evolution of Product Development Platforms

Historically, small molecule drugs and enzyme replacement therapy (ERT), or protein replacement therapy (PRT) more broadly (*i.e.* essentially the use of recombinant proteins for treating monogenic diseases with an aberrant or missing protein), have been key product development approaches and the mainstay of bringing orphan drugs to market. Additionally, product development platforms that help form the landscape for overall biopharmaceutical R&D have also contributed to orphan drugs – from traditional small molecule approaches to non-ERT large molecule platforms (*e.g.* monoclonal antibodies, peptides), especially in oncology.[3] A summary of orphan drug product development platforms is provided in Table 4.5. Given several product-profile drawbacks for 'traditional' ERT (*e.g.* intravenous administration, slow infusions, discomfort and logistical difficulty for patients), and the convergence of new science on drug development, orphan drug development is going through an evolution as newer (and even next-generational) product development platforms (*e.g.* gene therapy, oligomers and chaperones) become part of the treatment landscape. Even the latest wave of ERT has modifications upon the traditional ERT approach. For example, Protalix has created a unique plant biomanufacturing platform (using carrot and/or tobacco tissue culture cells) to develop the first plant-derived ERT in taliglucerase alfa to treat Gaucher disease.[26]

4.2.3 Gene Therapy

Gene therapy product development approaches, often for monogenic diseases, generally involve using viral vectors to infect target cells from a patient, which allows the patient to sustainably produce the aberrant or missing protein central to the disease. A key example is bluebird bio, a gene therapy company focused on rare diseases, which uses its non-replicating lentiviral vector

Table 4.5 Orphan drug product development platforms.

Platform	*Approach*	*Approvals*	*Examples*
1. Traditional small molecules[a]	Low MW organic compounds: based on disease-specific screens	Many	Imatinib, zoledronic acid
2. Large molecules[a]	High MW compounds: based on understanding molecular pathogenesis	Many	Epoetin alfa, somatropin
3. Monoclonal antibodies	Immunoglobulins: based on specificity for antigenic epitopes	Many	Infliximab, rituximab
4. Traditional ERT	Enzymes: replace deficient/aberrant enzymes (usually recombinant)	Many	Imiglucerase, laronidase
5. Next-generation ERT	Enzymes: ERT derived from novel platforms (*e.g.* non-mammalian)	1	Taliglucerase
6. Gene therapy	Vector-delivered gene sequences: replace deficient/aberrant gene products	1	Alipogene tiparvovec
7. RNA therapeutics[b]	Anti-sense oligonucleotides: modulate RNA biology and gene expression	1	Mipomersen
8. Pharmacological chaperones	Small molecules: stabilize and/or reshape misfolded proteins	0	N/A

[a]Small molecules are low molecular weight (<900 Daltons) organic compounds and have been the main molecule platform for drug development. Large molecules (naturally occurring, recombinant, synthetic) comprise a broad cross-section of compound classes (*e.g.* proteins, antibodies, peptides, oligonucleotides, genes), some of which are covered here as distinct orphan drug development platforms.

[b]Fomivirsen, a first-generation AON approved in 1998 and discontinued in 2004, is not considered within the scope of the current 'RNA therapeutics' platform.

platform to infect a patient's own bone marrow-extracted haematopoietic cells *ex vivo*, which are returned to patients to address the defective gene product, based on research in this field.[27,28] Currently, bluebird bio has key late-stage clinical development programmes in Lenti-D and LentiGlobin. Lenti-D, its lead programme, is an experimental treatment for the rare hereditary disorder childhood cerebral adrenoleukodystrophy, which affects young boys (world-wide population of ~3000), by replacing ABCD1, a peroxisomal protein responsible for the degradation of very long chain fatty acids in the brain. LentiGlobin, which uses a similar vector but replaces beta-globin, is being evaluated for the treatment of beta thalassaemia major and sickle cell disease. The lentiviral vector platform is also being applied in oncology (*e.g.* chronic lymphoid leukaemia), with bluebird bio's recent early-stage clinical collaboration with Celgene (another company focused on orphan diseases) to genetically modify a patient's own chimeric antigen receptor T-cells, for targeting and destroying cancer cells.[29] GSK, in collaboration with Fondazione Telethon and Fondazione San Raffaele, is using a retrovirus vector platform to address adenosine deaminase deficient severe combined immunodeficiency (ADA-SCID), which affects approximately 350 children in the world.[30] Although this indication is quite rare, the expectation is that validation of this gene therapy platform will facilitate application in other monogenic diseases (*e.g.* Wiskott–Aldrich syndrome).

uniQure's Glybera (alipogene tiparvovec), approved in Europe in 2012 for the orphan disease lipoprotein lipase deficiency, after previous rejection by the EMA, is the first example of a gene therapy approved in any country outside of China and is considered by many as a landmark milestone for gene therapy platforms.[31,32] The disease is characterised as an inefficiency or lack of absorbing dietary fats, resulting in inflammation in the pancreas. Alipogene tiparvovec's approval, based on clinical data from 27 patients, restricts prescribing to the subset of lipoprotein lipase deficiency patients who have suffered repeated pancreatitis, and requires ongoing monitoring by the company, to demonstrate long-term efficacy and safety.[33–35] Early indicators suggesting alipogene tiparvovec will probably be priced at more than €1.2 m per patient (or approximately €250 000 per patient per year), which would make it the world's most expensive drug, have stirred the ongoing debate about orphan drug pricing. There are a number of arguments for the proposed price: high pharmaceutical R&D costs in general, and alipogene tiparvovec's high-cost development programme in particular (*i.e.* it has been in development for about 10 years), are significant; R&D costs are to be recouped by sales to the very small patient population (*i.e.* it affects 1 per 1–2 million population); alipogene tiparvovec will reduce the high costs of emergency admissions and treatment; the lifetime cost of the single-administration alipogene tiparvovec is much lower than many other continuous administration orphan drugs, especially ERTs, which cost hundreds of thousands of euros per year. Interestingly, uniQure has proposed an 'annuity' approach to charging health systems for alipogene tiparvovec (*e.g.* payments distributed over 5 years, duration for which efficacy has been shown), which could contribute to the evolution of the pricing and reimbursement paradigm for orphan drugs.[34] The gene therapy field is definitely gaining momentum as other pharmaceutical companies continue to invest in this area (*e.g.* Novartis and Biomarin), and investor sentiment (notably bluebird bio's successful IPO in June 2013) demonstrates some validation of the field, which suffered severe scepticism only 5 years ago.[36] Nonetheless, the case for gene therapy has many appealing factors including high expectations for safety and efficacy given the use of patients' own cells, avoidance of patient compliance and adherence challenges given the one-time nature of the therapeutic intervention, and potential for reductions in the total cost of care (*e.g.* lifetime cost of care for a sickle cell anaemia patient approximates to $9 million).[37]

4.2.4 RNA Therapeutics

There is an exciting and fast-evolving group of novel therapeutic approaches involving the use of oligomers, or anti-sense oligonucleotides (AONs) more specifically – exploiting their inherent sequence-specific Watson–Crick RNA base pairing characteristics – to bring about therapeutic effect through a number of subcellular mechanisms that are central to RNA biology (*e.g.* transcription, splicing, translation and RNA degradation).[38–41] The key therapeutic approaches include downregulation of gene expression or post-transcriptional 'gene silencing' (*e.g.* mipomersen sodium's inhibition of apolipoprotein B's translation in homozygous familial hypercholesterolemia (HoFH)), and interference with an inherently abrogated transcription splicing process to re-establish normal function (*e.g.* drisapersen enables restoration of the DMD gene's normal reading frame through 'skipping' of exon 51 in Duchenne muscular dystrophy (DMD)).[39–43]

Mipomersen (Kynamro), an AON gapmer, co-developed by Genzyme and Isis and approved by the US FDA in late 2012 for HoFH, which affects ~400 people in the USA, is an example of a therapy that has validated the translation inhibition approach.[44,45] ISIS-TTR_{RX}, another example of this approach, inhibits translation of transthyretin (TTR) and is currently being developed by GSK and Isis for treating TTR amyloidosis and TTR-FAP (familial amyloid polyneuropathy), a disorder characterised by misfolded transthyretin protein and intracellular amyloid fibril aggregates.[46] ALN-TTR02 (Alnylam), yet another example of this approach, is also in clinical development for the same indication.[47]

Drisapersen, currently in late-stage clinical development for DMD in a collaboration between GSK and Prosensa and with FDA Breakthrough Therapy designation, takes a different AON approach: drisapersen enables restoration of the dystrophin/DMD gene's normal reading frame through 'skipping' of exon 51 in DMD.[39,42,43,48] This DMD exon-skipping programme is also significant for its innovative clinical development and regulatory approach. Duchenne's, an X-linked disease characterised by progressively debilitating natural history disease stages, has a spectrum of manifestations with important implications for selecting clinical end points and trial design, on a background of a wide array of exon-deletion abnormalities.[42] The first two stages, in younger-to-teenage boys, exhibit minimal loss of function and are not understood well enough to have validated end points. The latter two stages, affecting teenagers and older patients, exhibit more debilitating disease affecting cardiac, pulmonary and upper limb function. Prosensa's exon-targeting therapeutic approach, which would create a 'menu' of therapies for each exon-deletion abnormality, is influenced by the decreasing prevalence of the target exon (*e.g.* 13% for exon 51 through to 2% for exon 8).[49] This directly impacts the ability to recruit patients and obtain clinical data in all sub-populations, specific to each exon-deletion abnormality. Based on this background, the clinical development and regulatory approach will probably pursue a full development programme for compounds addressing the most prevalent target exon mutations (*e.g.* exon 51) and more targeted end points and clinical data sets (*e.g.* use of smaller unpowered studies; bridging and extrapolation; personalised end points) for compounds addressing the rarer target exon mutations (*e.g.* exon 8).[49]

Beyond AONs, there are small molecule investigational therapies within the emerging field of RNA therapeutics that target post-transcriptional control processes (to correct or compensate for a genetic defect), particularly for orphan and ultra-orphan disorders.[39–41] One company pursuing this approach is PTC Therapeutics, whose lead compound ataluren, an orally bioavailable, small molecule compound for the treatment of genetic disorders that arise from nonsense mutations, is currently in late-stage clinical trials for nonsense-mutation Duchenne's (nmDMD) and nonsense-mutation cystic fibrosis (nmCF).[50,51] Ataluren is believed to interact with the ribosomal machinery that decodes mRNA, enabling the ribosome to read through premature nonsense-mutation stop signals, thereby facilitating the production of a full-length, functional protein.[50,51]

4.2.5 Pharmacological Chaperones

Another category of products being developed for orphan diseases are pharmacological chaperones that help to reshape or restore function to misfolded proteins. Chaperones bring about therapeutic effect downstream of translation by 'protecting' their target proteins (*e.g.* mutant enzymes) from misfolding, and subsequent sequestration and degradation by the usual intracellular processes that manage misfolded proteins.[52,53] This allows these mutant enzymes to essentially function normally.

Amicus Therapeutics, arguably the company with the broadest portfolio of small molecule pharmacological chaperones, is leveraging its technology platform to develop orally bioavailable therapies to address lysosomal storage disorders including Fabry, Gaucher and Pompe diseases.[54] A potential benefit of this novel approach is the ability to enhance target enzyme activity in hard to reach tissues. Migalastat, Amicus' lead chaperone programme (currently in late-stage clinical development for Fabry disease), is an experimental oral therapy which stabilises alpha-galactosidase mutants and represents a collaborative development programme between Amicus and GSK.[55] Purported benefits of migalastat monotherapy, compared with traditional ERT, are its oral route of administration and enhanced efficacy.[55] Moreover, migalastat (and pharmacological chaperones in general) are being explored as part of combination therapy approaches with ERT, to overcome the inherent decreased activity of ERTs due to denaturing effects when ERTs are introduced into the bloodstream.[54] For example, current development programmes are investigating the co-formulation of chaperones with traditional ERTs (*e.g.* Fabrazyme, Lumizyme) for enhanced therapeutic effect in Fabry and Pompe diseases.[56]

Tafamidis (Vyndaquel), a small molecule chaperone therapy, is an example of the validation of this therapeutic approach and product development platform for orphan diseases, as outlined in Table 4.5.[57] It was approved in Europe in 2011 to treat the orphan disease TTR-FAP.[57] Tafamidis, originally developed by FoldRx which was acquired by Pfizer, stabilises transthyretin and prevents fibre growth.[58,59]

4.3 Orphan Drug Commercialisation Approaches: Pricing, Reimbursement and Market Access

Historically orphan drug commercialisation has been characterised by features consistent with niche markets – relatively high pricing, specialised distribution channels and targeted marketing. More so, orphan drug reimbursement, by private or public payer, has traditionally been generous, affording most patients in the small orphan disease communities with access to medicines, which are often life-saving and/or provide significant quality of life attributes. The availability of orphan drugs across Europe during the first decade after the passing of the 1999 EU orphan drug regulation was quite variable, with smaller countries exhibiting availability challenges.[22] More recently (*i.e.* within the last 5 years), ultimate patient access to and utilisation of orphan therapies have been subject to increasing restrictions, notably in the EU and the USA, due to several factors: higher pricing for newly launched orphan drugs; stricter reimbursement philosophies (due to greater scrutiny on overall drug costs); and more stringent payer management approaches and tactics.[60–62] That said, there has been an increase in the availability of patient access programmes, which counterbalance these limitations to address critical patient unmet need and facilitate utilisation.

4.3.1 Pricing

The higher pricing for orphan drugs compared to the broader pharmaceutical market is clear. Indeed, the 2010 list of the top ten priciest drugs, all of which treat ultra-orphan diseases, outlines costs that start at $200 000 per patient per year, and reported eculizumab (Alexion Pharmaceuticals) as first on the 2010 list at ~$400 000 per year[63] More recently, teduglutide (estimated pricing of ~$300 000 per patient per year), was approved by the FDA in 2012 for the treatment of short bowel syndrome.[64] The pricing of orphan drugs, albeit complex and in many ways non-transparent, is driven by a number of factors – the need to recoup high pharmaceutical R&D costs in general while deriving a reasonable profit margin, in the context of the high-risk nature of pharmaceutical R&D; the fact that product sales will be derived from a very small consumer market corresponding to the patient populations of individual orphan diseases; the value of the product to patients and their families, compared to the number and nature of alternative or incumbent health technologies, which may be non-existent for many orphan diseases; and the inherent monopoly market dynamics.[65–68] Orphan drug pricing is also influenced by the fact that oncology (as a therapeutic area) and large molecules (as a product modality), both of which attract higher pricing in the broader non-orphan drug marketplace due to market demand for these cutting-edge technologies, comprise the majority of orphan drug approvals.[11,60] As mentioned before, drugs for ultra-orphan or ultra-rare diseases (*i.e.* those with less than 20 patients per million of population) tend to attract even higher pricing within the orphan disease universe, due to the economics related to recouping significant R&D costs from sales to a very small consumer population; and an inverse correlation has been found between prevalence and price.[67,69,70] The list of the highest priced orphan drugs also illustrates this trend. That said, the 'effective' price for a health system or payer, which may be the result of negotiations and discounts, notably in the EU, influences ultimate adoption and market access. For example, the National Institute for Health and Care Excellence (NICE), the UK's health technology assessor (HTA), had considered dasatinib and nilotinib as equally effective in treating chronic myeloid leukaemia (CML). However, nilotinib was recommended (in addition to imatinib, the incumbent therapy) given that the manufacturer had already made an agreement with the Department of Health to offer nilotinib at a discounted price which improved its cost-to-quality adjusted life years (QALY) ratio compared with dasatinib.[71] Moreover, countries that regulate prices (*e.g.* Spain, Portugal) tend to have lower prices compared to countries with free-market pricing for pharmaceutical products, such as the USA or Germany.[67] The high prices of orphan drugs, especially ultra-orphan drugs, has become a very sensitive issue.[67,68,72] There are stakeholders who are of the opinion that prices are too high for the value delivered by many orphan drugs, but others have noted that this is difficult to evaluate.[67,68,72,73] As a result, there are ongoing proposals for price controls, which pose the risk of curbing innovation and ongoing investment in orphan drug R&D.[67,68,72]

4.3.2 Reimbursement

The high prices of orphan drugs have contributed to a longstanding debate on balancing the allocation of public funds to address the high-cost healthcare needs for severe diseases affecting a small minority *versus* maximising the health of the population as a whole.[68,74–76] Historically, there has been a high willingness to pay for drugs treating orphan diseases (especially ultra-orphan diseases) by patients, the population and payers, given disease seriousness, lack of alternatives and the identification with these patients.[72,77] Restricted payer budgets, for public and private payers alike, have resulted in stricter reimbursement philosophies in recent years.[60–62,67] The global economic recession and financial challenges since 2008, along with an increased focus on managing healthcare costs (especially pharmaceutical costs) and higher pricing of innovative new therapies, has led to a 'perfect storm' for overall pharmaceutical reimbursement, within which orphan drugs are also impacted.

NICE, which is evolving its assessment philosophy to value-based pricing (beyond the focus on cost-effectiveness), is becoming increasingly stringent with orphan drugs, and is in many ways taking a similar approach to the evaluation of orphan drugs as it does with non-orphan drugs. For example, some ultra-orphan drugs were reimbursed in the past, despite having relatively high incremental cost-effectiveness ratios (ICERs), even beyond NICE's acceptable range, as seen with therapies for Fabry and Gaucher disease.[78] However, recent NICE assessments suggest that orphan and ultra-orphan drugs are expected to have ICERs within the acceptable range and/or be accompanied by manufacturer-provided discounts, as mentioned earlier, as part of patient access schemes, to receive a favourable NICE recommendation.[71] NICE recently released a new final draft guidance decision to recommend ipilimumab and vemurafenib as treatment options for advanced malignant melanoma, supported by manufacturer-provided discounts.[79] In contrast, NICE recommended, in June 2013, against ruxolitinib for myelofibrosis (which has 0.4 cases per 100 000 each year in the UK), citing that the agent was not considered cost-effective, although deemed clinically effective.[80] Ruxolitinib, for which NICE had issued draft guidance with similar recommendations in February 2013, was not submitted with a patient access scheme, even during the follow-up interactions after the draft guidance, and was assessed by NICE to have an ICER of ~£150 000 (almost twice that estimated by the manufacturer).[81] Another indicator of the increased focus on orphan drug reimbursement decisions is the fact that NICE is taking more control on assessments for ultra-orphan drugs compared to regional UK health authorities, which were responsible for ultra-orphan drug assessments in the past.[82]

The shift towards 'stricter reimbursement' philosophies will be accentuated for orphan diseases where there are multiple agents within a disease category, which could be perceived by payers as therapeutic substitutes, even as disease management approaches evolve towards combination, cocktail and/or sequential therapy (*e.g.* melanoma or even Fabry disease).[60,62,78] Given concerns about unsustainable orphan drug pricing, many policy makers and healthcare managers will have an increasing tendency to restrict orphan drug reimbursement, but this must be weighed against the unintended consequences of limiting patient access, in many cases, to life-enabling therapies.[67,68] It is interesting that orphan drugs have been included in the scope of these tighter reimbursement philosophies given that, at a macro level, orphan drugs represent only a minor component of healthcare budgets.[61] Indeed, historically orphan drugs have represented 3.6–4.6% of national healthcare budgets and will probably not exceed 6.6% even with worst-case sensitivity analyses.[61] An emerging trend, central to the evolution of the reimbursement paradigm, is the increased focus among many HTAs on health-related quality of life (HRQoL) and patient-reported outcomes (PRO) parameters in pricing, market access and reimbursement decision making.[83] This will probably be considered as part of NICE's approach to implementing value-based pricing.[84] The environmental trends towards more stringent reimbursement will require orphan drug companies to demonstrate the value of their products in a holistic manner for key stakeholders (*e.g.* for patients, population, payers and the overall system), while highlighting product benefits relevant to the severity of the disease and compared to alternative therapies.

4.3.3 Payer Management

Within the environment of increasing restrictions on orphan drug reimbursement, several payers have started to employ benefit designs, management tools, approaches and tactics, previously reserved for other high-cost therapy categories (*e.g.* oncology, injectables, speciality drugs) to manage orphan drug utilisation and spend.[60,62] This has become most apparent in oncology, and for US commercial or private payers, where this trend has started to spill over from non-orphan to orphan oncology payer management. Some of these payer management tools, approaches and tactics include the use of restrictive tiers, prior authorisation, step therapy, increased patient coinsurance and/or co-payment, genetic testing (*i.e.* to ensure diagnostic and other parameters assessed in clinical trials are adhered to in the treatment setting; to restrict prescribing to drug label), case management and plan-imposed maximum limits/caps.[60,62] Clinical pathways – efforts among payers, practices and third-party groups to develop clinical protocols, which will standardise treatment and control spending – are also becoming more commonplace in the US oncology treatment arena.[85] While the majority of orphan drug access and utilisation is not yet covered by all these payer management approaches and tactics, there is considerable evidence that payer practices are trending in that direction. This is illustrated by the results of a 2008–09 survey of US payers: orphan drugs are under increased scrutiny for cost control; the majority of survey respondents would employ prior authorisation to manage orphan drugs; clinical data and overall cost exposure were the highest-ranked factors, among plans, driving benefit design and restrictions for orphan drugs.[60] Less than one-third of plans surveyed would tolerate orphan drug pricing greater than a $75 000 per patient per year threshold before being subject to scrutiny, which is interesting since only three of nine orphan drug agents in the survey were priced below that threshold.[60] Cost sharing can represent a significant burden for patients, as highlighted by 10–20% coinsurance benefit designs, which could translate to $48 500 to $88 200 per year in patient out-of-pocket spend, for cases where there is no plan-imposed maximum.[60] More recently, biological orphan drugs have been categorised as 'tier 4' medications in many private medical insurance plans in the USA, which require patients to cover up to one-third of overall drug costs.[86] Manufacturers will increasingly need to understand and address a payer's budget management goals, for the specific disease category as well as overall, as part of their commercialisation approach.

4.3.4 Patient Access Programmes

A number of approaches have emerged, predominantly driven by manufacturers, to enhance patient access to orphan therapies in response to challenges for orphan drug availability, affordability and reimbursement. Creative risk-sharing schemes, in addition to traditional patient access programmes and manufacturer discounts, are increasingly playing an important role in the provision of orphan drugs to patients.[22,73,87–90] Risk-sharing schemes are often designed on a premise where payers cover the costs, or the majority of costs, for cases where the drug is shown to have favourable outcomes, while the manufacturer covers the costs for cases where the drug doesn't work. This concept is taken further with performance-based risk-sharing agreements for ultra-orphan therapies, where price reductions can be entertained or negotiated if clinical outcomes are suboptimal or not compelling, which provides an approach to address the uncertainty regarding the long-term effectiveness of costly ultra-orphan drugs.[73,87,88]

Traditional patient access programmes provide free or significantly subsidised drugs to patients, often based on means testing of a patient's ability to pay, with drug costs being covered by pharmaceutical manufacturers and/or third-party organisations, such as non-profit groups like the National Organization of Rare Diseases (NORD) and Patient Access Network Foundation.[60,62,90] Other creative schemes consider the subsidisation of orphan drug costs at the expense of drugs for non-orphan diseases.[22,67] On balance, while patients may have fair 'access' to many orphan drugs that are covered under medical insurance plans, ultimate utilisation will depend on the combination of pricing and payer management (*e.g.* plan design, cost management tactics employed).[60,65] Even when there is access, there is a risk of under-utilisation, if reimbursement limitations and high cost-sharing burdens are placed on patients.

In summary, the key dimensions of commercialisation success around which companies must differentiate in order to win in the orphan drug market include understanding and exploiting orphan disease market fundamentals (*e.g.* 'community stickiness'), and partnering with PAGs; connecting and communicating with the community, through education (including scientific platforms), outreach, marketing and sales force efforts; maximising pricing, reimbursement and cost-sharing support; and optimising the product lifecycle (*e.g.* expanding to other indications and having multiple orphan indications).

4.4 Business Case for Developing Orphan Drugs

The favourable economics and generally positive investment case for orphan drug development and commercialisation has been fairly well characterised.[7,11] The favourable business case for orphan drugs is compelling at the macro level for the global orphan drug market, which has robust market growth fundamentals even when compared with the global pharmaceutical and non-orphan drug market, as discussed earlier in this chapter, as well as at the level of individual cases of orphan drug development. There are two key evaluations or reports that have investigated this topic – the *Drug Discovery Today* article 'Orphan Drug Development: An Economically Viable Strategy For Biopharma R&D' (published in 2012), and EvaluatePharma's 'Orphan Drug Report' (published in 2013).[7,11]

- *'Orphan Drug Development: An Economically Viable Strategy For Biopharma R&D' (2012):* this evaluation performed an analysis of the present value of revenues (PV) of 86 launched orphan drugs and compared them to a control sample set of 291 non-orphan drugs.[11] The total PV for orphan and non-orphan drugs was found to correspond to a mean per-year economic value of $406 million for an orphan drug, compared with $399 million for a non-orphan drug. This indicates that mean per-year economic values of the orphan and non-orphan drug cohorts were almost equal, which underscores the value-creation viability of orphan drugs. In addition, an analysis of the change in PV over time demonstrated that the mean PV of orphan drugs approximately doubled from $351 million in 2000 to $637 million in 2010, while the mean PV for non-orphan drugs remained constant at just over $600 million over the same time period, as in Figure 4.3.
- *'EvaluatePharma's Orphan Drug Report' (2013):* this report described an assessment of the revenue forecast for marketed and pipeline orphan drugs and indicates that the growth rate of the orphan drug market, for the period 2012–18 (*i.e.* 7.4%), will double the growth estimate for the

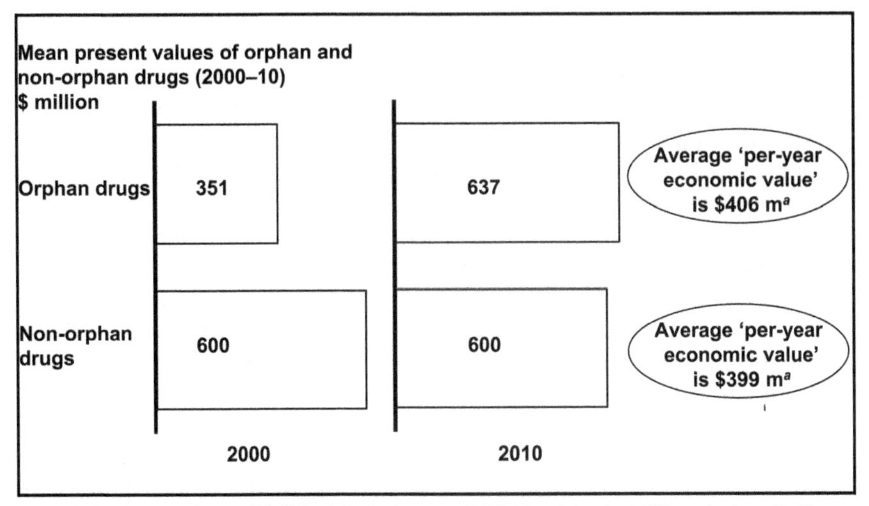
Mean present values of orphan and non-orphan drugs (2000–10)
$ million
Orphan drugs
351
637
Average 'per-year economic value' is $406 m[a]
Non-orphan drugs
600
600
Average 'per-year economic value' is $399 m[a]
2000
2010
[a]An analysis of the present value of revenues (PV) of 86 launched orphan drugs compared with that of a control sample set of 291 non-orphan drugs outlined the mean per-year economic values (over the 1990–2030 forecast period) for an orphan and a non-orphan drug.

Figure 4.3 Orphan *vs.* non-orphan economic value estimates.

overall pharmaceutical market (excluding generics), as outlined in Figure 4.1.[7] Moreover, Figure 4.2 illustrates that orphan drugs, which comprised 5.1% of the overall prescription drug market in 1998, will increase to a 15.9% share by 2018, corresponding to total orphan drug sales of $127 billion.[7] In addition, the expected return on investment (ROI) for the current late-phase R&D pipeline cohort was evaluated to be 10.3× for orphan drugs, *vs.* 6.0× for non-orphan drugs.[7] This translates to an ROI for orphan drugs which is 170% the value of that for non-orphan drugs and provides additional evidence for the value creation and economic viability of orphan drugs.[7]

4.4.1 Drug Development Drivers of Orphan Drug Economic Value

The factors underlying the positive investment case for orphan drugs, which are related to product development, include the following:

- *Tax credits*: the US orphan drug regulation grants tax credits amounting to 50% of R&D spend.[3,6] Interestingly, this incentive on its own may not be significant for many small orphan drug companies, given the requirement that the company must be profitable.
- *R&D grants*: there is annual grant funding in the USA to address the costs of qualified clinical testing expenses (*e.g.* $30 million per year for Phase I to III clinical trials for the fiscal years 2008–12).[7] In the EU, research grants are available at the levels of the European Community as well as member states.[9,22] In addition, EU and US regulations provide support to companies in protocol development and clinical study design, as shown in Table 4.1.[6,9,22]
- *Fee waivers*: US FDA Prescription Drug User Fee Act (PDUFA) filing or user fees are waived for companies with less than $50 million in worldwide revenues. PDUFA fees for an application requiring clinical data are currently approximately $2 million.[91] In the EU, there are fee reductions for protocol assistance (100%), marketing authorisation (50%), as well as pre-authorisation inspections, as outlined in Table 4.1.[9,22]
- *Favourable clinical development costs*: beyond the cost benefits already mentioned, there is strong evidence indicating that clinical development spend for orphan drugs is lower than that for non-orphan drugs. The average and median Phase III clinical trial costs for orphan drugs are 46% and 64% that of non-orphan drugs, respectively.[7] Practically, this could translate to the Phase III clinical trial costs for orphan drugs being one-quarter that of non-orphan drugs, when the 50% tax credit is taken into consideration.[7] The Phase III clinical trial size, a key factor of clinical development spend, is 4 (based on median) to 6 (based on average) times larger for non-orphan *vs.* orphan, based on data from EvaluatePharma and as depicted in Figure 4.4.[7] This is consistent with other reports which have highlighted that orphan drug trials require

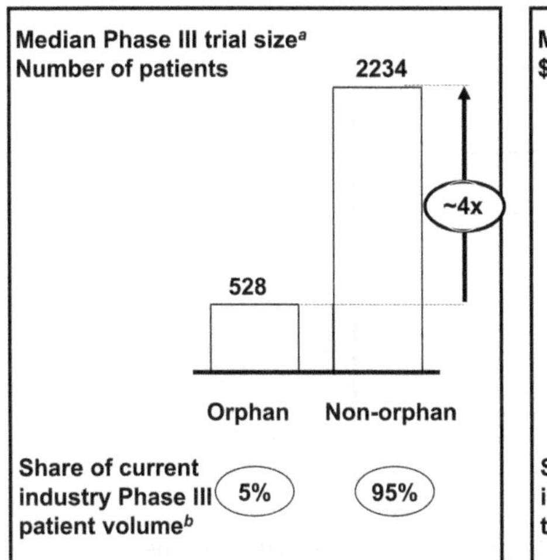

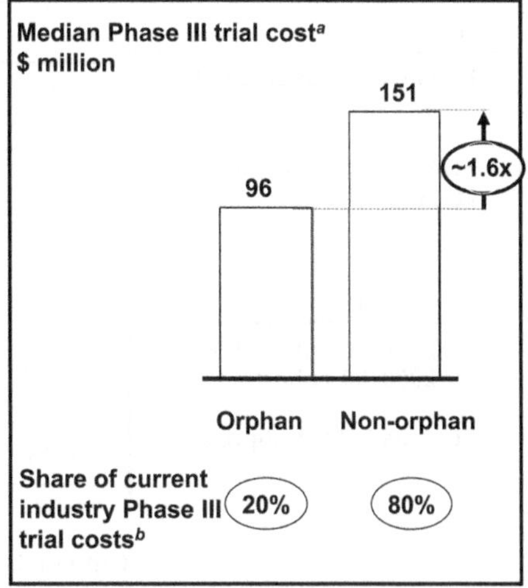

[a]EvaluatePharma's analysis of Phase III trial costs, for orphan drugs compared with non-orphan drugs, is based on all new drug products entering Phase III from January 2000. [b]EvaluatePharma's analysis of current cohort of Phase III or filed orphan and non-orphan drugs. EvaluatePharma also reported return on investment (ROI) estimates for orphan drugs of ×10.3 compared with ×6.0 for non-orphan drugs, for the current cohort of Phase III or filed drugs.

Figure 4.4 Orphan *vs.* non-orphan clinical development cost drivers.

fewer patients, including evidence of a 6.6-fold difference in the number of clinical trial participants in the largest trials.[92,93]

- *Regulatory approval success*: an analysis of the rate of regulatory success of orphan drugs, using data from CMR International for drugs approved between the years of 1997 and 2009, showed that filed orphan drugs demonstrated a 5% increased probability of regulatory success compared to the entire population (*i.e.* 93% *vs.* 88%).[11] The Tufts Center for the Study of Drug Development reported that 22% of pharmaceutical company orphan drug development programmes achieved approval in the 2000–09 period, compared to a 16% clinical approval success rate for 'mainstream' or non-orphan drug programmes.[12] Moreover, orphan drug designation status is associated with higher FDA first-cycle approval rates compared to non-orphan drugs (*i.e.* 60% for orphan drugs *vs.* 48% for non-orphan drugs), as illustrated in Figure 4.5.[94]
- *Efficient clinical development and regulatory approval timelines*: an analysis of clinical development timelines, using data from CMR International, reported an average Phase II-to-launch duration of 3.9 years for orphan drugs, compared to 5.42 years for non-orphan drugs (Figure 4.5).[11] A separate analysis, although finding no difference in Phase III trial length, showed a shorter FDA approval timeline for orphan compared to non-orphan drugs (*i.e.* 9.0 *vs.* 10.1 months), also illustrated in Figure 4.5.[7] Orphan drugs often secure fast track and/or priority review status from regulatory agencies, which contribute to shorter clinical development and regulatory review durations.[5,94] Indeed, orphan NMEs demonstrate a statistically significant greater use

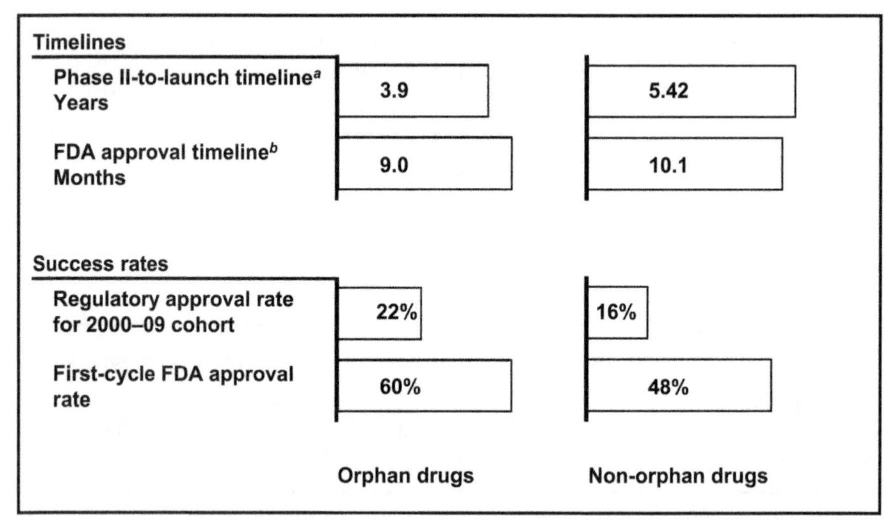

Figure 4.5 Orphan *vs.* non-orphan development timelines and approval success.

(compared with non-orphans) of any US FDA Fast Track or Subparts E/H review procedures (29.5 *vs.* 9.5%), as well as obtaining designation for priority review (83.5 *vs.* 35.4%).[5,94]

4.4.2 Commercialisation Drivers of Orphan Drug Economic Value

The factors underlying the positive investment case for orphan drugs, which are related to commercialisation, include the following:

- *Market exclusivity designations*: the USA and EU grant 7 year and 10 year market exclusivity from the date of approval respectively (Table 4.1), such that the market exclusivity benefit is not spent during clinical development before approval.[3,6,9,22] This market exclusivity benefit is incremental to patent exclusivity for drugs that are not patent-expired and can represent tremendous financial gains for the product. An analysis encompassing 99 orphan and 421 non-orphan drugs approved by the US FDA from 1983 to 2007 demonstrated that orphan drug market exclusivity extended the 'effective' life of the product by 0.8 to 0.9 years.[5] It is considered that the longer market exclusivity in the EU compared to the USA provides some balance to the lack of tax incentives at the level of the EU Commission, given that tax-related concessions are controlled by EU member states.[22]
- *Premium pricing*: despite being based on an often complex, and sometimes opaque, interplay of several underlying factors, it's clear that pricing pays an important role in driving the economic value of orphan,

and more so, ultra-orphan drugs, given that there is an inverse correlation between prevalence and price.[66–68,70] The top ten priciest drugs, all of which treat ultra-orphan indications, each cost more than $200 000 per patient per year.[63] Moreover, orphan drugs have demonstrated some of the most significant 'extraordinary price increases' (*i.e.* price increases greater than 100% at any one time), with examples of price increases greater than 3000% in extreme cases, and sometimes linked to the regulatory approval of prior off-label use.[72] The first high-priced orphan drug was imiglucerase (also the first ultra-orphan drug), an infusible ERT launched in 1994 for the treatment of Gaucher disease at a cost of approximately $300 000 per patient per year.[95] Currently, the most expensive drug in the world is eculizumab (Alexion Pharmaceuticals), which costs just under $500 000 per patient per year, and is approved for treating the ultra-orphan diseases PNH (paroxysmal nocturnal haemoglobinuria) and aHUS (atypical haemolytic-uremic syndrome).[68]

- *Product use expansion through follow-on indications*: while some of the top-selling orphan drugs are for single-indication use, the revenue-generating potential of orphan drugs is clearly enhanced by obtaining multiple follow-on indications, including additional orphan indications.[11] Indeed, results from 'Orphan Drug Development: An Economically Viable Strategy For Biopharma R&D', which evaluated the present value of revenues of a cohort of orphan drugs, indicate a correlation between multiple orphan indications and overall product value.[11] Among the top ten orphan drugs (ranked by PV), the total PV for those with multiple orphan indications was $34.3 billion compared to only $8.1 billion for those drugs with a single orphan indication. A separate analysis, in the same report, demonstrated a statistically significant greater trend for multi-indication orphan drugs to target initial approval in an orphan *vs.* non-orphan indication (75.5 *vs.* 24.5%), which could enable more favourable pricing.[11]

There are two additional factors that could be considered as commercialisation drivers of orphan drug economic value, albeit based more on directional evidence compared to those discussed above – rapid market uptake and lower marketing costs (relative to non-orphan drugs). A 2010 review of the US Orphan Drug Act (ODA) provides an indication of the rapid market uptake of orphan drugs, where 11 of the 18 single-indication orphan drugs which were 'blockbusters' (*i.e.* those with global annual sales exceeding $1 billion) achieved blockbuster status within the 7 year orphan drug market exclusivity period.[10] The case of a lower marketing cost requirement for orphan drugs, compared to non-orphans, is based on the fact that the orphan disease patient populations are smaller and often connected within a tight-knit community. Against that background, an orphan drug company may leverage compelling support from PAGs, strong relationships with treating physicians and the fast-growing impact of social media platforms to create awareness, rather than rely on the traditional pharmaceutical sales and marketing model, which would probably incur more cost.[95]

An important caveat is that the favourability of drug development and commercialisation drivers of orphan drug economic value, which have been outlined here, are based on analysis across the universe of orphan drugs. When the development plans for individual orphan drugs are being created, the cost, complexity, challenges and high-risk nature of pharmaceutical R&D in general should not be underestimated.[96]

4.5 Future Outlook

The favourable economics and investment case for orphan drug development and commercialisation are clear. The current trends in the orphan drug product development arena provide some interesting themes and an innovation imperative for influencing the evolution of the biopharmaceutical landscape – for orphan drug R&D specifically, as well as continual stimulation of biopharmaceutical R&D in general.

Orphan drug R&D will make important contributions to life sciences research, drug discovery and translational medicine, thereby enhancing therapeutic development approaches (*e.g.* genomic techniques, gene silencing, oligomer research, transcription science; epigenetic modification of DNA). Indeed, orphan drug R&D experiences will help to advance the development and use of personalised/stratified medicine approaches and targeted medicines.[11] Moreover, the crossover application of biological knowledge (*e.g.* disease targets and pathways) from orphan diseases will continue to inform and de-risk therapeutic approaches for non-orphan diseases, where orphan diseases may serve as a 'proof of concept' for non-orphan diseases.[13] In addition, orphan drug R&D will support the acceleration of new product concepts, product development technologies and platforms (*e.g.* next-generation antibodies, antibody drug conjugates, gene therapy, cell therapy, regenerative medicine) by evaluating their application to orphan diseases. Orphan drug R&D also has a key role in evolving clinical development paradigms (*e.g.* trial design, statistical modelling, patient selection, biomarker innovation, companion diagnostics, patient registries) and regulatory science (*e.g.* conditional approvals). It will be interesting to see the extent to which 'real world trials' are included as part of ongoing orphan drug development programme efforts to expand clinical data sets and update the risk–benefit profile of orphan drugs. The 'real world trial' paradigm, gathering efficacy and safety data across countries where feasible, could help encourage greater use of progressive (not only conditional) regulatory approval approaches for orphan drugs, which could help address concerns regarding the paucity of clinical data available at the time of marketing authorisation.[9]

Orphan drugs will also provide important contributions and innovative models for pharmaceutical commercialisation for pricing and reimbursement (*e.g.* risk-sharing schemes, lifetime or long-term payment models for one-time treatments such as gene therapy). Awareness, education, outreach and marketing approaches, for consumers and prescribers, will also be influenced by the degree to which orphan drugs embrace social media and community connectivity models.

References

1. FDA, *Fed. Regist.*, 1992, **57**, 62085.
2. M. A. Villarreal, *Orphan Drug Act: Background & Proposed Legislation in the 107th Congress, Congressional Research Services Report for Congress*, Congressional Research Services, The Library of Congress (US), 2001.
3. IOM, *Rare Diseases and Orphan Products: Accelerating Research and Development*, ed. M. J. Field and T. F. Boat, Institute of Medicine, The National Academies Press, Washington (DC), USA, 2010.
4. C. H. Asbury, *J. Am. Med. Assoc.*, 1991, **265**, 893–897.
5. E. Seoane-Vazquez, R. Rodriguez-Monguio, S. L. Szeinbach and J. Visaria, *Orphanet J. Rare Dis.*, 2008, **3**, 33.
6. FDA, *Developing Products for Rare Diseases & Conditions*, http://www.fda.gov/ForIndustry/DevelopingProductsforRareDiseasesConditions/default.htm, accessed 5 July 2013, United States Food and Drug Administration.
7. EvaluatePharma, *Orphan Drug Report 2013*, 2013.
8. M. M. Braun, S. Farag-El-Massah, K. Xu and T. R. Cote, *Nat. Rev. Drug Discovery,* 2010, **9**, 519–522.
9. K. Westermark, B. B. Holm, M. Soderholm, J. Llinares-Garcia, F. Riviere, S. Aarum, F. Butlen-Ducuing, S. Tsigkos, A. Wilk-Kachlicka, C. N'Diamoi, J. Borvendeg, D. Lyons, B. Sepodes, B. Bloechl-Daum, A. Lhoir, M. Todorova, I. Kkolos, K. Kubackova, H. Bosch-Traberg, V. Tillmann, V. Saano, E. Heron, R. Elbers, M. Siouti, J. Eggenhofer, P. Salmon, M. Clementi, D. Krievins, A. Matuleviciene, H. Metz, A. C. Vincenti, A. Voordouw, B. Dembowska-Baginska, A. C. Nunes, F. M. Saleh, T. Foltanova, M. Mozina, J. Torrent i Farnell, B. Beerman, S. Mariz, M. P. Evers, L. Greene, S. Thorsteinsson, L. Gramstad, M. Mavris, F. Bignami, A. Lorence and C. Belorgey, *Nat. Rev. Drug Discovery,* 2011, **10**, 341–349.
10. O. Wellman-Labadie and Y. Zhou, *Health Pol.*, 2010, **95**, 216–228.
11. K. N. Meekings, C. S. Williams and J. E. Arrowsmith, *Drug Discovery Today,* 2012, **17**, 660–664.
12. I. Melnikova, *Nat. Rev. Drug Discovery,* 2012, **11**, 267–268.
13. M. C. Fishman and J. A. Porter, *Nature,* 2005, **437**, 491–493.
14. T. R. Cote, K. Xu and A. R. Pariser, *Nat. Rev. Drug Discovery,* 2010, **9**, 901–902.

15. F. Braiteh and R. Kurzrock, *Mol. Cancer Ther.*, 2007, **6**, 1175–1179.
16. NIH, National Institutes of Health, 2013, vol. 2013.
17. U. S. Congress, in *Public Health Service Act (42 U.S.C. 281). H.R. 4013*, Congressional record, USA, 2002, vol. 148 (2002), vol. Public Law 107–280, pp. 1988–1991.
18. OIG, *The Orphan Drug Act – Implementation and Impact OEI-09-00-00380*, Office of Inspector General. United States Department of Health and Human Services, 2001.
19. EC, *Communication from the commission to the European parliament, the council, the European economic and social committee and the committee of the regions on rare diseases: Europe's challenges*, European Commission, Brussels, 2008.
20. P. Stolk, M. J. Willemen and H. G. Leufkens, *Bull. W. H. O.*, 2006, **84**, 745–751.
21. Orphanet, *The portal for rare diseases and orphan drugs*, http://www.orpha.net/consor/cgi-bin/index.php, accessed 5 July 2013.
22. B. Hughes, *Nat. Rev. Drug Discovery*, 2008, 7, 190–191.
23. NICE, *NICE Citizens Council Report: Ultra Orphan Drugs*, http://www.nice.org.uk/niceMedia/pdf/Citizens_Council_Ultraorphan.pdf, accessed 15 July 2013, National Institute for Health and Care Excellence, National Health Service (NHS), UK, London, 2004.
24. R. C. Griggs, M. Batshaw, M. Dunkle, R. Gopal-Srivastava, E. Kaye, J. Krischer, T. Nguyen and K. Paulus, P. A. Merkel and Rare Diseases Clinical Research Network, *Mol Genet Metab*, 2009, **96**, 20–26.
25. B. M. Buckley, *Lancet*, 2008, **371**, 2051–2055.
26. Y. Shaaltiel, D. Bartfeld, S. Hashmueli, G. Baum, E. Brill-Almon, G. Galili, O. Dym, S. A. Boldin-Adamsky, I. Silman, J. L. Sussman, A. H. Futerman and D. Aviezer, *Plant Biotechnol. J.*, 2007, **5**, 579–590.
27. bluebird, *Platform Overview*, http://www.bluebirdbio.com/platform-overview.php, accessed 10 July 2013.
28. N. Cartier, S. Hacein-Bey-Abina, C. C. Bartholomae, G. Veres, M. Schmidt, I. Kutschera, M. Vidaud, U. Abel, L. Dal-Cortivo, L. Caccavelli, N. Mahlaoui, V. Kiermer, D. Mittelstaedt, C. Bellesme, N. Lahlou, F. Lefrere, S. Blanche, M. Audit, E. Payen, P. Leboulch, B. l'Homme, P. Bougneres, C. Von Kalle, A. Fischer, M. Cavazzana-Calvo and P. Aubourg, *Science*, 2009, **326**, 818–823.
29. bluebird, *News Release: bluebird bio Announces Global Strategic Collaboration with Celgene to Advance Gene Therapy in Oncology*, http://www.bluebirdbio.com/pdfs/Celgene-bluebird-PR-32113.pdf, accessed 10 July 2013, 2013.
30. GSK, *News Release: GSK, Fondazione Telethon and Fondazione San Raffaele to collaborate on gene therapy for rare diseases*, http://us.gsk.com/html/media-news/pressreleases/2010/2010_pressrelease_10113.htm, accessed 10 July 2013, 2010.
31. A. Flemming, *Nat. Rev. Drug Discovery*, 2012, **11**, 664.
32. N. Moran, *Nat. Biotechnol.*, 2012, **30**, 807–809.

33. uniQure, *News Release: uniQure's Glybera first gene therapy approved by EC*, http://www.uniqure.com/news/167/182/, accessed 10 July 2013, 2012.
34. A. Jack, in *The Financial Times*, 2012.
35. S. Stovall, in *The Pink Sheet Daily*, 2012.
36. P. Bonanos, *START-UP, The Unlikely Renaissance Of Gene Therapy*, 2012.
37. S. D. Kotiah and S. K. Ballas, *Expert Opin. Invest. Drugs,* 2009, **18**, 1817–1828.
38. C. F. Bennett and E. E. Swayze, *Annu. Rev. Pharmacol. Toxicol.,* 2010, **50**, 259–293.
39. H. L. Lightfoot and J. Hall, *Nucleic Acids Res.,* 2012, **40**, 10585–10595.
40. R. Kole, A. R. Krainer and S. Altman, *Nat. Rev. Drug Discovery,* 2012, **11**, 125–140.
41. F. Muntoni and M. J. A. Wood, *Nat. Rev. Drug Discovery,* 2011, **10**, 621–637.
42. L. van Vliet, C. L. de Winter, J. C. van Deutekom, G. J. van Ommen and A. Aartsma-Rus, *BMC Med. Genet.,* 2008, **9**, 105.
43. G. J. van Ommen, J. van Deutekom and A. Aartsma-Rus, *Curr. Opin. Mol. Ther.,* 2008, **10**, 140–149.
44. FDA, *News Release: FDA approves new orphan drug Kynamro to treat inherited cholesterol disorder*, http://www.fda.gov/NewsEvents/Newsroom/PressAnnouncements/ucm337195.htm, accessed 10 July 2013, United States Food and Drug Administration, 2013.
45. P. Hair, F. Cameron and K. McKeage, *Drugs,* 2013, **73**, 487–493.
46. Isis, *News Release: Isis Initiates Phase 2/3 Study of ISIS-TTR Rx and Earns $7.5 Million Milestone Payment From GlaxoSmithKline*, http://ir.isispharm.com/phoenix.zhtml?c=222170&p=irol-newsArticle&id=1786054, accessed 10 July 2013, 2013.
47. Alnylam, *News Release: Alnylam Reports Positive Phase II Data for ALN-TTR02, an RNAi Therapeutic Targeting Transthyretin (TTR) for the Treatment of TTR-Mediated Amyloidosis (ATTR)*, http://phx.corporate-ir.net/phoenix.zhtml?c=148005&p=irol-newsArticle2&ID=1834040&highlight=, accessed 10 July 2013, 2013.
48. GSK, *News Release: GlaxoSmithKline's drisapersen (previously GSK2402968/PRO051) to receive Food and Drug Administration Breakthrough Therapy designation for potential treatment of patients with Duchenne Muscular Dystrophy*, http://www.gsk.com/media/press-releases/2013/glaxosmithkline_s-drisapersen–previously-gsk2402968-pro051–to-.html, accessed 10 July 2013, 2013.
49. N. M. Goemans, M. Tulinius, J. T. van den Akker, B. E. Burm, P. F. Ekhart, N. Heuvelmans, T. Holling, A. A. Janson, G. J. Platenburg, J. A. Sipkens, J. M. Sitsen, A. Aartsma-Rus, G. J. van Ommen, G. Buyse, N. Darin, J. J. Verschuuren, G. V. Campion, S. J. de Kimpe and J. C. van Deutekom, *N. Engl. J. Med.,* 2011, **364**, 1513–1522.
50. A. Opar, *Nat. Rev. Drug Discovery,* 2011, **10**, 479–480.
51. S. W. Peltz, M. Morsy, E. M. Welch and A. Jacobson, *Annu. Rev. Med.,* 2013, **64**, 407–425.
52. F. U. Hartl, A. Bracher and M. Hayer-Hartl, *Nature,* 2011, **475**, 324–332.

53. G. Parenti, *EMBO Mol. Med.*, 2009, **1**, 268–279.
54. R. E. Boyd, G. Lee, P. Rybczynski, E. R. Benjamin, R. Khanna, B. A. Wustman and K. J. Valenzano, *J. Med. Chem.*, 2013, **56**, 2705–2725.
55. Amicus, *News Release: Amicus Therapeutics Presents Additional 6-Month Results from Phase 3 Fabry Monotherapy Study at LDN World Symposium*, http://ir.amicustherapeutics.com/releasedetail.cfm?ReleaseID=740800, accessed 10 July 2013, 2013.
56. Amicus, *Amicus Clinical Studies*, http://www.amicusrx.com/clinical.aspx, accessed 10 July 2013.
57. G. Said, S. Grippon and P. Kirkpatrick, *Nat. Rev. Drug Discovery*, 2012, **11**, 185–186.
58. T. Coelho, L. F. Maia, A. Martins da Silva, M. Waddington Cruz, V. Plante-Bordeneuve, P. Lozeron, O. B. Suhr, J. M. Campistol, I. M. Conceicao, H. H. Schmidt, P. Trigo, J. W. Kelly, R. Labaudiniere, J. Chan, J. Packman, A. Wilson and D. R. Grogan, *Neurology*, 2012, **79**, 785–792.
59. S. M. Johnson, S. Connelly, C. Fearns, E. T. Powers and J. W. Kelly, *J. Mol. Biol.*, 2012, **421**, 185–203.
60. R. Hyde and D. Dobrovolny, *American Health and Drug Benefits*, 2010, **3**, 15–23.
61. C. Schey, T. Milanova and A. Hutchings, *Orphanet J. Rare Dis.*, 2011, **6**, 62.
62. E. Silverman, *Manag. Care*, 2013, **22**, 10–14.
63. M. Herper, *Forbes, The World's Most Expensive Drugs*, http://www.forbes.com/2010/02/19/expensive-drugs-cost-business-healthcare-rare-diseases.html, accessed 15 July 2013, 2010.
64. M. Herper, *Forbes, Inside The Pricing Of A $300,000-A-Year Drug*, http://www.forbes.com/sites/matthewherper/2013/01/03/inside-the-pricing-of-a-300000-a-year-drug/, accessed 15 July 2013, 2013.
65. J. C. Roos, H. I. Hyry and T. M. Cox, *BMJ*, 2010, **341**, c6471.
66. S. Aballea, M. Toumi, A. L. Vataire, A. Millier and M. Lamure, *Value Health*, 2010, **13**, A82.
67. S. Simoens, *Orphanet J. Rare Dis.*, 2011, **6**, 42.
68. E. Picavet, D. Cassiman and S. Simoens, *Orphan Drugs: Research and Reviews*, 2013, **3**, 23–31.
69. M. Zitter, *Manag. Care*, 2005, **14**, 52–67.
70. A. Messori, A. Cicchetti and L. Patregani, *BMJ*, 2010, **341**, c4615.
71. NICE, *News Release: NICE recommends nilotinib and standard dose imatinib for first line chronic myeloid leukaemia*, http://www.nice.org.uk/newsroom/pressreleases/DasatinibNilotinibImatinibForCMLGuidance.jsp, accessed 15 July 2013, National Institute for Health and Care Excellence, 2013.
72. T. A. Hemphill, *J. Bus. Ethics*, 2010, **94**, 225–242.
73. P. Kanavos and E. Nicod, *Value Health*, 2012, **15**, 1182–1184.
74. J. W. Dear, P. Lilitkarntakul and D. J. Webb, *Br. J. Clin. Pharmacol.*, 2006, **62**, 264–271.
75. R. Boy, I. V. Schwartz, B. C. Krug, L. C. Santana-da-Silva, C. E. Steiner, A. X. Acosta, E. M. Ribeiro, M. F. Galera, P. G. Leivas and M. Braz, *J. Med. Ethics*, 2011, **37**, 233–239.

76. S. Simoens, D. Cassiman, M. Dooms and E. Picavet, *Drugs,* 2012, **72**, 1437–1443.
77. A. Cote and B. Keating, *Value Health,* 2012, **15**, 1185–1191.
78. R. Brown, *UBC Evidence Matters Newsletter. Is There an Easier Path to Acceptance for Orphan Drugs?*, 2012.
79. NICE, *News Release: NICE plans to say yes to two breakthrough treatments for skin cancer*, http://www.nice.org.uk/newsroom/pressreleases/NICEPlansYesToBreakthroughTreatmentsSkinCancer.jsp, accessed 15 July 2013, National Institute for Health and Care Excellence, 2012.
80. NICE, *News Release: NICE publishes final guidance for a new treatment for enlarged spleen in adults with a rare blood cancer*, http://www.nice.org.uk/newsroom/pressreleases/NICEPublishesFinalGuidanceForANewTreatmentForEnlargedSpleenInAdultsWithARareBloodCancer.jsp, accessed 15 July 2013, National Institute for Health and Care Excellence, 2013.
81. NICE, *News Release: NICE consults on preliminary recommendations for a new treatment for enlarged spleen in adults with myelofibrosis*, http://www.nice.org.uk/newsroom/pressreleases/NICEConsultsOnPreliminaryRecommendationsForANewTreatmentForEnlargedSpleenInAdultsWithMyelofibrosis.jsp, accessed 15 July 2013, National Institute for Health and Care Excellence, 2013.
82. D. Risinger, *Morgan Stanley Analyst Report*. Pharma Europe, 2012.
83. A. Lloyd, D. Wild, K. Gallop and W. Cowell, *Expert Rev. Pharmacoecon Outcomes Res.,* 2009, **9**, 527–537.
84. NICE, *News Release: NICE "central" to value-based pricing of medicines*, http://www.nice.org.uk/newsroom/news/NICECentralToValueBasedPricingOfMedicines.jsp, accessed 15 July 2013, National Institute for Health and Care Excellence, 2013.
85. B. A. Feinberg, J. Lang, J. Grzegorczyk, D. Stark, T. Rybarczyk, T. Leyden, J. Cooper, T. Ruane, S. Milligan, P. Stella and J. A. Scott, *Am. J. Manag. Care,* 2012, **18**, SP159–SP165.
86. T. H. Lee and E. J. Emanuel, *N. Engl. J. Med.,* 2008, **359**, 333–335.
87. D. A. Hughes, B. Tunnage and S. T. Yeo, *QJM,* 2005, **98**, 829–836.
88. J. P. Cook, J. A. Vernon and R. Manning, *Pharmacoeconomics,* 2008, **26**, 551–556.
89. L. P. Garrison, A. Towse, A. Briggs, G. de Pouvourville, J. Grueger, P. E. Mohr, J. L. Severens, P. Siviero and M. Sleeper, *Value Health,* 2013, **16**, 703–719.
90. S. Villa, A. Compagni and M. R. Reich, *Int. J. Health Plann. Mgmt.,* 2009, **24**, 27–42.
91. FDA, *Fed. Regist.,* 2012, 77, 45639–45643.
92. E. Tambuyzer, *Nat. Rev. Drug Discovery,* 2010, **9**, 921–929.
93. R. Kneller, *Nat. Rev. Drug Discovery,* 2010, **9**, 867–882.
94. Parexel, *PAREXEL Biopharmaceutical Statistical Sourcebook 2012/2013*, ed. M.P. Mathieu, Parexel Intl Corp, Waltham, MA, USA, 2012.
95. M. I. Phillips, *Expert Opin. Orphan Drugs,* 2013, **1**, 1–3.
96. J. E. Davies, S. Neidle and D. G. Taylor, *Br. J. Cancer,* 2012, **106**, 14–17.

[illegible] and [illegible], Drugs, 2013, 72, [illegible]
77. A. Gregory [illegible], 2008, 18, [illegible]
78. [illegible] Bristol, [illegible]
[illegible]
[illegible]
[illegible]
[illegible], National Institute for Health and Care Excellence, 2013.
81. NICE, [illegible] on preliminary recommendations [illegible]
[illegible] Recommendations [illegible] National Institute for Health and Care Excellence, 2013.
87. [illegible]
88. [illegible]
[illegible]
[illegible], National Institute for Health and Care Excellence, 2013.
[illegible]
[illegible] 2013, [illegible]
96. [illegible]
97. [illegible], 2005, [illegible]
98. J. [illegible] Cook, [illegible] and [illegible], 2004, 26, [illegible]
99. [illegible], P. J. Mann, [illegible]
91. [illegible]
[illegible]
[illegible]
[illegible]
M. P. Mattson, [illegible], Waltham, MA, USA, 2013.
96. M. J. Phillips, [illegible] Drugs, 2013, [illegible]
97. [illegible]

THE ROLE OF ADVOCACY

CHAPTER 5

Disease Advocacy Organisations

SHARON F. TERRY*[a,b] AND CAROLINE KANT[b]

[a]Genetic Alliance, Washington, DC, USA; [b]EspeRare Foundation, Geneva, Switzerland
*E-mail: sterry@geneticalliance.org

5.1 What are Disease Advocacy Organisations?

Patient support groups, voluntary health organisations and disease advocacy organisations are just a few of the names by which advocacy and support for rare conditions is known. These organisations run the gamut from simple support for people affected by a condition to full-blown research entities that rival some pharmaceutical companies in financing and capacity. When specifically considering drug development for rare diseases, it is more likely that the organisation lies at the research entity end of the spectrum.

In determining what phrase to use to describe these entities, it should also be noted that there is a growing distaste in both umbrella bodies comprised of these organisations as well as among the individuals affected by rare conditions for the term 'patient'. Much of the lives of these individuals and their families are spent living with a chronic condition, and not in the care of a physician. The word 'patient' connotes the less than empowering position of being in the doctor–patient dyad and not in a position of power and participation. Biomedical research, and particularly drug development, lying as it does on the far end of the translational spectrum, requires participation of the individuals, families and communities it will benefit.[1]

Although phrases such as 'non-governmental organisation (NGO)' in the EU and Asia and 'voluntary health organisation (VHO)' in the USA are

RSC Drug Discovery Series No. 38
Orphan Drugs and Rare Diseases
Edited by David C Pryde and Michael J Palmer

Published by the Royal Society of Chemistry, www.rsc.org

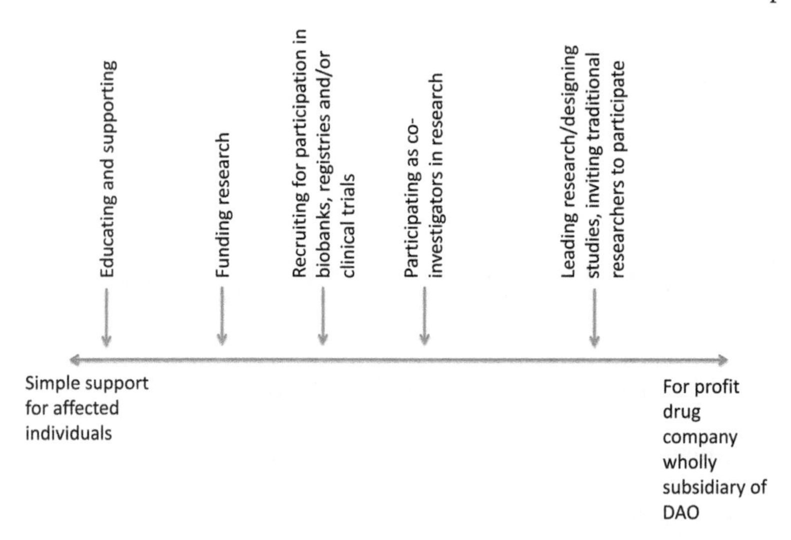

Figure 5.1 Continuum of influence and activity on rare disease research.

common, for the purposes of this chapter we will use the more specific term 'disease advocacy organisation' (DAO) to describe the non-profit organisations that are working to accelerate discovery and development of interventions. It would perhaps be more precise to say 'disease research organisations', but that would limit the discussion unnecessarily, because a substantial part of the acceleration of drug development in rare diseases comes from activity other than direct scientific research.

For the most part, DAOs have been somewhat invisible participants in the development of diagnostics and interventions, even in the common disease realm. Their participation is uneven and fragmented, thus not easily discernable or measured, although there are certainly extraordinary exceptions. Further, as will be discussed below, the overall system of drug development does not have a typical role for DAOs and hence their contributions are neither systematic nor obvious.[2–4] Figure 5.1 depicts the continuum of influence and activity by DAOs in rare disease research. These organisations span the continuum from providing simple support for affected individuals and families, to creating and operating full-blown for-profit pharmaceutical companies.

5.2 The Drug Development Process for Rare Diseases is Ripe for Change

Traditionally, drug discovery has been the domain of the pharmaceutical and (more recently) biotechnology industries. The profit-driven return on investment (ROI) business model used by the bio-pharmaceutical industry has led to enormous advances in health and to virtually all major innovations in pharmaceutical development over the past three to four decades. However, as described in detail in other chapters of this book, there is little incentive for these companies to invest in diseases with low or negligible commercial potential, because this business model is driven by very ambitious revenue and quarterly profit goals. Looking ahead and in the context of a quite generalised economic crisis, the commercial attractiveness of developing drugs for small disease populations is becoming increasingly 'uncertain'. Indeed six-digit treatment costs per year are less and less likely to remain bearable for strangled health care systems and the ethical pressure of providing access to health care to those suffering from rare and often debilitating diseases is putting strong pressure on treatment pricing.[5]

There is no lack of commentary on the problems in the drug development industry, and shining the spotlight on the concerns of individuals and families living with rare diseases highlights those problems. As described elsewhere, it takes up to 18 years, on average, for a drug to move from discovery to commercialisation.[6,7] Parents of children with rare diseases do not want to hear that solutions might advance in timelines measured in generations, perhaps after their child passes away, or the disease has taken its toll in the form of severe disability. Only five in 5000 compounds that enter preclinical testing make it to human testing, and only one of those five is ever approved for use. This is not an imaginable or acceptable failure rate for those affected by disease. Estimates for R&D costs for a single new drug (taking into account failed projects) can now range between $800 million to $1.5 billion. At this rate, the 7000 rare diseases will cost at least $5.6 trillion for funding just the successes. Because successes are scarce, failures are the norm, and costs continue to rise, the pharmaceutical industry is scrambling for answers to reduce the inherent risks of this endeavour and is attempting to shorten the timelines of the R&D process, while minimising attrition through stringent quality controls and de-risking strategies at all levels.

Despite all these efforts, the uncertainties and the financial risks remain high, translating into enormous pressure on pharmaceutical companies. In the last 5 years, most large pharmaceutical companies have cut their staffs substantially, most notably within the research workforce, and now focus principally on only the lowest risk and highest prevalence targets and diseases with proven or plausible commercial potential. A significant number of companies have added orphan diseases divisions to their structure, believing that there will be successes catalysed by market exclusivity, development incentives and generally large ROI because of traditionally high drug prices.[6–11] In the earliest days of DAO interactions with drug development, the work was largely focused on raising awareness and convincing pharmaceutical companies to participate in development and commercialisation. That is no longer the major concern – pharmaceutical and biotech companies have begun to see rare diseases as a solution to their woefully stressed business models.[9] One indicator of this shift is the number of rare disease drug development meetings around the world. As recently as 5 years ago there were one or two a year, mostly convened by DAOs or umbrella groups for them. Now large commercial conveners hold more than one a month around the globe. They are clearly aware that this is a lucrative business and will draw a substantial audience. Thus registration fees of more than $1000 per attendee are typical, and there is no lack of participation.

There has been a great deal of discourse about how to actually impact the translational research system to increase its effectiveness.[12] As common conditions are stratified using genetics and genomics, they are stratified into rare forms of common diseases, and though most are not reclassified as rare diseases per the technical definitions of various countries and regions,[13,14] many of the same challenges (small cohorts, low study power, lack of incentive to develop drugs for few customers) are relevant. Many pharmaceutical companies are opening rare disease therapeutic development divisions, hoping to capitalise on the generally high prices these drugs can garner, as well as attempting to discover new models to sustain and advance the industry.[6]

It is clear that the old drug development model described as a pipeline is not an effective one.[15] It is also clear that systemic change must occur if the pharmaceutical industry is going to produce interventions at the rate needed. Mindful of the fact that only about 5% of all rare conditions have treatments, and aware that organisations such as the International Rare Disease Research Consortium[16] declare an international goal of 200 new interventions by 2020, there is no doubt that a new model or models are critical. There simply is no pathway to accelerate drug development for rare (or any) diseases without transforming the current system.

5.3 DAOs as Indirect Participants in the Drug Development Process

It can be argued that drug development has both indirect and direct influences, catalysts and instigators. Indirect influences are at work on the ecosystem or environment, and are perhaps more influential on the system overall than is direct action. Direct engagements are the processes and activities that involve the nuts and bolts of drug development. DAOs are at work in both indirect and direct engagement. As mentioned previously, the involvement of DAOs in drug development isn't yet systematically defined. However, within the rare diseases landscape it has been increasingly recognised that a new system is being forged where the multi-faceted understanding of diseases of DAOs and their drive to act is becoming a central component of a DAO-empowered drug development model.

5.3.1 Policy Development

In the early years of DAO involvement in drug development the participation was largely indirect: DAOs influenced policy and culture. In the 1950s to the 1980s, DAOs spent much of their time and resources on supporting individuals and families affected by rare diseases and had very little to do with drug discovery and development. There were some early pioneers. The AFM (l'Association Française contre les Myopathies), founded in France in 1958, created a scientific committee and funding capabilities that supported research programmes and developed research institutes as early as 1981.[17] On the other side of the Atlantic, the Cystic Fibrosis Foundation was founded to directly impact research in 1955 by providing grants for research, funding for clinical trials and accreditations for clinical care reference centres. In 1989, Francis Collins, Lap-Chee Tsui and John R. Riordan, who were working with the Foundation at the time, discovered the gene that causes cystic fibrosis.[18] Following this major breakthrough the organisation created a non-profit drug development entity: the Cystic Fibrosis Foundation Therapeutics, Inc., which co-develops therapeutic interventions with pharmaceutical and biotech partners. By entering into contractual agreements with its commercial partners, it allows the Foundation to receive financial returns such as royalties after approval and/or sale of certain drugs that are developed as a result of the organisation's funding. This creates a financially sustainable model where financial R&D returns are reinvested into R&D to further fight for a cure. These DAO-driven R&D efforts have led to the approval of more than 10 therapeutic interventions, including Kalydeco, the first intervention targeting the underlying genetic cause of the disease in a subset of those patients.[19] Another pioneer, Abbey Myers, then housewife from the US state of Connecticut, rallied many small rare disease groups and advocated for a law in the US called the Orphan Drug Act of 1983.[20] Abbey's son was affected by Tourette syndrome, and she was a mother on a mission. She created an umbrella organisation called the National Organization of Rare Disorders (NORD).[21] In this first cross-disease effort, she rallied many DAOs, using community engagement, lobbying and the power of media such as the popular US television show 'Quincy', and she brought the faces and voices of approximately 25 million individuals affected by rare diseases to the attention of the US Congress. The US Orphan Drug Act of 1983 provides tax relief and some marketing exclusivity for companies that develop an orphan drug, and is credited with the explosion in drug approvals for rare diseases after 1983. Without question, the Orphan Drug Act of 1983 has worked exceedingly well in achieving the important purpose for which it was enacted – to provide incentives that would encourage drug research for rare diseases. Since 1995, more than 400 medicinal products have been approved to treat rare diseases,[22] compared to 108 in the decade before and fewer than 10 in the 1970s. While this acceleration is laudable, it is dwarfed by the scale of the rare disease problem, and at this rate, it will take hundreds of years to find therapies for all 7000 rare conditions.

Nevertheless, the work of Abbey Meyers marked the earliest significant involvement of a coalition of DAOs in drug development for rare diseases, particularly with regard to regulatory policy. In subsequent years, DAOs around the world have lobbied their respective governments for similar laws. Similar laws have been passed in many other countries (Table 5.1). These laws certainly have a great influence on the acceleration of drug development for rare diseases. There are criticisms of the US law and a sense that, in its

Table 5.1 Rare disease policies throughout the world.

Country	*Definition of rare disease*	*Name of policy*	*Date policy was enacted*
Argentina	Possible definition: prevalence of less than 1 in 2000	Expediente numero 128/09	3-Mar-09
Australia	Rare diseases to 2000 people in the country, or about 1 in 11 000 citizens	Therapeutic Goods Act	11-Jun-05
China	Proposed that a disease be classified as a rare disease if it is prevalent in fewer than 1 in 500 000 or has a neonatal morbidity of fewer than 1 in 10 000	Bill of Rare Diseases' is under consideration by the national People's Congress	In progress; disagreement over the definition
European Union (EU)	Life-threatening or chronically debilitating disease affects less than 5 in 10 000 of the general population	Regulation (EC) no. 141/2000	16-Dec-99
France	1 in 5000 people	French National Plan on Rare Diseases	20-Nov-04
Great Britain and Northern Ireland	EU's definition	In progress	In progress
Greece	EU's definition	Greek National Plan for Rare Disease	29-Feb-08
Japan	Affects less than 50 000 people	Pharmaceutical Affairs Law	25-Jul-02

Korea	Diseases that affect fewer than 20 000 people or diseases for which an appropriate treatment or alternative medicine has yet to be developed	Pharmacy Act 91	18-Dec-91
Peru	None	Law 29698	4-Jun-11
Portugal	EU's definition	Portuguese National Plan for Rare Diseases	12-Nov-08
Romania	EU's definition	Romanian National Plan for Rare Diseases	29-Feb-08
Singapore	Life-threatening and severely debilitating illness	Medicines (Orphan Drug Exemption) order	4-Nov-91
Slovenia	EU's definition	Work Plan for Rare Diseases in Slovenia	1-Sep-11
Switzerland	A prevalence of no more than 5 in 10 000 individuals	Therapeutic Drugs Act	15-Dec-00
Taiwan	Disease is rare if it's prevalent in 1 in 10 000 people	Rare Disease and Orphan Drug Act	1-Jan-00
USA	Any disease or condition that affects fewer than 200 000 persons in the USA	Orphan Drug Act of 1983	4-Jan-83

30th year, it needs to be updated,[23,24] but overall the fact that 35 new molecular entities for rare diseases were approved in 2012 in the USA is an indication that the Act is effective. Results for the EU are 10 in the same year. This lower rate cannot be explained because of population, but perhaps because the directive is more recent in Europe. Over the last decade the European Organisation for Rare Diseases (EURORDIS) developed an overarching/federated EU campaign to enact legislation, and in 1999 the Regulation on Orphan Medicinal Products was signed into law. EURORDIS has created EU-wide research initiatives tailored to the specific social economic and medical features of rare diseases (http://www.eurordis.org/content/promoting-orphan-drug-development).

Many other groups have taken up the charge that NORD laid down 30 years ago, and are working to change regulatory policy and practice with regard to how clinical trials for rare diseases are assessed. An excellent example of this is a white paper published by Parent Project Muscular Dystrophy, called 'Putting Patients First'.[25] The paper, a collaborative work of several organisations and consulting firms, lays out four recommendations for the US Food and Drug Administration, which while focused on Duchenne muscular dystrophy, are relevant to all rare diseases:

- Expand the use of accelerated approval for therapies intended to treat rare diseases, including Duchenne muscular dystrophy.
- Issue clear guidance outlining the level of evidence required for the use of surrogate endpoints in order to expand the scope of acceptable endpoints, including novel surrogate and intermediate clinical endpoints, used to approve drugs for serious or life-threatening diseases with unmet medical need.
- Pilot the use of adaptive approval for serious and life-threatening disorders with significant unmet medical need, using existing authority under current law.
- Give greater weight to the demonstrated benefit/risk preferences of patients, as well as caregivers in the case of paediatric illness, when making risk benefit determinations. Subpart D considerations must be evaluated here, yet benefit/risk should also be addressed within the context of patients living with Duchenne.

It is easy to see that these are well thought out recommendations designed to change the system at a significant junction. The work of DAOs in policy is increasing in its credibility and professionalism. The next level of this work will capitalise on collaborative activity and begin to reduce the divides between DAOs as they embark on the same quest.

Another indirect but significant effect DAOs have on drug development is culture change. This is not as concrete as policy change, but in fact precedes it in a foundational way. Recommended policy changes such as the ones above are easier to implement if the work of changing the culture that underlies the policy receives attention. For example, these recommendations would not even be considered a decade ago. Frequent dialogue between the FDA and many advocates has created a climate that is more ready to consider these changes. Importantly, overall culture change has resulted in pharmaceutical and biotech companies working with DAOs as partners rather than simply using them. This creates alliances that are effective in effecting change and reinforcing a culture of partnership.

5.4 Direct Impact Throughout the System

There is no longer any doubt that DAOs are, and will be, important to retooling the drug development system. There is substantial evidence that they have made a difference, to a degree, for a number of diseases.[2,26–39] There is also a great deal of potential that has not been realised, in part because these organisations can both cause system change and more fully influence the system once it is changed.

One major way that DAOs have affected all of the phases of drug development for rare diseases is funding. DAOs have raised hundreds of millions of dollars for rare disease research. Unfortunately there is no way to quantify this contribution. It is significant, however. For example, the Cystic Fibrosis Foundation raised $100 million in 2011 and dispersed $73 million of that in research grants.[40] In 2012, the CFF announced it would grant $53 million to the US pharmaceutical company Pfizer.[41] This money has been used for funding traditional science in academic and industry groups, for catalysing competitions, as well as for funding research that the DAO initiates and manages independent of any other entity. In all of the areas described below, DAOs have offered substantial funding.

The aforementioned AFM was a pioneer in philanthropic fundraising for rare disease research in Europe when, in 1987, it launched a concept imported from the USA, the Telethon, a fundraising television broadcast. This effort has been very successful over the years, raising about €1.8 billion[42] to fund a wide range of rare disease research and drug development projects as well as patient care programmes. As a fundraising concept, Telethon has become a successful franchise exported all around the globe. Although large rare disease organisations are very successful at raising money for their diseases, most smaller DAOs representing conditions with names difficult to pronounce, and that personally touch few people, are still struggling to create the necessary momentum for substantial fundraising that can fuel research and drug development.

Each of the following sections will describe further contributions, in addition to funding. We will orient this discussion of direct influence of DAOs on drug discovery and development around the Navigating the Ecosystem of Translational Science (NETS) model[15] (Figure 5.2), rather than a pipeline model. We will describe some of the impact of DAOs in relation to each of the areas in the Figure 5.1 NETS model. The NETS model allows visualisation of an integrated, interdependent, network model of key drug discovery and development stages.

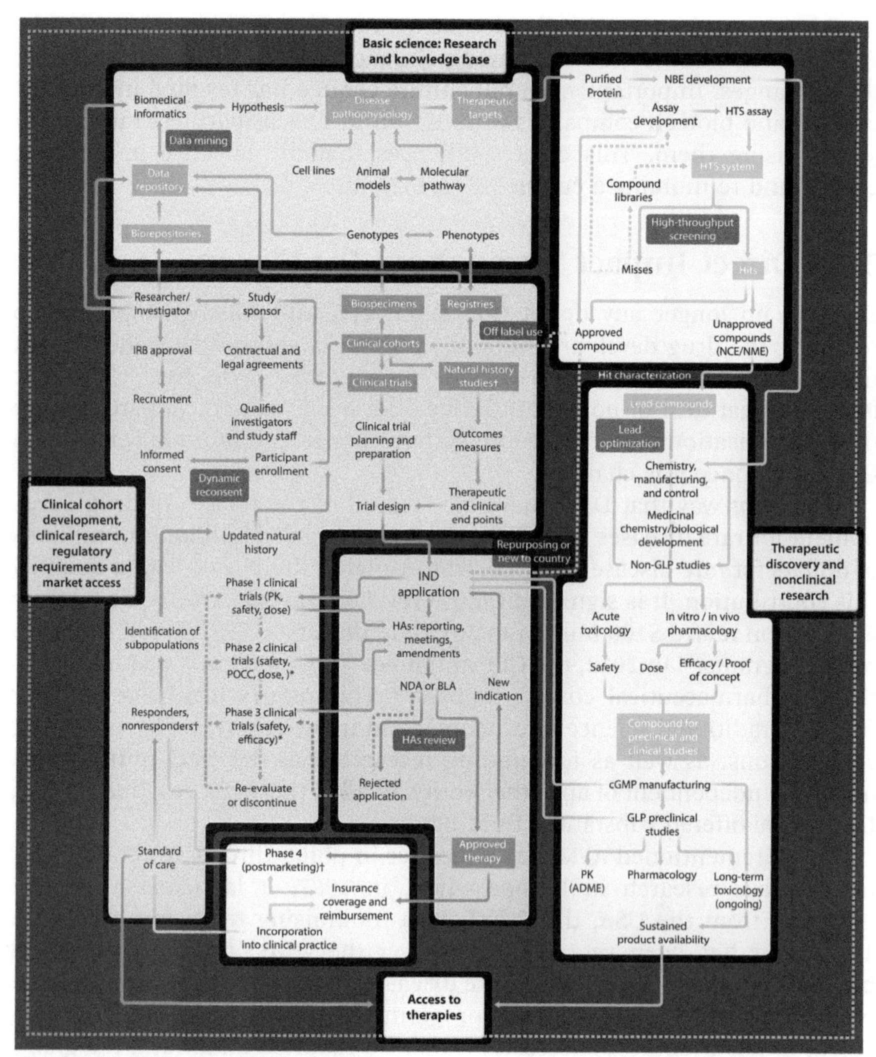

Figure 5.2 Navigating the Ecosystem of Translational Science (NETS) model.[15]

5.4.1 Basic Science and Therapeutic Target Discovery

5.4.1.1 Contributions Made

The orange and green sections of the NETS map describe some of the major steps in basic science and target discovery. This is an area in which DAOs have worked extensively. In the 1980s and 1990s, basic science, gene discovery, assay and animal model development were all supported by DAOs. PXE International, a DAO for the rare disease pseudoxanthoma elasticum

(PXE), was the first lay organisation to establish and manage a registry and biorepository, independent of any academic, government or industry involvement or support.[36,43] These biological samples were the material used in the gene discovery.[44,45] Further, the lay founders of PXE International learned sufficient bench science to discover the gene associated with PXE, called ABCC6, in 1999. These founders then patented the gene, turned the rights over to PXE International and retained the ability to steward its use in an open and equitable way.[46,47] In fact, it was so odd that lay individuals would discover and patent a gene that several erudite authors in credible publications continually misstated that the reason the founders were able to patent the gene is because they contributed biological materials to the search.[47,48] Other DAO founders, for example the Progeria Research Foundation, have also participated materially in the discovery of the associated gene and subsequent patenting.[29]

Many DAOs have worked alongside academic scientists to develop assays. The National Institutes of Health has offered technical assistance in assay development for some time in their molecular libraries programme.[49] The leadership from NIH has engaged some DAOs in this activity. Further, individual organisations have undertaken their own programmes that have successfully resulted in assays capable of high-throughput screening.

Developing animal models has become a major activity of DAOs. For example in 2008, the AKU Society, a patient-led group for the debilitating metabolic genetic disease alkaptonuria (AKU), worked with research teams at the Royal Liverpool University Hospital and the University of Liverpool (UoL) and started the findAKUre coalition, a scientist–patient partnership model to develop treatments for AKU. Over the past 9 years, this coalition single-handedly developed a major R&D programme. This started with the analysis of post-mortem data of an AKU patient,[50] funded through sponsored events, followed by the funding of a PhD programme that unbaled the generation of an *in vitro* model of AKU.[51] Thanks to support from the Big Lottery Fund, the AKU Society then funded a 4 year programme at UoL that successfully led to the development of an animal model of AKU, in which new therapies are being tested.

5.4.1.2 *Potential Areas for Further Development*

5.4.1.2.1 Assay Bootcamp. The leadership of NCATS and Genetic Alliance has discussed setting up an 'assay bootcamp'. This could be set up in such a way as to build on the strengths of the NIH, academic scientists and DAOs. For example, NCATS has a world-class facility and highly developed expertise for developing assays. The DAOs often work with academic scientists who are the intended recipients of the aforementioned Molecular Libraries Program at NIH. Still it is difficult for academic scientists to develop assays robust enough for high-throughput screening. In a bootcamp, DAOs could identify an academic scientist who would benefit from intensive training. They could sponsor the travel and lodging for the scientists with whom they collaborate to spend a week training at NIH, after doing requisite preparatory work on an assay. Returning to their labs, these scientists can continue to refine what they worked on at NIH, with follow-up as needed.

5.4.1.2.2 Open Access for All Molecular Libraries. DAOs can work together to establish a new norm for access to molecular libraries, commercial or not. Building on the work of the Molecular Libraries Small Molecule Repository at NIH and the example offered by Merck and Eli Lilly for Tuberculosis[52] the DAOs are in a unique position to work in concert with one another to ask that all molecular libraries be considered a pre-competitive resource, and they be open and available to scientists working on any disease. Certainly proprietary interests and ownership can be dealt with creatively, including novel licensing and profit-sharing arrangements.

5.4.1.2.3 Interoperable Registries. One area that is not dependent on leadership from governments or industry is the development of interoperable registries. As standards are developed and agreed upon, DAOs can ensure that the registries they construct are standards based and properly coded. This will allow for interoperability that will accelerate discovery, particularly in systems biology and common pathways. Even more dynamic, these registries can be federated and enable cross-disease research.

A successful example of DAO involvement in standardisation efforts of patient registries is the RARECARE initiative[53] that has been funded by the European Commission, with 15 organisational and institutional partners including the participation of the European Cancer Patient Coalition (ECPC). The assessment of validity, completeness and standardisation across rare cancer registries has set common criteria and rules to improve the quality and comparability in those registries. Beyond that, the project has produced an operational definition of rare cancers that establishes a list of conditions and an estimate of the health care burden of rare disease cancers in Europe. Building on the outcome of this project, in May 2013, the consortium started developing RARECARENet. This is a web-based information network that provides comprehensive information on rare cancers to the community at large (patients and their families, oncologists, general practitioners, researchers, health authorities).

5.4.2 Non-Clinical Research (Blue)

5.4.2.1 Current Work

Once a hit is identified, many DAOs participate in traditional ADME TOX (pharmacokinetic studies which assess absorption, distribution, metabolism, excretion and toxicity) work. They do this along a continuum from assisting in the laboratory to being part of work groups analysing results to actually developing in-house capacity to do the studies. A good example of this is the work of Parent Project Muscular Dystrophy to enable TOX studies when members of the community think they have discovered a potential solution to the progression of the disease.[54]

5.4.2.2 Areas for Further Development

DAOs could pool their resources and either set up or contract with facilities that can do these studies. They could also work together to get those who do ADME TOX work to share those results in a commons, thus pooling knowledge that will, in the simplest terms, keep others from repeating the work, and in more profound terms, create banks of knowledge that when mined for meta-analysis will reveal important information to accelerate pathway discovery. This sort of 'negative data' is given little attention, and could be very useful in advancing disease understanding and optimising drug discovery methods.

5.4.3 Regulatory Science (Purple)

5.4.3.1 Current Work

One area of intense activity in drug development for rare diseases is regulatory science. DAOs have been actively involved in discussions, deliberations, lobbying and guidance development in regulatory science. As mentioned above, the most significant work of NORD in the 1980s and 1990s was to drive Congress to establish the Orphan Drug Act. Since the Act passed more than 30 years ago, DAOs have been involved with the resulting Office of Orphan Products and also individuals in the drug and device offices at the FDA. Some of these, representing far more, were recognised in 2013 as 30th Anniversary Rare Disease Heroes,[55] and reading their bios helps to paint a picture of the diverse forms of interaction that the advocates have with the FDA. Further, as part of the current Prescription Drug User Fee Act (PDUFA V), the FDA is focusing on a number of disease areas to create what it is calling Patient Focused Drug Development. Focus groups and public meetings include the leadership of a number of DAOs. A concrete illustration of the agency's debut efforts to expand the role of the patient perspective within that initiative is to gather patients' perspectives on the impact of a condition on daily life and the available therapies to treat that condition. In addition, the FDA has publically announced that it supports and fosters the use of patient-reported outcome measures in clinical trials to support labelling claims in medical product development.[56] At the European Medicines Agency (EMA) there are guidelines for the involvement of DAOs with the agency.[57] These criteria are largely related to credibility, medical focus, appropriate legal structure and transparency. There are currently about 35 DAOs involved in EMA activities. In this context, the Patients' and Consumers' Working Party is providing recommendations to the EMA and its human scientific committees on all matters of interest to patients in relation to medicinal products. One of the roles of this party is to help the agency appropriately adapt the framework that enables the integration of DAOs in drug development. The role and implication of the DAOs in regulatory science is evolving and still being defined. However, regulatory agencies are actively exploring how to incorporate this new active stakeholder in assessing risk/benefit of medicinal products and beyond.

5.4.3.2 Areas for Further Development

Regulatory science is a difficult field. Government agencies have committed to working on improving this science, for example NIH and FDA announcing a joint initiative on it.[58] Some DAOs that support individuals and their families affected by rare and severely debilitating conditions feel that they have a higher risk tolerance when it comes to drug development in conditions where there is no disease-modifying treatment. More specifically, DAOs sometimes ask for more lenient regulation in the areas of biomarkers and statistical power of small studies without realising the risk in such approaches. The mantra of both the regulatory agency and the DAO is 'every disease is unique, nothing can be generalised'. At other times, DAOs are fully aware, have excellent examples and are very articulate about ways the regulatory agencies can be more useful to the rare disease drug approval process in general. As DAOs become more savvy at regulatory science, their voice is gaining more weight in the agencies' decisions. However, the regulatory agencies have to walk a careful line between engagement with the DAOs and undue influence on decisions.

5.4.4 Clinical Research (Pink)

When drugs are developed for rare diseases, thousands of individuals are often not available to power clinical trials. New methods for clinical trials must be developed and tested. Sometimes these are called '*N* of 1', or low power, trials. DAOs are a key partner in addressing some of the key challenges associated with '*N* of 1' trials.

5.4.1.1 Registries and Biorepositories

Biological samples from individuals affected by rare diseases, such as blood, tissue or DNA, are key to drug development and translational research. The analysis of these samples supports advancements in research and disease characterisation by uncovering molecular mechanisms and targets involved in the diseases as well as refining the understanding of the genotype–phenotype relationship. Clinical data analyses of these samples are also critical to establish new disease stratification approaches, through molecular profiling of omics data and biomarker discovery. This can enable enrolment of the right individuals for clinical trials on the basis of genetic inclusion criteria rather than more subjective criteria such as age, treatment history or stage of the disease. Hence, appropriate pharmacogenomics analysis of biobank samples can increase the chance of discovering predictive biomarkers and selecting the right clinical cohort that will have the best chance to respond to a particular investigational therapeutic product.

While in 1995 no other DAO had created a registry or biobank, these are now a relatively common activity among DAOs.[2,59] There are now hundreds of registries run by DAOs. It is clear that the trust community built by these organisations is an excellent basis for engaging people.

In 2003, Genetic Alliance created a cross-disease shared infrastructure platform built on the foundation of the PXE International Registry and BioBank (PIRB) which, in 1995, was the first lay owned and managed registry and biobank. Ten disease groups have used the Genetic Alliance Registry and BioBank (GARB) over the past 10 years and have gone on to accelerate research based on the resource. PXE International alone has 23 peer reviewed papers in a variety of clinical journals as a result of the clinical information and biological samples in the PIRB; a few are cited here.[60–63]

Registries have become considerably more sophisticated in recent years. One registry created by a collaboration of DAOs is Registries for All (Reg4ALL).[64] It is novel because it is:

- cross-disease, disease agnostic, and allows for participants having multiple conditions;
- engaging through gamification; and
- gives people control – they set their own sharing and data access preferences (Figure 5.3), which can be changed over time and context. Thus, individuals can be part of many projects, can easily be recontacted, and can even manage their loved ones.

This sort of registry will allow the vast amounts of data that should be associated with 'big data' on common and rare conditions to be shared according to individual preferences that can change over time. Comorbidities and other associations will allow researchers to more readily identify disease pathways.

5.4.4.2 *Clinical Cohorts*

DAOs are best positioned to efficaciously recruit the right individuals for clinical trials. The individuals in the community trust the leadership and the organisation. The organisation can vet the plethora of clinical trials available to the scarce number of affected individuals and help to determine which would be most beneficial to the community at large and to specific individuals in the community. The DAO can help researchers identify and enrol the correct sub-population in the disease to meet a desired end point. Perhaps the most dramatic example of this is in the rare disease Hutchinson–Gilford progeria. The Progeria Research Foundation has enrolled 103 children from 37 countries in clinical trials for the condition with an incidence of only 1 per 8 million live births per year.

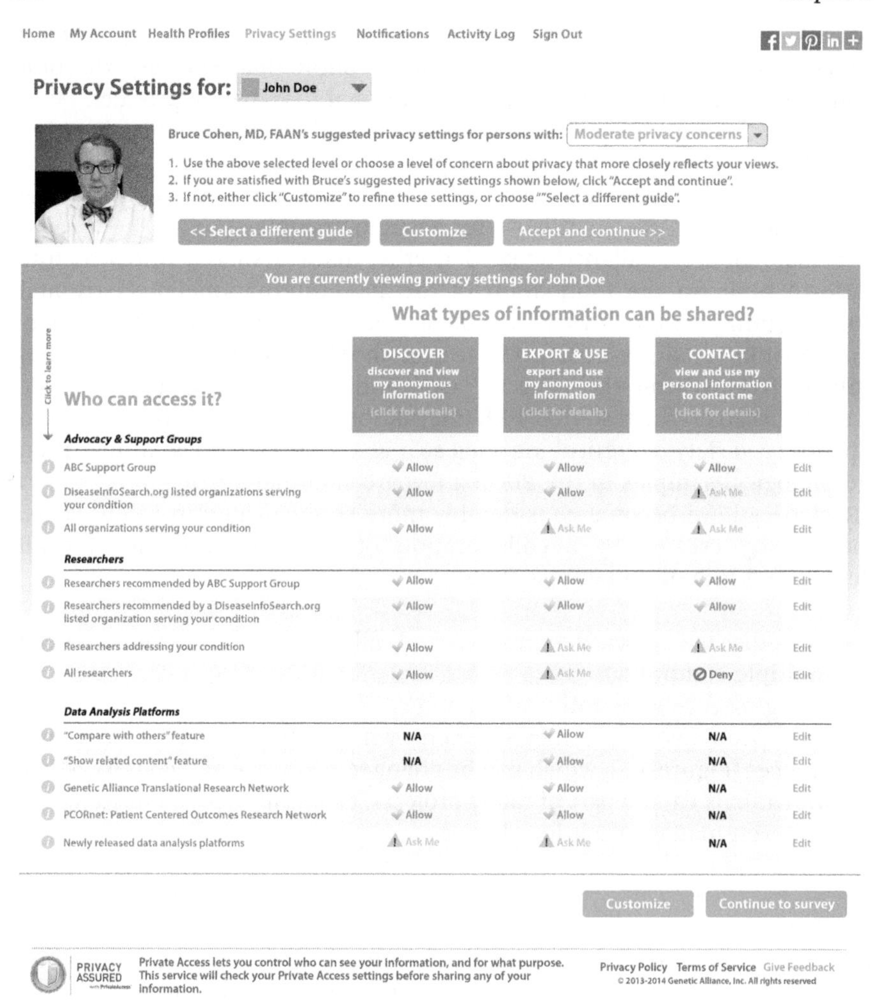

Figure 5.3 Privacy settings for Reg4ALL.

5.4.4.3 Natural History

Natural history studies are critical to the development of relevant clinical end points and outcome measures such as biomarkers. Disease natural history is an essential foundation of any clinical development programme and they are an important tool to understand the aetiology, range of manifestation and progression of diseases. Obtaining maximum value from drug development programmes depends on having natural history data of good quality. Unfortunately in rare diseases this type of information is often scarce. Some of the challenges are due to small populations that limit the opportunity for natural history study and replication, as well as high phenotypic diversity within those diseases (https://events-support.com/events/Natural_History_Studies).

Natural history is a particularly good place for DAOs to be engaged. There is no other participant in the research continuum that is motivated solely as an advocate for individuals living with a condition. As such, DAOs are given a unique window into the disease. In situations as simple as support group meetings, email lists and blogs, DAOs will 'hear' repetitive attributes that may in fact be part of the disease, and also will understand what aspects of the disease are of greatest concern. For example, in support group meetings in the late 1990s, women with PXE described the experience of being subject to a biopsy as a result of having a mammogram. PXE International collected mammograms from 51 affected women and 109 age-matched controls, and through a double-blind review, determined that mineralised arteries in the breast are common in PXE, and women should not be biopsied or scared about cancer.[65] This was important for clinical practice guidelines and quality of life for the women, alleviating both unnecessary anxiety and invasive procedures.

In order for natural history information to be used for drug development it is of critical importance to conduct well-controlled studies that have defined research goals, valid comparisons with control, appropriate subject selection and scientifically sound standardised data analysis methods. The information that can be generated out of these robust and well-designed studies on the natural course of the disease can be critical, especially in rare diseases where it can sometimes be unethical to conduct placebo-controlled studies. In these cases, the availability of those studies simply allow the investigation of potential treatment in those diseases. To increase the chance of success of drug development in rare diseases it is essential to start natural history studies early in the therapeutic development process. Under increasing regulatory authority expectations to have clinical studies compared to historical controls, drug development companies are initiating more and more industry-sponsored longitudinal studies. Here collaboration with DAOs can be critical for participant recruitment and design of an appropriate study. The Spinal Muscular Atrophy Foundation provides another example of how DAOs are sponsoring and spearheading natural history studies. They and the NIH funded a study conducted by the paediatric neuromuscular clinical research network that better characterised the natural history of II and III subtypes of this disease and collected data on clinical and biological outcomes for use in trial planning.[66] In addition to providing funds for the study, in this case the DAO was essential in enrolling sufficient participants to enable statistically significant study results.

5.4.4.4 *Therapeutic and Clinical End Points*

Parents of children with Duchenne muscular dystrophy have described that while clinical scientists chose the 6 minute walk test as an excellent indicator of the efficacy of a drug,[67] the boys themselves will declare that feeding themselves and using a computer to communicate is also important and that non-ambulatory boys are not eligible for trials if the 6 minute walk test is the only end point. Thus, when determining the right outcomes, or end points, it is important that the individuals who live with the condition are part of the considerations. This has become somewhat institutionalised in the previously mentioned FDA's Patient Focused Drug Development initiative as well as at the Patient-Centered Outcomes Research Institute (PCORI), in its work to produce evidence-based research about health outcomes. Over the coming years it will be funding a great deal of research, much of which will include DAOs.

5.4.4.5 Clinical Trials

There is no question that one of the strongest contributions DAOs make to rare disease drug development is in the area of clinical trials. On one hand, most DAOs inform patients about the goal of clinical trials, their process and how to enroll in them. Acting as an information hub for clinical development efforts in the disease, DAOs websites often reference ongoing clinical trials, their status and outcomes. Some also feature patient experiences of participating in those investigations. On the other hand there are numerous examples of a DAO defining inclusion and exclusion criteria, recruiting participants and screening them for clinical trials, managing their participation, supplementing the information and reimbursement and doing follow-up after the trial. For instance, PXE International's CEO is designated as a co-investigator in a PXE trial at Mt Sinai Hospital in NY. The staff perform all of the aforementioned activities.

5.4.4.6 Drug Repositioning

One approach to speeding up drug discovery and development and increase its chance of success is drug repositioning. This approach can be especially attractive in rare diseases, because cost of drug development can be reduced to compensate for a smaller market potential upon commercialisation. Drugs that have failed or been shelved due to lack of effectiveness or efficiency for common conditions can in some cases be repurposed for rare diseases. Notably between 2010 and 2012 half of the 48 approved drugs for rare diseases were repositioned drugs, illustrating the impact of this approach in bringing new therapeutic solutions to these patients. (Ref: Presentation of Anne Pariser, Director at CDER FDA.) Increasingly, testing drugs on smaller subsets of individuals can allow previously failed drugs to be used safely and effectively on these subsets of patients.[13] To more systematically uncover drug repositioning opportunities some researchers and organisations are examining molecular pathways and developing complex knowledge management approaches to match information about a disease pathophysiology and a drug mode of action. These advances have been promoted by some forward-thinking public–private partnerships, including those between patient support groups and government incentives.[68]

Many public–private partnerships and initiatives between pharmaceutical companies, DAOs and government have a long-term goal of maximising academic efforts by supporting collaborative repositioning efforts. The Progeria Research Foundation (PRF) is a good example of a DAO taking the lead in driving drug repositioning efforts to accelerate treatment for this debilitating paediatric disease. This DAO's efforts have bolstered translational research, moving from gene discovery to clinical treatment at an unprecedented pace. From 1999, when the organisation was founded, to today, the PRF has identified the genetic mutation that causes the disease, funded preclinical research and completed a ground-breaking trial on a drug repositioning opportunity. In 2012 the PRF and Merck & Co announced that Lonafarnib, a type of farnesyltransferase inhibitor (FTI) originally developed to treat cancer, has proven effective for progeria, with every child showing improvement in one or more of four ways: gaining additional weight, better hearing, improved bone structure and/or increased flexibility of blood vessels. This Phase II study was funded and coordinated by the DAO. The foundation was able to enrol 75% of all known progeria cases worldwide in this 2.5 year study.[69]

Despite therapeutic success of drug repositioning approaches, pharmaceutical companies often consider this approach as a less attractive approach than '*de novo* drug discovery', the main reasons being weak or lack of IP and lower potential for high pricing. Further collaborative and financial incentives are necessary to speed up the process and deliver more treatments to patients in need more quickly. One idea that emerged from an Institute of Medicine workshop on drug repurposing is a 'clearing house' that might be housed within a DAO and include information about shelved drugs, prior studies and so on.

There are several examples of DAOs enabling repurposing. The SMA Foundation initiated a programme at PTC Therapeutics in November 2011, in which PTC Therapeutics and the SMA Foundation signed a licensing agreement with Roche. Under the agreement, Roche gained an exclusive worldwide licence to PTC Therapeutics' SMA programme and PTC Therapeutics received $30 million upfront with more payments being made at various milestones. In August 2013, PTC Therapeutics chose a target and Roche paid another $10 million.

Perhaps the only non-profit organisation dedicated to repurposing for a broad range of rare diseases is the EspeRare Foundation. This non-profit organisation has developed an innovative venture philanthropic model that focuses on the translational validation (from preclinical to early clinical testing) of unexplored repositioning opportunities for rare diseases. With its patient-centred focus this organisation is looking to involve DAOs as early as possible in the drug development process.[70] For instance, EspeRare has uncovered the therapeutic potential of a shelved drug, previously developed for heart failure, to address Duchenne Muscular Dystrophy. Partnerships with Genetic Alliance and Duchenne Advocacy organizations such as Parent Project Muscular Dystrophy and l'Association de Myopathies Française allowed EspeRare to gain access to Duchenne research experts and the patient perspective that were key to the development of an effective validation strategy. Additionally, collaborations with those DAOs supported access to research grants to finance research studies necessary to develop this opportunity. With its non-profit drug development engine model, EspeRare has created a collaborative and financial way forward to explore the potential of repositioning opportunities for rare diseases that remained traditionally unexplored by commercial drug developers. However, it is clear that more work needs to be done to strengthen this Product Development Partnership model for rare diseases and make asset owners aware of this mechanism as a win-win pathway to accelerate drug development for rare conditions.

5.5 Beyond Disease Advocacy Organisations

5.5.1 Large NGOs in the Rare Disease Arena

Large NGOs, such as the Bill and Melinda Gates Foundation, spend enormous amounts of money on neglected disease research. One might ask why this attention has not been showered on rare diseases as well. We speculate that there are three reasons for this: (i) neglected diseases are phenomenally common major health burdens when one looks at health on a global scale; (ii) neglected diseases do not usually have advocacy organisations associated with them; no group champions the cause on a disease basis because these diseases usually occur in underdeveloped parts of the world, without the resources to create advocacy on a large scale; and (iii) they are not as fragmented as rare diseases are fragmented. Rare diseases affect small numbers of people, and there are thousands of them, sometimes with more than one group for a disease. Thus, it is difficult for large foundations to determine where they would have a major impact.

5.5.2 Systemic Change *vs.* One-Off Solutions

It is clear that continuing drug development in the current vein will not succeed, not for common conditions and certainly not for rare ones. Systemic change is badly needed. With the advances that new technologies, a networked age and a mature DAO culture offer, this should be possible. Of critical importance will be the integration of new processes and methods in the quest for interventions. In previous chapters it was clear that DAOs connect previously disparate research, invent new models for collaboration and use social media to integrate various components of the system. At the same time, this networked age will see a transformation of DAOs as they once were functioning in the world, because they too are experiencing the results of the social media explosion. In general, many organisations and communities are being born online, and they offer some tools that many of the older brick and mortar groups have not yet mastered. Online collaboration for profit models like PatientsLikeMe and the Inspire community compete for attention and energy. This is not necessarily detrimental to the needs of those living with rare diseases because these new offerings are powerful and in some cases transforming drug development more dramatically than anything else.[71]

Other non-profits that were established to create new more effective paradigms are also offering phenomenal tools for individuals to co-create solutions. Sage Bionetworks is a particularly robust example, having developed an open access database (Synapse), a 'Portable Legal Consent' tool which allows consent to be in the hands of the participant and to follow the data, and Bridge, a combination of collaborative competition and community around a disease or disease problem. Individuals can even get into the act without belonging to an organisation, and create their own clinical trials *via* Genomera's platform, or help to characterise tumour tissue on Cancer Research UK's website.

A unique collective has formed with the goal of 200 new therapies and genetic testing for all rare conditions by 2020. The International Rare Disease Research Consortium has embraced the skills and energy of DAOs and integrated them into the executive committee (umbrella groups) and work groups. This initiative, birthed by the EU Commission and the NIH, is now composed of 30 members from 11 countries. It is an example of the kind of collaboration that must power a revolution in rare disease research.

The previous purely competitive environment will no longer, if it ever did, sustain and advance the necessary research agenda. The pre-competitive space must be enlarged, and we have seen examples above, in drug repurposing, data sharing and collaborations where these experiments are being tried. It is time for DAOs to collaborate as well, and discover that by sharing resources, skills and energy, the solution will arrive much more quickly.[72]

DAOs were early pioneers in forging pathways into a very technical and difficult field. It is now evident, in hindsight, that the creative and innovative leaders of these organisations are the cutting edge of individuals leading research, as participants and citizen scientists. Crowdsourcing is not yet a proven pathway, but is certainly garnering interest and perhaps revealing some important lessons to the whole system. Research can no longer afford to ignore the participants, and especially for rare diseases, this may be a very important part of the catalyst for success.[73]

References

1. J. Kaye, L. Curren, N. Anderson, K. Edwards, S. M. Fullerton, N. Kanellopoulou, D. Lund, D. G. Macarthur, D. Mascalzoni, J. Shepherd, P. L. Taylor, S. F. Terry and S. F. Winter, *Nat. Rev. Genet.*, 2012, **13**, 371–376.
2. D. C. Landy, M. A. Brinich, M. E. Colten, E. J. Horn, S. F. Terry, R. R. Sharp, R. R. Sharp, D. C. Landy, S. F. Terry, P. F. Terry, K. A. Rauen, J. Uitto, L. G. Bercovitch, I. Van Hoyweghen, B. Penders, R. R. Sharp, M. Yarborough, J. W. Walsh, L. Psillidis, J. Flach, R. M. Padberg, S. Gupta, A. M. Bayoumi, M. E. Faughnan, A. Panofsky, B. W. Morrison, C. J. Cochran, J. G. White, J. Harley, C. F. Kleppinger, A. Liu, J. T. Mitchel, D. F. Nickerson, C. R. Zacharias, J. M. Kramer, J. D. Neaton, J. Jervell, J. Uitto, P. Birch, J. M. Friedman, I. S. Kohane, S. F. Terry, K. Zeitz, M. A. Majumder, P. F. Terry, A. R. Patterson,

H. Davis, K. Shelby, J. McCoy, L. D. Robinson, S. K. Rao, P. Banerji, G. E. Tomlinson, H. E. Briggs, N. M. Koroloff, T. Rolstad, G. Zimmerman, E. S. Nilsen, H. T. Myrhaug, M. Johansen, S. Oliver, A. D. Oxman, J. H. Platner, L. M. Bennett, R. Millikan, M. D. Barker, A. A. Lara and L. Salberg, *Genet. Med.*, 2012, **14**, 223–228.
3. R. R. Sharp and D. C. Landy, *Am. J. Med. Genet., Part A,* 2010, **152**, 3051–3056.
4. I. Chalamon, *Eur. J. Marketing,* 2011, **45**, 1736–1745.
5. C. McCabe, K. Claxton and A. Tsuchiya, *BMJ,* 2005, **331**, 1016–1019.
6. K. N. Meekings, C. S. Williams and J. E. Arrowsmith, *Drug Discovery Today,* 2012, **17**, 660–664.
7. J. P. Cohen, *Nat. Biotechnol.,* 2012, **29**, 751–756.
8. A. Sharma, *J. Pharm. BioAllied Sci.,* 2010, **2**, 290–299.
9. R. Numerof, *BioPharm Int.,* 2012, **25**, 66.
10. D. Weinstein, *Medical Marketing and Media,* 2012, **47**, 44.
11. *The global drug development process: what are the implications for rare diseases and where must we go?*, ed. S. F. Terry and J. Swanson, Greenleaf Publishing, Sheffield, UK, 2013.
12. F. Collins, *Sci. Transl. Med.,* 2011, **3**, 90cm17.
13. M. A. Hamburg and F. S. Collins, *N. Engl. J. Med.,* 2010, **363**, 301–304.
14. P. Deverka, *Personalized Medicine Coalition*, 2006.
15. K. Baxter, E. Horn, N. Gal-Edd, K. Zonno, J. O'Leary, P. F. Terry and S. F. Terry, *Sci. Transl. Med.,* 2013, **5**, 171cm171.
16. K. Baxter and S. F. Terry, *Genet. Test. Mol. Biomarkers,* 2011, **15**, 465.
17. l'Association Française contre les Myopathies, http://www.afm-telethon.com/, accessed 23 September 2013.
18. J. R. Riordan, J. M. Rommens, B. Kerem, N. Alon, R. Rozmahel, Z. Grzelczak, J. Zielenski, S. Lok, N. Plavsic, J. L. Chou, M. L. Drumm, M. C. Iannuzzi, F. S. Collins and T. Lap-Chee, *Science,* 1989, **245**, 1066–1073.
19. Cystic Fibrosis Foundation, http://www.cff.org/research/Drug DevelopmentPipeline/, accessed 23 September 2013.
20. Orphan Drug Act, 1983.
21. S. Putkowski, *NASN Sch. Nurse,* 2010, **25**, 38–41.
22. US Food and Drug Administration, http://www.fda.gov/ForIndustry/DevelopingProductsforRareDiseasesConditions/default.htm, accessed 23 September 2013.
23. C. Thorat, *Pediatrics,* 2012, **129**, 516.
24. F. Sasinowski, *Quantum of Effectiveness Evidence in FDA's Approval of Orphan Drugs*, Danbury, CT, 2011.
25. Putting Patients First, White Paper, Parent Project Muscular Dystrophy, Hackensack, NJ, 2013.
26. A. A. Lara and L. Salberg, *Pacing Clin. Electrophysiol.,* 2009, **32**(suppl 2), S83–S85.
27. M. L. Oster-Granite, M. A. Parisi, L. Abbeduto, D. S. Berlin, C. Bodine, D. Bynum, G. Capone, E. Collier, D. Hall, L. Kaeser, P. Kaufmann, J. Krischer, M. Livingston, L. L. McCabe, J. Pace, K. Pfenninger, S. A. Rasmussen, R. H. Reeves, Y. Rubinstein, S. Sherman, S. F. Terry,

M. S. Whitten, S. Williams, E. R. McCabe and Y. T. Maddox, *Mol. Genet. Metab.*, 2011, **104**, 13–22.
28. R. R. Sharp, M. Yarborough and J. W. Walsh, *Am. J. Med. Genet., Part A,* 2008, **146**, 2845–2850.
29. S. F. Terry, P. F. Terry, K. A. Rauen, J. Uitto and L. G. Bercovitch, *Nat. Rev. Genet.,* 2007, **8**, 157–164.
30. C. Silverman and J. P. Brosco, *Arch. Pediatr. Adolesc. Med.*, 2007, **161**, 392–398.
31. S. F. Terry, K. Zeitz, M. A. Majumder and P. F. Terry, *Genomics, Cancer Care & Advocacy*, Genetic Alliance, Washington DC, 2006.
32. J. O. Weiss, in *WikiAdvocacy*, ed. G. Alliance, Genetic Alliance, Washington, DC, 2004, vol. 2008.
33. A. Stockdale and S. F. Terry, in *The Double-Edged Helix*, ed. J. Alper, *et al.*, The Johns Hopkins University Press, Baltimore, 1st edn, 2002, pp. 80–101.
34. J. H. Platner, L. M. Bennett, R. Millikan and M. D. Barker, *Environ. Mol. Mutagen.*, 2002, **39**, 102–107.
35. J. Uitto, *Trends Mol. Med.*, 2001, 7, 182.
36. S. F. Terry and C. D. Boyd, *Am. J. Med. Genet.,* 2001, **106**, 177–184.
37. T. Rolstad and G. Zimmerman, *Dermatol. Clin.*, 2000, **18**, 277–285.
38. L. Psillidis, J. Flach and R. M. Padberg, *J. Womens Health,* 1997, **6**, 227–232.
39. H. E. Briggs and N. M. Koroloff, *Community Ment. Health J.*, 1995, **31**, 317–333.
40. Return of Organizations Exempt from Tax Return, 2011.
41. J. LaMattina, *Forbes Magazine*, 2013.
42. l'Association Française contre les Myopathies, http://www.afm-telethon.com/, accessed 23 September 2013.
43. Editorial, *Nat. Genet.*, 2006, **38**, 391.
44. O. Le Saux, Z. Urban, C. Tschuch, K. Csiszar, B. Bacchelli, D. Quaglino, I. Pasquali-Ronchetti, F. M. Pope, A. Richards, S. Terry, L. Bercovitch, A. de Paepe and C. D. Boyd, *Nat. Genet.,* 2000, **25**, 223–227.
45. A. A. Bergen, A. S. Plomp, E. J. Schuurman, S. Terry, M. Breuning, H. Dauwerse, J. Swart, M. Kool, S. van Soest, F. Baas, J. B. ten Brink and P. T. de Jong, *Nat. Genet.,* 2000, **25**, 228–231.
46. E. Marshall, *Science,* 2004, **305**, 1226.
47. P. Smaglik, *Nature,* 2000, **407**, 821.
48. J. F. Merz, D. Magnus, M. K. Cho and A. L. Caplan, *Am. J. Hum. Genet.,* 2002, **70**, 965–971.
49. National Institutes of Health, Molecular Libraries Programme, http://mli.nih.gov/mli/, accessed 23 September 2013.
50. T. R. Helliwell, J. A. Gallagher and L. Ranganath, *Histopathology,* 2008, **53**, 503–512.
51. L. Tinti, A. M. Taylor, A. Santucci, B. Wlodarski, P. J. Wilson, J. C. Jarvis, W. D. Fraser, J. S. Davidson, L. R. Ranganath and J. A. Gallagher, *Rheumatology,* 2011, **50**, 271–277.
52. Infectious Diseases Institute, http://www.idi-makerere.com/, accessed 23 September 2013.
53. RareCare, http://www.rarecare.eu/, accessed 23 September 2013.

54. M. Blaustein, *HT-100: Patient-Partnered Drug Development for DMD, Parent Project Muscular Dystrophy*, 2012.
55. US Food and Drug Administration, http://www.fda.gov/ForIndustry/DevelopingProductsforRareDiseasesConditions/OOPDNewsArchive/ucm341676.htm, accessed 23 September 2013.
56. R. Klein, FDA Voice, 14 June 2013.
57. European Medicines Agency, http://www.ema.europa.eu/ema/index.jsp?curl=pages/partners_and_networks/general/general_content_000231.jsp&mid=WC0b01ac0580035bee, accessed 23 September 2013.
58. C. Jackson and K. Riley, National Institute of Health, http://www.nih.gov/news/health/feb2010/od-24.htm, accessed 23 September 2010.
59. E. Horn, J. Bialick and S. Terry, *Biopreserv. Biobanking,* 2010, **8**, 115–117.
60. L. Bercovitch, T. Leroux, S. Terry and M. A. Weinstock, *Br. J. Dermatol.,* 2004, **151**, 1011–1018.
61. R. S. Bercovitch, J. A. Januario, S. F. Terry, K. Boekelheide, A. D. Podis, D. E. Dupuy and L. G. Bercovitch, *Radiology,* 2005, **237**, 550–554.
62. Y. Shi, S. F. Terry, P. F. Terry, L. G. Bercovitch and G. F. Gerard, *J. Mol. Diagn.,* 2007, **9**, 105–112.
63. O. Le Saux, K. Beck, C. Sachsinger, C. Treiber, H. H. Goring, K. Curry, E. W. Johnson, L. Bercovitch, A. S. Marais, S. F. Terry, D. L. Viljoen and C. D. Boyd, *Hum. Genet.,* 2002, **111**, 331–338.
64. Reg4ALL, https://www.reg4all.org/Pages/ToolBox/Home.aspx, accessed 23 September 2013.
65. L. Bercovitch, B. Schepps, S. Koelliker, C. Magro, S. Terry and M. Lebwohl, *J. Am. Acad. Dermatol.,* 2003, **48**, 359–366.
66. P. Kaufmann, M. P. McDermott, B. T. Darras, R. S. Finkel, D. M. Sproule, P. B. Kang, M. Oskoui, A. Constantinescu, C. L. Gooch, A. R. Foley, M. L. Yang, R. Tawil, W. K. Chung, W. B. Martens, J. Montes, V. Battista, J. O'Hagen, S. Dunaway, J. Flickinger, J. Quigley, S. Riley, A. M. Glanzman, M. Benton, P. A. Ryan, M. Punyanitya, M. J. Montgomery, J. Marra, B. Koo and D. C. De Vivo, *Neurology,* 2012, **79**, 1889–1897.
67. C. M. McDonald, E. K. Henricson, R. T. Abresch, J. Florence, M. Eagle, E. Gappmaier, A. M. Glanzman, R. Spiegel, J. Barth, G. Elfring, A. Reha and S. Peltz, *Muscle Nerve,* 2013, **38**, 343–356.
68. R. Muthyala, *Drug Discovery Today: Ther. Strategies,* 2011, **8**, 71–76.
69. L. B. Gordon, M. E. Kleinman, D. T. Miller, D. S. Neuberg, A. Giobbie-Hurder, M. Gerhard-Herman, L. B. Smoot, C. M. Gordon, R. Cleveland, B. D. Snyder, B. Fligor, W. R. Bishop, P. Statkevich, A. Regen, A. Sonis, S. Riley, C. Ploski, A. Correia, N. Quinn, N. J. Ullrich, A. Nazarian, M. G. Liang, S. Y. Huh, A. Schwartzman and M. W. Kieran, *Proc. Natl. Acad. Sci. U. S. A.,* 2012, **109**, 16666–16671.
70. EspeRare, http://www.esperare.org/, accessed 23 September 2013.
71. J. Frost, S. Okun, T. Vaughan, J. Heywood and P. Wicks, *J. Med. Internet Res.,* 2011, **13**, e6.
72. S. Terry, *Nat. Rev. Genet.,* 2010, **11**, 310–311.
73. S. F. Terry and P. F. Terry, *Sci. Transl. Med.,* 2011, **3**, 69cm63.

LYSOSOMAL STORAGE DISORDERS

CHAPTER 6

Pharmacological Chaperones to Correct Enzyme Folding, Cellular Trafficking and Lysosomal Activity

ROBERT E. BOYD* AND KENNETH J. VALENZANO

Amicus Therapeutics, 1 Cedar Brook Drive, Cranbury, NJ, USA
*E-mail: rboyd@amicusrx.com

6.1 Introduction

Glycosphingolipids (GSLs), glycoproteins, and a class of oligosaccharides known as glycosylated aminoglycans (GAGs) are important for membrane formation/integrity, signalling and various other cellular functions. Consequently, maintaining a balance in the production and degradation of these molecules is extremely important for cellular homeostasis. The synthesis of these molecules begins in the endoplasmic reticulum (ER), with additional modifications occurring as they traffic through the Golgi.[1] In contrast, their degradation occurs in an acidified organelle called the lysosome, and is facilitated by a variety of resident acid hydrolases, proteases and sulfatases. In addition glycogen, an available energy source especially for muscle tissues, can also be metabolised in the lysosome, as can cholesterol and small peptides. The catabolic sequences and responsible lysosomal enzymes for the metabolism of several important GSLs, glycogen and GAGs are shown in Schemes 6.1–6.3, respectively. Like many proteins, lysosomal enzymes are

RSC Drug Discovery Series No. 38
Orphan Drugs and Rare Diseases
Edited by David C Pryde and Michael J Palmer

Published by the Royal Society of Chemistry, www.rsc.org

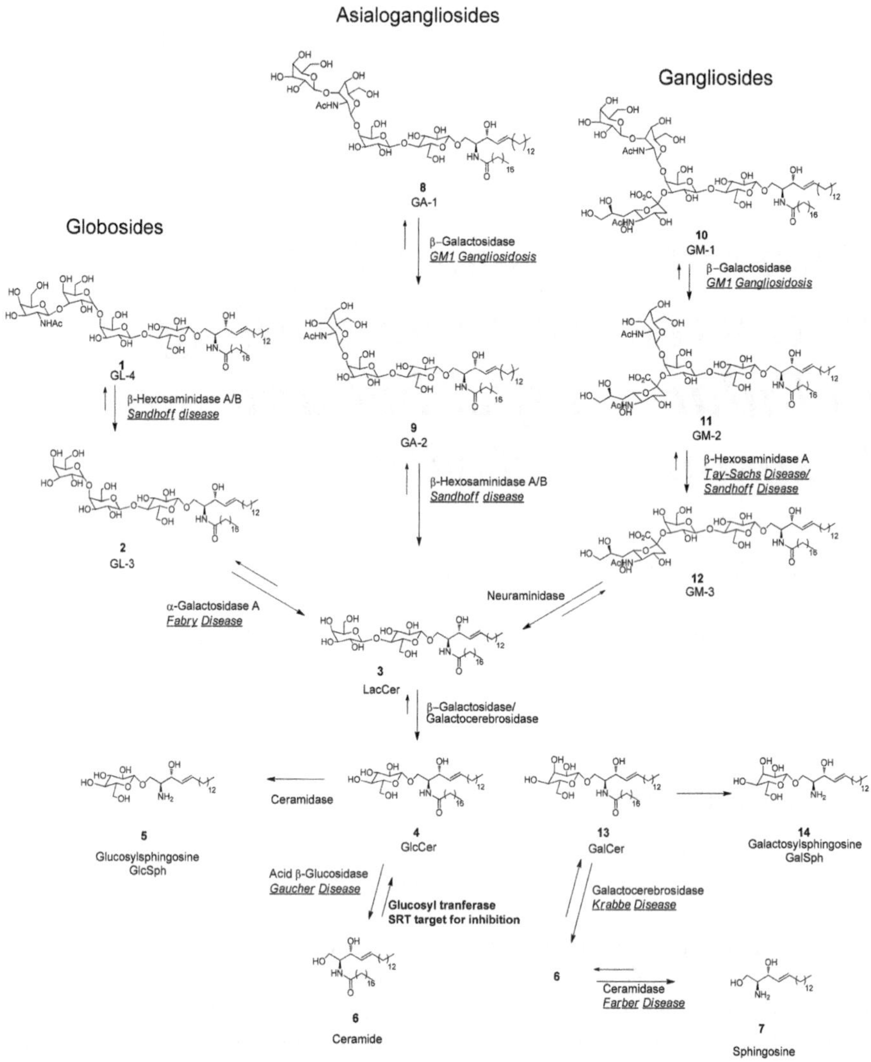

Scheme 6.1 Metabolic pathway for the synthesis and degradation of GSLs including: globosides (**GL-4** and **GL-3**), asialogangliosides (**GA-1** and **GA-2**) and gangliosides (**GM-1**, **GM-2** and **GM-3**). Anabolic enzymes (other than glucosyl transferase) are not shown. The disease resulting from the deficient enzymatic activity of any one of the degradation steps is shown in italics. SRT target enzyme shown in bold.

synthesised in the neutral pH environment of the ER (pH ≈ 7.2), trafficked through the secretory pathway, and delivered to the acidic environment of lysosomes (pH ≈ 5.2). Importantly however, unlike many other classes of proteins, lysosomal enzymes tend to be considerably less stable in a neutral pH environment (*e.g.* ER, blood) compared to an acidic pH environment.[2]

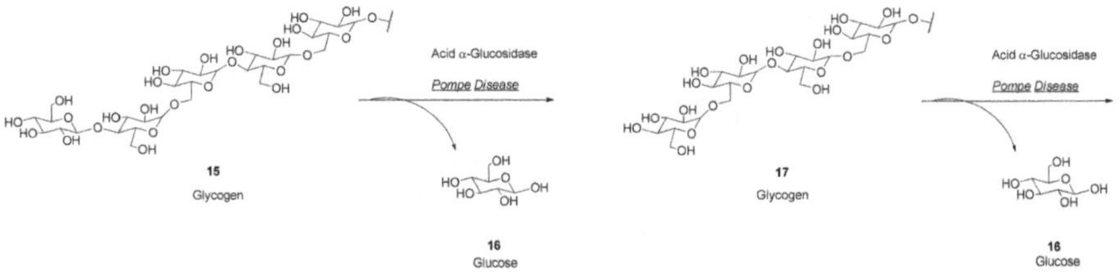

Scheme 6.2 Metabolism of glycogen 15/17. The α-1,4 and α-1,6 glycosidic linkages are cleaved to release glucose, which is an important energy source for cells.

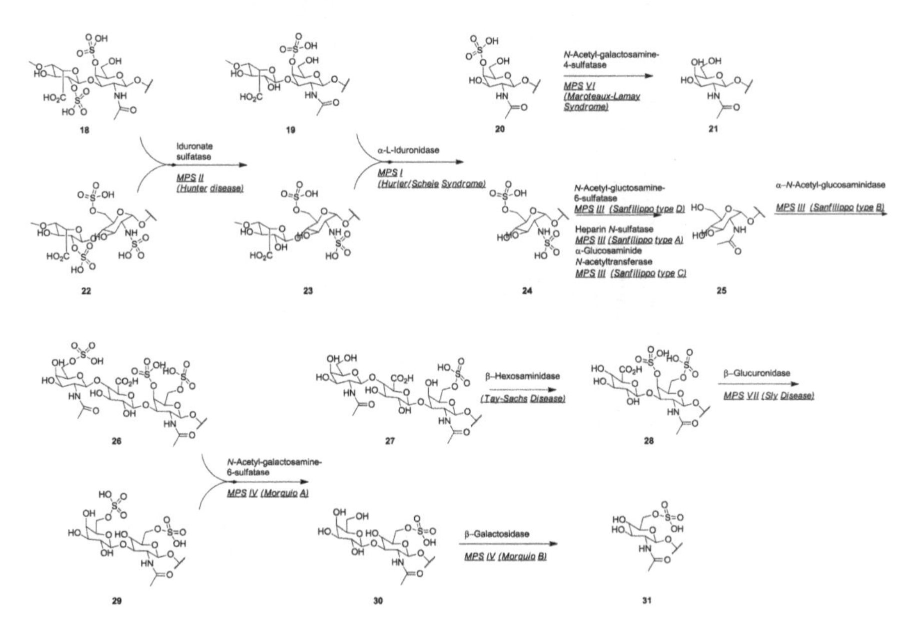

Scheme 6.3 Metabolic pathway of the glycosylated aminoglycans (GAGs) dermatan sulfate (**18**), heparin sulfate (**22**), chondroitin sulphate (**26**) and keratan sulfate (**29**). Anabolic enzymes are not shown. The disease resulting from the compromised enzymatic activity of any one of the degradation steps is shown in italics.

Substrate degradation in the lysosome occurs as sequential processes, with disruption of any specific step resulting in the accumulation of one or more substrate(s), cellular dysfunction and the manifestation of disease pathology. Table 6.1 identifies some of the more common lysosomal storage diseases (LSDs), along with the affected enzymes, the associated storage products, and their predominant clinical phenotypes. In addition, currently approved therapies as well as investigational drugs, both past and present, are presented. Tay–Sachs disease (TSD)/Sandhoff disease (SD) (collectively referred to as GM2 gangliosidosis),[3] Fabry disease (FD)[4] and Gaucher disease (GD)[5] all involve a disruption in the metabolism of GSLs and occur in individuals with

Table 6.1 Select lysosomal storage diseases (LSDs). Representative LSDs, including the specific protein deficiencies, primary storage material(s), and approved and/or investigational drugs, are presented.

Disease	*Deficiency*	*Storage product(s)*	*Typical phenotype*	*Approved drugs*[a]	*Clinical trials*[a]
Fabry	α-Galactosidase A (α-Gal A)	GL-3, Lyso-Gb_3[b]	Kidney disease; heart disease; peripheral neuropathy; angiokeritomas	*Fabrazyme®* *(agalsidase beta)* *Replagal™* *(agalsidase alfa)*	Galactose Migalastat HCl
Gaucher	Acid β-glucosidase (GCase)	GlcCer, GlcSph[c]	Hepatosplenomegaly; bone disease; low platelet; haemoglobin; CNS impairment (type II/III)	*Cerezyme®* *(imiglucerase)* *VPRIV™* *(velaglucerase alfa)* *Elelyso/Uplyso™* **Zavesca® (miglustat)**	Isofagamine Ambroxol **Eliglustat**
Tay–Sachs/ Sandhoff (GM2 gangliosidosis)	β-Hexosaminidase A/B (β-Hex A/B)	GM-2, GA-2	CNS impairment	N/A	Pyrimethamine
Morquio B (GM1 gangliosidosis)	β-Galactosidase	GM-1, GA-1, keratan sulfate	Highly varied/ mutation dependent	N/A	

Pompe (acid maltase deficiency)	Acid α-glucosidase (GAA)	Glycogen	Muscle disease/deterioration (heart/diaphragm/skeletal)	*Myozyme®* *Lumizyme®* *(both alglucosidase alfa)*	1-Deoxynojirimicin
Krabbe (globoid cell leukodystrophy)	Galactocerebrosidase	GalCer, Psychosine	Varied; mainly CNS-related disorders	N/A	
Hurler/Hurler-scheie (MPS I)	α-L-Iduronidase	Dermatan sulfate, heparan sulfate	Hepatosplenomegaly; skeletal abnormalities	*Aldurazyme® (laronidase)*	
Hunter (MPS II)	Iduronate sulfate sulfatase	Dermatan sulfate, heparan sulfate	Hepatosplenomegaly; CNS impairment; skeletal and connective tissue abnormalities	*Elaprase® (idursulfase)*	
Maroteaux–Lamay (MPS VI)	*N*-Acetylgalactosamine-4-sulfatase	Keratan sulfate, chondroitin sulfate	Heart disease; skeletal abnormalities	*Naglazyme® (galsulfase)*	
Niemann-Pick type C	Lysosomal/endosomal cholesterol transport	Cholesterol	CNS impairment	**Zavesca®**	**Cyclodextrin**

[a] *Enzyme replacement therapy*; **substrate reduction therapy**; pharmacological chaperone. N/A, no approved therapy.
[b] Deacylated form of GL-3.
[c] Deacylated form of GlcCer (Scheme 6.1).

genetic defects that result in insufficient β-hexosaminidase A/B (β-Hex A/B), α-galactosidase A (α-Gal A), and acid β-glucosidase (GCase) activity, respectively. Acid α-glucosidase (GAA) is responsible for the lysosomal degradation of glycogen, with compromised activity resulting in Pompe disease (PD; also referred to as acid maltase disease or glycogen storage disease type II).[6] It is important to point out that Table 6.1 represents only the most common LSDs; in fact, more than 50 of these genetic disorders have been identified and more comprehensive discussions have been previously published.[7–9] While the incidence of individual LSDs is extremely low, it is estimated that collectively they affect 1 : 1500 to 1 : 7000 individuals. Furthermore, the pathology associated with the accumulation of GSLs, GAGs or glycogen can vary dramatically; however, and somewhat surprisingly, clinical phenotypes and organ involvement can also be markedly different within LSD subsets such as those affecting GSL metabolism, and can even vary dramatically for individuals with the same disease. Presently there are two treatment options available for some LSD patients: enzyme replacement therapy (ERT)[10,11] and substrate reduction therapy (SRT).[12] The goal of each of these approaches is to reduce the levels of accumulated substrate in lysosomes. To achieve this, ERT relies on the intravenous infusion of a manufactured enzyme, generally on a weekly or bi-weekly basis to replace the compromised endogenous lysosomal activity. ERT is available for several LSDs (Table 6.1), and is generally quite effective at reducing substrate load and improving clinical outcomes; however, challenges do still exist with this therapeutic approach. The most significant is the fact that the first generation of these recombinant proteins are unable to cross the blood–brain barrier (BBB), and consequently are of little therapeutic value in treating the central nervous system (CNS) pathology associated with many LSDs.[13] Peptide tags that are covalently attached to ERT and theoretically help to target the modified proteins to the CNS are currently being investigated pre-clinically with several recombinant lysosomal enzymes.[14] In addition, intrathecal administration of Laronidase (Genzyme, Cambridge, MA), the ERT used to treat MPS I (Hurler syndrome; Scheme 6.3), is currently under clinical investigation. These revised ERT approaches may ultimately permit treatment of the CNS manifestations of some LSDs. In addition to the lack of CNS penetration of these first-generation ERT products, infusion of exogenous lysosomal proteins into the bloodstream often leads to immunological reactions that may limit efficacy and/or adversely affect tolerability.[15,16] Also, uptake of the infused enzymes into certain cell types and tissues is insufficient in certain cases, thereby limiting therapeutic efficacy.

In contrast to ERT, SRT involves the use of small molecules that inhibit the biosynthesis of the storage material or a key intermediate in the biosynthetic pathway of the storage material. Zavesca® (*N*-*n*-butyl-1-deoxynojirimycin), (Miglustat, Actelion, Allschwil, Switzerland) (**32a**), the only approved SRT, is an inhibitor of glucosyl transferase, a key enzyme in the biosynthesis of glucosylceramide (GlcCer; Scheme 6.1). It is currently indicated for GD, as well as to treat the CNS pathology associated with Niemann-Pick Type C (NPC). A second-generation SRT for GD, eliglustat tartrate (Genzyme, Cambridge, MA)

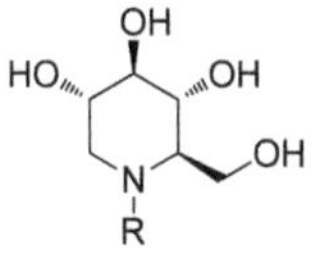

32a R = n-Bu
b R = H

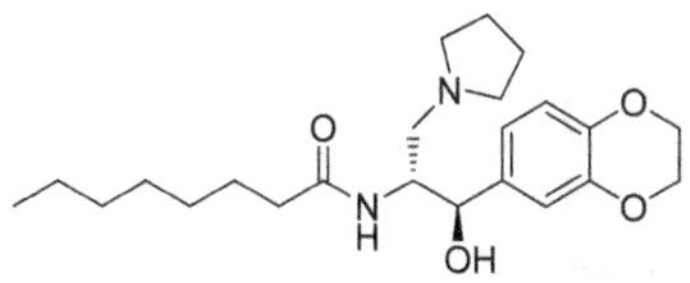

33

(**33**), is currently in Phase 3 clinical trials. While these SRT molecules directly inhibit only a single biosynthetic step in the production of GlcCer, the levels of all downstream GSLs may also be affected (Scheme 6.1); presently it is not known if or how reduced GSL levels may affect overall cellular function.

It is now understood that compromised lysosomal enzyme activity is frequently the result of mutations in the genes that encode these enzymes. While some of these mutations involve large insertions or deletions, frame shifts, or premature stop codons that lead to the synthesis of no enzyme or a catalytically inactive enzyme, some mutations are fairly subtle and lead to the production of enzymes that differ from wild type only by a single amino acid residue (*i.e.* missense mutations). Missense mutations in lysosomal enzymes can often exacerbate the low physical stability that many of these proteins demonstrate in a neutral pH environment, with consequently significantly less of the properly folded enzyme able to pass the quality control mechanisms of the ER and efficiently traffic to lysosomes. While most LSDs are inherited in an autosomal recessive fashion, some, such as FD, are X-linked, and while once thought to affect only males, it is now understood that both males and females are equally affected.

A pharmacological chaperone (PC) is a small molecule that can selectively bind and stabilise a specific protein early during biosynthesis, and allow it to pass the quality control of the ER. In the case of LSDs, a large percentage of the PCs that have been identified bind at their respective enzyme's active site and act as competitive inhibitors of the metabolism of endogenous substrate; consequently, after translocation to the lysosome, dissociation of the PC is necessary to allow substrate turnover. Ideally, a PC would bind with highest affinity in the neutral pH environment of the ER to facilitate stabilisation, and with much lower affinity at the acidic pH of lysosomes, thereby favouring dissociation and minimising inhibition of the target enzyme. Furthermore, the elevated lysosomal level of substrate is also an important factor in driving enzyme occupancy toward substrate binding/metabolism and minimising reassociation of the PC/enzyme complex. In principle, PC stabilisation of a lysosomal enzyme through binding at allosteric sites is also possible, with significant progress towards the identification of such compounds recently reported (*vide infra*).

The goal of this chapter is to highlight the key aspects of preclinical PC identification and characterisation. A general screening pathway is shown in Figure 6.1. Experimental protocols and important results for some of the more common LSDs previously mentioned will be used to exemplify various aspects of this overall strategy.

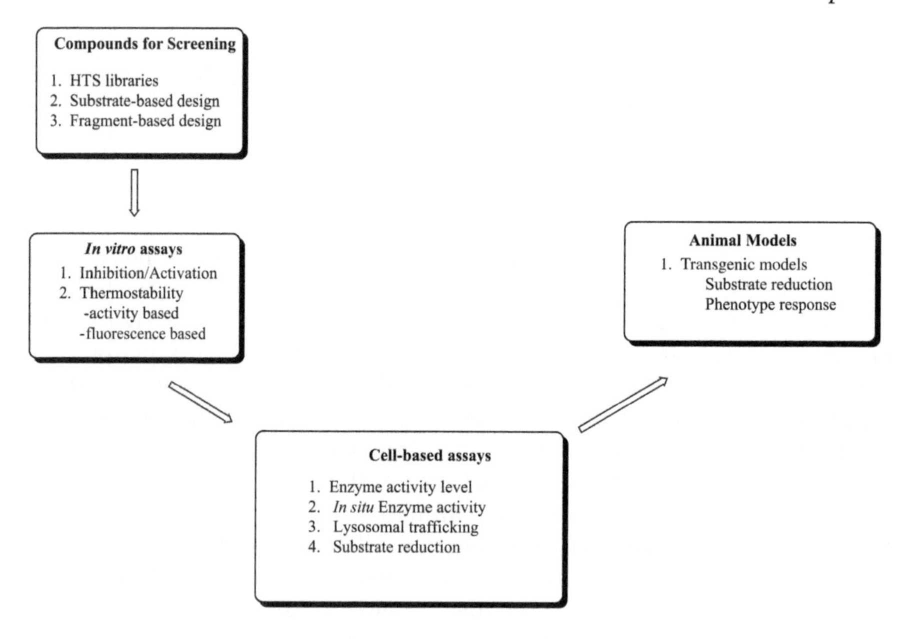

Figure 6.1 Identification and development pathway for pharmacological chaperones.

6.2 Identification of Pharmacological Chaperones (*In Vitro* Assays)

The challenges of identifying suitable PC lead structures for medicinal chemistry development has been addressed in several ways. Known glycosidase inhibitors such as iminosugars, azasugars and carbasugars have provided the basis for many of the first-generation PCs that have been identified. 1-Deoxynojirimycin (**32b**), isofagomine (IFG; **34**) and *N*-octyl-4-*epi*-

34 **35**

valienamine (**35**) bind with high affinity to the active sites of GAA, GCase and β-galactosidase, respectively, and are generally considered to be non-hydrolysable mimetics of their endogenous substrates. A large number of analogues of these inhibitors have also been synthesised and evaluated for their ability to bind and stabilise mutant lysosomal enzymes, many of which have recently been reviewed.[17] While the design of these molecules is often based on structural similarities with the terminal sugar residue of the natural substrate, this similarity is limited to the stereochemistry at the non-anomeric positions of the sugar. As a result, in some cases the selectivity (*e.g.* α- *versus* β-glucosidases, or lysosomal *versus* non-lysosomal glucosidases) of such compounds is only modest, with off-target activity often responsible for a variety of unwanted effects.

Fragment-based drug design (FBDD) is a relatively new approach that has been used successfully to identify lead structures for a number of therapeutic targets. It has been argued that small fragment libraries can more efficiently probe drug space for protein or receptor binding compared to larger drug-like molecules.[18] Furthermore, FBDD also affords the ability to target allosteric sites as a means to stabilise mutant lysosomal enzymes. On the other hand, the relatively weak binding (*i.e.* affinities in the high micromolar to millimolar range) for many low molecular weight fragments often requires detection using specialised techniques, including X-ray crystallography or ^{1}H NMR. Although this approach is quite new, some recent success in the identification of active leads for some non-lysosomal protein targets, and even a clinical candidate, has been reported. These successes demonstrate the viability of FBDD, particularly as an adjunct to more traditional screening approaches for some intractable protein targets. As evidence of this, Petsko *et al.* recently reported an initial fragment-based approach utilising both *in silico* and crystallographic techniques to identify suitable starting points for compounds that are capable of stabilising GCase.[19] In addition to binding sites at or near the enzyme's active site, several non-active site 'hotspots' were identified that could provide added stability to mutant GCase. It remains an opportunity for further research and drug discovery.

Finally, high-throughput screening (HTS) has been a mainstay in lead identification for over 20 years. A variety of HTS assays have identified several PCs for a number of LSDs (*vide infra*). The advantage in an HTS approach is the identification of new and unexpected chemotypes, which in turn can lead to a better understanding of binding interactions, and therefore to the identification of compounds with improved binding and selectivity profiles. In general, assays used for HTS have measured the ability of a small molecule to bind, and inhibit or stabilise, a target protein. As will be described below, many of these assays clearly distinguish active site *versus* allosteric binding, although in some cases follow-up assays are required to clearly elucidate the mechanism of action.

6.2.1 Inhibition/Activation Assays

Assays for inhibition of enzymatic activity have been developed for many lysosomal enzymes, typically using purified recombinant enzymes, or cell lysates that contain the activity of interest, together with the appropriate fluorogenic or chromogenic substrate, many of which are commercially available. These assays have been well characterised and are readily adaptable to 96-, 384- and 1536-well formats. Typically, these are end-point assays using a single concentration of test compound, although variations have been incorporated in certain cases.

For instance, Tropak *et al.* developed a fluorescence-based assay to identify inhibitors of β-Hex, the enzyme deficient in TSD and SD (see Table 6.1 and Scheme 6.1). Rather than the conventional end-point fluorescence determination that can produce a significant number of false-negatives and/or false-positives due to auto-fluorescence and fluorescence quenching, respectively, this modified assay technique relied on continuous monitoring of a fluorescent product over a 25 minute period.[20] Sixty-four active (*i.e.* inhibitory) compounds were identified from the initial screen of a 50 000-compound library; this set was further reduced to 24 compounds as confirmed inhibitors of β-Hex. The identification of three single-digit micromolar β-Hex inhibitors capable of increasing cellular activity (*vide infra*) in some SD/TSD patient-derived cell lines, provided validation for this screening approach. A second screen of 1040 FDA-approved compounds resulted in the identification of pyrimethamine (**36**), which was then quickly advanced into an open-label Phase 1/2 clinical trial for TSD.[21] While a 3- to 6-fold increase in enzyme activity was seen in peripheral blood mononuclear cells from a majority of enrolled patients, significant toxicity was observed at higher doses and the trial was suspended.

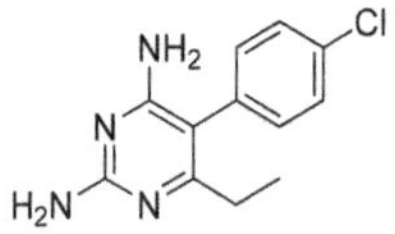

36 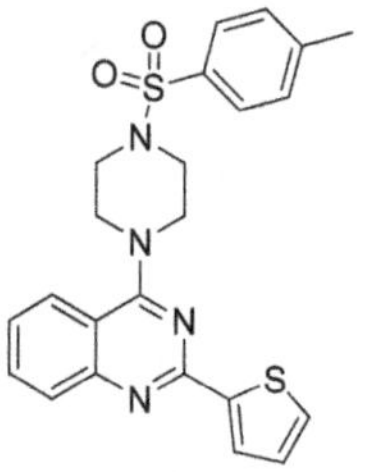 **37** 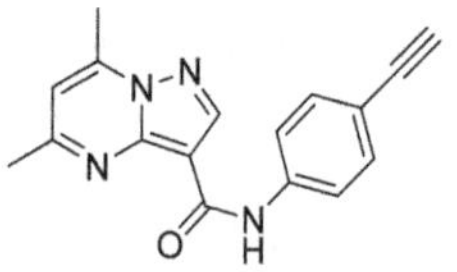 **38**

Marugan *et al.* also developed a variation of the traditional inhibition assay that used human spleen homogenates as the source for GCase.[22] This type of assay makes use of the enzyme's natural cofactors that are present in tissue, such as Saposin C, rather than utilising exogenous detergents and salts that can substitute for these cofactors *in vitro*. It is also worth noting that this assay was run using homogenates prepared from normal human spleen, as well as from GD patients homozygous for the common N370S mutant allele; purified, recombinant normal and N370S GCases were also included for comparison. Interestingly, it was discovered that assays run using tissue homogenates (normal or N370S) yielded dramatically different results compared to assays run with the purified recombinant enzymes. The reasons for this are not entirely clear, but do probably reflect the activity of the natural cofactors present in tissue homogenates. It should be pointed out that the different outcomes seen when using the purified enzyme *versus* the cell lysate versions of this assay are enzyme specific, because no differences were seen when a similar approach was utilised for GAA.[23] Furthermore, this assay was configured as a quantitative HTS approach, providing multi-point inhibition curves for a 50 000-compound library. A series of quinazoline structures were identified as confirmed hits; medicinal chemistry optimisation resulted in a series of active-site PCs for GCase, as exemplified by **37**. A second series of compounds was later found that *increased* enzyme activity in this assay; again medicinal chemistry led to the identification of **38** as the optimised lead for this series.[24] It is presumed that compounds such as **38** bind outside the active site of GCase (*i.e.* allosterically) to increase enzyme stability and promote proper translocation to the lysosome. While this series of compounds has produced the first evidence of enzyme stabilisation through allosteric binding, questions around their ability to reduce endogenous substrate levels remain unanswered. A similar approach has also led to the identification of GAA activators; unfortunately, this series has also been unable to clearly demonstrate any substrate reduction in cell-based or animal models.

6.2.2 Stability Assays

Enzyme stabilisation is believed to be the basic mechanism by which PCs are able to increase proper ER export and lysosomal translocation of mutant lysosomal enzymes. Multiple approaches have been used to demonstrate changes in the physical stability of lysosomal enzymes as a function of pH, temperature and/or small molecule binding. Kelly *et al.* used circular dichroism to monitor GCase stability, as measured by denaturation and unfolding during a thermal ramp.[25] These were some of the first studies to demonstrate the significant stabilisation of GCase when incubated with 5 to 50 μM concentrations of the *N*-substituted deoxynojirimycin analogue **39**. More recently, utilising differential scanning calorimetry (DSC), Petsko *et al.* demonstrated the ability of 1-deoxygalactonojirimycin (**40**) to stabilise α-Gal A, increasing its melting temperature (T_m) by 11.4 °C and 13.3 °C at pH 5.2 and pH 7.4, respectively. Although smaller in magnitude, a similar stabilisation and increase in T_m was shown when IFG (**34**) was incubated with GCase.

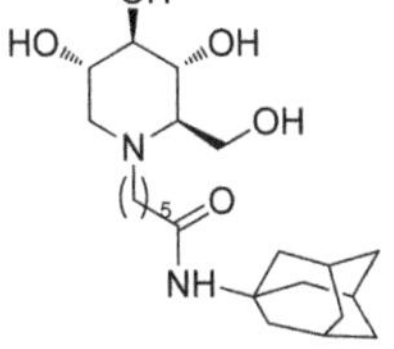

OH HO OH OH N O NH

39

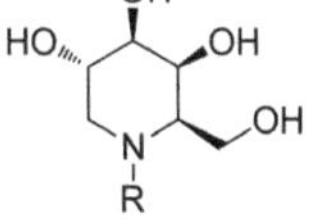

OH HO OH OH N R

40

NH2 OH Br N H Br

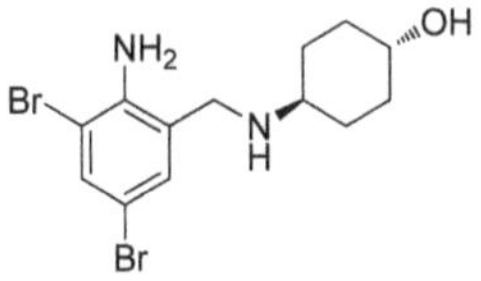

41

Fluorescent probes, such as SYPRO® Orange, have also been used to monitor protein stability/denaturation.[26] These molecules have a very low quantum yield in aqueous solution, but are highly fluorescent in many organic solvents, essentially providing the basis for the development of screening assays that measure the stabilising effect of potential PCs on lysosomal enzymes. To this end, thermal denaturation of many proteins causes significant changes in the tertiary structure, thereby exposing hydrophobic amino acid residues, which are normally confined to the inner core of these macromolecules, to the surrounding aqueous environment. The fluorescence of probes such as SYPRO® Orange increase as they bind to these non-polar amino acid residues. Consequently, these fluorophores can be used to evaluate protein stability (or melting) as a function of temperature. This concept has been miniaturised and validated for HTS, and was used to screen compound libraries to identify small molecules that bind and stabilise large proteins. Thermostability curves using **32a,b** and the lysosomal enzyme GAA demonstrate the concentration-dependent increase in the T_m of this enzyme, with temperature shifts of approximately 3 °C and 8 °C at 1 and 10 μM, respectively (Figure 6.2). A significant advantage of this approach is the ability to screen a wide variety of proteins/enzymes with a single assay set-up. Unfortunately, the number of false-negatives and false-positives that result from library compounds having excitation/emission spectra that overlap with SYPRO® Orange represent a limitation of this technique. However, alternative readouts for enzyme stability have been utilised and can help minimise this problem (*vide infra*).

A variation of the thermostability assay that uses an activity-based readout as the end point also can be used. For this approach, activity measurements of enzyme preparations that have been incubated at an elevated temperature for a given period of time are compared to the enzyme activities of control samples that have been maintained in an ice bath. Typically, the elevated temperatures lead to relatively rapid protein denaturation, which is measured as a loss of activity using a simple enzyme activity assay. This

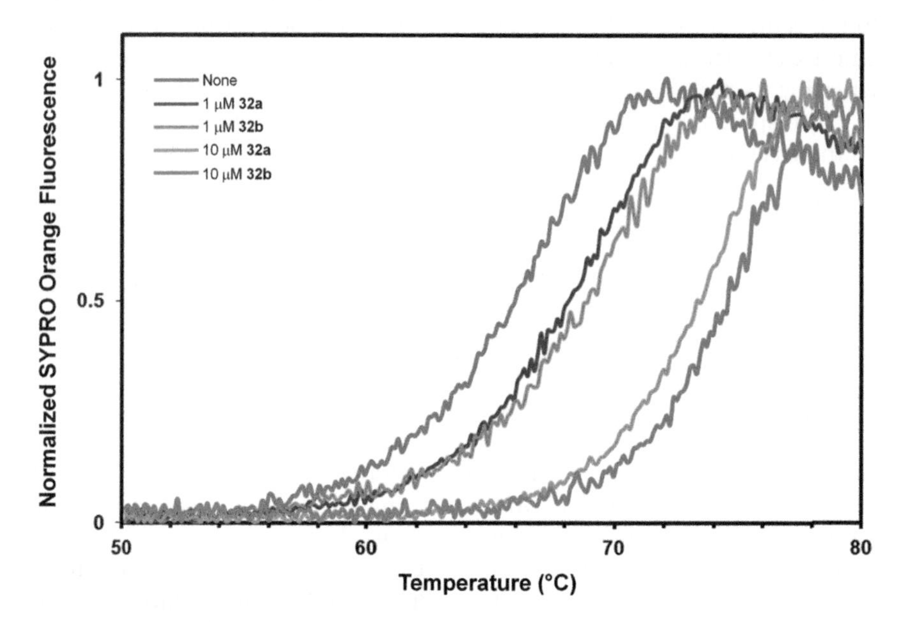

Figure 6.2 Thermal denaturation curves of GAA with increasing concentrations of **32a** and **32b**. Increased fluorescence of y-axis represents increased protein unfolding.

screening paradigm identified Ambroxol (**41**) from a collection of 1040 FDA-approved drugs as a PC capable of stabilising GCase.[27] This molecule has subsequently been evaluated in a pilot study with a small number of GD patients.[28] Ten of the 12 patients completed the study, with only one showing a minor adverse reaction. Haemoglobin level and platelet count as well as spleen and liver volume were monitored. Although results did not reach statistical significance, positive results were seen in some patients, suggesting a follow-up study is warranted.

6.3 Identification of Pharmacological Chaperones (Cell-based Assays)

As with any drug discovery programme, a number of secondary assays are necessary to help prioritise compounds for advancement through preclinical development. This is in addition to some of the standard requirements such as target selectivity, bioavailability, and in many cases, the ability to cross the BBB. In general, these assays tend to be more complex and difficult to utilise in large screening campaigns. They do, however, provide an effective way to prioritise compounds that have been identified *via* various screening strategies, and provide important information on mechanism of action.

6.3.1 Assays to Measure Total Cellular Enzyme Levels

Typically, cell-based assays have been widely used early in the discovery process to show that incubation with a PC can increase the total cellular levels of an enzyme of interest, indicative that the first steps of PC action have occurred (*i.e.* the PC has crossed the plasma and ER membranes, and has bound and stabilised the mutant enzyme in the ER). For these assays, primary fibroblasts or immortalised lymphoblastoid cell lines (LCLs) (often derived from LSD patients), as well as transfected cells that express wild-type or mutant enzymes, are incubated with a potential PC for a period of 3–6 days.[29,30] Cells are then washed and lysed, and changes in total cellular enzyme levels are measured using one of the two analytical techniques described below. Fibroblasts or LCLs derived from Fabry, Gaucher and other LSD patients have been used successfully to identify new PCs *via* HTS, and to support lead optimisation.[31,32] In the case of patient-derived cells that are homo- or hemizygous for the disease-causing mutation, the effect of the PC on the mutant enzyme can be readily determined, as only one mutant form is expressed.[30,33] This is more challenging with cells derived from patients that are heterozygous for two different mutant alleles, as it is difficult to interpret which mutant forms are expressed and responsive to the PC.[34] To circumvent this challenge, heterologous expression of individual mutant forms of lysosomal enzymes in host cells such as CHO or HEK-293 has been used when patient-derived cells that express only a single mutant form are not available. After transient transfection, cells can be incubated with a potential PC, with the effect on cellular levels of the mutant enzyme monitored. Expression of the mutant form needs to be sufficiently high, as these cell types also express wild-type forms of most lysosomal enzymes, which may mask the activity of the transiently expressed mutant enzyme and the effect of the PC. Heterologous expression has been used to characterise the activity of a number of PCs on several different mutant lysosomal enzymes.[35]

Two analytical methods have been used traditionally to assess enzyme levels in lysates from cells that were incubated with a PC. First, Western blot analysis is used to separate cellular proteins based on molecular weight, followed by detection and quantitation using chemiluminescence- or fluorescence-based techniques.[36,37] Increases in the total protein levels of specific lysosomal enzymes are indicative that the PC has stabilised the enzyme and led to greater cellular levels. Western blotting has been used to clearly show increased total cellular levels of mutant forms of α-Gal A, GCase, GAA, β-Hex, and others after incubation with PCs in a variety of different cell types.[20,30,33,38–41] In addition, for some lysosomal enzymes, Western blotting can also reveal processing of immature isoforms into more mature isoforms (*i.e.* post-translational modifications to the protein backbone and/or to the glycan structures), indirect evidence that the enzyme has exited the ER and entered the secretory pathway. It is important to point out that increased protein levels, even increases in the fully mature isoforms, do not necessarily indicate that the mutant enzyme is active *in situ* and able to metabolise accumulated lysosomal substrates; other assays are necessary for these purposes (*vide infra*).

The potential for a PC to have therapeutic efficacy in an LSD is at least partially determined by its ability to also increase the total cellular *activity* of the deficient enzyme. The assays used to measure this are sometimes referred to as 'enhancement assays', as they measure the ability of the PC to increase, or 'enhance', the cellular activity of a particular enzyme. The lysosomal enzyme activity in lysates from cells incubated with a PC has commonly been measured using artificial fluorogenic or chromogenic substrates as described above for the enzyme inhibition assays.[29–31,33,38,39] Enzyme activity is measured by mixing the cell lysate with exogenous substrate in a buffer that has been optimised for the catalytic function of the target enzyme (*e.g.* pH, salt composition, presence of detergents, *etc.*). As lysosomal hydrolases tend to have highest activity in an acidic environment, these assays typically utilise low pH buffers to minimise metabolism of the artificial substrates by related cellular hydrolases that have higher pH optima. Alternatively, parallel assays can be run in the presence of selective inhibitors that can help quantify the contribution of substrate metabolism by non-lysosomal enzymes.

As nearly all currently described PCs are reversible, competitive inhibitors of their target lysosomal enzymes, carry-over of residual PC into the lysed-cell enzymatic assay may interfere with the activity measurement, potentially masking the detection of increased enzyme activity. This phenomenon has been seen with wild-type and some mutant forms of α-Gal A, GAA and GCase following incubation with high concentrations of PC.[33,39–41] These hurdles have been overcome using procedures designed to reduce PC carry-over into the enzyme assay, including glycoprotein enrichment, enzyme immunocapture and extended incubation of cells in PC-free media prior to assay. While the simplest approach is to provide an extended incubation time in the absence of PC to achieve a more complete washout, this timeframe can be limited by the half-life of the chaperoned enzyme. Hence, the PC washout time must be sufficiently shorter than the time required for the increased enzyme level to return to baseline.[30,42]

It should be noted that these cell-based assays alone cannot distinguish whether the elevated cellular enzyme levels are caused by a PC-mediated mechanism of action or *via* an alternative pathway or mechanism, thus necessitating the use of parallel assays such as thermostability or enzyme inhibition to show direct interaction of the PC with the enzyme of interest.[31] Additionally, other assays are needed to verify that the 'enhanced' enzyme has been delivered to the lysosome and is competent to turn over substrate *in situ*.

6.3.2 Assays to Measure Lysosomal Enzyme Trafficking

The ability of the PC to promote lysosomal trafficking, and hence earn the name 'chaperone', also requires evaluation. These methods can utilise a direct assessment of the cellular location of the PC-stabilised lysosomal enzyme, or an indirect assessment based on changes in molecular weight due to protein and/or glycan modification, as well as reductions in endogenous levels of accumulated lysosomal substrates.

Subcellular fractionation is the classical method for monitoring protein trafficking. Cells are homogenised, organelle and membrane fractions are then isolated (*e.g.* by ultracentrifugation, magnetic beads, *etc.*), and protein content analysed by Western blotting and/or enzyme activity, coincident with established organelle-specific markers. Studies with mutant lysosomal enzymes have used this approach to examine defects in trafficking, as well as PC-mediated improvements.[29,39,43,44] However, subcellular fractionation often presents technical challenges, such as the requirement for large quantities of cells (>10^6 per sample) and the high potential for incomplete separation of cellular components.

As an alternative, and as discussed above, proteolytic processing of precursor proteins into mature forms can be used as an indirect marker for protein trafficking, provided that the processing is coupled to trafficking. Proteolytic processing en route to the lysosome has been seen for a number of lysosomal enzymes, including α-mannosidase,[45] GCase,[46] β-glucuronidase,[47] α-fucosidase,[48] β-Hex,[49] GAA[33,50–54] and others,[55–57] and thus could be used to monitor their trafficking in response to PC incubation. Glycan processing has also been used as a marker for protein trafficking. As glycosylated proteins traffic through the secretory pathway, their glycan chains are modified and remodelled by resident glycosyltransferases and glycosidases;[58] such changes can be detected by protein glycosylation analysis.[59] For example, the glycan chains of GCase undergo processing as the enzyme traffics through the secretory pathway, with clear changes in molecular weight that indicate their subcellular localisation and which are distinguishable by Western blot analysis.[44]

Perhaps the most robust method to monitor protein trafficking is imaging-based subcellular localisation, which utilises fluorescence microscopy to simultaneously monitor the target enzyme (*via* specific antibodies or genetically encoded tags) and organelle-specific markers such as the lysosome-resident transmembrane protein LAMP-1.[60] This approach has been used to examine aberrant trafficking of mutant enzymes and the subsequent restoration of normal trafficking following incubation of cells with potential PCs.[33,39,43,53,54,61–64]

6.3.3 Assays to Measure *In Situ* Lysosomal Activity

The cell-based assays highlighted above indicate the quantity and cellular location of mutant enzymes before and after incubation with a PC. However, an important concern in the successful selection of a PC is enzyme inhibition, specifically in the lysosome. In an ideal situation, PCs would stabilise mutant enzymes, correct folding defects, enhance trafficking, and stimulate lysosomal activity. Unfortunately, almost all PCs identified to date bind to the active sites of their target enzymes and act as reversible, competitive inhibitors (*vide supra*). Identifying compounds that are effective chaperones but weak lysosomal inhibitors would greatly aid in the development of good development candidates.

As discussed above, one approach to reducing chaperone-mediated inhibition in the lysosome is to select for compounds that show pH-dependent binding to their target enzyme; that is, PCs that have higher affinity for their target enzyme in the neutral pH environment of the ER, compared to a lower binding affinity in the acidic environment of the lysosome. An alternate and complementary approach is to select for compounds that rapidly leave the lysosome (and the cell) after the deficient enzyme has trafficked to the lysosome. An enzyme assay that measures activity in the lysosomes of intact, living cells (*i.e. in situ*) can be employed for both of these approaches. These assays utilise fluorogenic substrates that are selectively taken up into lysosomes, presumably by fluid phase endocytosis, or that measure changes in endogenous substrate levels following PC incubation.

In the first case, *in situ* assays can be used to measure the potency for enzyme inhibition in the lysosome by a PC. *In situ* assays have been described for measuring the lysosomal activity of GCase,[65–71] α-Gal A,[72,73] β-Hex[74] and β-galactosidase.[75,76] In the case of GCase, an *in situ* activity assay was used to characterise the potency of lysosomal inhibition by IFG (**34**).[35] Isofagomine inhibited lysosomal GCase activity with an IC_{50} value of approximately 300 nM. This was substantially lower affinity than when measured using purified enzyme in a cell-free inhibition assay (~44 nM at pH 5.2), suggesting the potential for even less IFG-mediated inhibition in an intact cell. Hence, screens for PC candidates can use *in situ* assays to select for compounds with reduced potency for lysosomal inhibition, with compounds showing lower potency taking priority.

In the second case, assays using these *in situ* fluorogenic substrates have been configured to measure the efflux rates of PCs from lysosomes. Here, cells are incubated with a substrate together with the PC (at concentrations that are high enough to provide near complete inhibition of the target enzyme). The PC is then washed away from the cells, and return of enzymatic activity over time as indicated by increased cellular fluorescence provides an indirect measure of the PC's efflux rate from the lysosome in an intact cellular environment. The rate of lysosomal efflux of IFG was assessed using the substrate 5-(pentafluorobenzoylamino)fluorescein-di-β-D-glucopyranoside (PFBF-β-glucose),[44] with the enzymatically-liberated PFBF fluorophore being rapidly conjugated to thiol groups[77] and remaining trapped inside the cell during the assay.[67] This trapping permits detection of activity within lysosomes by fluorescence microscopy, or within cells using fluorescence plate readers or flow cytometry. Similar assays have been developed for several other lysosomal enzymes including α-Gal A, β-Hex and β-galactosidase.[72–76] Taken together, *in situ* assays that utilise artificial, fluorogenic substrates can play an important role in identifying PCs that have low inhibitory potential and/or fast efflux rates from lysosomes, key properties for PCs that are active site inhibitors.

Lastly, *in situ* cell-based assays have been reported that monitor reduction of endogenous substrate levels in patient-derived cells. Unfortunately, the development of these assays presents significant challenges and only a few have been reported. Yam *et al.* described the use of immunofluorescence to show a reduction in globotriaosylceramide (GL-3; Scheme 6.1), the primary storage product associated with FD, after human Fabry fibroblasts harbouring the R301Q mutant form of α-Gal A were incubated with 20 μM 1-deoxygalactonojirimycin (**40**) for 60 days.[43] While these data were only semi-quantitative, they did provide the first example of substrate reduction in a cell-based model. In similar studies, total cellular GL-3 levels in Fabry fibroblasts were semi-quantified using infrared fluorescence imaging.[30] In fibroblasts harbouring R301Q or L300P α-Gal A, 7-day incubation with **40** followed by 3-day washout resulted in GL-3 reductions of 45% and 38%, respectively; in contrast, continuous 10-day incubation with **40** did not significantly reduce GL-3 levels.

In a similar way, Khanna *et al.* were able to demonstrate an approximate 25–50% decrease in GlcCer levels in a Gaucher patient-derived cell line that was homozygous for L444P GCase following 7-day incubation with 30 μM IFG (**34**) followed by 3-day washout.[39] Similarly, SRT with Zavesca (**32a**) led to an approximate 75% reduction in GlcCer levels in this assay. In contrast, incubation with **34** for 10 consecutive days resulted in a net increase of 15–35% in GlcCer levels. This work again provided further proof-of-concept for the ability of PCs to reduce endogenous substrate in cell-based models. Even more importantly, these experiments demonstrated the importance of allowing sufficient time for PC dissociation to allow substrate turnover.

Assays that show decreased levels of the endogenous substrate in enzyme-deficient cells after incubation with a PC indicate that the function of the enzyme in the lysosome has been successfully restored. In combination with supportive results from the assays described above, restoration of *in situ* lysosomal enzyme function adds further support that the molecule acts as a PC for its enzyme target. Such results would thus warrant further evaluation of the bona fide PC in preclinical animal models, and ultimately in patients with the LSD.

6.4 Animal Models

The use of animal models should play a critical role in any drug development programme aimed at identifying efficacious PCs. An understanding of the pharmacokinetic and pharmacodynamic (PK/PD) properties of the PC and the enzyme, respectively, can play a key role in identifying and optimising administration regimens that lead to a net increase in lysosomal enzyme activity and substrate reduction in disease-relevant tissues *in vivo*.[35] An ideal model would express a common human mutant form of the lysosomal enzyme under study, would accumulate substrate, and would display a pathology that closely resembles that found in humans. While few if any of the current animal models possess all of these properties, some valuable information can be obtained from those that are available.

For example, a transgenic mouse model that expresses a human recombinant R301Q missense mutant form of α-Gal A that is transcriptionally regulated by the human *GLA* promoter was shown to accumulate GL-3 in disease-relevant tissues. This model was used to demonstrate dose-dependent increases in enzyme activity as well as a reduction in substrate burden in heart, skin and kidney following daily oral administration of **40**.[78] These effects reached statistical significance in 4 weeks, with even greater reductions in GL-3 observed following 24 weeks of administration. However, more important was the observation that less frequent administration using a regimen comprised of 4 consecutive days with drug followed by 3 days without drug (*i.e.* a '4 ON/3 OFF' regimen) led to greater reductions in tissue GL-3 levels than daily administration. Importantly, this '4 ON/3 OFF' regimen reduced tissue substrate levels to those that were comparable to four, once-weekly administrations of the ERT Fabrazyme (Genzyme, Cambridge, MA). Similar results were obtained with a Pompe transgenic mouse model that expresses the human P545L GAA missense mutation following oral administration of the PC 1-deoxynojirimycin (**32b**).[79] Once again, less frequent administration regimens resulted in significantly greater glycogen reduction in disease-relevant tissues such as heart and skeletal muscle compared to daily administration.

A mouse model of GM-1 gangliosidosis that expresses a human R201C missense mutation in β-galactosidase not only accumulates GM-1 and GA-1, but also displays a neurological phenotype that is similar to that found in humans.[80] It was shown that **35** significantly decreased substrate (primarily

GM-1) levels in brain following 3–5 months of administration. This decrease was observed only in animals in which administration was initiated by 2 months of age; administration to older animals did not show the same positive effects. More importantly, a statistically significant improvement in neurological function was also observed following administration to 2 month-old mice. These studies provide further evidence that PCs have the potential to restore mutant enzyme activity in tissue to levels that are effective in reducing substrate, and that this reduction can modify the phenotypic presentation of the animal model.

6.5 Conclusions and Future Directions

Genetic diseases such as LSDs represent a tremendous burden on patients and families, and consequently represent a true unmet medical need, despite the fact that they are quite rare. While therapies developed over the last 10–15 years represent a major step forward, most individuals suffering with these diseases still have few, if any, choices for promising therapy. In addition, the cost of existing treatments is a financial burden to the overall healthcare system. To this end, research in the identification and development of effective small molecule PCs have yielded some impressive results pre-clinically, and have begun to also show promise in the clinical setting; however, much remains to be done. The discovery of PCs that potently stabilise mutant enzymes during synthesis in the ER, thereby promoting efficient translocation through the secretory pathway, with lower potency for interaction with the target enzyme in the lysosome to facilitate more efficient substrate turnover, would be a tremendous step forward. A more complete understanding of the PK/PD properties of the small molecule PCs is critical in the optimisation of both dose and administration regimen in order to maximise substrate turnover *in vivo*. The development of more efficient and higher throughput substrate reduction models, both *in vitro* and *in vivo*, would also be extremely helpful in this regard. Finally, new research programmes aimed at the identification of non-active site binding PCs that stabilise mutant enzymes and promote lysosomal trafficking may offer the potential of reduced substrate without the liability of enzyme inhibition.

References

1. M. E. Taylor and K. Drickamer, *Introduction to Glycobiology*, Oxford University Press Inc., New York, 3rd edn, 2011.
2. R. L. Lieberman, J. A. D'Aquino, D. Ringe and G. A. Petsko, *Biochemistry,* 2009, **48**, 4816.
3. R. Gravel, M. Kaback, R. L. Proia, K. Sandhoff, K. Suzuki and K. Suzuki, in *The Metabolic and Molecular Bases of Inherited Disease Online*, ed. C. Scriver, A. Beaudet, W. Sly and D. Valle, McGraw-Hill, New York, 2001, vol. 2006.

4. R. Desnick, Y. Ioannou and C. Eng, in *The Metabolic and Molecular Bases of Inherited Disease*, ed. C. Scriver, A. Beaudet, W. Sly and D. Valle, McGraw-Hill, New York, 2001, vol. 2006.
5. E. Beutler and G. Grabowski, in *The Metabolic and Molecular Bases of Inherited Disease*, ed. C. Scriver, A. Beaudet, W. Sly and D. Valle, McGraw-Hill, New York, 2001, vol. 2006.
6. R. Hirschhorn and A. J. J. Reuser, in *The Metabolic and Molecular Bases of Inherited Disease*, ed. C. Scriver, A. Beaudet, W. Sly and D. Valle, McGraw-Hill, New York, 2001, vol. 2006, p. 3389.
7. Y.-H. Xu, S. Barnes, Y. Sun and G. A. Grabowski, *J. Lipid Res.*, 2010, **51**, 1643.
8. O. Staretz-Chacham, T. C. Lang, M. E. LaMarca, D. Krasnewich and E. Sidransky, *Pediatrics*, 2009, **123**, 1191.
9. E. F. Neufeld and J. Muenzer, in *The Metabolic and Molecular Bases of Inherited Disease*, ed. C. Scriver, A. Beaudet, W. Sly and D. Valle, McGraw-Hill, New York, 2001, vol. 2006.
10. R. O. Brady, *Annu. Rev. Med.*, 2006, **57**, 283.
11. M. Rohrbach and J. T. Clarke, *Drugs*, 2007, **67**, 2697.
12. F. M. Platt and M. Jeyakumar, *Acta Paediatr.*, 2008, **97**, 88.
13. D. J. Begley, C. C. Pontikis and M. Scarpa, *Curr. Pharm. Des.*, 2008, **14**, 1566.
14. D. Wang, S. S. El-Amouri, M. Dai, C.-Y. Kuan, D. Y. Hui, R. O. Brady and D. Pan, *Proc. Natl. Acad. Sci. U. S. A.*, 2013, **110**, 2999.
15. P. S. Kishnani, P. C. Goldenberg, S. L. DeArmey, J. Heller, D. Benjamin, S. Young, D. Bali, S. A. Smith, J. S. Li, H. Mandel, D. Koeberl, A. Rosenberg and Y. T. Chen, *Mol. Genet. Metab.*, 2010, **99**, 26.
16. J. M. de Vries, N. A. M. E. van der Beek, M. A. Kroos, L. Özkan, P. A. van Doorn, S. M. Richards, C. C. C. Sung, J.-D. C. Brugma, A. A. M. Zandbergen, A. T. van der Ploeg and A. J. J. Reuser, *Mol. Genet. Metab.*, 2010, **101**, 338.
17. R. E. Boyd, G. Lee, P. Rybczynski, E. R. Benjamin, R. Khanna, B. A. Wustman and K. J. Valenzano, *J. Med. Chem.*, 2013, **56**, 2705.
18. M. Congreve, G. Chessari, D. Tisi and A. J. Woodhead, *J. Med. Chem.*, 2008, **51**, 3661.
19. M. Landon, R. Lieberman, Q. Hoang, S. Ju, J. Caaveiro, S. Orwig, D. Kozakov, R. Brenke, G.-Y. Chuang, D. Beglov, S. Vajda, G. Petsko and D. Ringe, *J. Comput.-Aided Mol. Des.*, 2009, **23**, 491.
20. M. B. Tropak, J. E. Blanchard, S. G. Withers, E. D. Brown and D. Mahuran, *Chem. Biol.*, 2007, **14**, 153.
21. J. T. R. Clarke, D. J. Mahuran, S. Sathe, E. H. Kolodny, B. A. Rigat, J. A. Raiman and M. B. Tropak, *Mol. Genet. Metab.*, 2011, **102**, 6.
22. J. J. Marugan, W. Zheng, O. Motabar, N. Southall, E. Goldin, W. Westbroek, B. K. Stubblefield, E. Sidransky, R. A. Aungst, W. A. Lea, A. Simeonov, W. Leister and C. P. Austin, *J. Med. Chem.*, 2011, **54**, 1033.
23. J. J. Marugan, W. Zheng, O. Motabar, N. Southall, E. Goldin, E. Sidransky, R. A. Aungst, K. Liu, S. K. Sadhukhan and C. P. Austin, *Eur. J. Med. Chem.*, 2010, **45**, 1880.

24. S. Patnaik and J. J. Marugan, *J. Med. Chem.*, 2012, **55**, 5734.
25. A. R. Sawkar, M. Schmitz, K. Zimmer, D. Reczek, T. Edmunds, W. E. Balch and J. Kelly, *ACS Chem. Biol.*, 2006, **1**, 235.
26. M. W. Pantoliano, E. C. Petrella, J. D. Kwasnoski, V. S. Lobanov, J. Myslik, E. Graf, T. Carver, E. Asel, B. A. Springer, P. Lane and F. R. Salemme, *J. Biomol. Screening*, 2001, **6**, 429.
27. G. H. B. Maegawa, M. B. Tropak, J. D. Buttner, B. A. Rigat, M. Fuller, D. Pandit, L. Tang, G. J. Kornhaber, Y. Hamuro, J. T. R. Clarke and D. J. Mahuran, *J. Biol. Chem.*, 2009, **284**, 23502.
28. A. Zimran, G. Altarescu and D. Elstein, *Blood Cells, Mol., Dis.*, 2013, **50**, 134.
29. J.-Q. Fan, S. Ishii, N. Asano and Y. Suzuki, *Nat. Med.*, 1999, **5**, 112.
30. E. Benjamin, J. Flanagan, A. Schilling, H. Chang, L. Agarwal, E. Katz, X. Wu, C. Pine, B. Wustman, R. Desnick, D. Lockhart and K. Valenzano, *J. Inherited Metab. Dis.*, 2009, **32**, 424.
31. M. B. Tropak and D. Mahuran, *FEBS J.*, 2007, **274**, 4951.
32. G.-N. Wang, G. Reinkensmeier, S.-W. Zhang, J. Zhou, L.-R. Zhang, L.-H. Zhang, T. D. Butters and X.-S. Ye, *J. Med. Chem.*, 2009, **52**, 3146.
33. J. J. Flanagan, B. Rossi, K. Tang, X. Wu, K. Mascioli, F. Donaudy, M. R. Tuzzi, F. Fontana, M. V. Cubellis, C. Porto, E. Benjamin, D. J. Lockhart, K. J. Valenzano, G. Andria, G. Parenti and H. V. Do, *Hum. Mutat.*, 2009, **30**, 1683.
34. G. A. Grabowski, *Genet. Test.*, 1997, **1**, 5.
35. K. J. Valenzano, R. Khanna, A. C. Powe, R. Boyd, G. Lee, J. J. Flanagan and E. R. Benjamin, *Assay Drug Dev. Technol.*, 2011, **9**, 213.
36. H. Towbin, T. Staehelin and J. Gordon, *Proc. Natl. Acad. Sci. U. S. A.*, 1979, **76**, 4350.
37. W. N. Burnette, *Anal. Biochem.*, 1981, **112**, 195.
38. S. Ishii, H. H. Chang, K. Kawasaki, K. Yasuda, H. L. Wu, S. C. Garman and J. Q. Fan, *Biochem. J.*, 2007, **406**, 285.
39. R. Khanna, E. R. Benjamin, L. Pellegrino, A. Schilling, B. A. Rigat, R. Soska, H. Nafar, B. E. Ranes, J. Feng, Y. Lun, A. C. Powe, D. J. Palling, B. A. Wustman, R. Schiffmann, D. J. Mahuran, D. J. Lockhart and K. J. Valenzano, *FEBS J.*, 2010, **277**, 1618.
40. S. H. Shin, S. Kluepfel-Stahl, A. M. Cooney, C. R. Kaneski, J. M. Quirk, R. Schiffmann, R. O. Brady and G. J. Murray, *Pharmacogenet. Genomics*, 2008, **18**, 773.
41. S.-H. Shin, G. J. Murray, S. Kluepfel-Stahl, A. M. Cooney, J. M. Quirk, R. Schiffmann, R. O. Brady and C. R. Kaneski, *Biochem. Biophys. Res. Commun.*, 2007, **359**, 168.
42. P. Lemansky, D. F. Bishop, R. J. Desnick, A. Hasilik and K. von Figura, *J. Biol. Chem.*, 1987, **262**, 2062.
43. G. H. Yam, C. Zuber and J. Roth, *FASEB J.*, 2005, **19**, 12.
44. R. A. Steet, S. Chung, B. Wustman, A. Powe, H. Do and S. A. Kornfeld, *Proc. Natl. Acad. Sci. U. S. A.*, 2006, **109**, 13813.
45. J. M. Richardson, N. A. Woychik, D. L. Ebert, R. L. Dimond and J. A. Cardelli, *J. Cell Biol.*, 1988, **107**, 2097.

46. J. M. Richardson, N. A. Woychik, D. L. Ebert, R. L. Dimond and J. A. Cardelli, *J. Cell Biol.,* 1988, **107**, 2097.
47. A. H. Erickson and G. Blobel, *Biochemistry,* 1983, **22**, 5201.
48. D. M. Leibold, C. B. Robinson, T. F. Scanlin and M. C. Glick, *J. Cell. Physiol.,* 1988, **137**, 411.
49. D. V. Quon, R. L. Proia, A. V. Fowler, J. Bleibaum and E. F. Neufeld, *J. Biol. Chem.,* 1989, **264**, 3380.
50. R. J. Moreland, X. Jin, X. K. Zhang, R. W. Decker, K. L. Albee, K. L. Lee, R. D. Cauthron, K. Brewer, T. Edmunds and W. M. Canfield, *J. Biol. Chem.,* 2005, **280**, 6780.
51. H. A. Wisselaar, M. A. Kroos, M. M. Hermans, J. van Beeumen and A. J. Reuser, *J. Biol. Chem.,* 1993, **268**, 2223.
52. A. Hasilik and E. F. Neufeld, *J. Biol. Chem.,* 1980, **255**, 4946.
53. T. Okumiya, M. A. Kroos, L. V. Vliet, H. Takeuchi, A. T. Van der Ploeg and A. J. Reuser, *Mol. Genet. Metab.,* 2007, **90**, 49.
54. G. Parenti, A. Zuppaldi, M. Gabriela Pittis, M. Rosaria Tuzzi, I. Annunziata, G. Meroni, C. Porto, F. Donaudy, B. Rossi, M. Rossi, M. Filocamo, A. Donati, B. Bembi, A. Ballabio and G. Andria, *Mol. Ther.,* 2007, **15**, 508.
55. E. F. Neufeld, *Annu. Rev. Biochem.,* 1991, **60**, 257.
56. S. Kornfeld, *J. Clin. Invest.,* 1986, 77, 1.
57. S. Kornfeld, *FASEB J.,* 1987, **1**, 462.
58. R. Kornfeld and S. Kornfeld, *Annu. Rev. Biochem.,* 1985, **54**, 631.
59. T. Merry, *Acta Biochim. Pol.,* 1999, **46**, 303.
60. M. Fukuda, *J. Biol. Chem.,* 1991, **266**, 21327.
61. R. L. Lieberman, B. A. Wustman, P. Huertas, A. C. Powe, Jr, C. W. Pine, R. Khanna, M. G. Schlossmacher, D. Ringe and G. A. Petsko, *Nat. Chem. Biol.,* 2007, **3**, 101.
62. G. H. B. Maegawa, M. Tropak, J. Buttner, T. Stockley, F. Kok, J. T. R. Clarke and D. J. Mahuran, *J. Biol. Chem.,* 2007, **282**, 9150.
63. M. B. Tropak, G. J. Kornhaber, B. A. Rigat, G. H. Maegawa, J. D. Buttner, J. E. Blanchard, C. Murphy, S. J. Tuske, S. J. Coales, Y. Hamuro, E. D. Brown and D. J. Mahuran, *Chem Bio Chem,* 2008, **9**, 2650.
64. B. Rigat and D. Mahuran, *Mol. Genet. Metab.,* 2009, **96**, 225.
65. V. Agmon, S. Cherbu, A. Dagan, M. Grace, G. A. Grabowski and S. Gatt, *Biochim. Biophys. Acta, Lipids Lipid Metab.,* 1993, **1170**, 72.
66. E. Kohen, C. Kohen, J. G. Hirschberg, R. Santus, G. Grabowski, W. Mangel, S. Gatt and J. Prince, *Cell Biochem. Funct.,* 1993, **11**, 167.
67. M. Lorincz, L. A. Herzenberg, Z. Diwu, J. A. Barranger and W. G. Kerr, *Blood,* 1997, **89**, 3412.
68. L. Madar-Shapiro, M. Pasmanik-Chor, T. Dinur, A. Dagan, S. Gatt and M. Horowitz, *J. Inherited Metab. Dis.,* 1999, **22**, 623.
69. B. Rudensky, E. Paz, G. Altarescu, D. Raveh, D. Elstein and A. Zimran, *Blood Cells, Mol., Dis.,* 2003, **30**, 97.
70. N. Sasagasako, T. Kobayashi, Y. Yamaguchi, N. Shinnoh and I. Goto, *J. Biochem.,* 1994, **115**, 113.

71. H. H. van Es, M. Veldwijk, M. Havenga and D. Valerio, *Anal. Biochem.*, 1997, **247**, 268.
72. M. A. Hölzl, M. Gärtner, J. J. Kovarik, J. Hofer, H. Bernheimer, G. Sunder-Plassmann and G. J. Zlabinger, *Clin. Chim. Acta*, 2010, **411**, 1666.
73. C. R. Kaneski, R. Schiffmann, R. O. Brady and G. J. Murray, *J. Lipid Res.*, 2010, **51**, 2808.
74. M. B. Tropak, S. W. Bukovac, B. A. Rigat, S. Yonekawa, W. Wakarchuk and D. J. Mahuran, *Glycobiology*, 2010, **20**, 356.
75. C. R. Kaneski, S. A. French, M. R. Brescia, M. J. Harbour and S. P. Miller, *J. Lipid Res.*, 1994, **35**, 1441.
76. S. Marchesini, L. Demasi, P. Cestone, A. Preti, V. Agmon, A. Dagan, R. Navon and S. Gatt, *Chem. Phys. Lipids*, 1994, **72**, 143.
77. S. Arttamangkul, M. K. Bhalgat, R. P. Haugland, Z. Diwu, J. Liu, D. H. Klaubert and R. P. Haugland, *Anal. Biochem.*, 1999, **269**, 410.
78. R. Khanna, R. Soska, Y. Lun, J. Feng, M. Frascella, B. Young, N. Brignol, L. Pellegrino, S. A. Sitaraman, R. J. Desnick, E. R. Benjamin, D. J. Lockhart and K. J. Valenzano, *Mol. Ther.*, 2010, **18**, 23.
79. R. Khanna, J. Flanagan, J. Feng, M. Frascella, R. Soska, Y. Lun, B. Ranes, D. Guillen, D. Lockhart and K. Valenzano, Manuscript provisionally accepted in *PLOS ONE*.
80. J. Matsuda, O. Suzuki, A. Oshima, Y. Yamamoto, A. Noguchi, K. Takimoto, M. Itoh, Y. Matsuzaki, Y. Yasuda, S. Ogawa, Y. Sakata, E. Nanba, K. Higaki, Y. Ogawa, L. Tominaga, K. Ohno, H. Iwasaki, H. Watanabe, R. O. Brady and Y. Suzuki, *Proc. Natl. Acad. Sci. U. S. A.*, 2003, **100**, 15912.

CHAPTER 7

Discovery and Clinical Development of Idursulfase (Elaprase®) for the Treatment of Mucopolysaccharidosis II (Hunter Syndrome)†

MICHAEL HEARTLEIN* AND ALAN KIMURA

Shire, 300 Shire Way, Lexington, MA 02421, USA
*E-mail: mheartlein@shire.com

7.1 Introduction

Heparan sulfate (HS) and dermatan sulfate (DS) are glycosaminoglycans (GAGs) comprised of linear sulfated chains of alternating uronic acid and hexosamine residues that are modified enzymatically to produce a complex pattern of sulfation linkages (*e.g.* N-S, 2-S, 6-S, 2-O, 3-O). The specific pattern and degree of sulfation on GAGs is essential for distribution and presentation of growth factors, chemokines and immune modulators, and also instrumental in roles as diverse as the viscoelastic properties of bronchial secretions and mucoadhesive properties of bacteriophage and other viruses.[1–3] HS-GAG is

†This chapter is dedicated to the memory of Professor Ed Wraith. It is a historical account of the discovery and clinical development of Elaprase. Up-to-date safety and efficacy information on Elaprase can be found in the current summary of product characteristics and prescribing information for Elaprase.

RSC Drug Discovery Series No. 38
Orphan Drugs and Rare Diseases
Edited by David C Pryde and Michael J Palmer

Published by the Royal Society of Chemistry, www.rsc.org

degraded by a series of exo-enzymes whose activity is sharply restricted to the acidic environment of the lysosome. Importantly, these enzymes are only able to hydrolyse linkages at the non-reducing end of HS and must therefore act sequentially.[4] Loss of activity due to mutations in a single exo-enzyme results in one of several disorders known collectively as the mucopolysaccharidoses (MPS). There are seven well described and characterised MPS diseases with a possible eighth, MPS IIIE, recently identified as a deficiency of arylsulfatase G.[5] The first step in the degradation of HS occurs with iduronate-2-sulfatase (IDS, EC 3.1.6.13).[4] When iduronate-2-sulfatase is missing or deficient, it causes MPS II, also known as Hunter syndrome. The specific enzyme deficiency was first described by Bach and co-workers.[6]

MPS II is a rare X-linked recessive disorder with an incidence of 1 in 100 000 to 160 000 live births.[7–9] It occurs predominantly in males, but some females with MPS II have been reported. The clinical disease is secondary to the iduronate-2-sulfatase deficiency and is a result of the chronic and progressive accumulation of excessive amounts of partly digested HS and DS GAG fragments in the lysosomes of nearly all cell types, tissues and organs of the body. GAG deposition in patients leads to developmental delay; upper airway obstruction, limited lung capacity and sleep apnoea; organomegaly (liver, spleen and heart) and valvular heart disease; joint stiffness and decreased range of motion; and short stature and severe skeletal deformities.[10] There is a wide spectrum of clinical severity among MPS II patients. In the most severe cases, central nervous system involvement leads to mental retardation and progressive neurologic decline. The clinical manifestations of MPS II generally lead to death in the first or second decade of life. In the less severe form of MPS II (attenuated form), death may occur in early adulthood, but some patients have survived into the fifth and sixth decades of life.[10]

Beginning in the 1970s various attempts to purify iduronate-2-sulfatase from human tissue were successful in identifying core properties of the enzyme, but it was not until 1990 that the laboratory of John Hopwood purified the enzyme to homogeneity.[11] The Hopwood group was then able to clone the iduronate-2-sulfatase gene,[12] which eventually allowed for the expression of recombinant iduronate-2-sulfatase (idursulfase) and the development of an enzyme replacement therapy (ERT) for MPS II by Shire. The name for the final drug product formulation of idursulfase for human use is Elaprase® (Shire Human Genetic Therapies, Inc., Lexington, MA).

7.2 Production and Non-Clinical Testing of Idursulfase

7.2.1 Production of Recombinant Iduronate-2-Sulfatase (Idursulfase) in Human Cells

When Shire (at that time Transkaryotic Therapies, Inc.) endeavoured to establish a new platform for protein production in the early 1990s there were several key factors that influenced the choice of cells for manufacturing human therapeutic proteins. These factors included the ability of the cells to grow continuously under manufacturing conditions, viral sequence profile and capacity for a full range of human post-translational modifications. The industry standard for the production of therapeutic proteins in the early 1990s was Chinese hamster ovary (CHO) cells. Derived in the late 1950s, the CHO cells used today were originally established from ovary tissue isolated from a Chinese hamster.[13] Twenty years later, the cells were mutagenised with gamma radiation and chemicals to inactivate the dihydrofolate reductase (DHFR) gene.[14] This enabled methotrexate stepwise selection for co-amplification of recombinant transgenes encoding therapeutic proteins. Over the years, the accumulated passaging and mutagenesis caused additional genetic changes that led to an immortal phenotype with biochemical deficiencies (*e.g.* proline auxotrophy). It was recognised after the initial isolation of CHO cells that the cell line harboured virus-like particles.[15] Although widely considered benign, these reverse transcriptase-active particles nevertheless required specific procedures to ensure their removal during product recovery.[16] A new host cell for protein production was selected by Shire because of the desire to start with a virus-free cell line, and concerns about the immunogenic potential of rodent glycoforms and the absence of beneficial human glycoforms. As an alternative to CHO cells, development of a new protein production platform was focused on protein production from human cell lines, for example HT-1080 cells derived from a fibrosarcoma of transformed fibroblast cells.[17] The basis of the transformation event in HT-1080 cells was reasonably well understood and was shown to be caused, at least in part, by a single point mutation that activated the *NRAS* gene.[18] Recently transformed cells also lacked the sequential manipulations performed in multiple laboratories over several decades, which were required to establish DHFR-deficient CHO cell lines. For example, some such cell lines were shown to be free of detectable viruses or viral sequences, and were capable of a full range of human post-translational modifications.

Protein manufacturing in human cell lines, as compared to non-human cell lines, can result in the production of proteins with human glycosylation.[19] These carbohydrate signatures are an inherent property of secreted proteins that are often species-specific and that can influence the biology of a protein. For example, carbohydrate chains on proteins are capable of binding to specific receptors present on cell surfaces where they can initiate a signal transduction cascade often similar to those observed with growth factors.[20] They can also bind carbohydrate-specific receptors (*e.g.* the mannose-6-phosphate [M6P] receptor) to trigger uptake and trafficking in the cell, which is a key property enabling ERT. Specific glycoforms can also modulate the immunogenic profile of proteins.[21] These and the other specific post-translational modifications that serve as intrinsic features of therapeutic proteins were the primary motivations for developing a fully human protein production system. This human cell platform has been used to successfully develop epoetin delta,[19] agalsidase alfa,[22] velaglucerase alfa[23] and idursulfase.

7.2.2 Cell Line Development and Manufacturing of Idursulfase (Elaprase)

Idursulfase produced by recombinant DNA technology in a human cell line provides a human glycosylation profile, which is analogous to the naturally occurring enzyme. The enzyme activity of idursulfase is also the same as the endogenous enzyme, such that it hydrolyses the 2-sulfate esters of terminal iduronate sulfate residues from DS and HS in the lysosomes of various cell types. To generate a cell line suitable for the production of idursulfase, an expression construct was introduced into a continuous human cell line. This was achieved by insertion of a plasmid-based cDNA sequence, encoding human iduronate-2-sulfatase along with regulatory sequences and a marker gene for drug selection, into the genomic DNA of the human cell. The insertion occurs as a result of integration of the transfected DNA fragment into the genome and results in the expression of active idursulfase. Cells secreting high levels of idursulfase were then selected. A Master and Working Cell Bank was manufactured and subjected to comprehensive viral safety testing. To support clinical testing and commercial manufacturing, a conventional chromatographic purification process was developed that includes a virus removal filter and the entire manufacturing process was scaled appropriately. The final drug product, Elaprase, is a formulation of idursulfase intended for intravenous (IV) infusion and is supplied as a sterile, non-pyrogenic clear to slightly opalescent colourless solution that must be diluted prior to administration in 0.9% sodium chloride injection, USP. Each vial contains an extractable volume of 3.0 mL with an idursulfase concentration of 2.0 mg mL^{-1} at a pH of approximately 6, providing 6.0 mg idursulfase, 24.0 mg sodium chloride, 6.75 mg sodium phosphate monobasic monohydrate, 2.97 mg sodium phosphate dibasic heptahydrate and 0.66 mg polysorbate 20.

7.2.3 Recombinant Idursulfase Properties and Mechanism of Action

Idursulfase is expressed as a single polypeptide chain of 550 amino acids which, after removal of a signal peptide, is secreted as a 525 amino acid glycoprotein with a molecular weight of approximately 76 kilodaltons (Figure 7.1). The enzyme contains eight N-linked glycosylation sites that are occupied by oligosaccharide chains, including complex, hybrid and high-mannose type oligosaccharide chains. The enzyme activity of idursulfase is dependent on the post-translational modification of a specific cysteine (C59) to formylglycine.[24] Idursulfase has a specific activity in excess of 40 U mg^{-1} of protein (one unit is defined as the amount of enzyme required to hydrolyse 1 μmole of heparin disaccharide substrate per hour under the specified assay conditions).

In MPS II, HS and DS GAGs progressively accumulate within cellular lysosomes of many cell types in the body, leading to cellular engorgement,

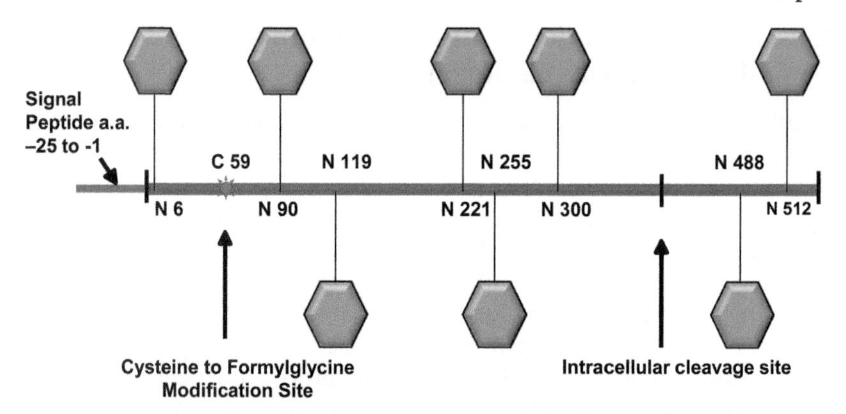

Figure 7.1 Structure of secreted idursulfase is presented schematically. Occupied N-linked glycosylation sites are depicted with hexagons. The post-translational modification of the active site cysteine-59 to formylglycine-59 is also marked. The amino acid sequence of idursulfase, predicted from the cDNA sequence, has been confirmed by peptide mass mapping and N-terminal sequence analysis.

organomegaly, tissue destruction and organ system dysfunction. Due to the intracellular compartmentalisation of lysosomes, the pharmacodynamic activity of idursulfase is dependent upon internalisation of the enzyme into the lysosomes. The binding of idursulfase *via* M6P moieties to cell surface M6P receptors provides a receptor-mediated uptake mechanism for this enzyme. M6P residues on the oligosaccharide chains of idursulfase allow specific binding of the enzyme to the M6P receptors on the cell surface, leading to cellular internalisation of the enzyme and subsequent targeting to intracellular lysosomes. Sialylation of the complex and hybrid oligosaccharide chains on idursulfase reduces uptake by hepatic asialoglycoprotein receptors and thereby prevents rapid elimination of idursulfase from the body.[25] Treatment of MPS II patients with Elaprase therefore provides exogenous enzyme for receptor-dependent cellular internalisation of the enzyme, targeting to intracellular lysosomes and subsequent catabolism of accumulated GAGs.

7.2.4 Non-Clinical Studies

An animal model of MPS II, the iduronate-2-sulfatase knockout (IKO) mouse, exhibits many of the physical characteristics of Hunter syndrome seen in humans, including coarse features, skeletal defects (including thickened digits), hepatomegaly and a reduced lifespan. Increased GAG levels are observed in urine and tissues throughout the body and widespread cellular vacuolisation is observed histopathologically.[26] The IKO model was used to evaluate the dose levels and dose regimen of idursulfase required to degrade stored GAGs in this animal model. The IV route of administration was generally used in animal studies to conform to the intended clinical route of administration of idursulfase.

A series of pharmacodynamic studies was conducted in which idursulfase was administered IV at weekly intervals to IKO mice.[27] The doses of idursulfase ranged from 0.1 to 5.0 mg kg^{-1}. These studies established that idursulfase caused a reduction in urinary and tissue (liver, spleen, kidney and heart) GAGs, indicating that idursulfase was active, reached target organs and was probably taken up into the lysosomes where catabolism of excess GAGs occurs. From these studies it was determined that doses as low as 0.1 mg kg^{-1} resulted in a measurable pharmacodynamic effect.

Another set of pharmacodynamic studies was performed to establish dose frequency. IKO mice were administered weekly, every other week or monthly IV injections of idursulfase. Clear reductions in urinary and tissue GAGs were observed after dosing regimens ranging from 8 weeks to approximately 6 months.[27] Long-term administration (12 and 24 weeks) of 1 mg kg^{-1} idursulfase administered weekly and every other week were both effective in reducing tissue GAG concentrations in various tissues, and were more effective than monthly dosing.

Studies to determine the pharmacokinetic properties of idursulfase in animal models (mice, rats and monkeys) have shown that idursulfase has a biphasic serum elimination profile with mean elimination half-lives of about 5–6 hours.[27] C_{max} values were proportional to dose for all species, however AUC values were not linearly proportional to dose. Based on pharmacokinetic studies in monkeys, it is likely that serum clearance mechanisms (*e.g.* cell surface M6P receptors, *etc.*) became saturated at doses of 0.5 mg kg^{-1} or higher.

Idursulfase was detected in all organs and tissues examined in ^{125}I-radiolabelled mouse[27] and rat biodistribution studies. Tissue half-lives were similar for the major organs and were approximately 1–2 days for liver, kidney, heart, spleen and bone (including marrow). The accumulation and retention of idursulfase in these organs and tissues was consistent with the distribution of M6P receptors in tissues and organs in mammals.

Non-clinical data revealed no special hazard for humans based on a conventional 6 month, repeat-dose toxicity study in cynomolgus monkeys, in which a no adverse effect level of at least 12.5 mg kg^{-1} idursulfase was established. Single-dose acute toxicology studies were also performed in rats and cynomolgus monkeys, establishing a no adverse effect level of at least 20 mg kg^{-1} for both species. A male fertility study was performed in rats, which revealed no evidence of impaired male fertility.

Overall, the development programme for idursulfase progressed from proof-of-principle pharmacodynamic studies, to dose and dose-frequency studies, to an analysis of relative tissue biodistribution of enzyme, and finally to pharmacokinetic and toxicological assessments. Information from these studies was supportive of the selected idursulfase therapeutic doses and regimens used in human clinical trials. Information gained from early non-clinical and Phase I/II clinical studies was also used to select the range of doses used in later toxicology studies.

7.3 Clinical Development of Elaprase

7.3.1 Background: Enzyme Replacement Therapies for Lysosomal Storage Diseases

As early as the 1960s, it was speculated that exogenous replacement of the deficient enzyme would provide therapeutic benefit to patients with lysosomal storage diseases (LSDs).[28–30] The identification of the specific enzyme deficiency for a number of LSDs,[29,31,32] together with recognition of the importance of glycosylation in targeting the enzyme to the appropriate cellular lysosomes,[33–36] led to the development of the first effective ERTs for LSDs.

In 1991, Ceredase® (Genzyme, Cambridge, MA), a placenta-derived β-glucocerebrosidase enzyme developed for type 1 Gaucher disease, became the first ERT to be approved for the treatment of an LSD. The licensure was based on an open-label trial conducted by Barton and colleagues,[37] in which the efficacy of Ceredase was demonstrated in 12 patients with type 1 Gaucher disease who received IV infusions of enzyme every 2 weeks for 9–12 months. Treatment with Ceredase was associated with clinical improvement, as evidenced by increases in haemoglobin concentration and platelet count, as well as reductions in hepatic and splenic volume.

Cerezyme® (Genzyme), a recombinant form of human β-glucocerebrosidase expressed in CHO cells, was subsequently developed and approved as a treatment for type 1 Gaucher disease in 1994. The approval of Cerezyme was based on a randomised double-blind trial comparing 15 patients treated with Cerezyme with 15 patients receiving Ceredase over a period of 6–9 months.[38]

The next generation of ERTs for LSDs focused on the development of recombinant human enzyme preparations of α-galactosidase A for the treatment of Fabry disease. Two recombinant forms of α-galactosidase A were developed for large-scale production: one was expressed in a CHO cell line (Fabrazyme®, Genzyme) and the other in a human cell line (Replagal®, Shire). Replagal and Fabrazyme were approved in Europe in 2001, with Fabrazyme subsequently receiving approval in the USA in 2003.

The pivotal trial for Fabrazyme licensure was conducted as a randomised double-blind study consisting of 58 Fabry patients randomised at a 1 : 1 ratio to receive IV infusions of either Fabrazyme or placebo every other week for 20 weeks.[39] The efficacy of Fabrazyme was demonstrated using a surrogate end point, which was clearance of microvascular endothelial deposits of globotriaosylceramide (Gb_3) from renal biopsy specimens (20 of 29 patients receiving Fabrazyme showed clearance of deposits after 20 weeks of treatment compared with none of the 29 placebo patients).

Two studies were conducted to demonstrate the efficacy of Replagal and to support its licensure.[40,41] These studies were conducted as randomised, double-blind, placebo-controlled trials consisting of 26 and 15 Fabry patients randomised equally to receive IV infusions of either Replagal or placebo every other week for 6 months. In contrast to the Fabrazyme pivotal study, the goal of these two studies was to demonstrate the efficacy of Replagal based on clinically important end points. Replagal treatment was associated with a reduction in neuropathic pain scores, improvement in renal pathology, increases in creatinine clearance, reductions in left ventricular mass, and reductions in plasma and cardiac Gb_3 levels.

The success of ERTs for Gaucher and Fabry diseases was soon matched by the development of an ERT for MPS I. In an initial Phase I open-label study, 10 MPS I patients were treated with weekly IV infusions of recombinant human α-L-iduronidase (Aldurazyme®, Genzyme) for 52 weeks.[42] The safety and efficacy of Aldurazyme was subsequently demonstrated in a randomised, double-blind, placebo-controlled trial in which 45 patients were administered either IV Aldurazyme or placebo weekly for 26 weeks.[43] The co-primary end points were changes in pulmonary function, as measured by forced vital capacity (FVC; per cent predicted of normal), and changes in distance walked in the 6 minute walk test (6MWT). Compared with placebo patients, patients treated with Aldurazyme showed improvements of 5.6 percentage points in per cent of predicted normal FVC ($p = 0.009$) and 38.1 metres walked in the 6MWT ($p = 0.066$). Treatment with Aldurazyme was also associated with significant reductions in liver volume and urinary GAG excretion, and improvements in sleep apnoea and shoulder flexion in a subset of more severely affected patients. The MPS I studies were not designed to evaluate the effect of Aldurazyme on neurological function, as most of the study patients did not have severe neurological impairment.

The development pathway for these first ERTs set many precedents for LSDs, as well as for the rare genetic disease field in general. It was in this setting that the clinical development programme for Elaprase was designed and executed by Shire. Consistent with many of its predecessors, the Elaprase clinical programme consisted of a single Phase I/II study, followed by a larger, pivotal Phase II/III licensure trial. The design and results of these studies will be described in Section 7.3.3.

7.3.2 Considerations in the Clinical Development of Therapies for Rare Genetic Diseases

As mentioned above, the first ERTs set many precedents as well as expectations for the clinical development of therapies for rare genetic diseases. Many of these were felt to be advantageous, the most significant of which were (i) abbreviated clinical programmes consisting of small numbers of patients participating in only one or two Phase I/II studies and one pivotal Phase III trial and (ii) the use of a surrogate end point (reduction of Gb_3 in renal biopsy specimens) that was more favourable for trials with small sample sizes. The shortened clinical development programmes and the potential use of surrogate end points could both be justified based on the rarity of LSDs and small patient numbers available for clinical studies. With a limited number of small trials to execute for licensure, it was also anticipated that the development time and cost to the companies would be significantly reduced, and delivery of the much needed therapy to the patients would be more rapid.

As discussed by other authors,[44–47] conducting studies with small sample sizes presents many challenges to the successful development of a new therapy. The challenges are even greater for LSDs, considering that the patient populations are highly heterogeneous with respect to disease manifestations and severity. Even though small clinical trials can be warranted for rare diseases, the early Phase I and II studies must still be of sufficient size, design and robustness to make the following critical decisions for Phase III development.

- The appropriate dose and schedule to be evaluated in Phase III.
- An optimal efficacy end point that is feasible, clinically meaningful for the patient population, and responsive to treatment.
- The patient population or subgroup of patients to target in Phase III.
- Duration of treatment to show clinical benefit.

Identification of an optimal efficacy end point that is feasible, clinically meaningful for the patient population and responsive to treatment is a significant undertaking for any clinical development programme, let alone one that is based on limited Phase I/II data in a small study population. Understanding how the clinical end point behaves over time in the population (*e.g.* renal function decline in Fabry patients) and its responsiveness to therapy are keys for the design of the Phase III study. A non-interventional study investigating the natural history of the clinical end point would be extremely helpful in this regard.

For LSDs, the relationship between responsiveness to ERT and the severity of the disease manifestation should be explored and considered in selecting the appropriate study population. With Aldurazyme treatment, improvements in sleep apnoea and shoulder flexion were observed in the more severely affected MPS I patients.[43] One must be cognisant, however, that the more severely affected patients with longstanding disease may have organ and tissue damage that is irreversible and therefore non-responsive to therapy.[43,48]

As stated above, the design of the Phase III study must rely on the results of the Phase I/II studies, specifically for determination of the dose and schedule of the treatment, and the optimal efficacy end points and patient population to be investigated. Even though small clinical trials can be justified for rare diseases, regulatory agencies still require convincing evidence of the safety and efficacy for licensure of a new medicinal product, which dictates a Phase III study designed with adequate statistical power to demonstrate clinical benefit.

The selection of an appropriate Phase III trial design is clearly critical for a successful outcome; the advantages and limitations of different trial designs for small clinical studies have been reviewed in several recent publications[46,47] and will not be discussed here. Some other considerations in the design of the Phase III study, however, are worth mentioning. If the optimal dose and schedule have not been completely defined in Phase I/II studies, the evaluation of different doses or schedules can still be performed in Phase III. This is an important approach if there is any uncertainty about dosing of the new study drug. If a clinically meaningful efficacy end point is not feasible or cannot be adequately powered, a surrogate end point can be considered. This end point would have to be justified as either predicting or reliably predicting clinical benefit. However, unless the surrogate is well established and understood, interpretation of the results and its clinical benefit may be difficult and could put the development programme at risk.

In summary, there are many challenges in the clinical development of therapies for rare genetic diseases. Rapid progression to a Phase III study following only one or two small Phase I/II clinical trials, coupled with rare diseases represented by very heterogeneous patient populations, makes it very challenging to design meaningful Phase III trials with high probabilities of success. One must identify the best ways to optimise the trials, not only in their design and statistical power, but also from a trial execution standpoint. Natural history studies could also help identify the optimal patient populations to target.

7.3.3 Clinical Development of Elaprase for the Treatment of MPS II

At the time, the successful development of ERTs for Gaucher, Fabry and MPS I diseases defined the therapeutic approach to the LSD field. This set the stage for the development of Elaprase as an ERT for MPS II. The clinical development of Elaprase began with the initiation of a Phase I/II study in 2001 and culminated in the completion of a Phase II/III pivotal trial in 2005. Similar to the MPS I studies, these studies focused on improvements of the systemic and not neurological manifestations of MPS II disease, as it was assumed that IV-administered Elaprase would not efficiently cross the blood–brain barrier.

7.3.3.1 Phase I/II Study

Elaprase was administered to MPS II patients for the first time in a Phase I/II trial conducted by Muenzer *et al.*[49] The goals of this first human study were (i) to investigate the safety of every other week IV infusions of certain doses of Elaprase; (ii) to explore the pharmacodynamics of Elaprase; and (iii) to explore the efficacy of Elaprase using a variety of clinical measures relevant to the disease. As it was the only planned Phase I/II study, combined with the anticipation that the results could be used to support efficacy in the future, the trial had a broad scope and was designed with rigour as a randomised, double-blind, dose-finding, placebo-controlled study. The duration of treatment was 24 weeks and an open-label extension was performed beyond this point. In retrospect, the small number of patients in the trial, combined with their marked disease heterogeneity, made interpretation of the results very challenging.

Twelve MPS II patients were enrolled sequentially as three groups of four patients; within each group, three patients were randomised in double-blind fashion to receive Elaprase and one patient was randomised to receive placebo. Elaprase was administered IV every other week at three dose levels: the first group received 0.15 mg kg^{-1} body weight, followed by escalation to 0.5 and 1.5 mg kg^{-1} body weight for the second and third groups of patients enrolled, respectively. As described earlier, the selection of the doses and the every other week regimen were based on the non-clinical data. The dose levels of Elaprase represented a 10-fold dose range, which was felt to be sufficiently broad for the testing of a protein therapeutic. After 24 weeks of the double-blind phase, all patients elected to continue in the open-label extension of the study; patients randomised to Elaprase remained on the dose of their treatment group, while patients randomised to placebo crossed over and were also given the dose of their treatment group. The analyses consisted of 48 weeks of treatment with Elaprase for all patients; for the placebo patients, this represented 72 weeks of participation in the trial, 24 weeks of placebo and 48 weeks of open-label Elaprase treatment.

As this was the first exposure of patients to Elaprase, close monitoring of safety was incorporated into the design and conduct of the study. The study started with the lowest Elaprase dose, initiating treatment in a single patient each week; progression to the next dose level was allowed only when all patients at the lower dose had been administered three infusions of study drug and were monitored for at least 7 days after the third dose.

7.3.3.1.1 Results. The age of patients ranged from 6 to 20 years. Baseline pulmonary function FVC ranged from severely compromised (15% of predicted) to normal (86% of predicted). Patients also had a very wide range of distances walked in 6 minutes. Each treatment group varied with respect to age and disease severity. The oldest patients (20 years of age) were in the 0.5 mg kg^{-1} mid-dose group; these patients had the most severe respiratory compromise (two of three patients had tracheostomies) and the lowest mean distance walked in the 6MWT at baseline.

All patients had abnormal levels of urine GAGs at baseline, which decreased markedly following initiation of Elaprase treatment. All dose groups experienced significant reductions in urine GAGs during the double-blind phase lasting 24 weeks, which was maintained through 48 weeks of the study. The majority of patients achieved near normal levels of urine GAGs. Based on mean urine GAG levels, as well as the percentage reduction from baseline, there was a trend for patients on the 1.5 mg kg^{-1} dose to have the highest treatment effect.

After 24 and 48 weeks of Elaprase infusions, liver and spleen volumes were significantly reduced in the overall treated population. The 0.5 mg kg^{-1} and

1.5 mg kg^{-1} dose levels resulted in earlier and higher reductions in liver volume, while all three dose levels appeared equally effective in decreasing spleen volume. Normalisation of liver volumes occurred in six of nine patients (67%) with hepatomegaly at baseline. All seven patients with splenomegaly at baseline had normal spleen volumes following 48 weeks of Elaprase treatment. These results, combined with the reduction in urine GAG levels, showed that IV-infused Elaprase was biologically active and could degrade accumulated GAGs in tissues and organs.

There was also evidence of clinical benefit associated with Elaprase treatment. After 48 weeks of treatment, patients in the mid- and high-dose groups had increases in walking distance of 17.7% (45 metres) and 27.9% (114 metres), respectively. Pooled results across the three dose groups at 48 weeks showed an increase in walking distance of 14.2% (47 metres) compared to baseline. Pulmonary function testing showed that 10 of the 12 patients (83%) had restrictive ventilatory disorder at baseline, with FVC measurements of less than 80% of predicted. After 48 weeks of Elaprase treatment, pooled results across the three dose levels showed a small increase in absolute FVC (0.071 L), with the high-dose group accounting for the greatest increases from baseline (0.173 L). Following 48 weeks of treatment, there also appeared to be a reduction in left ventricular mass across all three dose levels. Treatment effects for other cardiac parameters, however, were not observed. Finally, the study results also suggested improvements in some patients with sleep apnoea as well as certain joint range of motion measurements (*e.g.* wrist extension), but significant improvements overall were difficult to show because of the broad disease heterogeneity of the patients and small sample size.

Intravenous infusions of Elaprase were generally well tolerated. Infusion reactions occurred in patients receiving the mid- and high-dose levels; all patients were able to continue treatment by slowing the infusion rate (infusion time was extended from 1 to 3 hours) and by pre-medication with antihistamine and corticosteroids. No infusion reactions were associated with elevations of tryptase or complement activation.

Some patients at the higher dose levels developed IgG antibodies to Elaprase after exposure to three to six infusions. The induction of these antibodies did not appear to have an impact on either the biological or clinical activity of Elaprase.

7.3.3.1.2 Conclusions. This study was the first use of Elaprase in patients with MPS II. The study examined every other week infusions of three different dose levels of Elaprase in the blinded phase and all patients continued in the open-label extension. Infusion reactions were successfully managed by the combination of slowing the infusion rate and pre-medication. Reductions in urine GAGs, and liver and spleen size, showed that Elaprase was active and efficiently targeted to the cellular lysosomes of tissues where it could degrade stored GAGs. Demonstration of clear clinical benefit was more difficult, however, owing to the small sample size of the study and the heterogeneity of the disease of the patients, the latter resulting in variable treatment responses. Nonetheless, there was evidence of clinical benefit as many patients showed improvements in walking distance, pulmonary function and sleep apnoea, as well as a reduction in left ventricular mass.

Relevant to the design of the Phase III study, the results of the Phase I/II study indicated that (i) Elaprase was most effective when administered at the higher dose levels; (ii) consistent with the Aldurazyme experience in MPS I patients, walking distance in the 6MWT and pulmonary function were the end points with the highest likelihood of demonstrating clinical benefit in a trial of 6–12 months duration; and (iii) baseline disease severity affected patient responsiveness to therapy and, as such, imbalances in disease severity across treatment groups could decrease the power of the study, as well as complicate interpretation of the results.

7.3.3.2 Phase II/III Study

The goal was to design a Phase II/III study with the highest probability of success in demonstrating the safety, efficacy and clinical benefit of Elaprase. As discussed previously, this was a significant challenge due to the relatively small sample size of the study, disease heterogeneity of the MPS II patient population, and also the variability encountered in clinical testing. Moreover, regulatory approval would be based on the results of a single pivotal trial, requiring the trial to be conducted robustly and to provide firm evidence of safety and efficacy. The recent clinical trial experience with Aldurazyme in MPS I patients provided much insight into the design of the Phase II/III study for Elaprase, the main elements of which are discussed below.

- The 0.5 mg kg^{-1} dose administered every other week was initially selected for the Phase III study. The biodistribution studies in mice and rats, however, showed Elaprase to have a tissue half-life of 1–2 days, indicating that it would be eliminated from the tissues by the second week after the infusion. As a result, it was decided to investigate the 0.5 mg kg^{-1} dose administered weekly, in addition to the every other week schedule. The weekly administration would test the importance of having active enzyme continuously present in the tissues. The demonstrated efficacy of weekly administered Aldurazyme also supported this decision.[43] Because two schedules would be explored, the Elaprase trial was designated as a Phase II/III study.
- With two primary efficacy end points identified (walking distance in the 6MWT and FVC), in order to ensure adequate power it was decided that the primary efficacy outcome would be a composite score of the two end points and not co-primary end points. Other end points, including sleep apnoea, and liver and spleen size, were considered for the primary composite score but were eventually not used. It was felt that sleep studies would be highly variable between study sites and the clinical benefit of reducing liver and spleen size could not be justified, as there

was no clear injury to these organs due to GAG storage. Measurement of joint range of motion was also highly variable and responsiveness to therapy was difficult to show. Liver and spleen size and joint range of motion were, however, included as secondary outcomes; sleep studies were not performed during the study.

The two-component composite end point was clinically justified as it captured the effect of Elaprase treatment on respiratory function and physical functional capacity as measured by walking ability. As such, the clinical benefit of Elaprase treatment was determined across multiple organ systems adversely affected in MPS II, namely the lung, heart and joints.

The primary statistical analysis of the composite end point was performed by the global non-parametric rank-sum test as described by O'Brien.[50] In this procedure, every patient was ranked on the two outcomes, the ranks were summed, and the sums were compared between the treatment groups. The primary comparison of the composite variable was between the weekly Elaprase-treated group and the placebo group.

A sample size calculation was difficult to perform for the study due to the composite nature of the primary efficacy end point. The proposed sample size of 90 patients represented as large a number of patients as was feasible for the study. Based on this sample size, coupled with the composite score and its analysis for the primary efficacy end point, the power of the study was assumed to be sufficient and high. To obtain a general estimate of the power of the study, however, a sample size calculation was performed using FVC as an independent variable. Using an effect size and a variance estimator from the Phase I/II study, 28 patients per treatment group provided a power of at least 80% with a two-sided alpha = 0.05 level test.

- Because of its direct impact on responsiveness to treatment, a significant concern was a potential imbalance in disease severity between treatment groups. In general, MPS II disease progression and severity correlate with age. As such, the Phase II/III utilised dynamic randomisation stratified by age and disease severity (as determined by the 6MWT and FVC baseline results) to allocate patients to treatment group. It was hoped that this would improve the efficiency of the comparisons for a small study.
- To decrease the overall variability of the clinical measurements (pulmonary function, 6MWT), the number of sites performing the testing was kept to a minimum of six. These sites also administered study drug, but because of the burden of administering weekly infusions of study drug to 90 patients, other centres were recruited to perform study drug infusions but not clinical testing.
- Finally, it was decided that the study would require a 1 year treatment duration to show clinical benefit, based on the results of the Phase I/II trial.

The Phase II/III pivotal study for Elaprase was conducted as a randomised, double-blind, placebo-controlled trial, under the supervision of Muenzer and colleagues.[51] The study enrolled 96 patients with MPS II, aged 5–31 years. At the time, this was the largest and longest placebo-controlled study ever conducted for an LSD.

Patients were randomised to receive either weekly placebo IV infusions, weekly IV infusions of Elaprase at 0.5 mg kg^{-1}, or every other week IV infusions of Elaprase at 0.5 mg kg^{-1} (in the latter group, placebo was administered in the alternate week to maintain the blinding of the treatment regimens). The treatment duration was 53 weeks.

7.3.3.2.1 Results. As described earlier, the primary efficacy end point was a two-component score based on the change from baseline to week 53 in distance walked during the 6MWT and the change in per cent predicted of normal FVC. The primary end point showed the greatest statistically significant difference between the weekly Elaprase-treated group and the placebo group ($p = 0.0049$); the every other week Elaprase-treated group also reached significance compared to the placebo group ($p = 0.0416$). When the results of the individual components were examined, it was clear that weekly infusions of Elaprase provided significant clinical benefit based on improvements in the 6MWT distance and FVC compared to placebo. It was also evident that the weekly dosing regimen was superior to the every other week dosing regimen of Elaprase. The patients in the weekly dosing group experienced a 37 metre increase in the 6MWT distance ($p = 0.013$), a 2.7% increase in per cent predicted of normal FVC ($p = 0.065$), and a 160 mL increase in absolute FVC ($p = 0.001$) compared with patients in the placebo group. In contrast, the every other week dosing regimen did not reach statistical significance for any of these outcomes.

Consistent with the Phase I/II results, urine GAG levels, and liver and spleen size, were significantly reduced in both Elaprase-treated groups compared with the placebo recipients. Marked reductions were evident as early as the week 18 evaluation. In terms of joint range of motion, only an improvement in elbow mobility was detected between the weekly Elaprase and placebo groups.

Infusions of Elaprase were generally well tolerated, with no patients withdrawing from the 1 year study due to an infusion-related reaction. Anaphylactic-type reactions were observed in some patients receiving Elaprase. The majority of the other adverse events reported were consistent with those expected to be observed in an untreated MPS II population. Although certain patients in the Elaprase-treated groups developed IgG antibodies to Elaprase during the course of the study, there was no meaningful correlation of these antibodies with adverse events or clinical assessments. Reductions in urine GAG levels, however, were not as great in antibody-positive patients.

Data generated after completion of the Phase II/III study have provided additional insight into the immune response to Elaprase in MPS II patients.

The percentage of patients developing antibodies to idursulfase has ranged from 51% to 68%, with a proportion having antibodies that neutralise either the uptake of idursulfase into cells or enzymatic activity. Antibody-positive patients appear to have a higher incidence of hypersensitivity reactions and a reduced systemic exposure to idursulfase. Not surprisingly, an association between the patient's genotype and immunogenicity has been observed, indicating that the patient's endogenous expression of iduronate-2-sulfatase (lack of expression or expression of an altered protein) is an important factor in idursulfase immunogenicity. Regardless of antibody status, however, Elaprase treatment has continued to result in pharmacodynamic and clinical effects.

7.3.3.2.2 Conclusions. The Phase II/III results showed that infusions of Elaprase were generally well tolerated, and that the weekly dosing regimen provided significant clinical benefit as demonstrated by improvements in walking ability and pulmonary function. It was clear that the weekly dosing regimen, which allowed Elaprase to be continuously present in the tissues, was critical for demonstrating efficacy and clinical benefit. In addition, the composite scoring approach resulted in a powerful and sensitive analysis for the primary efficacy end point. On the basis of the Phase II/III data, Elaprase was approved in the USA in 2006 and subsequently in Europe in 2007.

7.4 Overall Conclusions

The development of Elaprase, as an effective therapy for MPS II patients, was part of a continuum of many significant scientific and medical advances in the field of rare genetic diseases. The production of Elaprase by recombinant DNA technology using a fully human cell line for protein expression was a unique approach taken by Shire. This process resulted in Elaprase having a human glycosylation profile, analogous to the naturally occurring enzyme.

The clinical development of Elaprase consisted of a Phase I/II study, followed by a Phase II/III pivotal trial. A clinical development programme consisting of only two small trials, coupled with the broad disease heterogeneity of the MPS II patient population, made the development of Elaprase very challenging. Moreover, regulatory approval would be based on the results of a single pivotal trial, requiring the trial to be conducted robustly and to provide firm evidence of safety and efficacy. As such, many features were carefully planned and incorporated into the design of the Phase II/III study to ensure its success. In the end, Elaprase has been widely approved worldwide including in the USA and Europe.

The clinical development of rare disease therapies must be thoughtfully optimised to be able to successfully demonstrate clinical benefit in small heterogeneous patient populations with therapeutics that are unique and complex.

References

1. J. D. Esko and S. B. Selleck, *Annu. Rev. Biochem.*, 2002, **71**, 435–471.
2. C. I. Gama, S. E. Tully, N. Sotogaku, P. M. Clark, M. Rawat, N. Vaidehi, W. A. Goddard, A. Nishi and L. C. Hsieh-Wilson, *Nat. Chem. Biol.*, 2006, **2**, 467–473.
3. M. Vlasak, I. Goesler and D. Blaas, *J. Virol.*, 2005, **79**, 5963–5970.
4. P. J. Meikle, M. Fuller and J. J. Hopwood, *Chemistry and Biology of Heparin and Heparan Sulfate*, ed. H. G. Garg, R. J. Linhardt and C. A. Hales, Elsevier Ltd, Kidlington, Oxford, 2005, ch. 10, pp. 285–311.
5. B. Kowalewski, W. C. Lamanna, R. Lawrence, M. Damme, S. Stroobants, M. Padva, I. Kalus, M. A. Frese, T. Lübke, R. Lüllmann-Rauch, R. D'Hooge, J. D. Esko and T. Dierks, *Proc. Natl. Acad. Sci. U. S. A.*, 2012, **109**, 10310–10315.
6. G. Bach, F. Eisenberg Jr, M. Cantz and E. F. Neufeld, *Proc. Natl. Acad. Sci. U. S. A.*, 1973, **70**, 2134–2138.
7. V. A. McKusick, *N. Engl. J. Med.*, 1979, **283**, 1466–1468.
8. I. D. Young and P. S. Harper, *Hum. Genet.*, 1982, **60**, 391–392.
9. P. J. Meikle, J. J. Hopwood, A. E. Clague and W. F. Carey, *J. Am. Med. Assoc.*, 1999, **281**, 249–254.
10. J. E. Wraith, M. Scarpa, M. Beck, O. A. Bodamer, L. De Meirleir, N. Guffon, A. M. Lund, G. Malm, A. T. Van der Ploeg and J. Zeman, *Eur. J. Pediatr.*, 2008, **167**, 267–277.
11. J. Bielicki, C. Freeman, P. R. Clements and J. J. Hopwood, *Biochem. J.*, 1990, **271**, 75–86.
12. P. J. Wilson, C. P. Morris, D. S. Anson, T. Occhiodoro, J. Bielicki, P. R. Clements and J. J. Hopwood, *Proc. Natl. Acad. Sci. U. S. A.*, 1990, **87**, 8531–8535.
13. T. T. Puck, S. J. Cieciura and A. Robinson, *J. Exp. Med.*, 1958, **108**, 945–956.
14. G. Urlaub and L. A. Chasin, *Proc. Natl. Acad. Sci. U. S. A.*, 1980, 77, 4216–4220.
15. D. N. Wheatle, *J. Gen. Virol.*, 1974, **24**, 395–399.
16. D. M. Strauss, S. Lute, K. Brorson, G. S. Blank, Q. Chen and B. Yang, *Biotechnol. Prog.*, 2009, **25**, 1194–1197.
17. S. Rasheed, W. A. Nelson-Rees, E. M. Toth, P. Arnstein and M. B. Gardner, *Cancer*, 1974, **33**, 1027–1033.
18. H. Paterson, B. Reeves, R. Brown, A. Hall, M. Furth, J. Bos, P. Jones and C. Marshall, *Cell*, 1987, **51**, 803–812.
19. Z. Shahrokh, L. Royle, R. Saldova, J. Bones, J. L. Abrahams, N. V. Artemenko, S. Flatman, M. Davies, A. Baycroft, S. Sehgal, M. W. Heartlein, D. J. Harvey and P. M. Rudd, *Mol. Pharmaceutics*, 2011, **8**, 286–296.
20. Y. Kariya, C. Kawamura, T. Tabei and J. Gu, *J. Biol. Chem.*, 2010, **285**, 3330–3340.
21. D. Ghaderi, R. E. Taylor, V. Padler-Karavani, S. Diaz and A. Varki, *Nat. Biotechnol.*, 2010, **28**, 863–867.

22. S. C. Garman and D. N. Garboczi, *J. Mol. Biol.*, 2004, **337**, 319–335.
23. B. Brumshtein, P. Salinas, B. Peterson, V. Chan, I. Silman, J. L. Sussman, P. J. Savickas, G. S. Robinson and A. H. Futerman, *Glycobiology*, 2010, **20**, 24–32.
24. J. U. Baenziger, *Cell*, 2003, **113**, 421–422.
25. G. Walsh and R. Jefferis, *Nat. Biotechnol.*, 2006, **24**, 1241–1252.
26. A. R. Garcia, J. Pan, J. C. Lamsa and J. Muenzer, *J. Inherited Metab. Dis.*, 2007, **30**, 924–934.
27. A. R. Garcia, J. M. DaCosta, J. Pan, J. Muenzer and J. C. Lamsa, *Mol. Genet. Metab.*, 2007, **91**, 183–190.
28. C. de Duve, *Fed. Proc.*, 1964, **23**, 1045–1049.
29. R. O. Brady, *N. Engl. J. Med.*, 1966, **275**, 312–318.
30. J. C. Fratantoni, C. W. Hall and E. F. Neufeld, *Science*, 1968, **162**, 570–572.
31. R. O. Brady, A. E. Gal, R. M. Bradley, E. Martensson, A. L. Warshaw and L. Laster, *N. Engl. J. Med.*, 1967, **276**, 1163–1167.
32. G. Bach, R. Friedman, B. Weissmann and E. F. Neufeld, *Proc. Natl. Acad. Sci. U. S. A.*, 1972, **69**, 2048–2051.
33. R. O. Brady, *Philos. Trans. R. Soc. London*, 2003, **358**, 915–919.
34. M. R. Natowicz, M. M.-Y. Chi, O. H. Lowry and W. S. Sly, *Proc. Natl. Acad. Sci. U. S. A.*, 1979, **76**, 4322–4326.
35. J. H. Grubb, C. Vogler and W. S. Sly, *Rejuvenation Res.*, 2010, **13**, 229–236.
36. F. Barbey, D. Hayoz, U. Widmer and M. Burnier, *Curr. Med. Chem.: Cardiovasc. Hematol. Agents*, 2004, **2**, 277–286.
37. N. W. Barton, R. O. Brady, J. M. Dambrosia, A. M. Di Bisceglie, S. H. Doppelt, S. C. Hill, H. J. Mankin, G. J. Murray, R. I. Parker, C. E. Argoff, R. P. Grewal and K.-T. Yu, *N. Engl. J. Med.*, 1991, **324**, 1464–1470.
38. G. A. Grabowski, N. W. Barton, G. Pastores, J. M. Dambrosia, T. K. Banerjee, M. A. McKee, C. Parker, R. Schiffmann, S. C. Hill and R. O. Brady, *Ann. Intern. Med.*, 1995, **122**, 33–39.
39. C. M. Eng, N. Guffon, W. R. Wilcox, D. P. Germain, P. Lee, S. Waldek, L. Caplan, G. E. Linthorst and R. J. Desnick, *N. Engl. J. Med.*, 2001, **345**, 9–16.
40. R. Schiffmann, J. B. Kopp, H. A. Austin, S. Sabnis, D. F. Moore, T. Weibel, J. E. Balow and R. O. Brady, *J. Am. Med. Assoc.*, 2001, **285**, 2743–2749.
41. D. A. Hughes, P. M. Elliott, J. Shah, J. Zuckerman, G. Coghlan, J. Brookes and A. B. Mehta, *Heart*, 2008, **94**, 153–158.
42. E. D. Kakkis, J. Muenzer, G. E. Tiller, L. Waber, J. Belmont, M. Passage, B. Izykowski, J. Phillips, R. Doroshow, I. Walot, R. Hoft and E. F. Neufeld, *N. Engl. J. Med.*, 2001, **344**, 182–188.
43. J. E. Wraith, L. A. Clarke, M. Beck, E. H. Kolodny, G. M. Pastores, J. Muenzer, D. M. Rapoport, K. I. Berger, S. J. Swiedler, E. D. Kakkis, T. Braakman, E. Chadbourne, K. Walton-Bowen and G. F. Cox, *J. Pediatr.*, 2004, **144**, 581–588.

44. Committee on Strategies for Small-Number-Participant Clinical Trials, *Institute of Medicine, Small Clinical Trials: Issues and Challenges*, ed. C. H. Evans, Jr and S. T. Ildstad, National Academy Press, Washington, DC, 2001, ch. 2, pp. 20–59.
45. CHMP: guideline on clinical trials in small populations (Doc. Ref. CHMP/EWP/83561/2005), London, 27 July 2007.
46. J. W. Gerss and W. Kopcke, *Adv. Exp. Med. Biol.*, 2010, **686**, 173–190.
47. S. Gupta, M. E. Faughnan, S. Gupta, G. A. Tomlinson and A. M. Bayoumi, *J. Clin. Epidemiol.*, 2011, **64**, 1085–1094.
48. R. O. Brady, *Annu. Rev. Med.*, 2006, **57**, 283–296.
49. J. Muenzer, M. Gucsavas-Calikoglu, S. E. McCandless, T. J. Schuetz and A. Kimura, *Mol. Genet. Metab.*, 2007, **90**, 329–337.
50. P. C. O'Brien, *Biometrics*, 1984, **40**, 1079–1087.
51. J. Muenzer, J. E. Wraith, M. Beck, R. Giugliani, P. Harmatz, C. M. Eng, A. Vellodi, R. Martin, U. Ramaswami, M. Gucsavas-Calikoglu, S. Vijayaraghavan, S. Wendt, A. Puga, B. Ulbrich, M. Shinawi, M. Cleary, D. Piper, A. M. Conway and A. Kimura, *Genet. Med.*, 2006, **8**, 465–473.

GENETIC DISORDERS

CHAPTER 8

Discovery and Development of Ilaris® for the Treatment of Cryopyrin-Associated Periodic Syndromes

HERMANN GRAM

Novartis Institutes of BioMedical Research, Forum 1, CH-4002 Basel, Switzerland
E-mail: hermann.gram@novartis.com

8.1 Introduction

Interleukin-1 (IL-1) is a pro-inflammatory cytokine produced by a variety of cell types, particularly mononuclear phagocytes, in response to injury, infection and inflammation. The biological activity of IL-1 is attributed to two distinct proteins, IL-1α and IL-1β. IL-1β is the main secreted form, while IL-1α appears mainly implicated in intracellular signalling events, although it can be released as a biologically active protein from necrotic cells.[1] The observed biological activities of IL-1β are mediated through the type I receptor (IL-1RI), which is ubiquitously expressed and binds both IL-1β and IL-1α with medium affinity.[2] IL-1 signalling requires the engagement of a second receptor chain, termed IL-1 receptor associated protein (IL-1RacP), which together with IL-1RI forms a high-affinity binding site for IL-1.[3] The bioactivity of IL-1β is tightly regulated by a high-affinity decoy receptor and IL-1Ra, an endogenous and structurally similar competitive antagonist binding to IL-1RI.[4–6] Signalling

RSC Drug Discovery Series No. 38
Orphan Drugs and Rare Diseases
Edited by David C Pryde and Michael J Palmer

Published by the Royal Society of Chemistry, www.rsc.org

downstream of the IL-1 receptor is facilitated by kinases IRAK-1 and IRAK-4, and amongst others, the adapter molecule myD88, leading to the activation of the NFκB, the MAP kinase, and the pI3kinase pathways. IL-1 signalling is thereby overlapping with the signalling through Toll-like receptors and further downstream, with signalling by TNF family members.

Impaired balance between IL-1β and its endogenous antagonist IL-1Ra have been described for many chronic inflammatory conditions, such as rheumatoid arthritis, renal, pulmonary and arterial diseases, and auto-inflammatory disorders,[7,8] which are characterised by mild to severe inflammatory phenotypes. IL-1β induces numerous downstream mediators implicated in the inflammatory processes and tissue remodelling, including chemokines, cytokines, cyclooxygenase 2, leukocyte adhesion molecules, as well as extracellular matrix components, fibronectin, and collagenases.[9–11] The most prominent physiological effect of IL-1β administered to animals or humans is its ability to induce fever.[12] Indeed, the first name of the identified IL-1 entity was pyrexin, based on its identified fever-inducing activity.[1]

The main source of IL-1β in acute and chronic inflammation is monocytes or macrophages, although IL-1β can also be produced by other cell types, such as fibroblasts, endothelial cells or epithelial cells. IL-1β is produced as an inactive precursor protein (pro-IL-1β) which remains intracellular. The production of pro-IL-1β mRNA is tightly regulated at the transcriptional level. Typically, IL-1β mRNA is induced by cytokines, interleukins or ligands of Toll-like receptors. The cleavage of the pro-IL-1β precursor protein by caspase I and the release of bioactive IL-1β from cells is regulated on the post-translational level by the inflammasome, a multi-protein complex which is able to respond to extracellular and intracellular stimuli.[13]

Because of the prominent role of IL-1β in a number of chronic inflammatory conditions, numerous attempts to target the IL-1 signalling pathway have been undertaken by the pharmaceutical industry. Notably, inhibitors of caspase I, believed to be an essential and indispensable protease in IL-1β production, were heavily pursued by many pharmaceutical companies. However, IL-1β can be processed by neutrophil proteases, such as proteinase 3 or elastase, making caspase I largely redundant in overt chronic inflammatory processes.[14,15] This may in part explain the at best very modest effects of caspase I inhibitors in clinical trials in rheumatoid arthritis.

In the mid-1990s the ability to produce human antibodies with high affinity for the target molecule by immunisation of mice carrying the human antibody repertoire was developed to maturity.[16] This technical development opened up the possibility of generating a high-affinity neutralising human antibody to human IL-1β, making such a pharmaceutical suitable for clinical use. Such pharmaceutical intervention towards IL-1β would circumvent the apparent redundancy in the production machinery for IL-1β and the lack of specificity in downstream signalling. An anti-IL-1β antibody would have to be a long-acting biological to avoid frequent injections, and it needed to neutralise IL-1β with potency at least similar to the endogenous antagonist, IL-1Ra, to achieve maximum clinical efficacy.

8.2 Development of Canakinumab

8.2.1 Pre-Clinical Development

In the late 1990s Novartis Pharma AG licensed the HuMab-Mouse™ technology from Medarex and started a programme aimed at the generation of therapeutic human antibodies targeting human IL-1β. HuMab mice carry part of the human antibody repertoire of the IgG1 heavy and light chain, giving rise to a human antibody response to administered antigens.[16,17] Such mice were immunised with recombinant human IL-1β, and subsequently, hybridomas secreting human anti-human IL-1β monoclonal antibody were generated to capture the antibody response to the human IL-1β antigen. Several different antibodies emerged from this endeavour, and two of them were progressed into pre-clinical development. The more potent antibody, termed ACZ885 and later canakinumab, a human IgG1/κ antibody, entered clinical development.

Canakinumab is produced for Phase I and Phase II clinical studies in recombinant NS0 cells and later in genetically engineered Sp2/0-AG14 cells for the commercial production process. Canakinumab binds to human IL-1β with an apparent affinity of about 40 pM [18] and neutralises the biological activity of IL-1β *in vitro* with an IC_{50} of about 43 pM, similar to the recombinant endogenous IL-1 antagonist IL-1Ra (Table 8.1). Canakinumab has high selectivity towards human IL-1β, it does not bind to other members of the IL-1 family, including the most related members IL-1α and IL-1Ra. Canakinumab also exhibits a very high degree of species specificity, inasmuch as it does not bind to mouse, rat or rabbit IL-1β, and not even to the highly related IL-1β from rhesus or cynomolgus monkeys. Elucidation of the X-ray structure of canakinumab in complex with IL-1β revealed Glu64 in human IL-1β as a critical residue for antibody–antigen interaction.[19] This residue is not conserved in macaque monkeys, rodents, canines, and many other mammalian species. Mutational analysis confirmed the importance of this residue for the binding of canakinumab and, for example, a Glu64Ala substitution in human IL-1β is not recognised by canakinumab.

This exquisitely high species selectivity posed a problem for the pre-clinical development of canakinumab, as the commonly used macaque non-human

Table 8.1 Inhibition of IL-1β-induced release of IL-6. IL-6 release from human dermal fibroblasts stimulated with human and marmoset IL-1β was measured in the presence of canakinumab and IL-1Ra. Inhibition of IL-6 production was determined and IC_{50} were derived by logistic curve fitting. The data represent the mean ± SEM for $n = 5$ experiments.

IC_{50}	*Canakinumab [pM]*	*IL-1Ra [pM]*
Human IL-1β	43.6 ± 5.5	109.2 ± 13.9
Marmoset IL-1β	40.8 ± 5.6	120.3 ± 15.3

primates, cynomolgus or rhesus monkeys, were not acceptable for toxicological evaluation due to the lack of target binding. Marmoset monkeys belong to the group of non-human primates, and breeding colonies for pharmacological testing exist for this species. IL-1β from marmoset shares 96% homology with human IL-1β and, most interestingly for the pre-clinical development of canakinumab, marmoset IL-1β has Glu64-like human IL-1β. Canakinumab exhibits full cross-reactivity to marmoset IL-1β, and the bioactivity of marmoset IL-1β is effectively neutralised by canakinumab (Table 8.1). Therefore, marmoset monkeys fulfilled the criteria of a relevant species for toxicological examination of canakinumab. However, toxicological studies required for the clinical development of antibodies in marmoset monkeys were never reported before, and reagents for immunophenotyping in this species were largely lacking. Further, population background data on fertility, pre- and postnatal survival, pre-term birth rate, and the incidence of tumours were scarce. Therefore, part of the toxicological examination required for market authorisation had to be performed in mice using an antibody specific for mouse IL-1β. A high-affinity monoclonal mouse anti-mouse IL-1β antibody had been generated at Ciba-Geigy, one of the predecessor companies of Novartis, about 5 years before starting the anti-IL-1β project at Novartis.[20] This mouse anti-mouse IL-1β antibody was similar in potency to canakinumab, and could therefore be used in mouse studies for toxicological evaluation of fertility, pre- and postnatal development, and immunotoxicity. Standard 4, 13 and 26 week toxicology for canakinumab and part of the embryo-foetal development programme was conducted in marmosets without revealing pre-clinical safety signals (Table 8.2).

8.2.2 Clinical Development

Clinical development of canakinumab was initiated in healthy volunteers and patients with mild asthma to assess tolerability, safety and pharmacokinetics at 0.3, 1, 3 and 10 mg kg^{-1} intravenous (i.v.) dose levels.[21] In parallel, a randomised, double-blind, placebo-controlled, dose escalation study was conducted to explore the safety, tolerability, pharmacokinetics and pharmacodynamics of canakinumab in patients with active rheumatoid arthritis (RA).[18] The proof of concept trial in patients with RA provided first evidence of efficacy. In this study, the patient group receiving 10 mg kg^{-1} on days 1 and 15 achieved an improvement in their disease activity score (DAS28) until day 43, and a statistically significant reduction in their C-reactive protein *versus* the placebo group.[18] A Phase II dose-ranging study suggested efficacy in RA with an unclear dose–response relationship.[22]

RA is a multi-factorial autoimmune and inflammatory disease responding to a number of biological treatments with different mechanisms of action. IL-1β is one of several important inflammatory drivers in this disease, and its contribution to pathology may vary widely amongst patients.[23] Furthermore, antibodies targeting the signalling pathways of other pro-inflammatory cytokines, such as TNF or IL-6, have demonstrated clinical benefit in this multi-factorial

Table 8.2 Preclinical studies.

Study type	*Species*	*Route*	*Antibody*	*Purpose*
Tolerability study	Marmoset	s.c.	Canakinumab	Repeated dose toxicity
13 week with an 8 week recovery	Marmoset	s.c.	Canakinumab	Repeated dose toxicity
13 week	Marmoset	s.c.	Canakinumab	Repeated dose toxicity
4 week with an 8 week recovery	Marmoset	i.v. slow bolus injection	Canakinumab	Repeated dose toxicity
26 week with a 6 week recovery	Marmoset	i.v. slow bolus injection	Canakinumab	Repeated dose toxicity
Fertility study	Mouse	s.c.	Anti-mouse IL-1β	Fertility and early embryonic development
Embryo foetal study	Marmoset	s.c.	Canakinumab	Embryo-foetal development
Embryo foetal study	Mouse	s.c.	Anti-mouse IL-1β	Embryo-foetal development
Peri- and postnatal development study	Mouse	s.c.	Anti-mouse IL-1β	Peri- and postnatal development study
Juvenile study	Mouse	s.c.	Anti-mouse IL-1β	Juvenile study
Immunotoxicity study 4 weeks	Mouse	s.c.	Anti-mouse IL-1β	Immunotoxicity

inflammatory disease. It is therefore demanding to support a therapeutic hypothesis in which IL-1 has a major role in the inflammation-related clinical symptoms in RA. Given the modest efficacy and variable responses in RA, it was felt difficult to establish the dose response and overall pharmacokinetic/pharmacodynamic relationship for canakinumab in a mid-size study before full development of this drug. Also, a decision not to continue with the development in RA was taken based on market considerations.[24] It was therefore decided to study canakinumab in a pathophysiological condition entirely dependent on the overproduction of IL-1β and thereby, specifically targeting the principal effector molecule in such disease to unequivocally assess the therapeutic utility of canakinumab in human pathology.

8.2.2.1 Clinical Development in CAPS

Targeted causative treatment of a disease requires an in-depth understanding of its pathophysiology and aetiology to match the underlying molecular pathophysiology with the drug's mode of action. The most straightforward understanding of the aetiology of a disease comes from studies on monogenetic diseases in which the physiology of the affected gene product is linked to known pathophysiological pathways.

More than a decade ago, the genetic cause of rare periodic fever syndromes, such as familial Mediterranean fever (FMF), Muckle–Wells syndrome (MWS), familial cold auto-inflammatory syndrome (FCAS) or TNF receptor associated periodic syndrome (TRAPS) was mapped to specific genes, such as MEFV for FMF, CIAS1 for MWS and FCAS, and TNFRSF1A for TRAPS. As these genetic diseases are characterised by fever attacks of different severity, duration and periodicity, IL-1β became a prime suspect to be involved in the pathophysiology of these syndromes.

MWS and FCAS are autosomal dominant hereditary diseases, and they belong together with the non-hereditary neonatal onset multisystem inflammatory disease/chronic infantile neurologic cutaneous and articular syndrome (NOMID/CINCA) to the family of cryopyrin-associated periodic syndrome (CAPS). CAPS is an extremely rare disease with an estimated prevalence of about 1 per 1 000 000 subjects.[25] The identified CIAS1 gene mutations in CAPS patients affect a protein termed NLRP3 (formerly NALP3).[26–28]

The clinical presentation of CAPS can be quite variable; it includes severe fatigue, periodic fever, influenza-like myalgia, anaemia and inflammation of the skin, eyes, bones, joints and meninges.[29] FCAS, the least severe form within the CAPS spectrum, is characterised by an urticaria-like rash that appears shortly after exposure to cold and can last for up to 12 hours. Attacks of FCAS commonly include fever, headache, fatigue, conjunctivitis and polyarthralgia. MWS shares many of the systemic manifestations of FCAS; however, attacks in MWS are typically not induced by temperature changes and can persist for up to 2 days. Patients may experience progressive sensorineural hearing loss (75% of patients) and amyloid A (AA) amyloidosis (25% of patients), leading to renal impairment.[30] CINCA/NOMID, the most severe of the CAPS diseases, is associated with urticarial rash, central nervous system inflammation and disabling arthritis. Progressive visual loss, sensorineural deafness, headaches, aseptic meningitis, developmental delay and mental retardation are common. If left untreated, the mortality rate in childhood is 20%.[30] Until recently, CAPS patients have used anti-inflammatory agents, such as high-dose corticosteroids and non-steroidal anti-inflammatory drugs to treat the symptoms of the disease, but treatments for the cause of the inflammation were lacking.

At about the same time as the discovery of mutations in the NLRP3 protein in CAPS, regulation of the production of mature and bioactive IL-1β was unravelled. Caspase I had been identified as the protease responsible for direct cleavage of pro-IL-1β,[31] but the regulation of caspase I activity was not known at that time. The discovery of a multi-protein complex interacting with caspase I, and its ability to induce processing of pro-IL-1β to the mature form led to the concept of the inflammasome as a regulator of caspase I activity and thus, IL-1β and IL-18 production.[32] NLRP3 is a member of a larger protein family which is characterised by the presence of a caspase recruiting domain (CARD) and is an integral part of the NLRP3 inflammasome.

NLRP3 mutations in CAPS patients are associated with constitutive secretion of IL-1β from peripheral blood monocytes demonstrating a regulatory role of NLRP3 for inflammasome activity.[13,26,33] Further biochemical and cellular studies led to the hypothesis that NLRP3 is a negative regulator of inflammasome activity which responds to intracellular Ca^{++} and cAMP.[34] While intracellular Ca^{++} stimulates inflammasome activity, cAMP appears to inhibit inflammasome activity by binding to the negative regulator NLRP3. The cAMP binding to NLRP3 is impaired by amino acid substitutions observed in CAPS patients, leading to inflammasome activity and deregulated production of mature IL-1β. Consequently, the treatment of CAPS patients with anakinra, a recombinant IL-1 receptor antagonist and specific blocker of IL-1 signalling, licensed for the treatment of RA, led to a rapid and complete clinical and serological response in a case study with two and three MWS patients, respectively[35,36] or a cohort of FCAS patients.[33] Clinical response in MWS is typically sustained by daily injections of anakinra.[37,38]

The molecular understanding of the affected cellular pathway by mutations observed in CAPS patients allows the application of a targeted medicine paradigm, *i.e.* specifically and exclusively targeting IL-1β, a highly potent inflammatory mediator closely downstream of the causative defect in NLRP3 and the inflammasome.

As expected, four MWS patients treated by i.v. infusion of canakinumab at 10 mg kg^{-1} had a fast and complete clinical, serological and biochemical response within 8 days.[39] However, the median duration of 185 days for clinical remission in these first patients was unexpected and surprising. Complete clinical responses were subsequently observed in the same patients upon i.v. infusion of 1 mg kg^{-1} canakinumab or s.c. injection of 150 mg canakinumab. The median time to clinical relapse in this study was 90.5 days (range 90–98 days) and 127 days (range 55–230 days) for the 1 mg kg^{-1} i.v. dose and the 150 mg s.c. dose, respectively. The long duration of clinical remission in CAPS patients is fully explained by the pharmacokinetics and potency of canakinumab, which has a serum half-life of 26 days in CAPS patients.[40]

Canakinumab is slowly absorbed in humans when given as a s.c. injection with a T_{max} of 7 days and an absolute bioavailability of about 70%. The volume of distribution of 8.33 $\pm$ 2.62 L in adult CAPS patients is consistent with a largely intravascular distribution. Canakinumab exhibits dose-proportional pharmacokinetics, both when given as an i.v. infusion (0.3–10 mg kg^{-1}) or when administered as a single s.c. injection of 150 or 300 mg. Maximum serum concentrations (C_{max}) reached by the marketed strength of 150 mg is about 16 μg mL^{-1} after s.c. injection in CAPS patients.[40]

A mathematical model describing the relationship between pharmacokinetics (PK), pharmacodynamics (PD), target binding and the production rate of IL-1β was developed in this first clinical Phase I/II study in CAPS.[39] Pharmacodynamic activity in healthy individuals and patients can be traced by measuring circulating IL-1β in complex to canakinumab in the serum. The elimination of free IL-1β from blood is high, and steady state concentrations are hardly measurable in the serum of healthy individuals or patients. IL-1β in complex to canakinumab has a much lower rate of elimination, and therefore, steady state serum concentrations of the complex can be determined in specific assays. Measurements of the rate of complex formation, steady state serum concentrations, and elimination rates of the canakinumab–IL-1β complex allowed for the generation of a pharmacokinetic and pharmacodynamic binding model, describing antibody–target complex formation and the distribution of canakinumab into extravascular and serum compartments. The resulting model predicted that the endogenous production rate of IL-1β in the four enrolled CAPS patients was about 31 ng per day, and thereby about 5-fold higher than the IL-1β production in healthy volunteers (6 ng per day). Interestingly, the increased production rate of IL-1β in CAPS patients decreased upon treatment and approached a normal rate after 4–6 weeks, whereas the endogenous production rate for IL-1β did not change in healthy individuals upon treatment with canakinumab.[39] This finding confirms the notion of increased IL-1β production by isolated blood mononuclear cells from CAPS patients *ex vivo*, and it is consistent with the existence of a positive feedback loop in CAPS patients in which IL-1β can stimulate its own production.[41]

A flare-probability model based on the PK/PD model was created to derive a dosing regimen for a pivotal Phase III study[39] (Figure 8.1). It was found that the duration of clinical remission, *i.e.* time to subsequent flare, is inversely correlated to the endogenous production rate of IL-1β. Based on the data obtained from seven patients, Monte Carlo simulations were run, and the derived flare-probability model predicted that a dose regimen of 150 mg s.c. every 8 weeks keeps the majority of CAPS patients in clinical remission.

This model-based dose regimen was tested in a subsequent Phase III clinical trial in CAPS patients.[42] In this three-part, 48 week, double-blind,

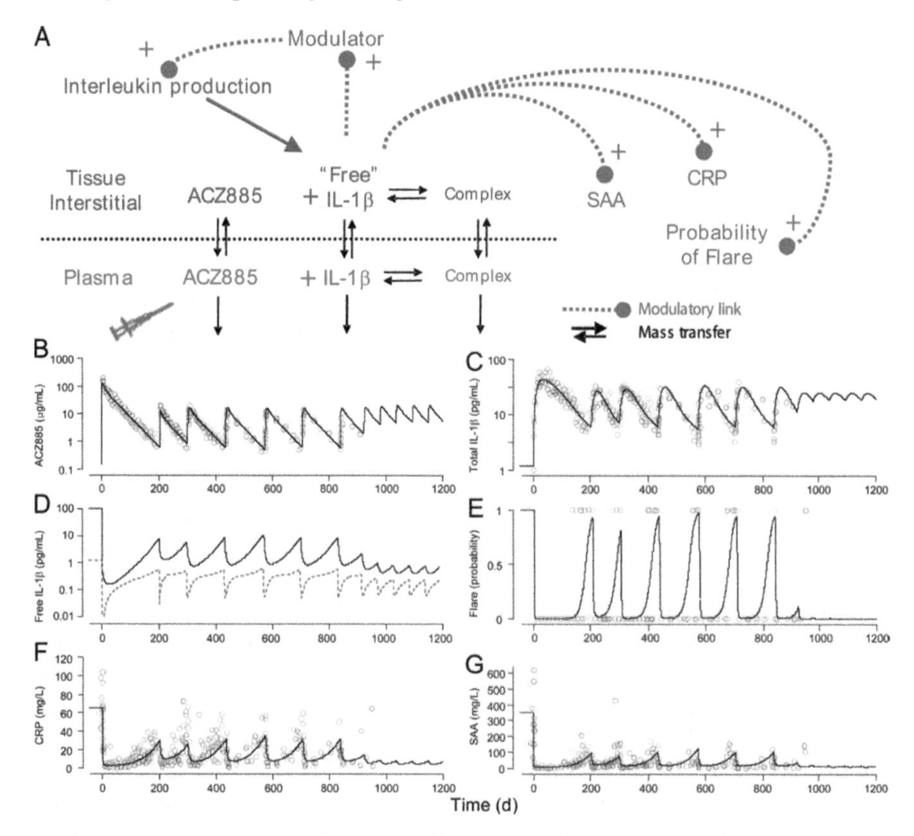

Figure 8.1 Prediction of clinical response in CAPS. (A) Pharmacokinetic/pharmacodynamic model flare probability model for clinical response in CAPS, (B) pharmacokinetics, (C) total IL-1β in plasma, (D) model derived free IL-1β in the peripheral (black) and intravascular (red) compartment, (E) observed flares, (F) CRP, (G) SAA. Reprinted from ref. 39 with permission.

placebo-controlled, randomised withdrawal study 35 patients received 150 mg of canakinumab s.c. Those patients with a complete response to treatment in part 1 entered part 2 and were randomly assigned to either 150 mg of canakinumab or placebo every 8 weeks for up to 24 weeks. After completion of part 2 or at the time of relapse, patients proceeded to part 3 and received at least two more doses of canakinumab in an open-label fashion (Figure 8.2). CAPS patients with mutations in the NLRP3 gene and aged between 4 years and 75 years were eligible for this study. Previous medication with canakinumab or anakinra was permitted, but enrolment in the open-label part 1 required a discontinuation of previous treatment and recurrence of disease. Thirty-four out of 35 patients who entered part 1 of the study had a complete response to a single dose of canakinumab. A complete clinical response was achieved in the majority of patients between day 8 and

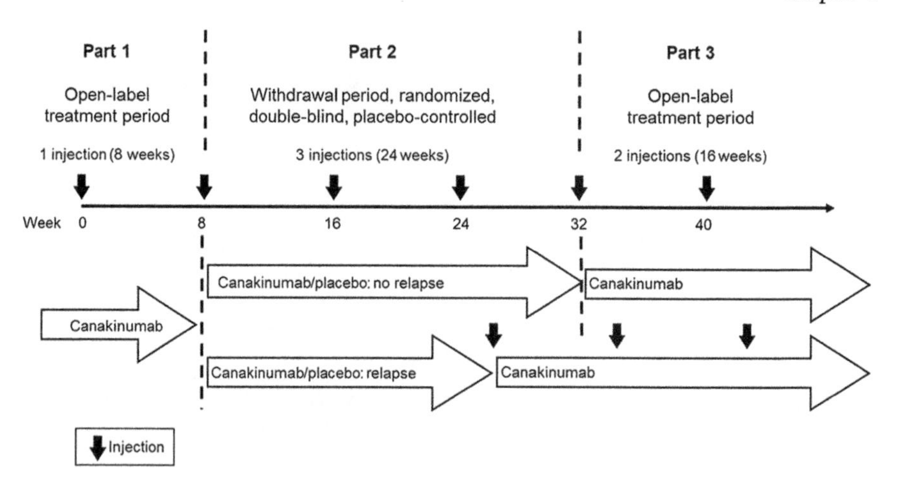

Figure 8.2 Design of the clinical phase III pivotal study in CAPS patients.

day 15 post-treatment. Thirty-one patients who maintained complete response during the 8 week period of part 1 were randomised to either placebo or a 150 mg s.c. dose of canakinumab, subsequently administered every 8 weeks. All 15 patients in the canakinumab group remained in remission during the 24 week time period of part 2. In contrast, 13 out of 16 patients (81%, $p < 0.001$) who received placebo had a disease flare with a median time to flare of 22 weeks post the initial canakinumab dose (Figure 8.3). Flares occurred in this group starting at 12 weeks after the first dose and throughout the 24 weeks of part 2. Inflammatory markers such as IL-6, C-reactive protein (CRP) or serum amyloid A (SAA) normalised upon treatment with canakinumab compared to pre-treatment and active disease state, and stayed mostly within the normal throughout continuous treatment with canakinumab, indicating sustained control of inflammation.[42,43] Clinical symptoms like skin rash, arthralgia, headache/migraine, conjunctivitis or fatigue/malaise were recorded in the enrolled patients. All symptoms improved rapidly, both by physician's and patient's assessments. Disease activity was judged absent or minimal in >85% of patients by day 8 and at the end of part 1, and clinical response was maintained until the end of study.[43] Health-related quality of life improved significantly from baseline to the end of part 1 and the end of part 3. Quality of life was examined by functional assessment of chronic illness therapy-fatigue (FACIT-F), health assessment questionnaire (HAQ), and 36-item Short-Form Health Survey mental and physical component summary (SF36 MCS/PCS) scores. All scores improved significantly from baseline to the end of part 1 or part 3. In particular, the FACIT-F and SF36 PCS scores improved drastically, reaching values in the range of the general US population.[43]

Results from the Phase III double-blind withdrawal study were confirmed in a larger open-label design study in which 166 patients with CAPS were

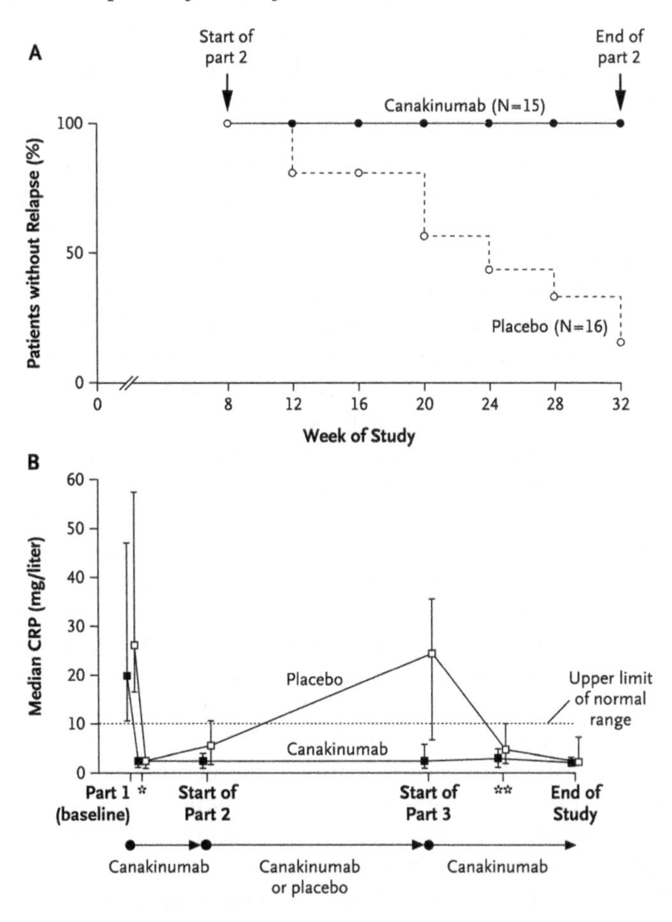

Figure 8.3 Response to canakinumab in pivotal phase III trial. (A) Percentage of patients without relapse in the placebo controlled part of the phase III study. (B) Median levels and interquartile range of CRP throughout the study. Reprinted from ref. 42 with permission.

enrolled and treated with a 150 mg s.c. dose of canakinumab every 8 weeks, or weight-based 2 mg kg^{-1} every 8 weeks in case the body weight was ≤40 kg.[44] Patients in this 2 year study represented the full spectrum of CAPS, with diagnosis for FCAS (n = 30), MWS (n = 103) or NOMID/CINCA (n = 32). Forty-seven were paediatric patients of 3 years to less than 18 years in age. NLRP3 mutations were observed in 94% of these patients. The median duration of treatment was 414 days (range 29–687 days) in the entire cohort. Dose adjustment up to 600 mg was permitted, and 40 patients (24.1%) required adjustment in dose or dose frequency. In most cases, doubling of the dose was sufficient to establish clinical response. Severe manifestation of CAPS and paediatric patients required upward dose adjustment more often than less severe phenotypes within the CAPS spectrum. Complete clinical response to the first dose of canakinumab was observed in 85 out of

109 (78%) patients. Another 23 patients showed a partial clinical response with the first 3 weeks of treatment. Available data from 141 patients showed that 90% of the patients had no relapse with the chosen 8 weekly dosing interval and the established dose. Control of clinical symptoms in the study population was also demonstrated by rapid improvement of the inflammatory serum parameters CRP and SAA which normalised within 8 days and stayed within the normal limits throughout the entire treatment period. Improvement in neurological manifestation was observed in 9 out of 20 patients with observed neurological abnormalities. Hearing normalised or improved in a fraction of patients during the 2 year study period.

In general, canakinumab was well tolerated and most adverse events were transient and mild in nature. Only occasional mild injection site reactions were noted. Reported adverse events did not cluster around a specific phenotype or age group, other than more infections reported in children.[44] Infections were commonly reported in patients treated with canakinumab. In the 2 year study the most common infection-related adverse events were bronchitis (event rate per patient-year 0.11), rhinitis (0.17), and upper respiratory tract infection (0.13). Also, severe adverse events (SAEs) were observed more frequently in children in the 2 year study (12.8% *vs.* 10.8%), concerning predominantly the upper respiratory tract. Amongst those, bronchitis and influenza strain H1N1 was reported in one NOMID patient, tonsillitis in two MWS patients, and pneumonia in one MWS patient. No SAEs were suspected to be treatment related.[44] None of the SAEs were fatal, and all infection-related SAEs responded to standard of care treatment. Three adults and one child with MWS developed vertigo on at least one occasion, which was self-limiting in all cases and did not lead to discontinuation of treatment with canakinumab.

Safety information from studies in patients with systemic juvenile idiopathic arthritis, gouty arthritis, RA, type I and type II diabetes suggest that canakinumab is overall well tolerated in adult and juvenile patients. The most common observed adverse effects are a mildly increased rate of infections, which is compatible with its mode of action. Although these infections are mostly upper respiratory tract or urinary infections, some cases of severe bacterial infections have been observed in the overall development programme for canakinumab. Mild, transient and asymptomatic cases of elevations of serum transaminases, bilirubin or triglycerides have been reported in clinical trials. Transient episodes of neutropenia have been observed under treatment with canakinumab.

Long-term safety and efficacy in CAPS patients is currently investigated in a number of studies, also including children of 2 years of age and older (study numbers: NCT0121364, NCT01302860, NCT01576367).

8.2.2.2 Regulatory Status

Canakinumab currently has orphan drug status for CAPS in the USA. The US Food and Drug Administration (FDA) granted market authorisation in June 2009 for the treatment of FCAS and MWS patients aged 4 years and older under the trade name Ilaris®. Ilaris® is given as a single injection of 150 mg s.c. for patients with body weight >40 kg, and 2 mg kg^{-1} for CAPS patients with body weight ≥15 kg and ≤40 kg; for children 15–40 kg with an inadequate response, the dose can be increased to 3 mg kg^{-1}.

The European Medicines Agency (EMA) approved Ilaris® for the treatment of CAPS, including NOMID/CINCA, in October 2009. In January 2013 EMA granted an extension of the label allowing the treatment of CAPS patients down to the age of 2 years and adjustment of the dose up to 600 mg.

Ilaris® is currently approved for the treatment of CAPS in about 65 countries worldwide. In addition, Ilaris® obtained approval for the treatment of gouty arthritis attacks in the EU, and it is approved for the treatment of systemic juvenile idiopathic arthritis (sJIA) in the USA and in the EU.

8.3 Outlook

Genetic and biochemical characterisation of other periodic fever syndromes such as FMF, TRAPS or HIDS suggests that IL-1β plays a prominent role in the pathophysiology of these diseases. These diseases are rare, whereas TRAPS and HIDS are mostly found in patients of Western European origin at a similar prevalence to CAPS. FMF is most common in individuals originating from around the Mediterranean.[45] The prevalence of FMF is higher than for TRAPS, CAPS or HIDS.[25,46]

Mutations in the MEFV gene identified in FMF lead to functional change in the pyrin protein, which interacts with components of the inflammasome.[47] Introduction of mutations identified in FMF patients into the pyrin protein leads to enhanced IL-1β secretion and auto-inflammation in an experimental mouse model.[48,49] These findings link mutations underlying the cause of FMF to the regulation of the inflammasome and IL-1β production, and therefore establish IL-1β blockade as a targeted treatment paradigm for FMF. Indeed, targeting IL-1 by anakinra or canakinumab in colchicine-resistant FMF patients resulted in a significant clinical response.[50] A recent small open-label study with colchicine-resistant FMF patients demonstrated that canakinumab largely prevented recurrent FMF flares and improved quality of life.[51] Mitroulis *et al.* reported improvement of arthritic symptoms and inflammation in response to canakinumab in a single case of colchicine-resistant FMF.[52]

TRAPS is a periodic fever disease associated with mutations affecting TNF receptor I (TNFSFR1A). The molecular pathophysiology is not well understood, but it appears that TNFSFR1A is defective in intracellular trafficking, which leads to sensitisation of phagocytes to inflammatory stimuli, enhanced production of reactive oxygen and, in turn, to the activation of the inflammasome and production of IL-1.[53] TRAPS patients paradoxically do not show a long-lasting and sustained response to treatment with TNF blocking agents, but appear to respond well to IL-1 targeted treatment.[54] A recent study in a cohort of 20 juvenile and adult TRAPS patients reported excellent efficacy of canakinumab.[55] In this study 19/20 patients achieved complete or almost complete remission of their disease activity by day 15 and a highly significant reduction of the inflammatory markers serum amyloid A and CRP close to the normal range.

HIDS is an autosomal recessive disease associated with defects in the mevalonate kinase (MVK) gene. MVK is a key enzyme in the cholesterol and isoprenoid synthesis pathway. Deficiency in this enzyme leads to accumulation of mevalonate, and further downstream in the pathway to a shortage of isoprenoids, like farnesyl- and geranylgeranylpyrophosphate.[56] Peripheral blood mononuclear cells obtained from HIDS patients secrete higher amounts of pro-inflammatory cytokines, and in particular, IL-1β [57] Activation of caspase I by mevalonate pathway inhibitors or MVK deficiency has been demonstrated,[56,58] and IL-1β appears therefore to play a prominent role in the disease pathology of HIDS. Not surprisingly, inhibitors of IL-1 signalling, including canakinumab, are clinically effective in alleviating disease activity in HIDS.[59,60]

Although the underlying genetic defects and the molecular aetiology and pathology of the FMF, TRAPS and HIDS is different from CAPS, pre-clinical and clinical research has demonstrated that the biochemical consequences of these defects converge on the activation of the inflammasome and dysregulation of IL-1β synthesis. The aetiology of Schnitzler's syndrome, another extremely rare auto-inflammatory disorder, is unknown, but excellent clinical responses to treatment with canakinumab or anakinra have been reported.[61,62] The encouraging initial findings with anakinra and canakinumab in a variety of auto-inflammatory disorders suggest that targeting IL-1 has clinical utility also in these rare and orphan diseases.

References

1. C. A. Dinarello, *Eur. J. Immunol.,* 2010, **40**, 599–606.
2. K. Matsushima, T. Akahoshi, M. Yamada, Y. Furutani and J. J. Oppenheim, *J. Immunol.,* 1986, **136**, 4496–4502.
3. E. B. Cullinan, L. Kwee, P. Nunes, D. J. Shuster, G. Ju, K. W. McIntyre, R. A. Chizzonite and M. a. Labow, *J. Immunol.,* 1998, **161**, 5614–5620.
4. S. A. Greenfeder, P. Nunes, L. Kwee, M. Labow, R. A. Chizzonite and G. Ju, *J. Biol. Chem.,* 1995, **270**, 13757–13765.
5. C. J. McMahan, J. L. Slack, B. Mosley, D. Cosman, S. D. Lupton, L. L. Brunton, C. E. Grubin, J. M. Wignall, N. A. Jenkins and C. I. Brannan, *EMBO J.,* 1991, **10**, 2821–2832.
6. D. E. Smith, R. Hanna, D. Friend, H. Moore, H. Chen, A. M. Farese, T. J. MacVittie, G. D. Virca and J. E. Sims, *Immunity,* 2003, **18**, 87–96.
7. W. P. Arend, *Cytokine Growth Factor Rev.,* 2002, **13**, 323–340.
8. C. A. Dinarello, A. Simon and J. W. M. van der Meer, *Nat. Rev. Drug Discovery,* 2012, **11**, 633–652.
9. D. Kessler-Becker, T. Krieg and B. Eckes, *Biochem. J.,* 2004, **379**, 351–358.
10. G. Fantuzzi and C. A. Dinarello, *J. Leukocyte Biol.,* 1996, **59**, 489–493.

11. M. Tsuzaki, G. Guyton, W. Garrett, J. M. Archambault, W. Herzog, L. Almekinders, D. Bynum, X. Yang and A. J. Banes, *J. Orthop. Res.*, 2003, **21**, 256–264.
12. P. Bodel, P. Ralph, K. Wenc and J. C. Long, *J. Clin. Invest.*, 1980, **65**, 514–518.
13. F. Martinon and J. Tschopp, *Cell*, 2004, **117**, 561–574.
14. G. Fantuzzi, G. Ku, M. W. Harding, D. J. Livingston, J. D. Sipe, K. Kuida, R. A. Flavell and C. A. Dinarello, *J. Immunol.*, 1997, **158**, 1818–1824.
15. L. a. B. Joosten, M. G. Netea, G. Fantuzzi, M. I. Koenders, M. M. A. Helsen, H. Sparrer, C. T. Pham, J. W. M. van der Meer, C. A. Dinarello and W. B. van den Berg, *Arthritis Rheum.*, 2009, **60**, 3651–3662.
16. N. Lonberg, *Nat. Biotechnol.*, 2005, **23**, 1117–1125.
17. D. M. Fishwild, S. L. O'Donnell, T. Bengoechea, D. V. Hudson, F. Harding, S. L. Bernhard, D. Jones, R. M. Kay, K. M. Higgins, S. R. Schramm and N. Lonberg, *Nat. Biotechnol.*, 1996, **14**, 845–851.
18. R. Alten, H. Gram, L. A. Joosten, W. B. van den Berg, J. Sieper, S. Wassenberg, G. Burmester, P. van Riel, M. Diaz-Lorente, G. J. M. Bruin, T. G. Woodworth, C. Rordorf, Y. Batard, A. M. Wright and T. Jung, *Arthritis Res. Ther.*, 2008, **10**, R67.
19. M. Blech, D. Peter, P. Fischer, M. M. T. Bauer, M. Hafner, M. Zeeb and H. Nar, *J. Mol. Biol.*, 2013, **425**, 94–111.
20. T. Geiger, H. Towbin, A. Cosenti-Vargas, O. Zingel, J. Arnold, C. Rordorf, M. Glatt and K. Vosbeck, *Clin. Exp. Rheumatol.*, 1993, **11**, 515–522.
21. J. Bonner, P. Lloyd, P. Lowe, G. Golor, R. Woessner and S. Pascoe, in *Annual Congress of the European Respiratory Society*, 2006, p. 748.
22. R. Alten, J. Gomez-Reino, P. Durez, A. Beaulieu, A. Sebba, G. Krammer, R. Preiss, U. Arulmani, A. Widmer, X. Gitton and H. Kellner, *BMC Musculoskeletal Disord.*, 2011, **12**, 153.
23. S. B. Cohen, L. W. Moreland, J. J. Cush, M. W. Greenwald, S. Block, W. J. Shergy, P. S. Hanrahan, M. M. Kraishi, A. Patel, G. Sun and M. B. Bear, *Ann. Rheum. Dis.*, 2004, **63**, 1062–1068.
24. I. Demin, B. Hamrén, O. Luttringer, G. Pillai and T. Jung, *Clin. Pharmacol. Ther.*, 2012, **92**, 352–359.
25. E. Lainka, U. Neudorf, P. Lohse, C. Timmann, M. Bielak, S. Stojanov, K. Huss, R. von Kries and T. Niehues, *Klinische Pädiatrie*, 2010, **222**, 356–361.
26. E. Aganna, F. Martinon, P. N. Hawkins, J. B. Ross, D. C. Swan, D. R. Booth, H. J. Lachmann, A. Bybee, R. Gaudet, P. Woo, C. Feighery, F. E. Cotter, M. Thome, G. A. Hitman, J. Tschopp and M. F. McDermott, *Arthritis Rheum.*, 2002, **46**, 2445–2452.
27. I. Aksentijevich, M. Nowak, M. Mallah, J. J. Chae, W. T. Watford, S. R. Hofmann, L. Stein, R. Russo, D. Goldsmith, P. Dent, H. F. Rosenberg, F. Austin, E. F. Remmers, J. E. Balow, S. Rosenzweig, H. Komarow, N. G. Shoham, G. Wood, J. Jones, N. Mangra, H. Carrero, B. S. Adams, T. L. Moore, K. Schikler, H. Hoffman, D. J. Lovell,

R. Lipnick, K. Barron, J. J. O'Shea, D. L. Kastner and R. Goldbach-Mansky, *Arthritis Rheum.*, 2002, **46**, 3340–3348.
28. H. M. Hoffman, J. L. Mueller, D. H. Broide, A. A. Wanderer and R. D. Kolodner, *Nat. Genet.*, 2001, **29**, 301–305.
29. A. Simon and J. W. M. van der Meer, *Am. J. Physiol.: Regul., Integr. Comp. Physiol.*, 2007, **292**, R86–R98.
30. R. L. Glaser and R. Goldbach-Mansky, *Curr. Allergy Asthma Rep.*, 2008, **8**, 288–298.
31. N. A. Thornberry, H. G. Bull, J. R. Calaycay, K. T. Chapman, A. D. Howard, M. J. Kostura, D. K. Miller, S. M. Molineaux, J. R. Weidner and J. Aunins, *Nature*, 1992, **356**, 768–774.
32. F. Martinon, K. Burns and J. Tschopp, *Mol. Cell*, 2002, **10**, 417–426.
33. H. M. Hoffman, S. Rosengren, D. L. Boyle, J. Y. Cho, J. Nayar, J. L. Mueller, J. P. Anderson, A. A. Wanderer and G. S. Firestein, *Lancet*, 2004, **364**, 1779–1785.
34. G.-S. Lee, N. Subramanian, A. I. Kim, I. Aksentijevich, R. Goldbach-Mansky, D. B. Sacks, R. N. Germain, D. L. Kastner and J. J. Chae, *Nature*, 2012, 10–15.
35. P. N. Hawkins, H. J. Lachmann, E. Aganna and M. F. McDermott, *Arthritis Rheum.*, 2004, **50**, 607–612.
36. P. N. Hawkins, H. J. Lachmann and M. F. McDermott, *N. Engl. J. Med.*, 2003, **348**, 2583–2584.
37. R. Goldbach-Mansky, N. J. Dailey, S. W. Canna, A. Gelabert, J. Jones, B. I. Rubin, H. J. Kim, C. Brewer, C. Zalewski, E. Wiggs, S. Hill, M. L. Turner, B. I. Karp, I. Aksentijevich, F. Pucino, S. R. Penzak, M. H. Haverkamp, L. Stein, B. S. Adams, T. L. Moore, R. C. Fuhlbrigge, B. Shaham, J. N. Jarvis, K. O'Neil, R. K. Vehe, L. O. Beitz, G. Gardner, W. P. Hannan, R. W. Warren, W. Horn, J. L. Cole, S. M. Paul, P. N. Hawkins, T. H. Pham, C. Snyder, R. A. Wesley, S. C. Hoffmann, S. M. Holland, J. a. Butman and D. L. Kastner, *N. Engl. J. Med.*, 2006, **355**, 581–592.
38. P. N. Hawkins, H. J. Lachmann, E. Aganna and M. F. McDermott, *Arthritis Rheum.*, 2004, **50**, 607–612.
39. H. J. Lachmann, P. Lowe, S. D. Felix, C. Rordorf, K. Leslie, S. Madhoo, H. Wittkowski, S. Bek, N. Hartmann, S. Bosset, P. N. Hawkins and T. Jung, *J. Exp. Med.*, 2009, **206**, 1029–1036.
40. A. Chakraborty, S. Tannenbaum, C. Rordorf, P. J. Lowe, D. Floch, H. Gram and S. Roy, *Clin. Pharmacokinet.*, 2012, **51**, e1–e18.
41. E. V. Granowitz, E. Vannier, D. D. Poutsiaka and C. A. Dinarello, *Blood*, 1992, **79**, 2364–2369.
42. H. J. Lachmann, I. Kone-Paut, J. B. Kuemmerle-Deschner, K. S. Leslie, E. Hachulla, P. Quartier, X. Gitton, A. Widmer, N. Patel and P. N. Hawkins, *N. Engl. J. Med.*, 2009, **360**, 2416–2425.
43. I. Koné-Paut, H. J. Lachmann, J. B. Kuemmerle-Deschner, E. Hachulla, K. S. Leslie, R. Mouy, A. Ferreira, K. Lheritier, N. Patel, R. Preiss and P. N. Hawkins, *Arthritis Res. Ther.*, 2011, **13**, R202.

44. J. B. Kuemmerle-Deschner, E. Hachulla, R. Cartwright, P. N. Hawkins, T. A. Tran, B. Bader-Meunier, J. Hoyer, M. Gattorno, A. Gul, J. Smith, K. S. Leslie, S. Jiménez, S. Morell-Dubois, N. Davis, N. Patel, A. Widmer, R. Preiss and H. J. Lachmann, *Ann. Rheum. Dis.,* 2011, **70**, 2095–2102.
45. H. M. Hoffman and A. Simon, *Nat. Rev. Rheumatol.,* 2009, **5**, 249–256.
46. C. Fonnesu, C. Cerquaglia, M. Giovinale, V. Curigliano, E. Verrecchia, G. de Socio, M. La Regina, G. Gasbarrini and R. Manna, *Joint, Bone, Spine: Revue du Rhumatisme,* 2009, **76**, 227–233.
47. J. J. Chae, G. Wood, S. L. Masters, K. Richard, G. Park, B. J. Smith and D. L. Kastner, *Proc. Natl. Acad. Sci. U. S. A.,* 2006, **103**, 9982–9987.
48. J. J. Chae, Y.-H. Cho, G.-S. Lee, J. Cheng, P. P. Liu, L. Feigenbaum, S. I. Katz and D. L. Kastner, *Immunity,* 2011, **34**, 755–768.
49. S. Papin, S. Cuenin, L. Agostini, F. Martinon, S. Werner, H.-D. Beer, C. Grütter, M. Grütter and J. Tschopp, *Cell Death Differ.,* 2007, **14**, 1457–1466.
50. U. Meinzer, P. Quartier, J.-F. Alexandra, V. Hentgen, F. Retornaz and I. Koné-Paut, *Semin. Arthritis Rheum.,* 2011, **41**, 265–271.
51. A. Gul, *et al.*, *Arthritis Rheum.,* 2012, **64**, s322.
52. I. Mitroulis, P. Skendros, A. Oikonomou, A. G. Tzioufas and K. Ritis, *Ann. Rheum. Dis.,* 2011, **70**, 1347–1348.
53. L. Cantarini, O. M. Lucherini, I. Muscari, B. Frediani, M. Galeazzi, M. G. Brizi, G. Simonini and R. Cimaz, *Autoimmun. Rev.,* 2012, **12**, 38–43.
54. N. Ter Haar, H. Lachmann, S. Ozen, P. Woo, Y. Uziel, C. Modesto, I. Koné-Paut, L. Cantarini, A. Insalaco, B. Neven, M. Hofer, D. Rigante, S. Al-Mayouf, I. Touitou, R. Gallizzi, E. Papadopoulou-Alataki, S. Martino, J. Kuemmerle-Deschner, L. Obici, N. Iagaru, A. Simon, S. Nielsen, A. Martini, N. Ruperto, M. Gattorno and J. Frenkel, *Ann. Rheum. Dis.,* 2012, **72**, 678.
55. M. Gattorno, *et al.*, *Arthritis Rheum.,* 2012, **64**, s322.
56. S. Normand, B. Massonnet, A. Delwail, L. Favot, L. Cuisset, G. Grateau, F. Morel, C. Silvain and J.-C. Lecron, *Eur. Cytokine Network,* 2009, **20**, 101–107.
57. R. van der Burgh, N. M. Ter Haar, M. L. Boes and J. Frenkel, *Clin. Immunol.,* 2012, **86**, 21–23.
58. B. Massonnet, S. Normand, R. Moschitz, A. Delwail, L. Favot, M. Garcia, N. Bourmeyster, L. Cuisset, G. Grateau, F. Morel, C. Silvain and J.-C. Lecron, *Eur. Cytokine Network,* 2009, **20**, 112–120.
59. C. Galeotti, U. Meinzer, P. Quartier, L. Rossi-Semerano, B. Bader-Meunier, P. Pillet and I. Koné-Paut, *Rheumatology,* 2012, **51**, 1855–1859.
60. E. J. Bodar, L. M. Kuijk, J. P. H. Drenth, J. W. M. van der Meer, A. Simon and J. Frenkel, *Ann. Rheum. Dis.,* 2011, **70**, 2155–2158.
61. H. D. de Koning, J. Schalkwijk, J. van der Ven-Jongekrijg, M. Stoffels, J. W. M. van der Meer and A. Simon, *Ann. Rheum. Dis.,* 2012, 1–6.
62. K. Krause, K. Weller, R. Stefaniak, H. Wittkowski, S. Altrichter, F. Siebenhaar, T. Zuberbier and M. Maurer, *Allergy,* 2012, **67**, 943–950.

CHAPTER 9

Discovery and Development of Tafamidis for the Treatment of TTR Familial Amyloid Polyneuropathy

RICHARD LABAUDINIÈRE

Labaudinière Consulting, LLC, Medfield, MA 02052, USA
E-mail: labaudiniere@gmail.com

9.1 Rare Genetic Diseases and Drug Development

Half of the drugs under development are still failing in the clinic at the proof of concept stage for insufficient efficacy.[1] The reasons for this lack of clinical efficacy are certainly multiple but it cannot be denied that since the incorporation and quasi-exclusive reliance on high-throughput technologies in the pharmaceutical industry and despite the recent emphasis, following the completion of the human genome project, on translational medicine, particularly at the NIH, pharmaceutical research as a whole has focused much more on so-called druggable targets (screenable target classes with positive track record for lead generation) than on therapeutically validated targets, increasing the number of drug candidates searching in the clinic for the right therapeutic indication. The importance of the identification and validation of a direct clinical link between the biological function of a target and a disease process has been neglected as a key success factor for a positive

RSC Drug Discovery Series No. 38
Orphan Drugs and Rare Diseases
Edited by David C Pryde and Michael J Palmer

Published by the Royal Society of Chemistry, www.rsc.org

clinical outcome. The field of rare monogenic diseases constitutes a unique opportunity to develop drugs on genetically validated targets. Capitalising on the aftermath of the successful completion of the human genome project, the increasing numbers of genetic and functional genomics studies, and the favourable regulatory environment for rare diseases in the major pharmaceutical markets (USA, EU and Japan), rare monogenic diseases represent an attractive opportunity for drug development with a significantly increased chance of positive clinical outcome because of the genetic validation of the biological target (specific mutations clinically and functionally linked to disease state) and the access to a genetically selected patient population for clinical evaluation. Positive target engagement with the appropriate safety profile should guarantee a successful clinical development and translate into patients with the desired disease-modifying therapeutic effect.

Nevertheless, the field of rare genetic diseases is still largely an uncharted territory for drug development. Most of the genetically validated targets are non-druggable targets or pathways, not always easily amenable to high-throughput discovery technologies and without a track record for lead generation. Natural history studies for these diseases are scarce and validated clinical end points are lacking for most of them.

Among the rare genetic diseases, the field of protein misfolding diseases witnessed several successful drug development stories in the past two decades (Table 9.1). Genetic protein misfolding diseases are caused by mutations in genes causing protein misfolding, leading to protein trafficking defects and loss of function as observed in lysosomal storage diseases and cystic fibrosis or formation of amyloid or aggregates resulting in a toxic gain of function as in TTR amyloidosis and Huntington's disease (Figure 9.1). Since 1994 and the approval of Ceredase and Cerezyme for the treatment of Gaucher disease, treatments have been identified in several rare genetic diseases caused by protein misfolding. Initially, enzyme replacement therapy was successfully used in several lysosomal storage diseases. More recently, small-molecule pharmacological chaperones reached the market for the treatment of cystic fibrosis and phenylketonuria and tafamidis was approved for the treatment of transthyretin familial amyloid polyneuropathy (TTR-FAP), the neurological form of TTR amyloidosis (Table 9.1).

The purpose of this chapter is to present the story of the identification and development of tafamidis, the first drug approved for the treatment of a rare genetic form of amyloidosis due to the misfolding of a protein called TTR.

9.2 Familial Amyloid Polyneuropathy, an Autosomal Dominant Amyloidosis

The amyloidoses are diseases induced by accumulation of various insoluble fibrillar proteins (amyloid) in the tissues in amounts sufficient to impair normal function. Transthyretin (TTR) is one of 27 proteins known to form fibrillar deposits in human amyloidoses.

Table 9.1 Drug development success stories in rare genetic misfolding diseases.

Disease	*Causative gene*	*Defect*	*Drug*	*Mode of action*[a]	*Year of first approval*[b]	*Company*
Lysosomal storage diseases		Mistrafficking – loss of function				
Gaucher disease	β-Glucocerebrosidase		Alglucerase (Ceredase)	ERT	1994 (EU)	Genzyme
			Imiglucerase (Cerezyme)	ERT	1994 (USA)	Genzyme
			Miglustat (Zavesca)	Substrate reduction	2002 (EU)	OGS/actelion
			α-Taliglucerase (Elelyso)	ERT	2012 (USA)	Protalix/Pfizer
			α-Velaglucerase (Vpriv)	ERT	2010 (USA)	Shire
Fabry disease	α-Galactosidase		β-Agalsidase (Fabrazyme)	ERT	2003 (USA)	Genzyme
			α-Agalsidase (Replagal)	ERT	2001 (EU)	TKT/Shire
Pompe disease	α-Glucosidase		α-Alglucosidase (Myozyme)	ERT	2006 (EU)	Genzyme
Cystic fibrosis	CFTR	Trafficking and gating defects – loss of function	Ivacaftor (Kalydeco)	Chemical chaperone[c]	2012 (USA)	Vertex
Phenylketonuria	Phenylalanine hydroxylase	Degradation – loss of function	Sapropterin (Kuvan)	Chemical chaperone	2007 (USA)	Biomarin
Familial amyloid polyneuropathy	TTR	Amyloid formation – gain of toxic function	Tafamidis (Vyndaqel)	Native state stabiliser	2011 (EU)	FoldRx/Pfizer

[a]ERT = enzyme replacement therapy.
[b]Territory of first approval.
[c]Correct gating defect.

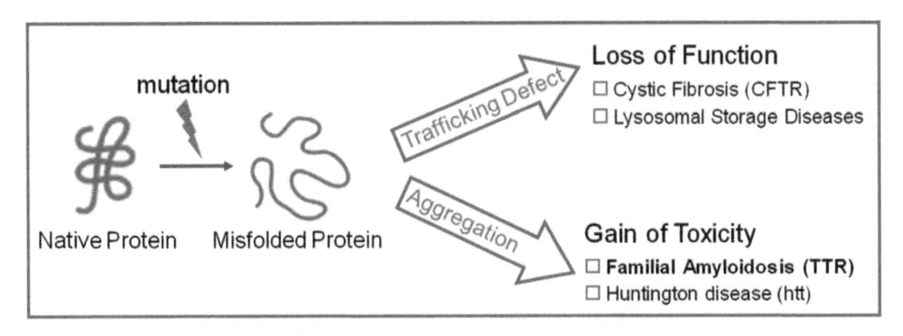

Figure 9.1 Genetic misfolding diseases. The folding and maintenance of proteins in a correctly folded active form is essential to normal cellular function. Protein misfolding, due to mutations or to defects in cellular quality control mechanisms, leads to the accumulation of proteins with insufficient activity to perform their function (loss of function) or results in the formation of toxic misfolded intermediates that themselves lead to pathology (toxic gain of function). In several disorders, such as cystic fibrosis and several lysosomal storage diseases (Gaucher, Fabry and Pompe diseases), misfolded proteins are not trafficked to their intended cellular location, in others such as Huntington's disease and familial amyloidosis, misfolded proteins aggregate with accumulation of toxic misfolded intermediates.

TTR, a 127 amino acid, 55 kDa protein, primarily synthesised in the liver, is a secreted protein present in the blood and cerebrospinal fluid and is a carrier of thyroxine and retinol–binding protein–retinol (vitamin A) complex.[2,3] In its native state, TTR exists as a homotetramer with two C2 symmetric funnel-shaped thyroxine binding sites located at the central dimer–dimer interface (Figure 9.2).[4–6]

Both natural sequence TTR and mutated variants of TTR are involved in amyloid disease, but a mutation in TTR accelerates the process of TTR fibrillogenesis and is the most important risk factor for TTR amyloidosis.[2,3] TTR amyloidosis (ATTR) is the most prevalent form of familial amyloidosis with an estimate of 8000 to 10 000 patients worldwide. In the USA, the gene frequency is estimated to be 1 in 100 000 in Caucasians, corresponding to a potential US patient population of around 1000 people.[7] More than 100 TTR single site variants have been thus far identified and associated with TTR amyloidosis.

In the pathogenesis of ATTR, an amyloidogenic mutation can lead to tetramer dissociation into monomers, and the folded monomers undergo partial denaturation to produce alternatively folded monomeric amyloidogenic intermediates.[8,9] These intermediates then misassemble into soluble oligomers, profilaments, filaments and amyloid fibrils (Figure 9.3). The rate-limiting step for TTR amyloid formation is tetramer dissociation.[10] All disease-associated mutations characterised thus far destabilise the TTR tetramer and many influence the velocity of rate-limiting tetramer dissociation.[8] Several studies are consistent with the misfolded monomer and/or the

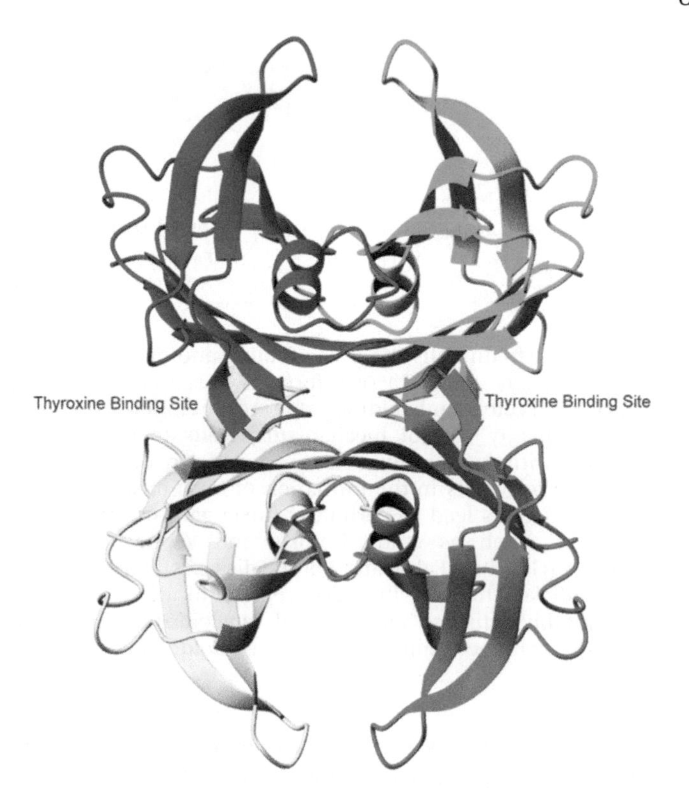

Figure 9.2 Tetrameric structure of transthyretin. TTR, a 127 amino acid, 55 kDa protein, primarily synthesised in the liver, is a secreted protein present in the blood and cerebrospinal fluid and is a carrier of thyroxine and retinol-binding protein–retinol (vitamin A) complex. In its native state, TTR exists as a homotetramer with a 222-symmetry (*e.g.* three perpendicular C2 symmetry axes) featuring a pair of funnel-shaped, C2-interconvertable thyroxine binding sites located at the dimer–dimer interface.

early soluble aggregates being the major neurotoxic species along the TTR amyloid cascade.[11–14]

TTR amyloidosis develops primarily in the heart and along the peripheral nerves, as well as in the gastrointestinal tract, vitreous and elsewhere.[15] ATTR is invariably progressive and fatal, patients usually dying 5–10 years from initial symptoms. Depending on the primary site of deposition, the disease has been termed familial amyloid cardiomyopathy (TTR-FAC) or familial amyloid polyneuropathy (TTR-FAP).

In TTR-FAC patients, TTR amyloid fibrils infiltrate the myocardium, leading to diastolic dysfunction progressing to restrictive cardiomyopathy.[16] Several single point mutations in TTR have been primarily associated with TTR-FAC: V20I, P24S, A45T, Gly47Val, Glu51Gly, I68L, Gln92Lys, L111M and

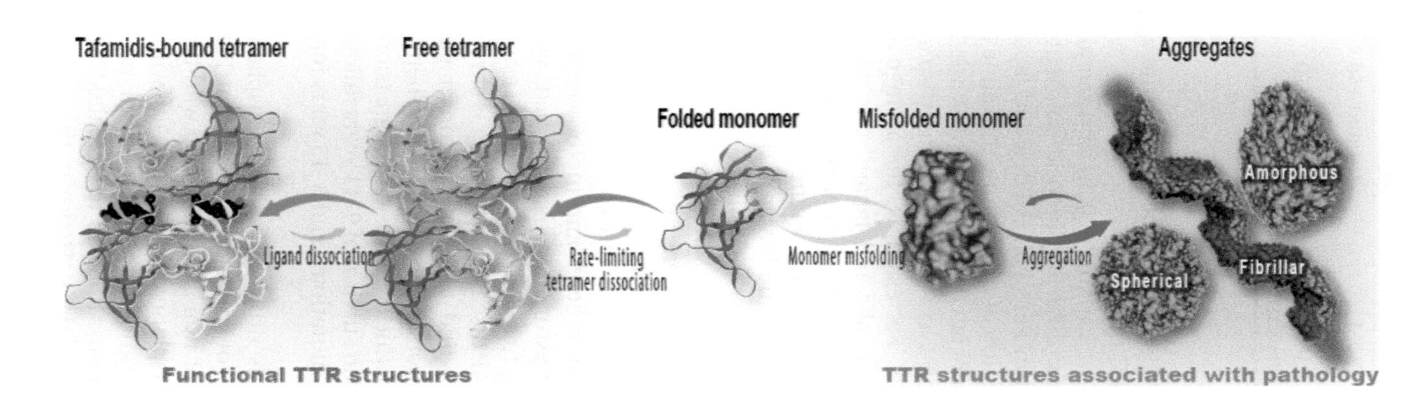

Figure 9.3 TTR amyloid cascade. In the pathogenesis of TTR amyloidosis, the rate-determining step for TTR amyloid formation is tetramer dissociation. Tetramers dissociate into folded monomers that undergo partial denaturation to produce alternatively folded monomeric amyloidogenic intermediates. These intermediates then misassemble into soluble oligomers, profilaments, filaments and amyloid fibrils. The misfolded monomer and/or the early soluble aggregates are suspected to be the major neurotoxic species inducing the neurodegenerative process observed in TR-FAP. Tafamidis binding to TTR stabilizes the tetrameric native state and slows tetramer dissociation, thereby efficiently inhibiting aggregation.

Table 9.2 Epidemiology of TTR-FAP.

Region	*Countries*	*Variant*	*Age of onset*	*Penetrance*
Endemic	Portugal	V30M	30s and >50	>80%
	Sweden	V30M	50	20–25%
	Japan	V30M	30s and >50	>80%
Non-endemic	ROW	80+ variants	>50	Unknown

V122I.[17–19] It is usually late onset (symptoms occur after the age of 50). No treatment is available for TTR-FAC patients.

Deposition of amyloid fibrils of variant TTR in peripheral nerve tissue produces the condition called familial amyloidotic polyneuropathy (TTR-FAP). Large foci of TTR-FAP exist in Portugal, Sweden and Japan (Table 9.2). The V30M variant with a substitution of methionine for valine at position 30 is the most common mutation associated with TTR-FAP. While TTR-FAP is an autosomal dominant disease, the average age of onset displays some variability within and across variants. For example, the average age at disease onset in endemic regions of Portugal[20] and Japan[21,22] is approximately 32 years. However, in Sweden disease onset generally occurs in the fifth decade,[12] as in non-endemic cases in Japan,[23] France[24,25] and Italy,[26] in which symptom onset usually occurs later in life (Table 9.2).

Clinical manifestations of the disease are similar regardless of age of onset and nature of mutation. The classic presentation is sensory neuropathy starting in the lower extremities and evidence of motor neuropathy follows within a few years.[20,24,27,28] Axonal degeneration involves small fibres with loss of pain and temperature sensations, large fibres with loss of touch, vibration and position sensations, abnormal reflexes and motor weakness with difficulty ambulating. Autonomic dysfunction is observed with dizziness, gastrointestinal disorders leading to severe malnutrition, sexual dysfunction and urinary incontinence. TTR-FAP disease progression is similar across TTR mutations, geographic locations, and early *vs.* late onset patients and is characterised by a progressive worsening of both peripheral (particularly motor) and autonomic neuropathies. Rates of progression are similar between TTR mutations, with mean survival of 9–11 years following symptom onset for patients with the V30M mutation[20] and 3–15 years for patients with non-V30M mutations.[29]

The only treatment available is orthotopic liver transplantation, which removes the main production site of amyloidogenic mutant TTR protein and replaces it with the production of non-amyloidogenic wild-type TTR. More than 2000 patients have been transplanted since the 1990s, with a 5 year post-transplant survival rate of 77% and a 10 year survival rate of 71%.[30,31] Progression of neuropathy symptoms appears to be halted in a significant proportion of patients (limited set of data). However, this invasive procedure is associated with significant short- and long-term morbidity, the first year mortality post-transplant averaging approximately 10%.

9.3 TTR Tetramer Stabilisation, a Genetically Validated Approach for Disease-Modifying Treatment

T119M, a TTR variant with a substitution of methionine for threonine at position 119 of the polypeptide chain, is a non-pathogenic variant. When this T119M mutation is present on one TTR allele with an amyloidogenic variant such as V30M on the other TTR allele, it has been observed to afford a protective effect. Nine compound heterozygous carriers of V30M/T119M belonging to five different kindreds have been described in the Portuguese population.[32] Among them, only one developed signs of the disease, albeit with a particularly benign course. The other carriers of the two mutations were asymptomatic well after the mean age of onset of their affected siblings (who were heterozygous for the V30M mutation). The simultaneous occurrence of T119M and V30M in the same individual seems to delay or prevent the deleterious effects of TTR V30M, suggesting a protective effect of T119M on the pathogenic effects of V30M.[33-36] Recently, a new TTR variant, TTR R104H, has been reported in a Japanese family.[37] The variant was detected in heterozygous and compound heterozygous individuals who were also carriers of V30M. Similar to T119M, R104H seems to be non-pathogenic and confers protective clinical effects in the compound heterozygous carrier.

Stability studies demonstrate an increased resistance of tetrameric TTR to dissociation into monomers when the T119M or R104H variant is present.[34] The resistance conferred by these different variants towards dissociation have been demonstrated under various denaturation conditions (urea or acidic) (Table 9.3).[36,38,39] Furthermore, subunit exchange experiments show that homotetrameric T119M TTR self-exchanges subunits slower than both wild-type and V30M TTR,[36] demonstrating that the energetic barrier for T119M dissociation, by destabilisation of the dissociative transition state, is considerably higher than those characterising wild-type and V30M.[8,39] Thus, tetramer stabilisation by increasing the kinetic barriers for tetrameric

Table 9.3 Stabilization properties identified for protective variants in TTR-FAP.[a]

Properties	*T119M*	*R104H*	*TTR binder/ tafamidis*
TTR tetrameric stabilisation under urea (or guanidinium thiocyanate) denaturation conditions	Yes[33,38]	Yes[38]	**Yes**[39,54]
Inhibition of fibril formation under acidic conditions	Yes[36]	N/A	**Yes**[39,54]
Higher kinetic barriers for tetramer dissociation	Yes[8,36]	N/A	**Yes**[39,54]
Clinical protective effects	Yes[32,37]	Yes[32,38]	**Yes**[76]

[a]N/A = not available.

dissociation, the rate-limiting step of TTR amyloidosis, is the likely mechanism for the intragenic trans-suppression and protective effect observed in the presence of T119M variant.[36,39]

By stabilisation of the tetrameric native state of TTR, small-molecule binders to TTR at the thyroxine binding site similarly increase the activation barrier associated with tetramer dissociation.[39] In fact binding of a TTR tetramer stabiliser to only one thyroxine binding site seems to be sufficient to stabilise tetrameric TTR.[40] Hence, small-molecule binding to TTR has been associated with increased resistance towards tetramer dissociation, the rate-limiting step in TTR fibril formation, mimicking the overall stabilisation effect shown by the intragenic trans-suppressors (Table 9.3).

Therefore, stabilisation of tetrameric TTR with a small-molecule binder represents a genetically validated therapeutic target for TTR amyloidosis and TTR-FAP, with such a TTR stabiliser drug expected to stop progression of the disease. Specific binding to TTR is not expected to significantly affect thyroxine metabolism and function. TTR is a tertiary carrier of thyroxine in the blood, responsible for only 10–15% of protein-bound plasma thyroxine and less than 1% of circulating TTR in the blood carries thyroxine, the primary carrier in blood being the thyroxine-binding globulin.[41,42] In TTR-null mice, despite a 50% reduction in total circulating thyroxine levels, the free thyroxine levels remained normal and no alteration was observed in circulating T3 and TSH levels.[43] Humans carrying TTR mutations that lead to increased or decreased thyroxine affinity do not show signs of altered thyroid hormone function. For example, although V30M TTR homotetramers have severely reduced thyroxine binding, V30M homozygous patients are euthyroid.[44]

9.4 Discovery of Tafamidis, a Potent TTR Binder

Denaturation stress and/or acidic conditions are sufficient to induce TTR tetramer dissociation. Incubation of TTR in low pH environments is all that is required to initiate the fibrillogenesis reaction.[10] Searches for inhibitors of TTR fibril formation under acidic conditions were conducted by Jeff Kelly in the 1990s. A limited screening of FDA-approved compounds identified several hits, most of them with NSAID activities such as flufenamic acid and diflunisal.[45] X-ray structures were generated to show the importance of the carboxylic acid moiety and the presence of aromatic rings substituted with halogens to optimally occupy the pockets formed at the TTR dimer–dimer interaction surface to accommodate thyroxine, the natural ligand of TTR. Using the X-ray structures of the early TTR binders as a model, a diverse set of pharmacophores was generated by the Kelly team[46] with good activity as inhibitors of fibril formation in acid conditions: N-substituted anthranilic acids,[47,48] *N*-phenyl-phenoxazine dicarboxylic acids,[49] benzoxazole carboxylic acids,[50] biphenyl carboxylic acids,[51] dibenzofuran carboxylic acids[52] and bisaryloxime carboxylic acids[53] (Scheme 9.1).

An ideal TTR tetramer stabiliser for the treatment of TTR amyloidosis would be a drug with a good safety profile, lacking the NSAID activity,

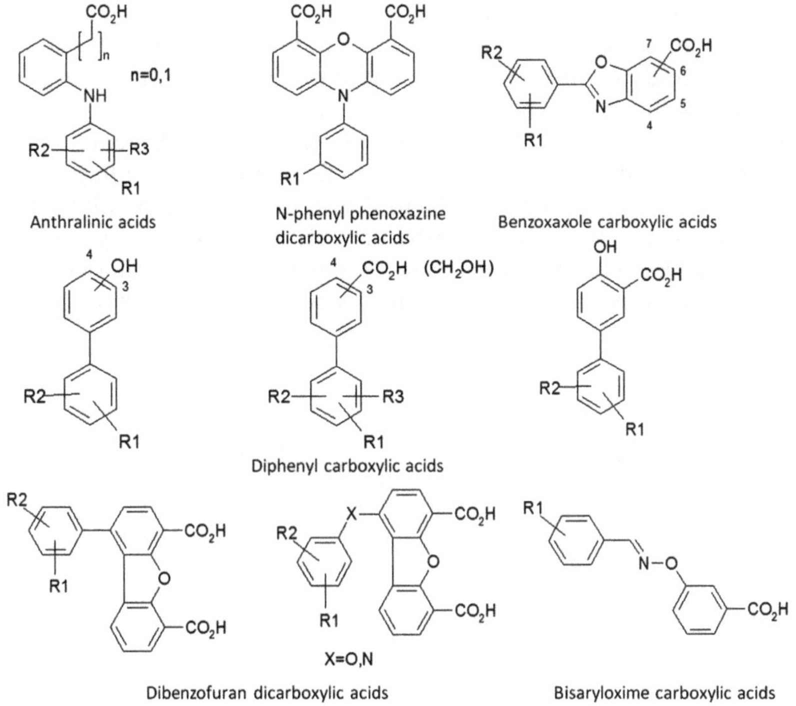

Scheme 9.1 Pharmacophores for TTR fibril formation inhibition identified by the Kelly team at SCRIPPS at the time of initiation of the FoldRx TTR programme.

contraindicated in patients with cardiomyopathy and gastrointestinal disorders, with sufficient efficacy to stabilise the tetrameric form of TTR and therefore inhibit fibril formation, with good absorption, low clearance and significant half-life to give sufficient blood exposure to occupy at least one binding site on all circulating tetrameric TTR (normal TTR plasma concentration range of 3–5 μM).

In early 2004, at the initiation of the TTR programme at FoldRx, a start-up company co-founded by Jeff Kelly, around 60 representative compounds were selected from the six major pharmacophores identified at that time by the Kelly group at SCRIPPS (Scheme 9.1). Analogues with good fibril inhibition activity (>50% inhibition at a molar ratio of 1) were tested for their lack of cyclooxygenase activity (COX1 and COX2). Compounds with less than 10% inhibition at 10 μM were further tested for lack of inhibition for key cytochrome P450 enzymes (CYP450 3A4, 2D6 and 2C9) (<25% at 10 μM) and for hERG channel activity (<10% at 10 μM). The best analogues remaining from three pharmacophores (benzoxazole carboxylic acids, biphenyl carboxylic acids and dibenzofuran dicarboxylic acids) were tested for plasma exposure after a single oral dose in rats. A better *in vitro* profile and superior plasma exposure were observed with the benzoxazole carboxylic acids. This series was therefore selected for further analoguing and profiling to identify an orally bioavailable and potent inhibitor of TTR tetramer dissociation as a clinical candidate for the treatment of TTR-FAP. The benzoxazole-6-carboxylic acid analogue with the 3,5-dichlorophenyl moiety, tafamidis (Scheme 9.2), was selected in December 2004 as a clinical candidate for IND profiling.

Tafamidis binds with high affinity to TTR at its thyroxine binding sites, as demonstrated by the crystal structure of TTR bound by tafamidis (Figure 9.4).[54] Each binding site located at the dimer–dimer interface can be divided into an inner and outer binding cavity, which are further defined by

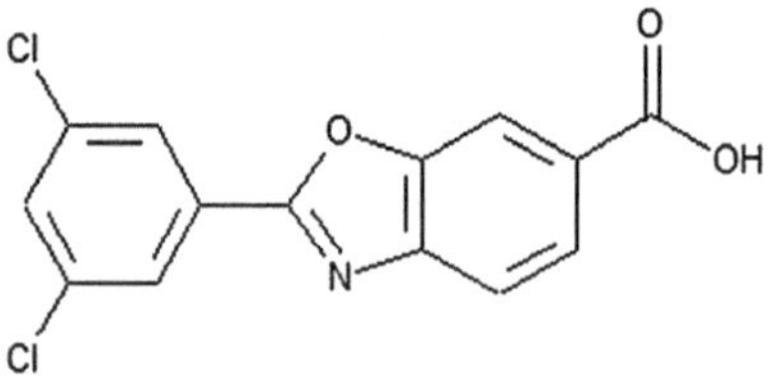

Scheme 9.2 Structure of tafamidis (2-(3,5-dichloro-phenyl)-benzoxazole-6-carboxylic acid).

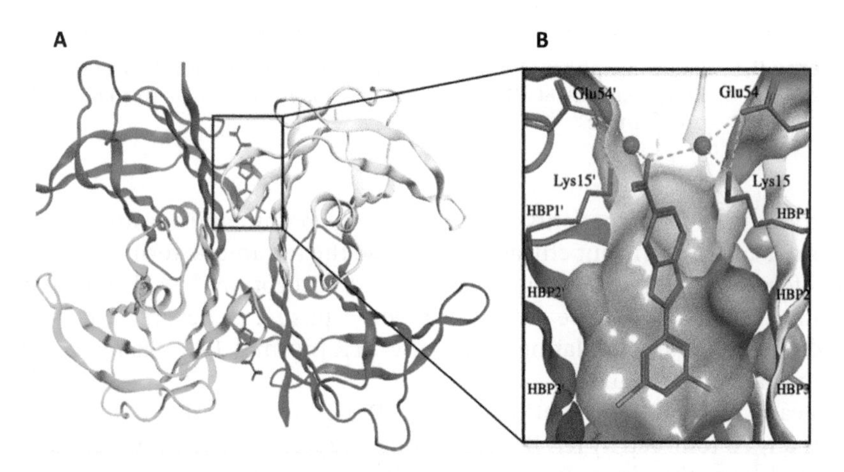

Figure 9.4 Crystal structure of tafamidis bound to TTR. (A) Three-dimensional ribbon diagram depiction of the TTR tetramer with tafamidis bound. The four TTR monomers are individually coloured. (B) Magnified image of tafamidis bound in one of the thyroxine binding sites. Connolly analytical surface representation (grey, hydrophobic; purple, polar) depicts the hydrophobicity of the binding site. The 3,5-chloro groups are placed in the halogen binding pockets (HBPs) 3 and 3′ making hydrophobic interactions, while the carboxylate function of tafamidis interacts with the Lys15/15′ residues of TTR through bridging water interactions (represented as dotted lines) at the edge of halogen binding pockets 1 and 1′.

three symmetry related pairs of binding pockets, called halogen-binding pockets (HBPs) because they accommodate the iodines of thyroxine.[55] HBP1 and 1′ are in the outer cavity, HBP2 and 2′ are at the interface between the inner and outer cavity and HBP3 and 3′ are in the inner cavity. The 3,5-dichloro substituents on the phenyl ring of tafamidis occupy the two symmetrical HBPs 3 and 3′, in the inner binding cavity (Figure 9.4B). The bis-meta chloro substitution permits simultaneous binding to both dimers constituting the weaker dimer–dimer interface, a superior feature for TTR kinetic stabilisers as compared to those having mono-meta substitution.[56,57] The benzoxazole ring of tafamidis is positioned in the hydrophobic environment of HBP2 and 2′ and HBP1 and 1′. In this orientation, the meta-carboxylate substituent on the benzoxazole ring extends out into the periphery of the thyroxine binding site, where it engages in bridging hydrogen bonds through ordered water molecules with Lys15/15′ (Figure 9.4B). This combination of hydrophobic and ionic interactions appears to bridge adjacent subunits and to kinetically stabilise the TTR tetramer.[53,56,58]

Isothermal titration calorimetry (ITC) was employed to determine the binding constants of tafamidis to TTR at pH 8.0 (25 °C).[49] Tafamidis binds to TTR with negative cooperativity, with $K_{d1} = 3$ nM and $K_{d2} = 278$ nM. Binding at the first site in TTR is 93 times stronger than binding at the second site; this negatively cooperative binding behaviour of ligands to TTR is typical and thought to arise from subtle conformational alterations after kinetic stabiliser binding at the first site.[54]

9.5 Preclinical Profiling of Tafamidis as a Potent Tetramer Stabiliser *In Vitro* and *Ex Vivo*

Three major success factors were identified to warrant clinical efficacy of tafamidis as an inhibitor of TTR tetramer dissociation:

- Demonstrated efficacy to inhibit TTR tetramer dissociation in both denaturating and non-denaturating conditions and across amyloidogenic variants.
- A pharmacological assay to assess biological activity in plasma and provide a measure of target engagement in the clinic.
- A safety margin appropriate for chronic treatment in TTR-FAP.

9.5.1 Inhibition of Tetramer Dissociation with Tafamadis

Most of the *in vitro* profiling of tafamidis for tetramer dissociation has been performed with the recombinant wild-type TTR. Because none of the 67 sites of mutations associated with TTR amyloidosis alter the key thyroxine[59] or tafamidis[54] binding residues, it was expected that tafamidis tetramer dissociation activity and potency would be similar between wild-type TTR and TTR amyloidogenic variants. So far, all subsequent findings using amyloidogenic variants have confirmed this hypothesis.

Tetramer dissociation is the rate-limiting step in the production of toxic species in the TTR amyloid cascade; therefore it was important to demonstrate the consistent biological activity of tafamidis in different biological conditions leading to tetramer dissociation: fibril formation under acidic conditions,[39] unfolding under urea[39] and subunit exchange under physiological conditions.[36] Incubation of wild-type or amyloidogenic variant TTR at pH of 4.4–4.5 induces tetramer dissociation, partial monomer denaturation and misassembly into amyloid fibrils. Tetrameric TTR does not denature in urea; hence dissociation to monomer is required for urea-induced denaturation. Therefore, at high urea concentrations, the rate of tetramer dissociation is linked irreversibly to fast monomer unfolding, easily monitored by far UV circular dichroism spectroscopy.[8,39,56] Folded or unfolded TTR monomers are difficult to detect under physiological conditions. However, under non-denaturing conditions, the rate of subunit exchange between wild-type TTR homotetramer and wild-type TTR homotetramer tagged with an N-terminal acidic flag tag is dictated by the rate of tetramer dissociation to its monomeric subunits prior to reassembly.[60] Upon mixing of the untagged with the tagged tetramer, time-dependent subunit exchange between them could be monitored by anion exchange chromatography.[56,60,61] When subunit exchange is complete, the monomers are randomly distributed across the five possible tetramers containing zero, one, two, three or four flagged monomers. Because each FLAG-tag adds six negative charges to each TTR subunit, the more FLAG-tag TTR subunits in the tetramer, the longer the retention time on an anion exchange column, allowing separation of all five tetramers and quantification of the extent of exchange as a function of time (Figure 9.5A).

Tafamidis was found to be a potent inhibitor of tetramer dissociation under both denaturating and physiological conditions, mimicking the overall tetramer stabilisation effect observed with the intragenic trans-suppressors, T119M and R104H. In all conditions, tafamidis dose-dependently inhibits tetramer dissociation. A tafamidis : TTR molar ratio of 1 consistently produced inhibition of around 70%: 68% inhibition of TTR amyloidogenesis over 72 hours under acidic conditions, 67% inhibition of tetramer dissociation in urea over 72 hours, and 70% inhibition of subunit exchange under non-denaturing conditions over 48 hours (Figure 9.6). Stabilisation over 48–72 hours is highly relevant because the TTR half-life in plasma is reported to be around 2 days. Based on binding constants of 3 and 278 nM for the first and second binding sites, respectively, a molar ratio tafamidis : TTR of 1 corresponds to the presence of more than 80% of TTR bound to one single molecule of tafamidis. Therefore, the occupancy of one binding site by tafamidis seems sufficient to reduce tetrameric dissociation by 70%, consistent with reports that binding to only one thyroxine-binding site is sufficient to kinetically stabilise tetrameric TTR.[40]

Under physiological conditions, negligible subunit exchange (<5%) was observed in the presence of 1.5 tafamidis : TTR molar ratio after 96 hours of incubation (Figure 9.5B), the time necessary for complete exchange in the absence of an inhibitor of tetramer dissociation.[54]

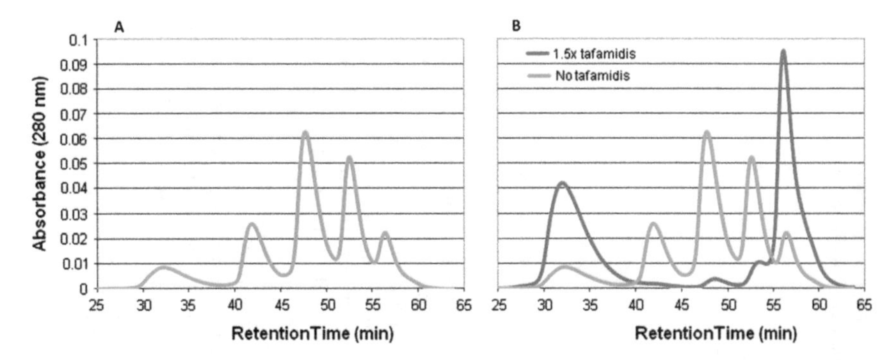

Figure 9.5 Tafamidis inhibits subunit exchange under physiological conditions. Solutions of wild-type homotetrameric TTR (1.8 μM) and wild-type homotetrameric TTR tagged with an N-terminal acidic flag-tag (1.8 μM) were pre-incubated with or without an inhibitor of TTR tetramer dissociation and mixed to initiate subunit exchange. The exchange reaction (25 °C; pH 7.0) was monitored by anion exchange chromatography every day for 7 days. Because the flag-tag provides about six negative charges to each tagged TTR subunit, the more flag-tag TTR subunits in the tetramer, the longer the retention time on the anion exchange column. (A) HPLC trace of subunit exchange after 96 hours of incubation in the absence of tafamidis. Predicted statistical distribution (1 : 4 : 6 : 4 : 1) of the five tetramers was achieved. (B) HPLC traces of subunit exchange after 96 hours of incubation in the presence of 0 (green line) or 1.5 (red line) equivalent of tafamidis.

As expected, we showed that tafamidis inhibited acid-induced amyloidogenesis of the two most clinically significant amyloidogenic variants, V30M for TTR-FAP and V122I for TTR-FAC, with potency and efficacy comparable to that of wild-type TTR (EC_{50} of 2.7–3.2 μM, corresponding to a tafamidis : TTR molar ratio of 0.75–0.9) (Figure 9.7).[54]

A modelling analysis of the subunit exchange time courses was used to calculate binding constants of tafamidis to TTR.[54] From this analysis, tafamidis was determined to bind to TTR with $K_{d1} = 2$ nM and $K_{d2} = 154$ nM, largely in agreement with the binding constants obtained by the ITC method.

9.5.2 Tetramer Stabilisation in Human Plasma with Tafamidis

The therapeutic concept behind the use of a TTR stabiliser to treat TTR amyloidosis is to specifically bind to circulating TTR for a constant stabilisation of its tetrameric form. It was essential for the successful drug development of tafamidis to establish a pharmacological marker to assess TTR tetramer stability directly in human plasma. This pharmacological marker would be used in the clinic to confirm target engagement directly in patients receiving tafamidis treatment, particularly useful for dose selection in the context of an ultra-orphan disease such as TTR-FAP.

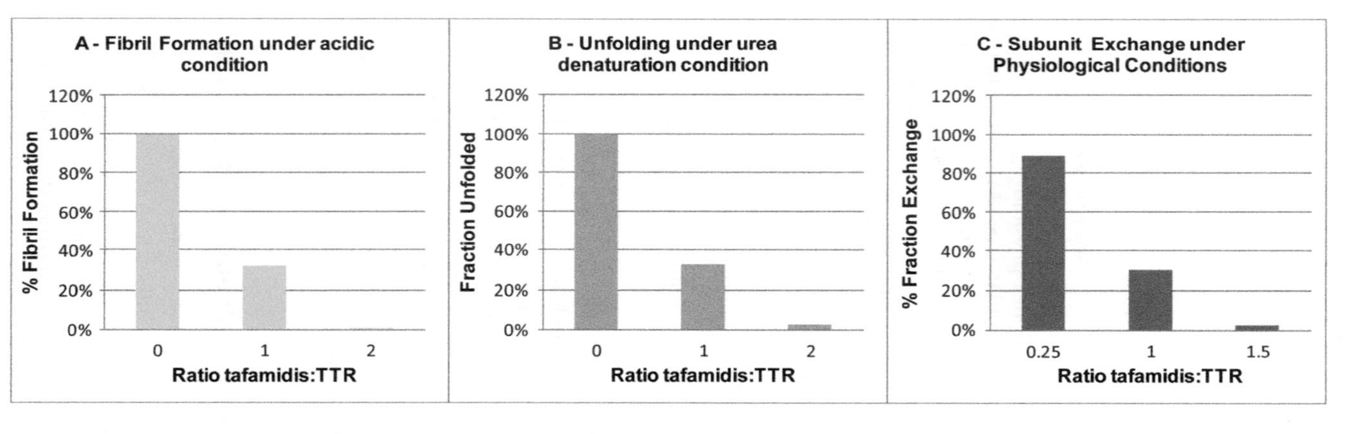

Figure 9.6 Tafamidis inhibits tetramer dissociation under denaturation and physiological conditions. (A) Fibril formation under acidic conditions: TTR amyloid fibril formation was initiated by acidification of TTR (3.6 μM) to a final pH value of 4.4. After incubation with 0, 1 or 2 equivalents of tafamidis for 72 hours (37 °C), turbidity was measured at 350 and 400 nm using an ultraviolet (UV)-vis spectrometer. All amyloid fibril formation data was normalised to wild-type TTR amyloid fibril formation in the absence of stabiliser, assigned to be 100% fibril formation. (B) Unfolding under denaturation conditions: TTR denaturation was initiated by adding urea to a solution containing TTR (1.8 μM) and 0, 1 or 2 equivalents of tafamidis to yield a final urea concentration of 6.5 M. After 72 hours, circular dichroism spectra were collected at time 0 and 72 hours (maximum denaturation). CD spectra were collected between 220 and 213 nm, with scanning every 0.5 nm and an averaging time of 10 seconds. Each wavelength was scanned once. The values for the amplitude were averaged between 220 and 213 nm to determine the extent of β-sheet loss throughout the experiment. Results were expressed as percentage of fraction unfolded relative to the amount observed in the absence of tafamidis. (C) Solutions of wild-type homotetrameric TTR (1.8 μM) and wild-type homotetrameric TTR tagged with an N-terminal acidic flag-tag (1.8 μM) were pre-incubated with different concentrations of tafamidis (tafamidis : TTR molar ratios: 0.25, 1, 1.5) and mixed to initiate subunit exchange. The exchange reaction (25 °C; pH 7.0) was monitored by anion exchange chromatography every day for 7 days. After 48 hours, the extent of exchange was calculated by dividing the integration of the tetramer with two flags by the total integration determined for all the tetramers.

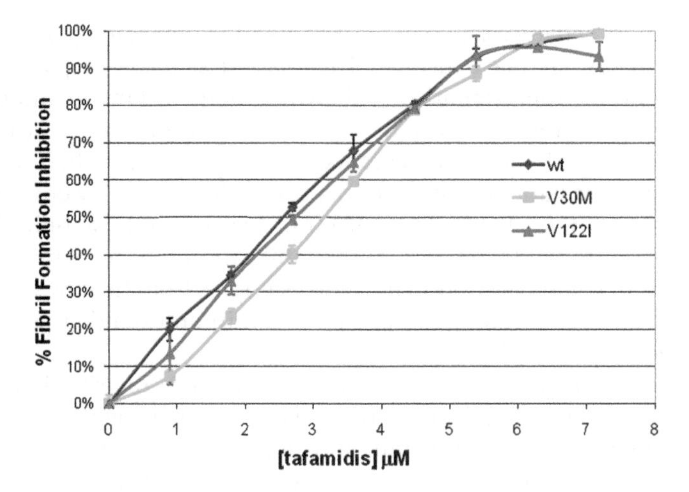

Figure 9.7 Tafamidis inhibits wild-type and amyloidogenic variant TTR fibril inhibition under acidic conditions. Fibril formation under acidic conditions: TTR amyloid fibril formation was initiated by acidification of purified TTR tetramers (wild-type, V30M or V122I at 3.6 μM) to a final pH value of 4.4 (wild type and V30M) or 4.5 (V122I) to allow comparable kinetics of fibril formation for the three alleles. After incubation with tafamidis at final concentrations of 0, 0.9, 1.8, 2.7, 3.6, 4.5, 5.4, 6.3 and 7.2 μM for 72 hours (37 °C), the turbidity was measured at 350 and 400 nm using an ultraviolet (UV)-vis spectrometer. All amyloid fibril formation data was normalised to TTR amyloid fibril formation in the absence of stabiliser, assigned to be 100% fibril formation. Therefore, 5% fibril formation corresponds to a compound inhibiting 95% of TTR fibril formation after 72 hours. Error bars represent the minimum and maximum values from three replicates.

Urea denaturation of human plasma samples has been used classically to assess TTR binder activity on tetramer dissociation.[62] After 3–4 days of incubation with or without a TTR binder and subsequent crosslinking with glutaraldehyde to crosslink the remaining tetrameric TTR, plasma samples are analysed by SDS/PAGE and Western blotting using anti-TTR serum to measure the remaining amount of tetramer.[54] Using this method, the efficacy of tafamidis to stabilise WT, V30M and V122I TTR was confirmed *ex vivo* in human plasma from healthy volunteers (WT), TTR-FAP (V30M) or TTR-FAC (V122I) patients. Dose-dependent tetramer stabilisation was observed at tafamidis : TTR molar ratios of 1, 1.5 and 2.[54]

Western blotting is not easily transferrable to the clinical setting for automated and accurate analysis of multiple clinical samples to support clinical development. It was decided to develop a more clinical-friendly method based on the urea denaturation method but using immunoturbidimetry as a readout to measure remaining tetrameric TTR, a readout used routinely in the clinic to measure plasma TTR levels. Using this methodology, dose-dependent stabilisation of patient plasma samples was observed with tafamidis, similar to that observed with Western blotting.[54]

To confirm its activity for the vast majority of TTR amyloidogenic variants, TTR stabilisation has been measured in plasma samples of patients harbouring mutations representative of the wide range of kinetic and/or thermodynamic stabilities observed in TTR amyloidosis.[63] Tafamidis significantly reduced the disappearance of TTR tetramer when added to plasma samples of patients with variants displaying either lower thermodynamic stability (V30M, Y69H, F64S, I84S) or lower kinetic stability (V122I) when compared to wild-type TTR, with stabilisation efficacy similar to the one observed in wild-type (WT) TTR plasma samples from healthy volunteers (Figure 9.8). Similar efficacy has been observed in an extended panel of 30 amyloidogenic variants.[54]

The pharmacokinetic studies have shown a selective distribution of tafamidis after oral administration in the plasma compartment in rat, small volume of distribution in both rats and dogs, and half-life in plasma of approximately 29–43 hours in rats and 55–62 hours in dogs,[64] all important attributes recognised as critical for optimum TTR tetrameric stabilisation (see Section 2.3.4). Long half-life observed in animals was confirmed in humans (~58 hours).[64]

It is worth noting that at the time of the preclinical development of tafamidis, an animal model of ATTR-PN that reflects the clinical disease or resembles the pathology was not available.

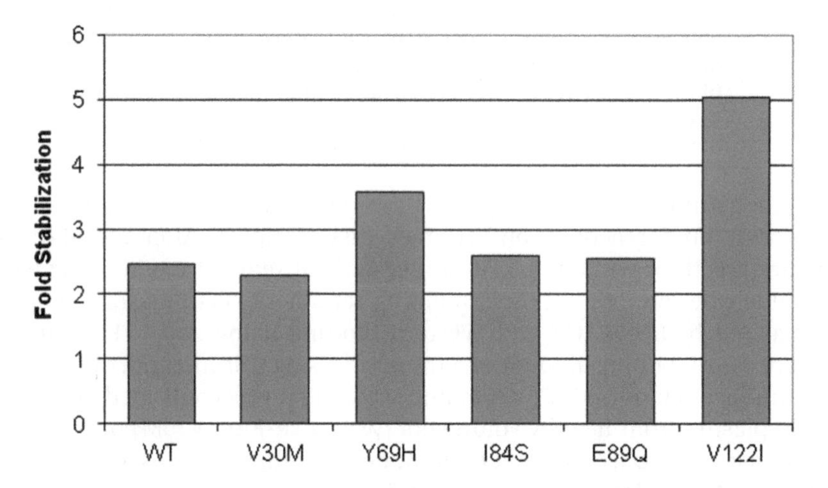

Figure 9.8 Tafamidis stabilises tetrameric TTR in plasma samples from patients with kinetically and/or thermodynamically destabilised TTR variants. Plasma samples from TTR amyloidosis patients with the indicated mutations and plasma samples from healthy volunteers were treated with urea in the presence of vehicle or 7.2 μM tafamidis and the amount of TTR tetramer present pre- and post-denaturation (48 hours) was determined by immunoturbidity. The results are presented as fold of stabilisation = fraction of initial for tafamidis/fraction of initial for DMSO. Fraction of initial = TTR tetramer (mg dL^{-1}) post-denaturation/TTR tetramer (mg dL^{-1}) pre-denaturation.

9.5.3 Safety Profile

To support chronic treatment in the clinics, toxicology studies were centred on two toxicity studies, a 6-month repeat dose toxicity study in rats and a 9-month repeat dose toxicity study in dogs. Tafamidis was considered to be well tolerated at exposure ratios of at least 24-fold and 9–11-fold above expected therapeutic human exposure, in rat and dog respectively.[64]

9.6 Clinical Strategy

In addition to the obstacle of discovering drug candidates acting on non-druggable but genetically validated targets, a major common challenge in the field of rare genetic diseases is the design of the clinical studies to prove clinical efficacy in a reasonable amount of time, with a limited pool of diagnosed patients available for enrolment. Natural disease histories are lacking for most of the rare diseases. Genotype–phenotype relationships are not well known and disease progression is not well understood. It is very common to be faced with a lack of clinical evaluation tools that could be used as clinical end points in a controlled study to support drug approval.

In the case of TTR-FAP, an ultra-orphan disease, at the time of initiation of the FoldRx clinical programme in September 2005, there was no pharmacotherapeutic agent approved for TTR-FAP and tafamidis was the first drug candidate to be studied in this ultra-orphan disease. No previous clinical studies or extensive literature on the natural disease history were available to guide trial design, to select suitable outcome measures, study duration and appropriate statistical analyses to demonstrate drug efficacy. In the context of a drug like tafamidis, designed to control disease progression, there had been no instrument studied clinically to measure progression of any aspect of the neurodegenerative process that characterised TTR-FAP. Disease progression in TTR-FAP has only been categorised in three main stages, with walking ability and ambulation status being the key determinants of stage of disease: Stage 1 for patients with no assistance required for ambulation, Stage 2 for patients needing assistance and Stage 3 for wheelchair bound or bedridden patients.[20]

9.6.1 Choice of Clinical End Points to Assess Disease Progression

In the absence of validated or even exploratory outcome measures for TTR-FAP, existing validated instruments assessing disease progression of other peripheral neuropathies were considered. It was important to select instruments assessing the progression of peripheral neuropathies and potentially useful in understanding the multifaceted nature of this disease. Specifically, the end points were intended to detect the consequences of the axonal degeneration, including increasing neurophysiological dysfunction, and worsening neurological impairment leading to deteriorating quality of life (QOL), and nutritional status.

The Neuropathy Impairment Score (NIS) is a structured, graded neurological examination that has been developed and validated in population-based studies to measure the evolution of diabetic polyneuropathy (DPN), another progressive, symmetric ascending axonal degenerative peripheral neuropathy.[65] The NIS-LL is a component of the NIS that focuses on impairment in the lower limbs (LL), the site affected first and most severely in DPN and in TTR-FAP. In fact, the NIS-LL has been used in several registration trials, especially in DPN patients.[66] The NIS-LL quantifies the findings of the neurological examination by attributing a score (ranging from 0 [normal] to 88 [total impairment]) to the clinical abnormalities noted in the physical assessment of sensation, muscle weakness and reflexes in the lower limb.

In concert with the objective measure of the NIS-LL, a patient-reported outcome assessment was selected to evaluate the patient's perception of neuropathy and its impact on QOL. The Norfolk QOL-DN, validated in DPN,[67,68] discriminates the presence of neuropathy using 35 scored questions encompassing five domains that comprise a total quality of life score (TQOL), with a range of −2 (best possible QOL) to 138 (worst possible QOL).[69]

Objective measures of nerve fibre impairment have been used to detect disease progression in other neuropathies such as DPN.[70,71] The summated seven nerve tests normal deviate (Σ7 NTs nds) is a composite score that summates normal deviates from seven nerve tests of primarily large-fibre function and the summated three nerve tests normal deviate (Σ3 NTSF nds) is a composite score that summates normal deviates from three nerve tests of small-fibre function.

TTR-FAP is characterised by severe gastrointestinal disturbances and overall wasting. Modified body mass index (mBMI), reflective of general nutritional status, has been reported to be associated with survival post liver transplant in TTR-FAP, with lower scores (600–700) associated with a poorer outcome.[72–74] The mBMI utilised the BMI calculation (a function of weight and height) multiplied by serum albumin ($g\ L^{-1}$) to correct for the effects of oedema on BMI.

9.6.2 Choice of Duration of Pivotal Trial and Patient Population

The objective of FoldRx's clinical programme was to design and conduct a single Phase 2/3 pivotal study to demonstrate the disease-modifying effect of tafamidis in TTR-FAP patients.

One challenge was to select a trial duration sufficient to observe a significant tafamidis effect on neurological outcomes, such as NIS-LL. An expert consensus report concluded that a 2-point change in the NIS-LL score is clinically meaningful and the smallest change that is recognisable by a physician.[75] In DPN, an axonal degenerative peripheral neuropathy deemed less severe and with a much slower rate of progression than TTR-FAP, the average annual change in NIS-LL is 0.85. It corresponds to a more than 2 year timeframe to observe a 2-point disease progression. Therefore, an 18-month duration was expected to be sufficient in TTR-FAP patients to observe significant progression in the placebo-treated group and to detect the expected disease-modifying effect of tafamidis.

For an ultra-orphan disease like TTR-FAP, enrolment is always a major challenge and it is important to keep this constraint in mind when designing a pivotal study. Because of the high prevalence in TTR-FAP of a single mutation, V30M (80% of liver transplantations have been performed in V30M patients[31]) and the presence of V30M endemic regions with an existing pool of identified families and diagnosed patients, it was decided to conduct the pivotal study in the V30M TTR-FAP population. Efficacy from the pivotal study in V30M patients was expected to translate to other TTR-FAP variants. As mentioned in Section 7.2, TTR-FAP disease progression is similar across TTR mutations and geographic locations, and the neurodegenerative process is caused, across all amyloidogenic variants, by tetramer dissociation. Tafamidis with demonstrated efficacy in stabilising all 30 variants so far tested was expected to be similarly active on V30M and non-V30M TTR-FAP variants. An open-label Phase II study in non-V30M has been conducted to confirm tafamidis efficacy in the non-V30M TTR-FAP population.[64]

Therefore, the pivotal Phase II/III study was a randomised (1 : 1), double-blind, placebo-controlled, multicentre international study designed to evaluate the efficacy during 18 months of tafamidis in patients with early stage ATTR-PN, to assess whether tafamidis impacted the rate of progression of the disease in patients diagnosed with ATTR-PN. The first patient was enrolled in January 2007.

9.6.3 Choice of Statistical Analysis

Most of the outcome measures considered to assess disease progression in TTR-FAP were intended to be used as a continuous variable (*i.e.* change over time). In the case of NIS-LL, it could be analysed either as a continuous variable or as a categorical variable, defining a 'NIS-LL responder' as one with less than a 2-point increase at a given time point, the smallest change recognisable by a physician.[75] In the absence of natural history and prior experience of using NIS-LL to assess disease progression in TTR-FAP, it was decided to choose the categorical analysis as the primary analysis, this analysis having recently been used in a DPN registration trial.[66]

The primary efficacy end points at month 18 were NIS-LL response to treatment ('NIS-LL responders' with an increase from baseline in NIS-LL of less than 2 points) and the change from baseline in the Norfolk QOL-DN total scores in the intent-to-treat (ITT) population defined as all randomised patients who received at least one dose of medication and who had at least one post-baseline evaluation or who were discontinued due to liver transplantation.[76] Patients discontinued due to liver transplantation were categorised as NIS-LL non-responders. Analyses of the co-primary end points were performed in an efficacy-evaluable (EE) population consisting of ITT patients who completed the study per protocol. This EE population was pre-specified as it was anticipated that the majority of patients enrolled would be on the liver transplant list and that many would undergo liver transplantation during the study if a donor organ became available.[76]

Multiple secondary end points were used to assess the efficacy of tafamidis, including change from baseline at months 6, 12 and 18 in NIS-LL, NORFOLK QOL-DN, $\Sigma 7$ NTs nds, $\Sigma 3$ NTSF nds and mBMI. Analyses of the secondary end points were conducted in the ITT population. Only observed values were used.[76]

Sample size, based on the co-primary end points, assumed two-sided tests, $\alpha = 0.05$, 90% power, and a discontinuation rate of <10%. Response rates of 20% for placebo and 50% for tafamidis were anticipated for NIS-LL. A true difference of 0.6 SD was assumed between the groups in NORFOLK QOL-DN. Each group required 58 patients.[76]

9.6.4 Dose Selection

Traditional dose-ranging studies are always difficult to perform in rare diseases, especially in ultra-orphan diseases such as TTR-FAP, with a limited population of diagnosed patients. Furthermore, for a disease-modifying approach to TTR-FAP, requiring long-term treatment, the absence of a surrogate marker of clinical efficacy did not allow short-duration, dose-ranging studies.

A pharmacodynamic approach was used to select the dose to be used in the Phase II/III pivotal study. Tetramer stabilisation in plasma (see Section 9.5.2) was assessed in plasma samples from the single- and multiple-dose ascending Phase I study conducted in healthy volunteers. Because tetramer dissociation is the rate-limiting step in TTR amyloid deposition and the cause of TTR-FAP (see Section 9.2), it is pertinent to use this TTR-specific pharmacodynamic marker to select a dose sufficient for optimum target engagement. Tafamidis inhibits tetramer dissociation by binding to the two thyroxine binding sites of circulating TTR; therefore it is expected to observe a saturable stabilisation effect with increased exposure to tafamidis in plasma samples of healthy volunteers. The use of wild-type TTR samples from the Phase I study is relevant because tafamidis has been shown to be equally potent in stabilising wild-type and variant TTR tetramers.

Analysis of plasma samples from the single- and multiple-ascending Phase I study demonstrates the presence of a plateau for tetramer stabilisation (range of difference from baseline 114–234) at tafamidis : TTR molar ratio above 1.2–1.4 (Figure 9.9).[64] It is consistent with tafamidis preclinical data (see Section 9.5) and with the fact that binding to only one thyroxine-binding site is sufficient to kinetically stabilise tetrameric TTR.[40] Extrapolated data from all Phase I pharmacokinetic data indicates that a daily dose of 20 mg will result, at steady state, in sufficient plasma concentration to reach a tafamidis : TTR molar ratio above 1.2. Therefore, a dose of 20 mg of tafamidis was selected to conduct the pivotal efficacy study.[64]

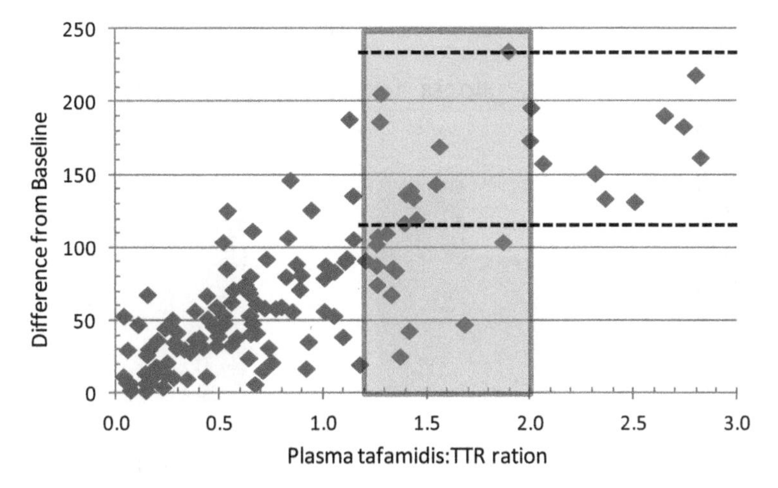

Figure 9.9 Pharmacodynamic marker and dose selection. Plasma samples from the single- and multiple-dose ascending Phase I study in healthy volunteers were incubated in 4.8 M urea to measure remaining TTR tetramer by immunoturbidimetry and to calculate the difference from baseline defined as: (fraction of initial − fraction of initial at baseline) × 100/ fraction of initial at baseline; fraction of initial: per cent remaining TTR after urea denaturation. Levels of tafamidis and TTR were also measured for each analysed plasma sample to calculate tafamidis : TTR molar ratio. The figure shows the difference from baseline as a function of molar ratio. Stabilisation plateau effect is delineated by horizontal black dotted lines. Tafamidis range of exposure predicted at steady state at a chronic daily dose of 20 mg is delineated by the pink box.

9.7 Clinical Results in Pivotal Phase 2/3 Study

One hundred and twenty-eight patients were randomised to tafamidis ($n = 65$) and placebo. Ninety-one patients completed the 18 month study, 47 in the tafamidis group and 44 in the placebo group. Thirteen patients in each group (21%) discontinued treatment to undergo liver transplantation.[76]

In the ITT population at month 18, more NIS-LL responders were observed in the tafamidis group than in the placebo arm (45.3% *vs.* 29.5%) and a 5.2-point difference in Norfolk QOL total score was observed in favour of the tafamidis group, without reaching statistical significance (Figure 9.10).[76] Statistical significance was achieved in the EE population with 60% of NIS-LL responders (no measurable disease progression) in the tafamidis group *versus* 38.1% in the placebo group ($p = 0.041$) and with a 8.8-point difference in QOL score in favour of tafamidis ($p = 0.045$) (Figure 9.10).[76] Because of the higher than anticipated discontinuation rate due to liver transplantation (21% observed *vs.* 10% estimated) and the *a priori* designation of the

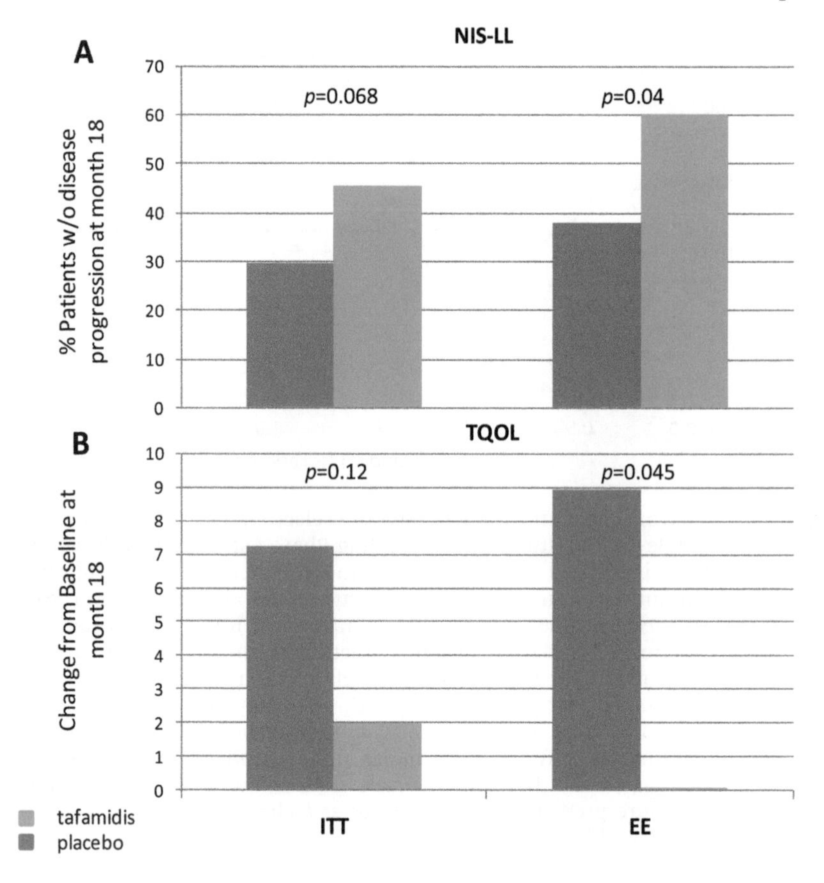

Figure 9.10 Tafamidis reduces neurological impairment and maintains quality of life in TTR-FAP patients. (A) Percentage of patients in each treatment group with less than 2 points increase in NIS-LL overall score (NIS-LL responder or no detectable disease progression) at month 18 in both the ITT (intent-to-treat) and the EE (efficacy evaluable) populations. (B) Change from baseline in each treatment group at month 18 in the total quality of life (TQOL) from the Norfolk QOL-DN in both the ITT and the EE populations.

liver-transplanted patients as NIS-LL non-responders in the ITT population, the study was certainly underpowered to demonstrate a statistical difference. Nevertheless, the results in patients completing the 18-month treatment per protocol (EE population) provide an accurate measure of the treatment effects of tafamidis over that period of time.[76] In the NIS-LL analysis as a continuous variable (*i.e.* change from baseline), a significant reduction of 52% in neurophysiological deterioration was observed in the ITT population at month 12 and 18,[76] with a difference of 3 NIS-LL points observed at the end of the 18-month treatment period (2.81 *vs.* 5.83; $p = 0.027$) (Figure 9.11A).

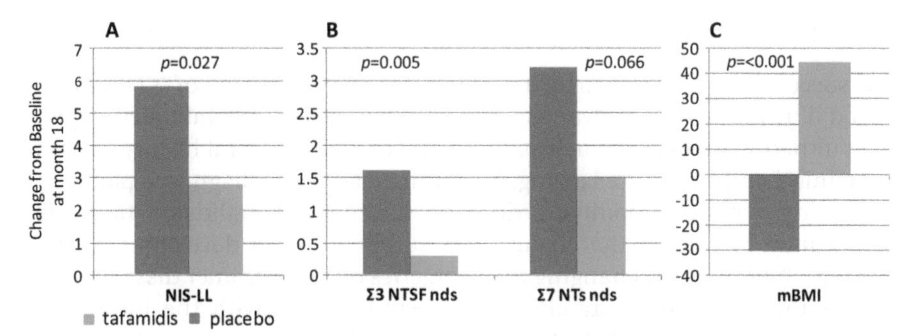

Figure 9.11 Tafamidis significantly slows the neurological process, preserves nerve function and improves nutritional status. Changes from baseline in each treatment group at month 18: (A) for the neurological impairment as measured by NIS-LL score; (B) for nerve function as assessed by $\sum 3$ NTSF nds and by $\sum 7$ NTs nds for small and large fibre function, respectively; (C) for nutritional status as assessed by modified body mass.

As mentioned previously, in a multifaceted disease like TTR-FAP it was important to show efficacy on multiple end points measuring different aspects of the disease and to demonstrate consistency of results across the chosen end points.

The significant reduction of neurophysiological deterioration noticed with tafamidis was confirmed by the preservation of nerve function observed in the tafamidis-treated patients: 54.3% ($p = 0.066$) and 83.8% ($p = 0.005$) preservation of function as compared to placebo in large and small fibres, respectively (Figure 9.11B). Nutritional status at month 18 was significantly improved (mBMI increase from baseline of +39.3) in tafamidis-treated patients compared with a worsening mBMI in placebo patients (−33.8; $p < 0.0001$) (Figure 9.11C).[76]

It is worth noting that TTR stabilisation at 18 months was demonstrated in 98% of tafamidis-treated patients and none of the patients who received placebo ($p < 0.001$).[76]

In an extension study in which all patients, having completed the pivotal study, received oral tafamidis 20 mg once daily for 12 months, it was shown that the effect of tafamidis in slowing neurological progression and preserving QOL is sustained over 30 months and that delayed introduction of tafamidis significantly slowed the rates of change in NIS-LL, Norfolk TQOL and mBMI compared with placebo.[77]

Eighteen months of tafamidis treatment was well tolerated with an adverse event profile similar to placebo.[76]

Based on the review of this data package, the Committee for Medicinal Products for Human Use of the EMA considered that the risk–benefit balance of tafamidis in the treatment of TTR-FAP to delay peripheral neurological impairment was favourable and therefore recommended the granting of the marketing authorisation in September 2011.[64]

9.8 Conclusion

The successful development of tafamidis for the treatment of TTR-FAP is a good illustration of the advantage provided by a genetically validated target. In an ultra-orphan disease such as TTR-FAP, for which natural history studies and clinical end points are lacking, with limited access to diagnosed patients, developing a disease-modifying drug on a genetically validated target is significantly increasing the chance of a successful drug development programme. Tafamidis, by changing the energetics of TTR tetramer dissociation, mimics the all tetramer stabilisation effects observed with the intragenic trans-suppressors (Table 9.3). Following dose selection based on pharmacodynamic studies in Phase I, clinical efficacy was demonstrated at that dose, in a single pivotal Phase II/III study in 128 TTR-FAP patients. Reminiscent of the protective effect observed in the presence of intragenic trans-suppressors, tafamidis was able to slow the neurodegenerative process, preserve neurophysiological function, maintain QOL and improve nutritional status.

It is worth noting that tafamidis is the first example of a disease-modifying therapy for any amyloid disease. It validates the amyloid hypothesis, demonstrating that the amyloid cascade actually causes the neurodegenerative process and that its inhibition halts the course of the disease, paving the way for other success stories in the field of amyloidosis.

Acknowledgements

It is a pleasure to acknowledge the key contributions of collaborators in the tafamidis project: JW Kelly and his team at SCRIPPS, all FoldRx employees, especially J. Packman, D. Grogan, C. Adams, C. Bulawa, J. Fleming, M. DeVit, C. Weigel and all investigators and co-workers involved in the Phase II/III pivotal study, especially T. Coelho, L. Maia, M. Waddington Cruz, V. Planté-Bordeneuve, P. Lozeron, O. Suhr, J. Campistol, I. Conceição, H. Schmidt and P. Trigo.

References

1. J. Arrowsmith, *Nat. Rev. Drug Discovery,* 2011, **10**, 328.
2. C. C. Blake, M. J. Geisow, S. J. Oatley, B. Rerat and C. Rerat, *J. Mol. Biol.,* 1978, **121**, 339.
3. H. L. Monaco, M. Rizzi and A. Coda, *Science,* 1995, **268**, 1039.
4. S. F. Nilsson, L. Rask and P. A. Peterson, *J. Biol. Chem.,* 1975, **250**, 8554.
5. R. A. Pages, J. Robbins and H. Edelhoch, *Biochemistry,* 1973, **12**, 2773.
6. M. J. Saraiva, *Hum. Mutat.,* 2001, **17**, 493.
7. M. D. Benson, Amyloidosis, in *The Metabolic and Molecular Bases of Inherited Diseases*, ed. C. R. Scriver, A. L. Beaudet, W. Sly and D. Valle, McGraw-Hill, New York, NY, 2001, vol. 4, pp. 5345–5378.
8. P. Hammarström, X. Jiang, A. R. Hurshman, E. T. Powers and J. W. Kelly, *Proc. Natl. Acad. Sci. U. S. A.,* 2002, **99**, 16427.

9. A. Quintas, D. C. Vaz, I. Cardoso, M. J. Saraiva and R. M. Brito, *J. Biol. Chem.,* 2001, **276**, 27207.
10. W. Colon and J. W. Kelly, *Biochemistry,* 1992, **31**, 8654.
11. N. Reixach, S. Deeshongkit, X. Jiang, J. W. Kelly and J. N. Buxbaum, *Proc. Natl. Acad. Sci. U. S. A.,* 2004, **101**, 2817.
12. M. Sousa, I. Cardoso, R. Fernandes, A. Guimaraes and M. J. Saraiva, *Am. J. Pathol.,* 2001, **159**, 1993.
13. K. Andersson, A. Olofsson, E. H. Nielsen, S.-E. Svehag and E. Lundgren, *Biochem. Biophys. Res. Commun.,* 2002, **294**, 309.
14. X. Hou, M.-I. Aguilar and D. H. Small, *FEBS J.,* 2007, **274**, 1637.
15. D. R. Jacobson and J. N. Buxbaum, *Adv. Hum. Genet.,* 1991, **20**, 69.
16. P. Nihoyannopoulos, *Curr. Opin. Cardiol.,* 1987, **2**, 371.
17. L. H. Connors, A. M. Richardson, R. Theberge and C. E. Costello, *Amyloid,* 2000, 7, 54.
18. D. R. Jacobson, M. Ittmann, J. N. Buxbaum, R. Wieczorek and P. D. Gorevic, *Tex. Heart Inst. J.,* 1997, **24**, 45.
19. P. D. Gorevic, F. C. Prelli, J. Wright, M. Pras and B. Frangione, *J. Clin. Invest.,* 1989, **83**, 836.
20. P. Coutinho, A. Martins da Silva, J. Lopes Lima and A. Resende Barbosa, in *Amyloid and Amyloidosis*, ed. G. G. Glenner, P. Pinho e Costa and A. Falcao de Freitas, Excerpta Medica, Amsterdam, 1980, pp. 88–98.
21. Y. Ando, M. Nakamura and S. Araki, *Arch. Neurol.,* 2005, **62**, 1057.
22. T. Hattori, Y. Takei, J. Koyama, M. Nakazato and S. Ikeda, *Amyloid,* 2003, **10**, 229.
23. S. Ikeda, M. Nakazato, Y. Ando and G. Sobue, *Neurology,* 2002, **58**, 1001.
24. V. Plante-Bordeneuve, T. Lalu and M. Misrahi, *Neurology,* 1998, **51**, 708.
25. V. Plante-Bordeneuve, A. Ferreira and T. Lalu, *Neurology,* 2007, **69**, 693.
26. G. Di Iorio, G. Sanges, A. Cerracchio, S. Sampaolo, V. Sannino and V. Bonavita, *Ital. J. Neurol. Sci.,* 1993, **14**, 303.
27. E. Hund, R. P. Linke, F. Willig and A. Grau, *Neurology,* 2001, **56**, 431.
28. M. D. Benson, *Best Pract. Res., Clin. Rheumatol.,* 2003, **17**, 909.
29. C. Rapezzi, C. C. Quarta and L. Riva, *Nat. Rev. Cardiol.,* 2010, 7, 398.
30. Familial Amyloidotic Polyneuropathy World Transplant Registry (FAPWTR), http://www.fapwtr.org, update of 8 February 2013.
31. G. Herlenius, H. E. Wilczek, M. Larsson and B. G. Ericzon, *Transplantation,* 2004, 77, 64.
32. T. Coelho, R. Chorao, A. Sousa, I. Alves, M. F. Torres and M. J. Saraiva, *Neuromuscular Disord.,* 1996, **6**, 27.
33. S. L. McCutchen, Z. Lai, G. J. Miroy, J. W. Kelly and W. Colon, *Biochemistry,* 1995, **34**, 13527.
34. I. Longo-Alves, M. T. Hays and M. J. Saraiva, *Eur. J. Biochem.,* 1997, **249**, 662.
35. M. P. Sebastiao, V. Lamzin, M. J. Saraiva and A. M. Damas, *J. Mol. Biol.,* 2001, **306**, 733.
36. P. Hammarström, F. Schneider and J. W. Kelly, *Science,* 2001, **293**, 2459.

37. H. Terazaki, Y. Ando, S. Misumi, M. Nakamura, E. Ando, N. Matsunaga, S. Shoji, M. Okuyama, H. Ideta, K. Nakagawa, T. Ishizaki, M. Ando and M. J. Saraiva, *Biochem. Biophys. Res. Commun.*, 1999, **264**, 365.
38. M. R. Almeida, I. L. Alves, H. Terazaki, Y. Ando and M. J. Saraiva, *Biochem. Biophys. Res. Commun.*, 2000, **270**, 1024.
39. P. Hammarström, R. L. Wiseman, E. T. Powers and J. W. Kelly, *Science*, 2003, **299**, 713.
40. R. L. Wiseman, S. M. Johnson, M. S. Kelker, T. Foss, I. A. Wilson and J. W. Kelly, *J. Am. Chem. Soc.*, 2005, **127**, 5540.
41. L. Bartalena and J. Robbins, *Clin. Lab. Med.*, 1993, **13**, 583.
42. H. N. Rosen, A. C. Moses, J. R. Murrell, J. J. Liepnieks and M. D. Benson, *J. Clin. Endocrinol. Metab.*, 1993, **77**, 370.
43. J. A. Palha, M. T. Hays, E. G. De Morreale, V. Episkopou, M. E. Gottesman and M. J. Saraiva, *Am. J. Physiol.*, 1997, **272**, E485.
44. G. Holmgren, S. Bergström, U. Drugge, E. Lundgren, C. Nording-Sikström, O. Sandgren and L. Steen, *Clin. Genet.*, 1992, **41**, 39.
45. P. W. Baures, S. A. Peterson and J. W. Kelly, *Bioorg. Med. Chem.*, 1998, **6**, 1389.
46. T. Klabunde, H. M. Petrassi, V. B. Oza, P. Raman, J. W. Kelly and J. Sacchettini, *Nat. Struct. Biol.*, 2000, **7**, 312.
47. V. B. Oza, H. M. Petrassi, H. E. Purkey and J. W. Kelly, *Bioorg. Med. Chem. Lett.*, 1999, **9**, 1.
48. V. B. Oza, C. Smith, P. Raman, E. K. Koepf, H. A. Lashuel, H. M. Petrassi, K. P. Chiang, E. T. Powers, J. Sachettinni and J. W. Kelly, *J. Med. Chem.*, 2002, **45**, 321.
49. H. M. Petrassi, T. Klabunde, J. Sacchettini and J. W. Kelly, *J. Am. Chem. Soc.*, 2000, **122**, 2178.
50. H. Razavi, S. K. Palaninathan, E. T. Powers, R. L. Wiseman, H. E. Purkey, N. N. Mohamedmohaideen, S. Deechongkit, K. P. Chiang, M. T. A. Dendle, J. Sachettinni and J. W. Kelly, *Angew. Chem., Int. Ed.*, 2003, **42**, 2758.
51. S. L. Adamski-Werner, S. K. Palaninathan, J. C. Sacchettini and J. W. Kelly, *J. Med. Chem.*, 2004, **47**, 355.
52. H. M. Petrassi, S. M. Johnson, H. E. Purkey, K. P. Chiang, T. Walkup, X. Jiang, E. T. Powers and J. W. Kelly, *J. Am. Chem. Soc.*, 2005, **127**, 6662.
53. S. M. Johnson, R. L. Wiseman, Y. Sekijima, N. S. Green, S. L. Adamski-Werner and J. W. Kelly, *Acc. Chem. Res.*, 2005, **38**, 911.
54. C. E. Bulawa, S. Connelly, M. Devit, L. Wang, C. Weigel, J. A. Fleming, J. Packman, E. T. Powers, R. L. Wiseman, T. R. Foss, I. A. Wilson, J. W. Kelly and R. Labaudinière, *Proc. Natl. Acad. Sci. U. S. A.*, 2012, **109**, 9629.
55. C. C. Blake and S. J. Oatley, *Nature*, 1977, **268**, 115.
56. T. R. Foss, R. L. Wiseman and J. W. Kelly, *Biochemistry*, 2005, **44**, 15525.
57. S. M. Johnson, S. Connelly, I. A. Wilson and J. W. Kelly, *J. Med. Chem.*, 2008, **51**, 6348.
58. S. Connelly, S. Choi, S. M. Johnson, J. W. Kelly and I. A. Wilson, *Curr. Opin. Struct. Biol.*, 2010, **20**, 54.

59. V. Cody and A. Wojtczak, in *Recent Advances in Transthyretin Evolution, Structure and Biological Functions*, ed. S. J. Richardson and V. Cody, Springer-Verlag, Berlin, Heidelberg, 2009, pp. 1–21.
60. F. Schneider, P. Hammarström and J. W. Kelly, *Protein Sci.*, 2001, **10**, 1606.
61. R. L. Wiseman, N. S. Green and J. W. Kelly, *Biochemistry,* 2005, **44**, 9265.
62. Y. Sekijima, M. A. Dendle and J. W. Kelly, *Amyloid,* 2006, **13**, 236.
63. Y. Sekijima, R. L. Wiseman, J. Matteson, P. Hammerström, S. Miller, A. R. Sawkar, W. E. Balch and J. W. Kelly, *Cell,* 2005, **121**, 73.
64. European Medicines Agency Committee for Medicinal Products for Human Use (2011) Tafamidis Meglumine (Vyndaqel) assessment report, 22 September 2011. EMA/729083/2011, procedure no. EMEA/H/C/002294.
65. D. J. Dyck, J. L. Davies, W. J. Litchy and P. C. O'Brien, *Neurology,* 1997, **49**, 229.
66. S. C. Apfel, S. Schwartz, B. T. Adornato, R. Freeman, V. Biton, M. Rendell, A. Vinik, M. Giuliani, J. C. Stevens, R. Barbano and P. J. Dyck, *J. Am. Med. Assoc.*, 2000, **284**, 2215.
67. E. J. Vinik, R. P. Hayes, A. Oglesby, E. Bastyr, P. Barlow, S. L. Ford-Molvik and A. I. Vinik, *Diabetes Technol. Ther.*, 2005, **7**, 497.
68. A. I. Vinik, J. Ullal, H. Parson and C. Casellini, *Nat. Clin. Pract. Endocrinol. Metab.*, 2006, **2**, 269.
69. E. J. Vinik, R. P. Hayes, A. Oglesby, E. Bastyr and A. I. Vinik, *J. Diabetes Sci. Technol.*, 2008, **2**, 1075.
70. P. J. Dyck, W. J. Litchy, J. R. Daube, C. M. Harper, J. Davies and P. C. O'Brien, *Muscle Nerve,* 2003, **27**, 202.
71. P. J. Dyck, P. C. O'Brien, W. J. Litchy, C. M. Harper and C. J. Klein, *Diabetes Care,* 2005, **28**, 2192.
72. P. L. Bittencourt, C. A. Couto, A. Q. Farias, P. Marchiori, P. C. Bosco Massarollo and S. Mies, *Liver Transplant.*, 2002, **8**, 34.
73. O. Suhr, A. Danielsson, G. Holmgren and L. Steen, *J. Intern. Med.*, 1994, **235**, 479.
74. O. Suhr, S. Friman and B. G. Ericzon, *Amyloid,* 2005, **12**, 233.
75. Diabetic polyneuropathy in controlled clinical trials: Consensus Report of the Peripheral Nerve Society, *Ann. Neurol.*, 1995, **38**, 478.
76. T. Coelho, L. F. Maia, A. Martins da Silva, M. Waddington Cruz, V. Planté-Bordeneuve, P. Lozeron, O. B. Suhr, J. M. Campistol, I. M. Conceição, H. H. Schmidt, P. Trigo, J. W. Kelly, R. Labaudinière, J. Chan, J. Packman, A. Wilson and D. R. Grogan, *Neurology,* 2012, **79**, 785.
77. T. Coelho, L. F. Maia, A. Martins da Silva, M. Waddington Cruz, V. Planté-Bordeneuve, O. B. Suhr, I. M. Conceição, H. H. Schmidt, P. Trigo, J. W. Kelly, R. Labaudinière, J. Chan, J. Packman and D. R. Grogan, *J. Neurol.*, 2013, **260**, 2082.

CHAPTER 10

Small Molecules that Rescue F508del CFTR as Cystic Fibrosis Therapies

MARKO J. PREGEL

Rare Disease Research, Pfizer Inc., 610 Main Street, Cambridge, MA 02139, USA
E-mail: marko.pregel@pfizer.com

10.1 Cystic Fibrosis and the Cystic Fibrosis Transmembrane Conductance Regulator (CFTR)

Cystic fibrosis (CF) is an autosomal recessive monogenic disease affecting mainly Caucasians.[1] The hallmarks of CF are excessive mucus secretion, infection and inflammation in the lungs, pancreatic insufficiency, and elevated sweat chloride concentration. While CF is a multisystem disease, affecting the lungs, pancreas, gastrointestinal, hepatobiliary and reproductive tracts, most of the morbidity and mortality is caused by chronic lung disease.[2] There are an estimated 60 000–70 000 patients worldwide and the median age of survival is into the fourth decade of life.[3] For most patients there is a high burden of care for therapies that do not address the root cause of the disease. Supportive therapies include physical airway clearance techniques, inhaled medications (mucolytics, antibiotics and hypertonic saline) and oral anti-inflammatory drugs, as well as pancreatic enzyme replacements and nutritional supplements.[3]

RSC Drug Discovery Series No. 38
Orphan Drugs and Rare Diseases
Edited by David C Pryde and Michael J Palmer

Published by the Royal Society of Chemistry, www.rsc.org

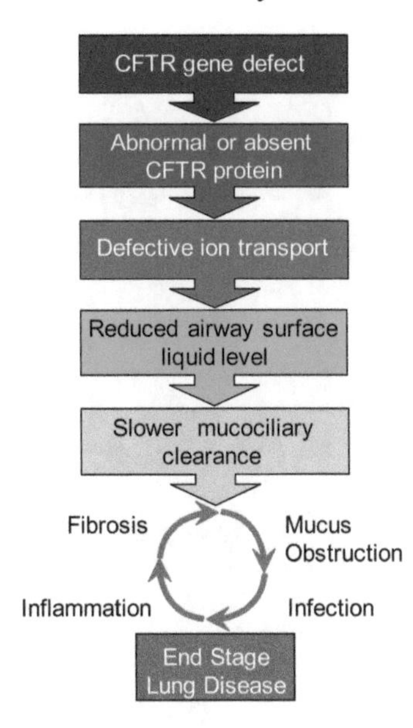

Figure 10.1 The cystic fibrosis pulmonary disease cascade.

CF is caused by mutations in the gene for CFTR (cystic fibrosis transmembrane conductance regulator), an ion channel involved in epithelial ion transport. CFTR is present mainly at the apical membrane of epithelial cells in the airways, intestine, pancreas, sweat glands and other tissues.[1] Mutation of *CFTR* leads to mucus obstruction in the airways, pancreas, intestine and vas deferens.[4] In the lungs, loss of CFTR function also decreases the levels of airway surface liquid, compromising the clearance of mucus from the airways and leading to cycles of infection, inflammation and fibrosis, ultimately degrading lung function (Figure 10.1).[2]

CFTR is a member of the ATP-binding cassette (ABC) transporter superfamily. It is an ion channel that conducts chloride and bicarbonate ions as well as other anions.[5] CFTR is composed of two membrane-spanning domains that make up the channel pore, two nucleotide binding domains that govern channel gating *via* ATP binding and hydrolysis, and a regulatory domain that must be phosphorylated to allow channel function.[6] Structures of related channels and experiments with CFTR have suggested a 'domain swapped' configuration in which the first membrane-spanning domain (MSD1) interacts in part with the second nucleotide binding domain (NBD2) and MSD2 interacts in part with NBD1 (Figure 10.2).[7]

10.1.1 CFTR Mutations

CF disease-causing mutations in CFTR have been classified into six types: (i) premature termination due to deletions, nonsense or frameshift mutations, (ii) defective trafficking out of the endoplasmic reticulum due to

Figure 10.2 Model of the CFTR structure.[42] Ribbon diagram showing the amino acid backbone of CFTR with MSD1 in purple, MSD2 in yellow, NBD1 in green, NBD2 in cyan, and the regulatory domain in grey. The intracellular loops of MSD1 and MSD2 are coloured in blue.

improper folding, (iii) improper gating, (iv) reduced conductance due to changes in the channel pore, (v) reduced production of channel due to altered splicing, and (vi) increased endocytosis from the plasma membrane.[8,9]

Many different mutations in CFTR cause CF. While the count of distinct mutations is now nearing 2000, only a handful of mutations affect a significant proportion of patients. Deletion of Phe508 of CFTR (F508del) occurs in approximately 70% of CFTR alleles.[3,10] Consequently, about 50% of patients are F508del homozygotes and about 40% are heterozygotes. Cumulatively at least one copy of F508del is present in about 90% of patients, making it by far the most common mutation: only four other mutations occur in more than 1% of sequences and none of these exceeds about 5%.[11]

10.1.2 F508del CFTR Defects

The F508del mutation causes loss of CFTR function in a number of ways. Channel density at the apical membrane is reduced due to protein misfolding. The misfolded CFTR is recognised by cellular quality control mechanisms and degraded through ER-associated degradation (ERAD), leaving only a tiny fraction of the translated CFTR protein to traffic to the plasma membrane (class 2 defect).[12] The channel density of F508del CFTR is also reduced because the mutant protein that does reach the plasma membrane undergoes rapid endocytosis but slow recycling back to the plasma membrane (class 6 defect).[13] F508del function is further reduced because it has a lower channel open probability (gating defect, class 3).[14] In addition, recent work has shown that once activated by phosphorylation, the F508del CFTR rapidly inactivates at physiological temperature.[15,16] The F508del mutation therefore causes a complex set of defects. However, small molecules have been shown to reverse some of these defects and recent clinical trials suggest that they may be capable of restoring sufficient function to benefit patients. It is thought that restoring approximately 10% of normal function should provide benefit to patients because this level of residual function is associated with mild disease. Restoration of 50% of normal function should provide substantial benefit because this level of function is present in CF carriers (heterozygotes) with no disease.

10.2 Correctors, Potentiators and Activators

Small molecules called 'correctors' have been shown to reverse the class 2 folding/trafficking defect of F508del CFTR.[17–19] Correctors increase the quantity of CFTR channels at the plasma membrane. Because the other F508del defects, including the gating defect, remain, a corrector alone is likely to provide only a small fraction of normal function. Corrector efficacy can be assessed functionally using a variety of ion channel assays. Western blotting to monitor the increase in the high molecular weight post-Golgi form of CFTR known as band C is also a commonly used approach.[20]

Several different modes of action can be envisaged for correctors. Correctors can be thought of as acting as transcriptional activators, pharmacological chaperones or proteostasis modulators.

Inhibitors of histone deacetylase 7 (HDAC7) have been shown to alter the transcriptional level of a number of CFTR-related genes, possibly including a modest effect on the transcription of CFTR itself, that either act alone or together to reverse the F508del CFTR trafficking defect.[21] The transcriptional activation effect resulted in restoration of activity in F508del primary airway epithelia to 28% of the wild-type level, demonstrating the potential of this mode of correction.

A pharmacological chaperone is a compound that directly binds and stabilises a misfolded protein in such a way that the protein achieves a more native fold. Pharmacological chaperones can be orthosteric (active site) or allosteric (non-active-site) binders. Numerous examples of pharmacological chaperones exist in the literature, especially in the field of G-protein-coupled receptors, where the concept first originated.[22]

Proteostasis modulators change the folding environment by changing the level or activity of pro-folding elements of the secretory pathway.[23] Selective up- or downregulation of specific trafficking proteins can also be envisaged. Because a low level of F508del CFTR escapes the ER and reaches the plasma membrane,[12] mechanisms that increase this 'leak' or allow accumulation of the 'leaked' protein might lead to an increase in the steady-state levels of CFTR at the cell surface. Such mechanisms could include modulating interactions with chaperone proteins to increase folding or decrease degradation, allowing more time for co-translational folding, promoting forward trafficking of CFTR from ER to Golgi or Golgi to plasma membrane, as well as slowing endocytosis from the plasma membrane or increasing recycling from early endosomes back to the plasma membrane. While there has been progress characterising the mode of action of some correctors, the molecular target(s) of corrector compounds have to date not been defined. In practice, correctors require time to achieve their effect, typically 12–48 hours to see a maximal response.

'Potentiators' are compounds that increase the channel open probability of mutant CFTR, reversing the class 3 gating defect. Potentiators are generally thought to influence gating through a direct interaction with CFTR, either at the transmembrane domains[24] or the nucleotide binding domains.[25] Direct interaction with CFTR has been inferred from excised patch experiments and from experiments in which purified CFTR protein is reconstituted into artificial bilayer membranes.[24,26,27] Potentiators can exert their effects almost instantly so potentiator assays usually involve acute additions or short incubation times of the order of minutes or tens of minutes.

While certain chemical scaffolds span both potentiator and corrector activities,[28,29] this has been more the exception than the rule and may arise more commonly for compounds that bind to proteins promiscuously. Correction and potentiation activities appear to arise from distinct mechanisms and optimising compounds with both activities will be challenging because the overlap between their structural requirements is likely to be small.

A third class of F508del modulators termed 'activators' has recently been identified.[30] These compounds act acutely to increase the activity of F508del and wild-type CFTR. Unlike potentiators, activators do not require the CFTR regulatory domain to be phosphorylated in order to increase CFTR function. Further work will be required to gauge the therapeutic utility of activator compounds.

10.3 Correctors and Potentiator Assays

10.3.1 Phenotypic Assays

Cell-based phenotypic assays have been used in high-throughput screening (HTS) to discover corrector compounds. Typically, recombinant F508del CFTR is overexpressed in a cell line under the control of a strong non-native promoter. Due to the F508del trafficking defect, these cell lines have low basal levels of CFTR function and cell-surface CFTR. The cells are treated with test compounds for 16–48 hours and the increase in CFTR function or channel density at the cell surface is measured. CFTR function has been assayed in HTS assays using fluorescent sensors of ion flux or membrane potential, while cell-surface CFTR has been assayed by antibody detection of epitope-tagged CFTR.

CFTR function has been assessed using a halide-sensitive yellow fluorescent protein sensor (hsYFP) in which YFP is engineered to allow quenching by iodide ion,[31] an anion that is channelled by CFTR. In this assay format, CFTR is activated by using a cyclic AMP agonist such as forskolin (to induce regulatory domain phosphorylation) and a potentiator compound such as genistein (to increase open probability). As iodide ions enter cells expressing hsYFP through CFTR, YFP fluorescence decreases. The magnitude of the change and the rate of change in fluorescence is proportional to the level of CFTR. In the case of membrane-potential-sensitive dyes, cAMP agonist and potentiator addition causes a change in membrane potential which induces redistribution of dye and in turn results in a change in fluorescence.[18]

For cell-surface CFTR assays, F508del CFTR has been engineered to include epitope tags in extracellular loops. In intact cells, the epitope tag is exposed only when CFTR is at the plasma membrane, allowing quantitation using an antibody in an ELISA format.[32] There are advantages and disadvantages to both approaches: functional assays can detect fluorescent false-positive compounds while cell-surface assays require changing the CFTR structure and can detect compounds that increase the surface density of non-functional channels. While both approaches have been used to successfully discover new correctors, only a small fraction of screening 'hits' show corrector activity in primary airway cells from CF patients.[17,19]

Potentiator discovery has used a variation of the cell-based functional assay described above. F508del CFTR can be corrected by incubation at reduced temperature (often around 27 °C) or by treatment with a corrector for 16–24 hours. Test compounds are applied just before measurement together with a cyclic AMP agonist such as forskolin. Iodide-containing buffer is added, and YFP fluorescence is followed (hsYFP iodide flux assay).[17] Alternatively, membrane potential dye fluorescence may be monitored after addition of test compounds with a cAMP agonist. Because most potentiator compounds appear to act by binding to CFTR, there is good correspondence between activity in recombinant cell lines and in patient primary cells.

10.3.2 Structure-Based Approaches

Biochemical structure-based approaches to HTS for correctors and potentiators have so far not been reported. The F508del mutation resides in the first nucleotide binding domain (NBD1) of CFTR. As described above, the F508del mutation causes CFTR to misfold and also causes a number of other defects.

Isolated F508del NBD1 domain is likewise misfolded and destabilised relative to wild-type NBD1. 'Suppressor' mutations that partially reverse the effect of F508del on stability and folding of NBD1 have been identified. Suppressor mutations that reverse the folding defect in isolated F508del NBD1 domain also correct the folding of full-length F508del CFTR protein, promoting trafficking and thereby increasing channel density at the plasma membrane.[33,34] Therefore, it is reasonable to suppose that compounds that bind to F508del NBD1 might, like suppressor mutations, stabilise the domain and correct folding in the context of full-length protein. To date, examples of NBD1-stabilising compounds have been rare[35,36] and HTS for F508del NBD1 stabilisers has not been reported.

A high-resolution crystal structure of full-length CFTR has also not yet been reported but several crystal and NMR structures of isolated NBD1 are available.[37–39] In the absence of a crystal structure of full-length CFTR, computational modelling has been employed.[40–42] EPIX Pharmaceuticals constructed a virtual structure incorporating information from the structure of Sav1866 transmembrane domains, NBD crystal structures, and distance constraints from intramolecular cysteine crosslinking experiments.[43] The CFTR structural model was then used to perform a 'virtual screen' by computational docking of test structures to a number of putative binding pockets at the interfaces between NBD1/NBD2, NBD1/ICL4 and NBD1/NBD2/ICL2/ICL4. The approach led to the identification of a number of new scaffolds with correction activity, potentiation activity and both correction and potentiation activities in the Fischer rat thyroid (FRT) cell line expressing recombinant F508del CFTR. Functional assay data from CF patient-derived airway cultures has so far not been reported. As knowledge of the CFTR 3D structure improves, this approach is likely to receive further attention.

10.4 Potentiators and Correctors in the Clinic

10.4.1 Potentiator Clinical Trials

As explained above, potentiator compounds will be needed to address the F508del CFTR gating defect. Clinical evaluation of potentiators is more straightforward in the G551D mutation, a pure class 3 mutation that causes only a gating defect. G551D is present in one copy in approximately 4% of CF patients.[3] The mutated channel is present at the plasma membrane at normal levels but the channel open probability is substantially reduced relative to wild-type. Kalydeco™ (ivacaftor, VX-770, Figure 10.3) is a marketed potentiator compound that improves the gating characteristics of G551D. It has been shown to improve lung function (absolute change in % predicted FEV_1 of +11–13%; FEV_1 or forced expiratory volume in 1 second is a measure of lung function), allow weight gain, and reduce the frequency of pulmonary exacerbations.[44,45] It is the first marketed disease-modifying therapy for G551D patients and appears to be providing a transformational benefit. Approximately 1100 US patients have the G551D mutation and are likely to

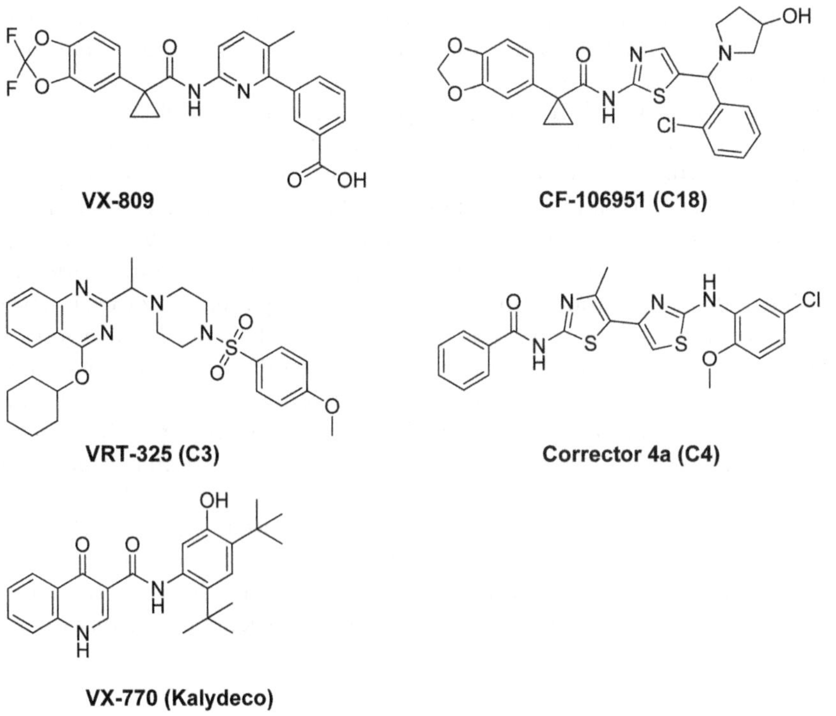

VX-809

CF-106951 (C18)

VRT-325 (C3)

Corrector 4a (C4)

VX-770 (Kalydeco)

Figure 10.3 Structures of VX-770 (Kalydeco™), VX-809, C18, corrector 4a and VRT-325.

benefit from Kalydeco™ therapy.[3] Kalydeco™ has been shown to potentiate other CFTR mutations that affect channel gating and to increase function in mutants with reduced conductance.[46,47] Clinical trials directed at testing efficacy in gating, conductance and other mutants are under way (http://www.clinicaltrials.gov).

Because most of the patients in these trials were compound heterozygotes (one G551D CFTR allele and a second mutant allele not responsive to Kalydeco™), the strong clinical response suggests that Kalydeco™ substantially or completely repairs the G551D defect. The magnitude of this clinical improvement through an effect on one copy of CFTR provides a useful benchmark for evaluating investigational therapies on the F508del mutation, discussed below.

10.4.2 Corrector Clinical Trials

The F508del mutation causes multiple defects including the trafficking defect addressed by correctors and the gating defect addressed by potentiators. Monotherapy with Kalydeco™ did not lead to any appreciable clinical improvement in F508del patients.[48] Similarly, the corrector VX-809 showed no statistically significant positive effects when administered alone, consistent with the fact that F508del causes folding and gating defects that must both be addressed.[49]

In trials on F508del homozygote patients, corrector plus potentiator combinations have so far shown improvements in lung function that are smaller than the responses achieved by Kalydeco™ in G551D compound heterozygotes (G551D and a second mutant allele). The corrector VX-661 plus Kalydeco™ caused a net absolute change in % predicted FEV_1 of +4–5%.[50] The clinical development of these corrector plus potentiator combinations is ongoing and such a combination is likely to benefit patients. However, based on FEV_1 responses it appears that their efficacy in F508del homozygotes is substantially lower than the efficacy of Kalydeco™ in G551D compound heterozygotes. This contrast suggests that the efficacy of the combination acting on two copies of F508del is much less than the efficacy of Kalydeco™ monotherapy on one copy of G551D. As a consequence, we may expect that greater corrector efficacy would provide further benefit to F508del patients.

Clinical responses to VX-661 or VX-809 combined with Kalydeco™ in F508del compound heterozygotes were smaller than in homozygotes and did not reach statistical significance (http://investors.vrtx.com/releasesArchive.cfm). Therefore greater corrector efficacy will be required to benefit patients with one copy of F508del.

10.4.3 Biomarkers

As will be outlined below, CF patient primary cells are a useful tool for optimising correctors and potentiators. In the case of Kalydeco™, robust responses to drug in patient-derived G551D bronchial epithelial cultures increased the confidence of success in the clinic at a time when CFTR modulation had not yet been shown to benefit patients. Retrospectively, the magnitude of the potentiation response in G551D bronchial epithelial cultures as a fraction of wild-type function closely paralleled the clinical response in nasal potential difference measurements of G551D patients, validating the use of CF primary airway cultures.[51]

One outcome of the clinical research on Kalydeco™ was a demonstration of the utility of sweat chloride concentration measurements. Kalydeco™ reduced sweat chloride concentration in G551D heterozygotes from ~100 mM (high) to below 60 mM (the diagnostic threshold for CF).[44] The changes were large and statistically significant after short periods of treatment (as few as 3 days) and in patient cohorts as small as 8. Because sweat chloride is readily measured by pilocarpine iontophoresis, it has become a useful clinical biomarker of potentiator efficacy in G551D patients, notwithstanding the fact that sweat chloride concentration responses do not correlate with improvements in lung function on a patient-by-patient basis.[52]

Clinical testing of VX-809 or VX-661 alone or in combination with Kalydeco™ in F508del homozygotes has to date shown smaller effects on sweat chloride concentration (change of 16 mM or less).[50,53] It is possible that the combinations are not efficacious enough at the exposures achieved so far and that larger responses will be seen once an efficacy threshold has been exceeded. It is also conceivable that the F508del sweat gland responds differently and more weakly to correctors than the G551D sweat gland responds to potentiators.

Even in the favourable situation in which compound responses in patient-derived airway cells can be measured and clinical biomarkers such as nasal potential difference and sweat chloride concentration measurements can be used, advancement of new compounds into the clinic would be facilitated by the ability to measure the same kind of response in primary cell cultures, animals and patients.

A translational biomarker, *i.e.* a biomarker that measures the effect of a corrector or potentiator both *in vitro* and *in vivo* would greatly improve the certainty of selecting appropriate doses for clinical trials. Understanding exposure–response relationships in rodents would facilitate selection of appropriate clinical doses. Biomarker response in small pilot clinical studies could define the exposure of a candidate drug that would be likely to provide benefit as well as the dosing regimen that would lead to the appropriate level and duration of exposure. Unfortunately, a true translational biomarker for correctors or potentiators has not yet been identified, although this is an area of active research.

10.5 The Goal of Combination Therapy

Clinical testing of a combination of a corrector and a potentiator has shown that modest efficacy is achievable in F508del patients. Recent *in vitro* work has shown how even greater efficacy may be attained. Screening combinations of FDA-approved drugs for correction activity showed that two-corrector combinations tested in the presence of a potentiator could deliver higher efficacy (8-fold over vehicle control) than the most efficacious single corrector plus potentiator (3.5-fold over vehicle control) in the FRT cell line expressing recombinant F508del CFTR.[54] Similar results were seen with combinations of published correctors tested in the FRT cell line.[55] Significantly, combinations of correctors also gave greater efficacy when tested in CF primary cells. Addition of corrector 4a to VX-809 gave a response double that of VX-809 in CF primary airway epithelial cells (Figure 10.4).[19] Two-corrector combinations gave responses 2- to 2.7-fold greater than VX-809 in F508del primary intestinal organoids.[56]

These findings demonstrate that combining two correctors and a potentiator can deliver substantially greater F508del rescue than a single corrector plus a potentiator. VX-809 plus Kalydeco™ has been shown to give approximately 35% of wild-type activity in CF patient primary airway cells[19] (Figure 10.4). As noted above, addition of a second corrector can double responses in patient-derived primary cells. This level of efficacy would be more than 50% of normal, the level of function associated with CF carriers who have no disease. Therefore, a combination of two correctors and

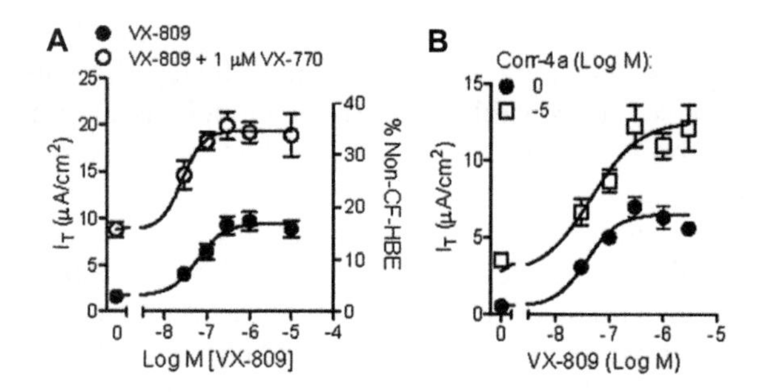

Figure 10.4 (A) Concentration response curve for VX-809 in the absence (•) and presence of 1 μM VX-770 (○) in F508del hBE from a single donor. Activity expressed in terms of short-circuit current and as a fraction of wild-type hBE activity. (B) Additive effects of VX-809 and corrector 4a indicated concentrations on CFTR-mediated chloride transport in cultured F508del hBE isolated from a single donor bronchus ($n = 4$). (Images are from ref. 19, used with permission.)

a potentiator can be expected to provide substantially greater benefit to patients than the combination of one corrector such as VX-809 or VX-661 with Kalydeco™.

The key to the three-compound combination is a synergistic combination of correctors. Synergy can be understood as efficacy greater than either individual corrector (Loewe additivity) that arises when two correctors with different modes of action are combined.[57,58] Synergy could conceivably arise from a combination of a pharmacological chaperone and a proteostasis modulator, between two proteostasis modulators with complementary modes of action, or between two pharmacological chaperones that bind to different sites on CFTR, among other possibilities. Two proteostasis modulators that affect parallel pathways that both lead to correction would be expected to combine favourably, as would compounds that increase flux through two steps in the same pro-folding pathway.[59]

Work from several groups has shown how two CFTR pharmacological chaperones might combine synergistically. Suppressor mutations that reverse the folding defect in isolated F508del NBD1 also correct the folding of full-length F508del CFTR.[33,34] NBD1 has been found to make important contacts with membrane-spanning domains (MSDs), especially intracellular loop 4 (ICL4) of MSD2. Suppressor mutations that repair F508del NBD1–ICL4 interactions also restore the folding and trafficking of full-length F508del CFTR. NBD1 suppressors and NBD1–ICL4 suppressors have modest effects alone, but when they are combined substantial levels of CFTR folding, trafficking and function are restored[33,34] (Figure 10.5). This provides a genetic rationale for targeting small molecules to improve NBD1 folding and to promote NBD1–ICL4 binding which when combined would deliver synergistic correction.

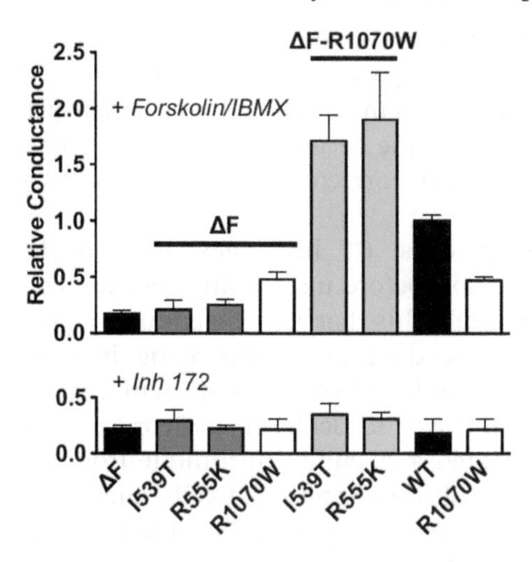

Figure 10.5 Correction of the F508del NBD1–ICL4 interface defect is synergistic with correction of the F508del NBD1 folding yield defect. CFTR-dependent trans-epithelial conductance experiments with wild-type, F508del, NBD1 and ICL4–NBD1 interface mutants. Monolayers of FRT cells transiently expressing CFTR with the indicated mutations were stimulated with 10 μM Forskolin + 100 μM IBMX and trans-epithelial conductance measured (top, ±SEM, $n = 3$, average conductance for wild-type = 1.24 mS). The CFTR dependence of conductance is indicated by the sensitivity to 20 μM Inh-172 (bottom). (Image from ref. 33, used with permission.)

10.6 Use of CF Patient Primary Cells for Corrector Discovery

10.6.1 CF Primary Airway Cell Culture

High-throughput screening (HTS) for correctors and potentiators has so far relied on overexpression of mutant CFTR in immortalised cell lines. While potentiators give broadly comparable activity in clinical trials, in patient primary cells and in cell lines, correctors that show activity in the cell lines used for HTS often fail to show activity in CF patient primary cells.[60] Consequently, optimisation of corrector drug candidates requires measurements of efficacy and potency in CF primary airway epithelia.

Primary cells can be harvested from CF lungs when patients undergo transplants. Epithelial cells can be harvested, expanded, then seeded onto filters and cultured at an air/liquid interface (ALI) to induce differentiation into a tissue-like epithelial layer.[61–63] These cultures contain tall columnar epithelial cells, goblet cells and basal cells arranged with polarised pseudo-stratified arrangement that closely resembles native tissue. Like tissues, the cultures spontaneously secrete mucus and surface liquid and contain beating cilia. CF airway epithelial cultures show reduced chloride secretion due to lack of CFTR function, reduced levels of airway surface liquid, and increased inflammatory responses. Measurements of corrector activity in well-differentiated CF human bronchial epithelial (CF hBE) cells are considered to have high value because this system is biologically similar to native tissue and clinical correctors and potentiators show appropriate responses.[19,64]

Like other primary cells, CF hBE cells are capable of only limited passaging and expansion before undergoing senescence. A new technique promises to push back this barrier. Co-culture techniques have been employed to extend the duration of passaging in other fields, such as embryonic stem cells, induced pluripotent stem cells and keratinocytes. The laboratory of Professor R. Schlegel at Georgetown University (Washington DC, USA) adapted a co-culture technique from keratinocyte culture and applied it to the culture of epithelial cells, including airway epithelial cells, to allow extended passaging with the capacity to differentiate appropriately in ALI culture.[65]

Airway epithelial cells are cultured in the presence of irradiated fibroblasts (the J2 clone of the Swiss 3T3 cell line) in the presence of a Rho kinase inhibitor (ROCKi) Y-27632. While typical hBE cultures may be used up to passage 5,[62] co-cultured human bronchial epithelial cells have been carried beyond passage 40 with no apparent loss of capacity to differentiate appropriately in ALI culture after removal of fibroblasts and ROCKi.[66] While this approach has great promise, it has not yet been widely applied or characterised, so the potential pitfalls are not yet known.

10.6.2 CF Primary Airway Cell Assays

There are several approaches to assess and characterise the activity of F508del corrector compounds in fully differentiated CF hBE cells. Optimisation of corrector activity involves iterative structural modification and activity testing to reveal structure–activity relationships. This requires assays with reasonable throughput and good data quality to profile new compounds and to generate dose–response curves.

Traditionally, corrector activity in CF hBE cells has been assessed using short-circuit current measurements in the Ussing chamber apparatus.[62] Compounds are applied to cells on filters for 24–48 hours, then the filters are mounted in Ussing chambers in which electrodes sense current and voltage across the epithelium through agar bridges. Voltage is clamped to zero and after inhibiting ENaC epithelial sodium channel activity with amiloride, CFTR is activated by sequential addition of forskolin and a potentiator. The cumulative forskolin plus potentiator response provides a measure of the extent of correction. While this method works well, its throughput is limited by the time-consuming maintenance of the chambers, electrodes and agar bridges.

The equivalent current electrophysiology assay and apparatus developed by Professors R. J. Bridges and W. Van Driessche allow measurement of correction activity in differentiated CF hBE cells in 24-well filter plates.[67] Rather than measuring the short-circuit current, this method measures trans-epithelial voltage and resistance under open circuit conditions and an 'equivalent' current is calculated. Maintenance is reduced dramatically because solid silver or silver chloride electrodes are used instead of agar bridges and the measurements are performed directly in disposable filter plates. A robot arm is used to cycle a 24-position electrode array across several 24-well filter plates allowing concurrent measurements over multiple plates. The assay has been implemented to provide high-quality data with throughput sufficient to support medicinal chemistry optimisation. A 96-well apparatus in development promises to further increase throughput.

10.6.3 Intestinal Organoid Culture and Assays

CF is a multi-organ disease in which the gastrointestinal tract is a significant source of morbidity. Between 6% and 20% of CF newborns present with the small intestinal obstruction known as meconium ileus and distal intestinal obstruction syndrome affects approximately 16% of patients. CF patients also experience increased frequency of malabsorption and gastro-oesophageal reflux disease.[68] Some of these issues arise because CFTR is present in enterocytes in intestinal crypts and mutation of CFTR reduces fluid flow into the intestinal lumen. Intestinal crypt cells are therefore a clinically relevant primary cell type.

Long-term culture of primary intestinal crypt cells was not possible until the lab of Professor H. Clevers at the Hubrecht Institute (Utrecht, the Netherlands) identified conditions that allowed long-term expansion of organ-like structures (organoids) from human colon.[69] In the presence of a defined set of growth factors and small molecules in Matrigel, the crypt cells formed organoids that could be passaged up to 100 times without measureable changes in phenotype or karyotype. Organoids could be stored frozen and in many other ways treated like cell lines. Importantly, the organoids were shown to represent a physiologically-relevant, differentiated and polarised collection of primary cells.[70]

Intestinal current measurements on rectal biopsies have been used to confirm diagnosis in CF for some time.[71] Crypt cells from rectal biopsy samples have been used to culture intestinal organoids to form polarised crypt-like structures in which CFTR secretes ions into the lumen.[72] Activation of CFTR with forskolin caused the crypts to expand due to anion secretion and osmotic water flow into the lumen. When CF intestinal organoids were pre-treated with corrector compounds, the rate of expansion was significantly greater than for untreated controls.[72] This functional response in organoids in microtitre plates could be readily quantified by simple image analysis techniques. The CF intestinal organoid assay has the potential to be miniaturised into 384-well format to allow screening for correctors directly in primary cells, avoiding some of the known pitfalls of HTS using recombinant cell lines.

10.6.4 Pluripotent Stem Cells

An alternative to using patient-derived primary cells is being actively investigated: there have been exciting developments in directing human pluripotent stem cells to differentiate into airway epithelia. Precisely timed treatment of embryonic stem cells (ESCs) or induced pluripotent stem cells (iPSCs) with specific growth factors allowed their conversion into lung progenitor cells.[73,74] Immature lung cells were cultured at an ALI to induce further differentiation, leading to the formation of epithelia containing basal cells, goblet cells and ciliated cells.[74] Comparison of epithelia derived from normal ESCs and CF iPSCs showed functional halide flux in normal but not CF cultures. Treatment of F508del CFTR CF iPSCs with corrector C18 (an analogue of VX-809) induced production of the mature C band of CFTR and relocalisation of CFTR to the plasma membrane.[74] Functional responses to C18 correction were not yet detectable, indicating the need for further optimisation of the method.

This work holds great promise for generation of cellular models of CF. Patient-derived iPS cells could be expanded to a much larger extent than primary cells and banked for future use. Testing of candidate drugs across arrays of patient epithelia could identify high responders in a personalised medicine approach. Because iPS cells can be much more readily genetically modified than primary cells, it should be possible to make stable genetic changes to delete or overexpress genes at will, allowing the construction of customised cell models.

10.6.5 Utility of CF Primary Cell Assays for Ranking Correctors

Corrector and potentiator activity can be assayed in cell lines as well as patient-derived primary cells. Potentiators that target CFTR directly show broadly similar efficacy and potency across different cell lines and in primary cells. In contrast, correction activity can differ significantly among different cell lines expressing F508del CFTR and between cell lines and primary cells.[60] Comparison of a set of 30 correctors in the FRT and A549 cell lines expressing recombinant F508del CFTR revealed only nine compounds with comparable activity in both cell lines, with most of the rest showing bias for one cell line or the other.

The difference in behaviour between cell lines and between cell lines and primary cells is difficult to understand and is a significant obstacle to identifying and optimising efficacious correctors. Multiple factors may be in play and no one factor appears to explain the difference.

CFTR is normally present at low levels in most tissues. Overexpression of recombinant CFTR in cell lines may flood the secretory pathway with much more CFTR than is usual, altering the stoichiometric balance between CFTR and the chaperone proteins. Overexpression might also induce a greater level of ERAD or activation of the unfolded protein response.

In tissues, the epithelial cells expressing CFTR are polarised, with clearly distinct apical and basolateral membranes. In the majority of tissues, CFTR is normally present at the apical membrane as a result of regulated trafficking processes.[75] Primary cells can be cultured in a well-differentiated state in which cell polarity and apical CFTR localisation is maintained but cell lines grown on plastic do not allow polarisation.

Many non-human cell lines, such as the FRT epithelial line, the baby hamster kidney (BHK) line and the NIH 3T3 rat fibroblast line, are commonly used in corrector discovery.[17,19,32] Species differences could contribute to the disconnect between corrector responses in these cell lines and human primary cells, especially for proteostasis modulators. In cell lines, strong non-native promoters are used to drive CFTR expression, a factor that could affect the response to transcriptional modulators.

There have been several attempts to overcome these limitations by generating immortalised cell lines from CF patient airway epithelial cells.[76] Unfortunately these efforts have not yet yielded lines that give corrector responses that parallel the primary cells. The difference in corrector behaviour between cell lines and primary cells means that the efficacy and potency of corrector molecules should ideally be confirmed using patient-derived primary cells at an early stage.

Many corrector compounds have been identified using cell lines expressing recombinant F508del CFTR. Side-by-side dose–response comparisons in well-differentiated CF hBE cultures using an electrophysiology readout such as short-circuit or equivalent current would efficiently rank these compounds according to efficacy and potency. Corrector responses can vary among primary cells from different patients and culture conditions are known to influence the magnitude of CFTR responses.[62] A reasonable solution would be to compare corrector responses to benchmark compounds such as those in the CFTR Chemical Compound Distribution Program from Cystic Fibrosis Foundation Therapeutics Inc. (CFFT) (described below) or a commercially available clinical corrector.

Because cell lines can provide a misleading picture of correction responses of small molecules, it is reasonable to suppose that the effects of genetic modulation of F508del CFTR trafficking may also differ between cell lines and primary cells. As for correctors, performing confirmatory experiments in patient primary cells would add significant weight to any conclusions derived from knocking down or overexpressing genes that regulate F508del CFTR trafficking in cell lines.

Transfection of siRNA oligonucleotides and lentiviral delivery of shRNA in polarised and well-differentiated CF airway epithelial cultures have both failed to substantially knock down target genes. Recent progress suggests that using Dicer-substrate siRNA (DsiRNA) chemically modified to increase stability in combination with reverse transfection at the time of seeding onto filters can give substantial knockdown.[77] For example, CFTR function could be reduced by >90% in polarised primary human airway epithelial cultures at 14 days post-seeding using DsiRNA reverse transfection.

10.7 Probe Compounds and Mechanisms of Action

A 'toolkit' collection of CFTR modulators from the literature has been assembled and made available to the CF research community through the CFTR Chemical Compound Distribution Program.[78] Commonly used corrector tool compounds include corrector 4a (C4),[17] VRT-325 (C3),[18] VX-809 (not part of the CFTR Chemical Compound Distribution Program)[19] and C18 (CF-106951, an analogue of VX-809). VX-809 and C18 are significantly more efficacious than C4 or C3 in CF primary airway cells.[19] This collection also includes potentiator compounds (but not Kalydeco™) and CFTR inhibitors. While the modes of actions of the correctors are under active investigation, the molecular targets of these compounds have so far not been defined.

10.7.1 Mechanisms of Action

The mode of action of VX-809 is the most studied of the correctors owing to its use in clinical trials and broad efficacy across cell lines and primary cells. It has been proposed to act as a CFTR pharmacological chaperone and has been reported variously to act on MSD1 of CFTR,[79,80] the interface between NBD1, ICL4 and ICL1,[56] and postulated to act at the NBD1–ICL4 interface.[81,82]

Definitive evidence of VX-809 action through direct binding to CFTR would require crosslinking studies, which have so far not been reported. Reversal of the thermal inactivation defect of an F508del CFTR construct containing additional mutations (R29K and R555K) by VX-809 and C18 in planar lipid bilayer electrophysiology studies was consistent with direct interaction,[56] while a similar study of an F508del CFTR construct containing a single additional mutation (I539T) did not show reversal of thermal inactivation.[81] Further studies will be required to resolve these apparent discrepancies.

Potentiator compounds are expected to act directly on CFTR to influence channel gating. Mechanistic studies of Kalydeco™ have shown that it increases the function of CFTR purified and reconstituted into liposomes (confirming binding to CFTR in a system free of other proteins) and that its effect is dependent on channel phosphorylation but not on the presence of ATP.[26] An elegant electrophysiology study of Kalydeco™ confirmed the lack of ATP dependency and gave detailed insight into its unique mode of action involving decoupling of channel gating from ATP hydrolysis.[24]

10.7.2 Probe Compounds

Corrector compounds have so far been discovered using the phenotypic screening approaches described above. The phenotypic approach is a powerful method to find new chemical matter when the identity of specific targets is not known and when disease mechanisms are imperfectly understood. Phenotypic screening has been shown retrospectively to have had a higher success rate for the discovery of first-in-class drugs than target-based approaches.[83] The phenotypic approach also has its drawbacks: assays are more complex and lower in throughput; without target knowledge, optimising compound potency and selectivity is more challenging. In some cases, phenotypic screening hits have been used to identify their molecular targets to facilitate compound optimisation. Such approaches can yield 'tool' or 'probe' compounds which are useful for validating the target and further understanding disease mechanism. Guidelines for good small-molecule probe compounds that have been proposed include: well-characterised chemical identity, potency (activity at <100 nM in biochemical assays or at 1–10 μM in cellular assays), selectivity in broad pharmacology panels (panels of assays for inhibition or activation of G-protein-coupled receptors, nuclear receptors, ion channels, kinases, phosphatases, proteases and ubiquitin ligases that are used to assess drug selectivity during drug development), and context (fit-for-purpose in a given system).[84]

There are no published data on toolkit correctors from broad pharmacology panels but the investigational clinical corrector VX-809 and the marketed potentiator Kalydeco™ have presumably undergone and passed extensive pharmacology and safety profiling. C3 and C4 have been shown to influence the trafficking of misfolded proteins other than F508del CFTR such as the G601S hERG channel and G268V P-glycoprotein while VX-809 did not.[19] Therefore, VX-809 and Kalydeco™ are likely to be useful as probe compounds while C3 and C4 may not be optimal because their selectivity has not been characterised and existing results suggest that their effects are broad.

Structure–activity relationships (SARs) of probe compounds can be used to powerfully discriminate between on-target and off-target mechanisms.[84] For a given chemical scaffold, if two structurally similar analogues are available, one with high potency and efficacy and the other inactive in the phenotypic assay, the pair can be tested in other assays designed to identify their molecular target. Correlation between the phenotypic assay responses and responses in a second assay are consistent with (but do not prove) the hypothesis that similar mechanisms may be operating. Conversely, lack of correlation suggests different modes of action and this approach can be used to rapidly eliminate potential targets/mechanisms. If extensive SAR information is available, a small collection of compounds with high, medium, low and no efficacy can be used as a probe set to triangulate on the target.[84] Application of this approach to F508del correctors would be expected to yield substantial benefits but awaits the availability of medium-active and inactive analogues of VX-809.

Corrector discovery efforts are hampered by the lack of broadly validated targets other than CFTR itself. Testing of probe molecules from other fields would be a quick route to discover new correction targets. If a potent and selective inhibitor of a given target was shown to cause rescue of F508del CFTR function in CF primary cells, it would be a strong indication that the target was implicated in correction. The chemical genetic result could be confirmed by classical genetic knockdown (siRNA, shRNA, or dominant negative approaches). Screening for new correctors could then use an HTS-compatible biochemical assay based on the target. With the wider use of CF primary cell assays, this approach can now be applied by many academic and industry groups. In summary, development of a corrector probe set could be used to identify the molecular targets for correction while use of probe sets from other fields could be used to find new putative correction targets.

10.8 Role of the Cystic Fibrosis Foundation

The Cystic Fibrosis Foundation is a non-profit donor-supported organisation dedicated to developing new drugs to fight the disease and improve quality of life for patients. The contributions span support of basic research, drug discovery and development, clinical care, a patient registry, and a therapeutics development network (http://www.cff.org). The organisation stands out among patient advocacy groups for the breadth and the amount of this support.

The CF Foundation funds 11 basic research centres at universities across the USA and provides research and training grants and awards to academic researchers through a competitive process for studies at the subcellular, cellular, animal or patient levels. Through its non-profit drug discovery and development affiliate, Cystic Fibrosis Foundation Therapeutics, Inc. (CFFT), the Foundation supports CF drug discovery, development and clinical evaluation. CFFT has organised several academic research consortia focused on CFTR folding, CFTR 3D structure, mucociliary clearance and biomarkers.

The Foundation's patient registry tracks the health and treatments of over 27 000 patients at Foundation-accredited care centres. Data from the patient registry permits studies of the effects of treatments, clinical care guideline development and clinical trial designs. The CF Foundation also manages a therapeutics development network of nearly 80 clinical research centres working to promote quality, safety and efficiency in CF clinical trials by centralising and standardising the research process.

In 1999, CFFT initiated funding for HTS for the discovery of corrector and potentiator compounds, leading to fruitful work in academic laboratories and in industry.[17,18,85,86] One of the industry laboratories was Aurora Biosciences (now Vertex Pharmaceuticals). The CFFT–Vertex collaboration was expanded and has led to the successful clinical programme which has so far produced Kalydeco™, VX-809, VX-661 and VX-983. In the process, CFFT has pioneered a business model that has been termed 'venture philanthropy', in which CFFT provided financial support, access to US CF patients and caregivers, and help with measurement of clinical outcomes.[87] This successful model is currently being employed to accelerate CF drug discovery and development research at Vertex Pharmaceuticals, Pfizer, Genzyme-Sanofi and Proteostasis Therapeutics, among others.

10.9 Concluding Thoughts

A number of developments have made the present time the most hopeful period to date for CF patients. Corrector plus potentiator combinations are being evaluated in F508del patients in late-stage clinical trials. Experiments in patient primary cells suggest that a combination therapy composed of two correctors and a potentiator could deliver efficacy at least double that of existing single corrector plus single potentiator combinations. Emerging data from pig and ferret models of CF are improving our understanding of disease physiology.[88,89] Existing and new primary cell systems are providing realistic and rigorous assays with which to evaluate new correctors at an early stage. CFTR folding and trafficking are becoming better understood and researchers are using the clinical agents as tools to develop new approaches to discover the next generation of correctors. Numerous academic laboratories are involved not just in basic research on CFTR but in the application of HTS for corrector discovery using a range of phenotypic and structure-based approaches. The pharmaceutical and biotechnology industries, recognising the high unmet need in orphan indications such as CF, have joined in the hunt for new therapies.

Many challenges remain. F508del CFTR has complex defects that may not allow complete rescue by small-molecule drugs. Advancing two- or three-drug combination therapies to market will be complicated and time-consuming. Nevertheless, the time when most CF patients have a therapy that prolongs and improves their lives by addressing the basic defects caused by CFTR mutations appears to be within sight. When that time arrives it will be due to the collaborative efforts of patient advocacy groups, academic researchers, and the pharmaceutical and biotechnology industries.

Acknowledgements

The support of the Cystic Fibrosis Foundation for the author's research is gratefully acknowledged, especially the leadership of Bob Beall, Preston Campbell, Melissa Ashlock, Diana Wetmore and Elizabeth Joseloff. The author wishes to thank the many colleagues, past and present, who contributed to the discussion and learnings summarised here, in particular Seng Cheng, Canwen Jiang, Richard Labaudinière, Chris Adams and Christine Bulawa. Thanks to Rajiah Aldrin Denny (Pfizer Worldwide Medicinal Chemistry) for preparing the CFTR structure model in Figure 10.2.

References

1. G. R. Cutting, F. Accurso, B. W. Ramsey, and M. J. Welsh, in *The Online Metabolic & Molecular Bases of Inherited Disease*, McGraw-Hill Global Education Holdings, 2013.
2. R. D. Coakley, in *Cystic Fibrosis*, ed. M. Hodson, D. Geddes and A. Bush, Edward Arnold, 3rd edn, 2007, pp. 59–68.
3. *Cystic Fibrosis Foundation Patient Registry 2011 Annual Data Report to the Center Directors*, Cystic Fibrosis Foundation, Bethesda, Maryland, 2012.
4. S. M. Rowe, S. Miller and E. J. Sorscher, *N. Engl. J. Med.*, 2005, **352**, 1992–2001.

5. R. A. Frizzell and J. W. Hanrahan, *Cold Spring Harbor Perspect. Med.*, 2012, **2**, a009563.
6. J. R. Riordan, *Annu. Rev. Biochem.*, 2008, **77**, 701–726.
7. J. F. Hunt, C. Wang and R. C. Ford, *Cold Spring Harbor Perspect. Med.*, 2013, **3**, a009514.
8. M. J. Welsh and A. E. Smith, *Cell*, 1993, **73**, 1251–1254.
9. P. A. Sloane and S. M. Rowe, *Curr. Opin. Pulm. Med.*, 2010, **16**, 591–597.
10. J. L. Bobadilla, M. Macek, Jr, J. P. Fine and P. M. Farrell, *Hum. Mutat.*, 2002, **19**, 575–606.
11. C. R. Scriver, *The metabolic & molecular bases of inherited disease*, McGraw-Hill, New York, 8th edn, 2001.
12. C. L. Ward and R. R. Kopito, *J. Biol. Chem.*, 1994, **269**, 25710–25718.
13. D. M. Cholon, W. K. O'Neal, S. H. Randell, J. R. Riordan and M. Gentzsch, *Am. J. Physiol.: Lung Cell. Mol. Physiol.*, 2010, **298**, L304–L314.
14. W. Dalemans, P. Barbry, G. Champigny, S. Jallat, K. Dott, D. Dreyer, R. G. Crystal, A. Pavirani, J. P. Lecocq and M. Lazdunski, *Nature*, 1991, **354**, 526–528.
15. A. A. Aleksandrov, P. Kota, L. A. Aleksandrov, L. He, T. Jensen, L. Cui, M. Gentzsch, N. V. Dokholyan and J. R. Riordan, *J. Mol. Biol.*, 2010, **401**, 194–210.
16. W. Wang, G. O. Okeyo, B. Tao, J. S. Hong and K. L. Kirk, *J. Biol. Chem.*, 2011, **286**, 41937–41948.
17. N. Pedemonte, G. L. Lukacs, K. Du, E. Caci, O. Zegarra-Moran, L. J. Galietta and A. S. Verkman, *J. Clin. Invest.*, 2005, **115**, 2564–2571.
18. F. Van Goor, K. S. Straley, D. Cao, J. Gonzalez, S. Hadida, A. Hazlewood, J. Joubran, T. Knapp, L. R. Makings, M. Miller, T. Neuberger, E. Olson, V. Panchenko, J. Rader, A. Singh, J. H. Stack, R. Tung, P. D. Grootenhuis and P. Negulescu, *Am. J. Physiol.: Lung Cell. Mol. Physiol.*, 2006, **290**, L1117–L1130.
19. F. Van Goor, S. Hadida, P. D. Grootenhuis, B. Burton, J. H. Stack, K. S. Straley, C. J. Decker, M. Miller, J. McCartney, E. R. Olson, J. J. Wine, R. A. Frizzell, M. Ashlock and P. A. Negulescu, *Proc. Natl. Acad. Sci. U. S. A.*, 2011, **108**, 18843–18848.
20. M. Benharouga, M. Sharma and G. L. Lukacs, *Methods Mol. Med.*, 2002, **70**, 229–243.
21. D. M. Hutt, D. Herman, A. P. Rodrigues, S. Noel, J. M. Pilewski, J. Matteson, B. Hoch, W. Kellner, J. W. Kelly, A. Schmidt, P. J. Thomas, Y. Matsumura, W. R. Skach, M. Gentzsch, J. R. Riordan, E. J. Sorscher, T. Okiyoneda, J. R. Yates, 3rd, G. L. Lukacs, R. A. Frizzell, G. Manning, J. M. Gottesfeld and W. E. Balch, *Nat. Chem. Biol.*, 2010, **6**, 25–33.
22. V. Bernier, D. G. Bichet and M. Bouvier, *Curr. Opin. Pharmacol.*, 2004, **4**, 528–533.
23. E. T. Powers, R. I. Morimoto, A. Dillin, J. W. Kelly and W. E. Balch, *Annu. Rev. Biochem.*, 2009, **78**, 959–991.
24. K. Y. Jih and T. C. Hwang, *Proc. Natl. Acad. Sci. U. S. A.*, 2013, **110**, 4404–4409.

25. O. Moran, L. J. Galietta and O. Zegarra-Moran, *CMLS, Cell. Mol. Life Sci.*, 2005, **62**, 446–460.
26. P. D. Eckford, C. Li, M. Ramjeesingh and C. E. Bear, *J. Biol. Chem.*, 2012, **287**, 36639–36649.
27. Z. Cai, Y. Sohma, S. G. Bompadre, D. N. Sheppard and T. C. Hwang, *Methods Mol. Biol.*, 2011, **741**, 419–441.
28. N. Pedemonte, V. Tomati, E. Sondo, E. Caci, E. Millo, A. Armirotti, G. Damonte, O. Zegarra-Moran and L. J. Galietta, *J. Biol. Chem.*, 2011, **286**, 15215–15226.
29. P. W. Phuan, B. Yang, J. M. Knapp, A. B. Wood, G. L. Lukacs, M. J. Kurth and A. S. Verkman, *Mol. Pharmacol.*, 2011, **80**, 683–693.
30. W. Namkung, J. Park, Y. Seo and A. S. Verkman, *Mol. Pharmacol.*, 2013, **84**, 384–392.
31. L. J. Galietta, P. M. Haggie and A. S. Verkman, *FEBS Lett.*, 2001, **499**, 220–224.
32. R. Robert, G. W. Carlile, C. Pavel, N. Liu, S. M. Anjos, J. Liao, Y. Luo, D. Zhang, D. Y. Thomas and J. W. Hanrahan, *Mol. Pharmacol.*, 2008, **73**, 478–489.
33. J. L. Mendoza, A. Schmidt, Q. Li, E. Nuvaga, T. Barrett, R. J. Bridges, A. P. Feranchak, C. A. Brautigam and P. J. Thomas, *Cell*, 2012, **148**, 164–174.
34. W. M. Rabeh, F. Bossard, H. Xu, T. Okiyoneda, M. Bagdany, C. M. Mulvihill, K. Du, S. di Bernardo, Y. Liu, L. Konermann, A. Roldan and G. L. Lukacs, *Cell*, 2012, **148**, 150–163.
35. H. M. Sampson, R. Robert, J. Liao, E. Matthes, G. W. Carlile, J. W. Hanrahan and D. Y. Thomas, *Chem. Biol.*, 2011, **18**, 231–242.
36. F. Liang, I. S. Kwok, P. Bhatt, S. Kaczmarek, R. Valdez, E. Layer, R. J. Fitzpatrick and A. F. Kolodziej, *Pediatr. Pulmonol.*, 2012, **47**, 224–225.
37. H. A. Lewis, X. Zhao, C. Wang, J. M. Sauder, I. Rooney, B. W. Noland, D. Lorimer, M. C. Kearins, K. Conners, B. Condon, P. C. Maloney, W. B. Guggino, J. F. Hunt and S. Emtage, *J. Biol. Chem.*, 2005, **280**, 1346–1353.
38. L. S. Pissarra, C. M. Farinha, Z. Xu, A. Schmidt, P. H. Thibodeau, Z. Cai, P. J. Thomas, D. N. Sheppard and M. D. Amaral, *Chem. Biol.*, 2008, **15**, 62–69.
39. V. Kanelis, P. A. Chong and J. D. Forman-Kay, *Methods Mol. Biol.*, 2011, **741**, 377–403.
40. J. Dalton, O. Kalid, M. Schushan, N. Ben-Tal and J. Villà-Freixa, *J. Chem. Inf. Model.*, 2012, **52**, 1842–1853.
41. A. W. Serohijos, P. H. Thibodeau and N. V. Dokholyan, *Methods Mol. Biol.*, 2011, **741**, 347–363.
42. L. He, A. A. Aleksandrov, A. W. Serohijos, T. Hegedus, L. A. Aleksandrov, L. Cui, N. V. Dokholyan and J. R. Riordan, *J. Biol. Chem.*, 2008, **283**, 26383–26390.
43. O. Kalid, M. Mense, S. Fischman, A. Shitrit, H. Bihler, E. Ben-Zeev, N. Schutz, N. Pedemonte, P. J. Thomas, R. J. Bridges, D. R. Wetmore,

Y. Marantz and H. Senderowitz, *J. Comput.-Aided Mol. Des.*, 2010, **24**, 971–991.
44. B. W. Ramsey, J. Davies, N. G. McElvaney, E. Tullis, S. C. Bell, P. Drevinek, M. Griese, E. F. McKone, C. E. Wainwright, M. W. Konstan, R. Moss, F. Ratjen, I. Sermet-Gaudelus, S. M. Rowe, Q. Dong, S. Rodriguez, K. Yen, C. Ordonez, J. S. Elborn and V. X. S. Group, *N. Engl. J. Med.*, 2011, **365**, 1663–1672.
45. J. C. Davies, C. E. Wainwright, G. J. Canny, M. A. Chilvers, M. S. Howenstine, A. Munck, J. G. Mainz, S. Rodriguez, H. Li, K. Yen, C. L. Ordonez, R. Ahrens and V. X. S. Group, *Am. J. Respir. Crit. Care Med.*, 2013, **187**, 1219–1225.
46. H. Yu, B. Burton, C. J. Huang, J. Worley, D. Cao, J. P. Johnson, Jr, A. Urrutia, J. Joubran, S. Seepersaud, K. Sussky, B. J. Hoffman and F. Van Goor, *J. Cystic Fibrosis*, 2012, **11**, 237–245.
47. F. Van Goor, H. Yu, B. Burton and B. Hoffman, *J. Cystic Fibrosis*, 2014, **13**, 29–36.
48. P. A. Flume, T. G. Liou, D. S. Borowitz, H. Li, K. Yen, C. L. Ordonez, D. E. Geller and V. X. S. Group, *Chest*, 2012, **142**, 718–724.
49. J. P. Clancy, S. M. Rowe, F. J. Accurso, M. L. Aitken, R. S. Amin, M. A. Ashlock, M. Ballmann, M. P. Boyle, I. Bronsveld, P. W. Campbell, K. De Boeck, S. H. Donaldson, H. L. Dorkin, J. M. Dunitz, P. R. Durie, M. Jain, A. Leonard, K. S. McCoy, R. B. Moss, J. M. Pilewski, D. B. Rosenbluth, R. C. Rubenstein, M. S. Schechter, M. Botfield, C. L. Ordonez, G. T. Spencer-Green, L. Vernillet, S. Wisseh, K. Yen and M. W. Konstan, *Thorax*, 2012, **67**, 12–18.
50. S. Donaldson, J. Pilweski, M. Griese, Q. Dong and D. Rodman, in *36th European Cystic Fibrosis Conference*, Lisbon, Portugal, 2013.
51. S. M. Rowe and A. S. Verkman, *Cold Spring Harbor Perspect. Med.*, 2013, **3**, a009761.
52. A. G. Durmowicz, K. A. Witzmann, C. J. Rosebraugh and B. A. Chowdhury, *Chest*, 2013, **143**, 14–18.
53. M. P. Boyle, in *26th Annula North American Cystic Fibrosis Conference*, Orlando, Florida, 2012.
54. S. Lin, J. Sui, S. Cotard, B. Fung, J. Andersen, P. Zhu, N. El Messadi, J. Lehar, M. Lee and J. Staunton, *Assay Drug Dev. Technol.*, 2010, **8**, 669–684.
55. R. J. Bridges, A. Thakerar, Y. Cheng, M. Mense, D. Wetmore, N. A. Bradbury, Y. Jia and W. van Driessche, *Pediatr. Pulmonol.*, 2010, 119–120.
56. T. Okiyoneda, G. Veit, J. F. Dekkers, M. Bagdany, N. Soya, H. Xu, A. Roldan, A. S. Verkman, M. Kurth, A. Simon, T. Hegedus, J. M. Beekman and G. L. Lukacs, *Nat. Chem. Biol.*, 2013, **9**, 444–454.
57. G. R. Zimmermann, J. Lehar and C. T. Keith, *Drug Discovery Today*, 2007, **12**, 34–42.
58. J. Lehar, B. R. Stockwell, G. Giaever and C. Nislow, *Nat. Chem. Biol.*, 2008, **4**, 674–681.

59. J. Lehar, G. R. Zimmermann, A. S. Krueger, R. A. Molnar, J. T. Ledell, A. M. Heilbut, G. F. Short, 3rd, L. C. Giusti, G. P. Nolan, O. A. Magid, M. S. Lee, A. A. Borisy, B. R. Stockwell and C. T. Keith, *Mol. Syst. Biol.*, 2007, **3**, 80.
60. N. Pedemonte, V. Tomati, E. Sondo and L. J. Galietta, *Am. J. Physiol.: Cell Physiol.*, 2010, **298**, C866–C874.
61. S. H. Randell, M. L. Fulcher, W. O'Neal and J. C. Olsen, *Methods Mol. Biol.*, 2011, **742**, 285–310.
62. T. Neuberger, B. Burton, H. Clark and F. Van Goor, *Methods Mol. Biol.*, 2011, **741**, 39–54.
63. P. H. Karp, T. O. Moninger, S. P. Weber, T. S. Nesselhauf, J. L. Launspach, J. Zabner and M. J. Welsh, *Methods Mol. Biol.*, 2002, **188**, 115–137.
64. F. Van Goor, S. Hadida, P. D. Grootenhuis, B. Burton, D. Cao, T. Neuberger, A. Turnbull, A. Singh, J. Joubran, A. Hazlewood, J. Zhou, J. McCartney, V. Arumugam, C. Decker, J. Yang, C. Young, E. R. Olson, J. J. Wine, R. A. Frizzell, M. Ashlock and P. Negulescu, *Proc. Natl. Acad. Sci. U. S. A.*, 2009, **106**, 18825–18830.
65. X. Liu, V. Ory, S. Chapman, H. Yuan, C. Albanese, B. Kallakury, O. A. Timofeeva, C. Nealon, A. Dakic, V. Simic, B. R. Haddad, J. S. Rhim, A. Dritschilo, A. Riegel, A. McBride and R. Schlegel, *Am. J. Pathol.*, 2012, **180**, 599–607.
66. F. A. Suprynowicz, G. Upadhyay, E. Krawczyk, S. C. Kramer, J. D. Hebert, X. Liu, H. Yuan, C. Cheluvaraju, P. W. Clapp, R. C. Boucher, Jr, C. M. Kamonjoh, S. H. Randell and R. Schlegel, *Proc. Natl. Acad. Sci. U. S. A.*, 2012, **109**, 20035–20040.
67. R. J. Bridges, A. Thakerar, Y. Cheng, S. H. Randell, J. M. Pilewski, C. M. Penland and W. van Driessche, *Pediatr. Pulmonol.*, 2011, 279–279.
68. I. Gooding and D. Westaby, in *Cystic Fibrosis*, ed. M. Hodson, D. Geddes and A. Bush, Edward Arnold, 3rd edn, 2007, pp. 209–224.
69. T. Sato, D. E. Stange, M. Ferrante, R. G. Vries, J. H. Van Es, S. Van den Brink, W. J. Van Houdt, A. Pronk, J. Van Gorp, P. D. Siersema and H. Clevers, *Gastroenterology*, 2011, **141**, 1762–1772.
70. T. Sato and H. Clevers, *Science*, 2013, **340**, 1190–1194.
71. M. J. Hug, N. Derichs, I. Bronsveld and J. P. Clancy, *Methods Mol. Biol.*, 2011, **741**, 87–107.
72. J. F. Dekkers, C. L. Wiegerinck, H. R. de Jonge, I. Bronsveld, H. M. Janssens, K. M. de Winter-de Groot, A. M. Brandsma, N. W. de Jong, M. J. Bijvelds, B. J. Scholte, E. E. Nieuwenhuis, S. van den Brink, H. Clevers, C. K. van der Ent, S. Middendorp and J. M. Beekman, *Nat. Med.*, 2013, **19**, 939–945.
73. H. Mou, R. Zhao, R. Sherwood, T. Ahfeldt, A. Lapey, J. Wain, L. Sicilian, K. Izvolsky, K. Musunuru, C. Cowan and J. Rajagopal, *Cell Stem Cell*, 2012, **10**, 385–397.
74. A. P. Wong, C. E. Bear, S. Chin, P. Pasceri, T. O. Thompson, L. J. Huan, F. Ratjen, J. Ellis and J. Rossant, *Nat. Biotechnol.*, 2012, **30**, 876–882.

75. W. B. Guggino and B. A. Stanton, *Nat. Rev. Mol. Cell Biol.*, 2006, **7**, 426–436.
76. D. C. Gruenert, M. Willems, J. J. Cassiman and R. A. Frizzell, *J. Cystic Fibrosis*, 2004, **3**(suppl. 2), 191–196.
77. S. Ramachandran, S. Krishnamurthy, A. M. Jacobi, C. Wohlford-Lenane, M. A. Behlke, B. L. Davidson and P. B. McCray, Jr, *Am. J. Physiol.: Lung Cell. Mol. Physiol.*, 2013, **305**, L23–L32.
78. http://www.cff.org/research/ForResearchers/ResearchTools/CFTRAntibodiesModulators/#CFTR_Modulators, administered by Prof. R.J. Bridges, Rosalind Franklin University (North Chicago, IL, USA) on behalf of the Cystic Fibrosis Foundation.
79. T. W. Loo, M. C. Bartlett and D. M. Clarke, *Biochem. Pharmacol.*, 2013, **86**, 612–619.
80. H. Y. Ren, D. E. Grove, O. De La Rosa, S. A. Houck, P. Sopha, F. Van Goor, B. J. Hoffman and D. M. Cyr, *Mol. Biol. Cell*, 2013, **24**, 3016–3024.
81. L. He, P. Kota, A. A. Aleksandrov, L. Cui, T. Jensen, N. V. Dokholyan and J. R. Riordan, *FASEB J.*, 2013, **27**, 536–545.
82. C. M. Farinha, J. King-Underwood, M. Sousa, A. R. Correia, B. J. Henriques, M. Roxo-Rosa, A. C. Da Paula, J. Williams, S. Hirst, C. M. Gomes and M. D. Amaral, *Chem. Biol.*, 2013, **20**, 943–955.
83. D. C. Swinney and J. Anthony, *Nat. Rev. Drug Discovery*, 2011, **10**, 507–519.
84. P. Workman and I. Collins, *Chem. Biol.*, 2010, **17**, 561–577.
85. L. V. Galietta, S. Jayaraman and A. S. Verkman, *Am. J. Physiol.: Cell Physiol.*, 2001, **281**, C1734–C1742.
86. B. H. Hirth, S. Qiao, L. M. Cuff, B. M. Cochran, M. J. Pregel, J. S. Gregory, S. F. Sneddon and J. L. Kane, Jr, *Bioorg. Med. Chem. Lett.*, 2005, **15**, 2087–2091.
87. M. A. Ashlock and E. R. Olson, *Annu. Rev. Med.*, 2011, **62**, 107–125.
88. L. S. Ostedgaard, D. K. Meyerholz, J. H. Chen, A. A. Pezzulo, P. H. Karp, T. Rokhlina, S. E. Ernst, R. A. Hanfland, L. R. Reznikov, P. S. Ludwig, M. P. Rogan, G. J. Davis, C. L. Dohrn, C. Wohlford-Lenane, P. J. Taft, M. V. Rector, E. Hornick, B. S. Nassar, M. Samuel, Y. Zhang, S. S. Richter, A. Uc, J. Shilyansky, R. S. Prather, P. B. McCray, Jr, J. Zabner, M. J. Welsh and D. A. Stoltz, *Sci. Transl. Med.*, 2011, **3**, 74ra24.
89. X. Sun, H. Sui, J. T. Fisher, Z. Yan, X. Liu, H. J. Cho, N. S. Joo, Y. Zhang, W. Zhou, Y. Yi, J. M. Kinyon, D. C. Lei-Butters, M. A. Griffin, P. Naumann, M. Luo, J. Ascher, K. Wang, T. Frana, J. J. Wine, D. K. Meyerholz and J. F. Engelhardt, *J. Clin. Invest.*, 2010, **120**, 3149–3160.

RARE MUSCLE DISORDERS

CHAPTER 11

Drug Discovery Approaches for Rare Neuromuscular Diseases

GRAHAM M. WYNNE AND ANGELA J. RUSSELL*

Chemistry Research Laboratory, University of Oxford, 12 Mansfield Road, Oxford OX1 3TA, UK
*E-mail: angela.russell@chem.ox.ac.uk

11.1 Introduction

Rare neuromuscular diseases encompass many diverse and debilitating disorders, ranging from ultra-orphan conditions such as various forms of autosomal dominant limb-girdle muscular dystrophy (which affect only a few families) and adult-onset proximal spinal muscular atrophy, autosomal dominant (which has an occurrence of approximately 1 in 1 000 000), to the so-called 'common' orphan diseases like Duchenne muscular dystrophy (DMD) which is an X-linked genetic disordering occurring in approximately 1 in 3500 male births (an occurrence of $\sim$3.7 per 100 000 population).[1]

The term neuromuscular disease is often used in a descriptive sense for all these conditions, although strictly speaking this term encompasses two rather more distinct types of disease. The first of these classes can be categorised as one that *directly* affects the functional mechanics of muscle operation, *i.e.* there is a structural or physiological defect in one or more muscle components, for example the lack of functional dystrophin in DMD. The second classification is a disease that affects either the nerves or the neuromuscular junctions (NMJs) rather than having a direct impact on the muscle tissue itself, *i.e.* the muscle function is reduced through impaired or

RSC Drug Discovery Series No. 38
Orphan Drugs and Rare Diseases
Edited by David C Pryde and Michael J Palmer

Published by the Royal Society of Chemistry, www.rsc.org

ablated transmission of electrical impulses. A prominent example of the latter is spinal muscular atrophy (SMA), where the lack of survival motor neuron (SMN) protein has a direct effect on motor neurones, thereby affecting muscle strength, tone, *etc.*

Together these diseases place a significant patient care and economic burden on society, both on the sufferers and their families, as well as the various national healthcare systems. It has been estimated that within the UK alone around 70 000 people suffer from some form of muscular dystrophy or a related disorder.[2]

From a treatment perspective, most current therapies for rare muscle disorders are aimed at alleviating the disease's symptoms rather than addressing the underlying primary cause. Because many of these diseases have a genetic basis and are often poorly characterised, current symptomatic treatments have met with limited success, and curative approaches have so far not proven to be generally viable. In recent years, however, several factors have combined to give renewed hope to sufferers of rare neuromuscular disease. The first of these is an enhanced understanding of the underlying mechanisms, be they genetic, biochemical or physiological, at the heart of the disease, although it should be noted that this does not necessarily mean that a unique molecular target has been identified for a particular disease. The second is resurgence in the use of phenotypic screening techniques, *i.e.* those that do not attempt to use a specific enzyme in a drug screen; rather they use cell-based or even model organism-based readouts, typically optimised for screening larger numbers of compounds than were previously possible. The third and perhaps most significant change that has proved pivotal is a paradigm shift within the drug discovery industry (*i.e.* pharmaceutical and biotechnology companies) away from the so-called 'blockbuster' drug strategies targeting mass patient groups. This has arisen largely due to an increasing awareness of the heterogeneity of most diseases, and a resultant move towards stratified (and more 'personalised') medicine, relying on the characterisation and treatment of smaller patient sub-populations, often utilising specific biomarkers. Incentives such as orphan designation, FDA fast-tracking and breakthrough classifications have also helped stimulate growth in the area; these are discussed elsewhere.

Two diseases which serve to illustrate the current state of the art for drug discovery towards rare neuromuscular disorders are DMD and SMA, and the remainder of this chapter will describe the status of research and development directed towards new therapies for these conditions. While the focus of this review is on the development of small-molecule therapeutic agents (*i.e.* small organic compounds with a molecular weight typically less than around 600 Daltons), a number of biological or macromolecular agents are also included where their application is important to illustrate mechanism, or provide proof-of-principle target validation.

A search of the literature gives an indication of this emergent trend towards rare diseases, using DMD and SMA as examples (Figure 11.1–11.4).[3]

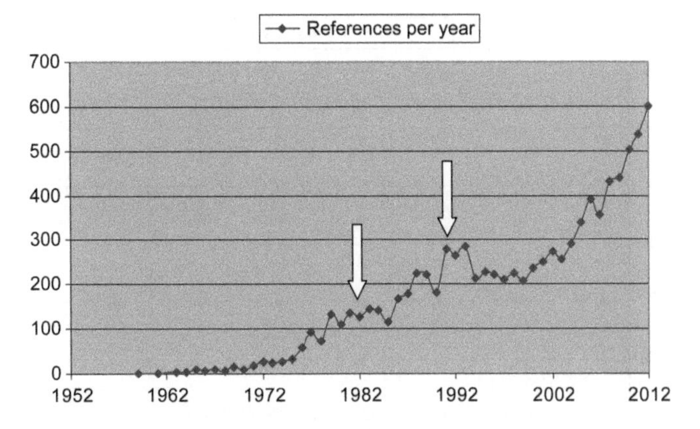

Figure 11.1 Total number of references per year containing the term 'Duchenne muscular dystrophy'. Arrows mark the dates of first publications relating to dystrophin (1982)[4] and its autosomal homologue utrophin (1992).[5]

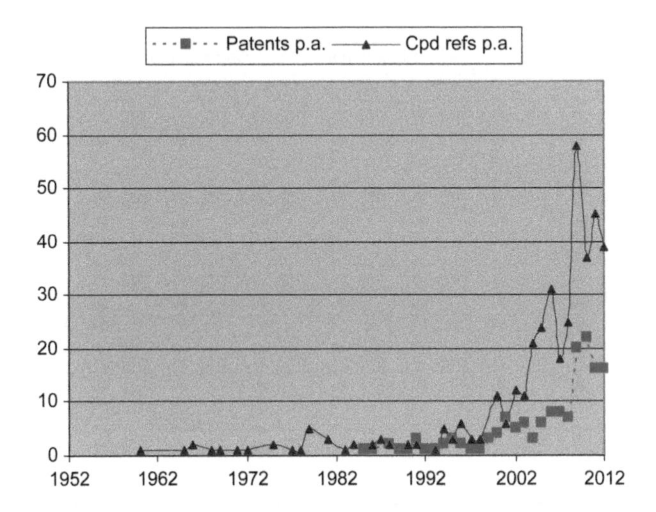

Figure 11.2 Total number of references per year containing the term 'Duchenne muscular dystrophy'; refined by document type 'patent' or which contain the additional term 'compound'.

From these data it can be seen that there is an smooth, although not exponential, increase in the total number of references published year-on-year for DMD, aside from a small spike in the early 1990s (Figure 11.1). Analysis of the data in more detail would be expected to establish why this flurry in activity occurred, but it may well be connected with the fact that much of the work relating to the identification of the dystrophin gene, and the protein product itself, occurred only a few years beforehand in the later 1980s, as well as the identification of utrophin, the autosomal homologue of dystrophin in

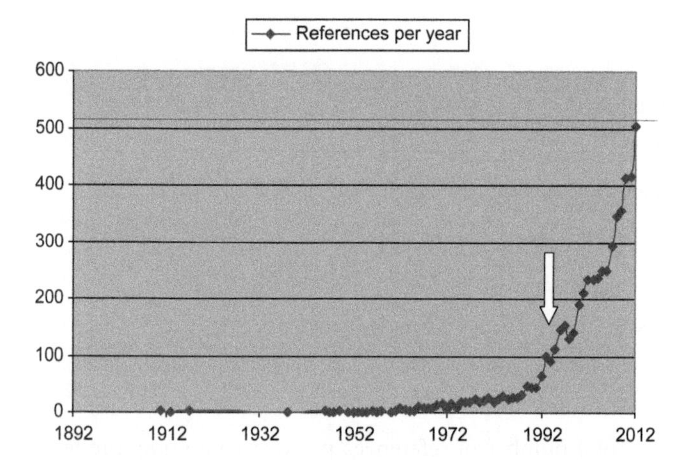

Figure 11.3 Total number of references per year containing the term 'spinal muscular atrophy'. An arrow marks the date of the first publication identifying the link between the SMN protein and SMA pathology.[6]

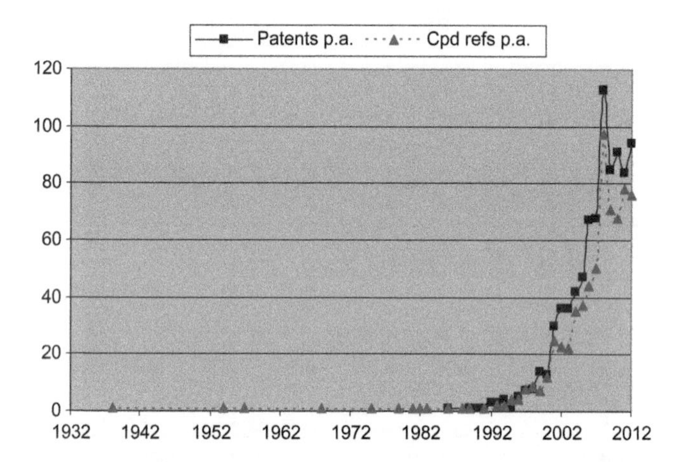

Figure 11.4 Total number of references per year containing the term 'spinal muscular atrophy'; refined by document type 'patent' or which contain the additional term 'compound'.

1992. Until 2000 or so there was virtually no (10 or less per annum) patent filings or publications relating to compounds/chemical matter for DMD (Figure 11.2). Given the genetic nature of the disease, its relatively poorly understood nature from a biochemical/molecular perspective and (as a result) fewer specifically defined molecular targets which could be considered for pharmacological intervention, this paucity is not entirely surprising.

By contrast, publication metrics plotted for references containing the term 'spinal muscular atrophy' have a much steeper curve (Figure 11.3 and 11.4). The major inflexion point again appears to take place around 1990, which was when much of the seminal work describing the genetic basis for the disease was published. Interestingly this curve shape is largely mirrored by the patent application/publication and compound disclosure metrics. Although the key biological discoveries took place at approximately the same time as the corresponding work on DMD, and since, in the main, similar therapeutic approaches (and molecular strategies) are being undertaken on both diseases, it is interesting to speculate as to why drug discovery research did not take place on DMD at a similar pace to SMA. It may be due to the more homogeneous genetic nature of SMA compared to DMD (see later), or perhaps to the sociological impact which follows from the fact that a significant proportion of severe SMA sufferers (and fatalities) are young babies.

Although disorders such as DMD and SMA are rare compared to (for example) cardiovascular disease and cancer, they have received little or no attention from the drug discovery industry to date, yet represent important potential sources of revenue for the industry at a time when there is considerable pressure on companies from financiers, healthcare providers and the public for new, innovative therapies rather than what may be seen as 'me too' drugs or diseases with less apparent unmet medical need. Indeed, the global market for muscular dystrophy therapeutics is significant, and has been estimated as potentially reaching levels in excess of $1 bn, assuming pricing models used in other orphan disease indications are applied.[7] For example, if a price point of $300–500k per patient per year is assumed, the market opportunity for a drug that would treat 2000 patients in the USA alone, such as an exon-skipping therapy, would equate to $600–1000 m of revenue.[8] Even a more conservatively priced therapeutic of $50k per patient per year would generate $100 m for the same population size. In an increasingly competitive industry it is therefore easy to appreciate the continued shift of the pharmaceutical industry towards orphan and rare diseases.

11.2 Duchenne Muscular Dystrophy

11.2.1 Introduction

Duchenne muscular dystrophy (DMD) is an X-linked neuromuscular disorder, which affects approximately 1 in 3500 births.[9,10] Despite being classified as a rare disease, DMD is one of the most common genetic diseases, with only cystic fibrosis (CF) being more prevalent. It has been estimated that there are likely to be in excess of 15 000 sufferers in the USA,[11] and in the UK alone there are around 2400 boys living with DMD, with 100 or so new diagnoses occurring every year.[12] Due in part to the relatively high incidence of new mutations, there is no effective genetic screening available at the current time, and so diagnosis often does not occur until boys are a few years old, when the physical symptoms present. Sufferers are afflicted by progressive muscle degeneration, and as a result are usually confined to a wheelchair before their early teenage years. Because DMD affects all muscles including the heart and diaphragm, as well as a loss of ambulation, as the disease progresses cardiac and respiratory problems start to manifest themselves. By their late teens and early 20s DMD boys are likely to require assistance in breathing, and will also be taking drugs intended to improve the cardiac symptoms, such as ACE inhibitors.

There is currently no effective treatment for DMD, although much improved symptomatic treatments and supportive care has resulted in an extended life for patients. Some now live to their late 20s, whereas in the past survival into the third decade of life was rare.

DMD is a devastating disease with a significant unmet medical need, and new therapies are desperately sought by families, support groups and charities, with much interest being seen as details of new ideas and approaches are published.

The genetic defect which is the underlying cause of DMD results in a lack of the key structural protein dystrophin within the muscles. The DMD gene which codes for dystrophin was originally identified in 1987 and is one of the largest genes in the human genome. Mutations, deletions or other rearrangements in this gene account for the vast majority of transcriptional errors,[13] with around 65% of patients suffering from DMD having a genetic mutation which leads to premature termination of gene translation. The dystrophin protein is coded by an mRNA which consists of 79 exons.[4] In-frame or out-of-frame mutations in the DMD gene for one or more of these exons can result in production of a truncated or non-functional protein. Accordingly, strategies intended to correct a mutation in one particular exon will only be applicable to a subset of DMD patients. For example, it has been estimated that exon-skipping therapies targeting an exon 51 mutation will treat around 13% of DMD boys. Likewise, premature stop codon (read-through) therapies will have applicability limited to a specific patient sub-population for a similar reason. In this manner DMD, like many other diseases, can be viewed as heterogeneous in nature, and the total patient group would therefore need to be treated either with multiple drugs to address the various factors causing the dystrophin lesion (*i.e.* a combination of medications), or a drug which acts in a mutation independent manner (*i.e.* it does not aim to restore dystrophin production).

DMD differs from the closely related Becker muscular dystrophy (BMD) in that BMD patients, while having mutations in the DMD gene, do still produce some dystrophin. This is a truncated form which arises due to in-frame deletions or mutations, but critically it still retains sufficient function to allow a reasonably normal lifespan, with some sufferers living until their 60s.[14]

The human body contains a large number of skeletal muscles, around 600 in total, and when cardiac muscle is also considered it is apparent that these tissues constitute a significant percentage of the human body.[15] Dystrophin is a critical component of all these, and consequently any widespread structural defects would be expected to have significant and far-reaching consequences for the individuals concerned. Dystrophin is a crucial structural constituent of skeletal muscle, being part of the dystrophin-associated protein complex (DAPC), where the C-terminal section forms interactions with a range of

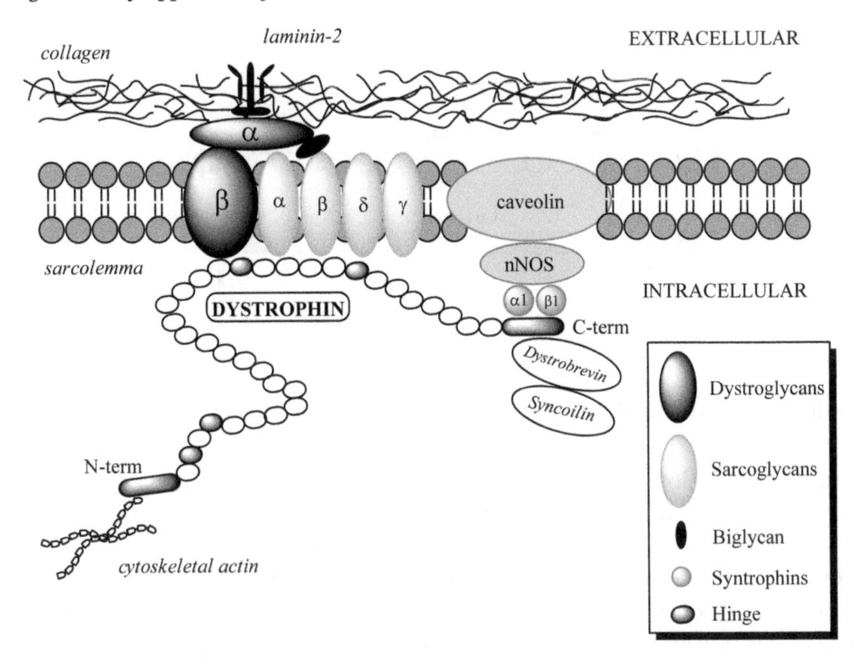

Figure 11.5 Detailed role of dystrophin in maintaining muscle fibre integrity – the DAPC (dystrophin-associated protein complex).

macromolecular glycoproteins including dystroglycans and sarcoglycans (Figure 11.5). In this role, as a kind of molecular 'shock absorber', it connects the external cell membrane (called the sarcolemma) to the internal actin cytoskeleton and provides protection from the mechanical stresses placed upon muscle during exercise-induced contraction and extension.[16]

Aside from a small amount of dystrophin produced in so-called 'revertant fibres' resulting from alternative processing of the DMD gene, production of functional dystrophin does not take place in DMD boys. As a result of this the dystrophin structural link between the sarcolemma and the internal cytoskeletal components of the muscle is absent; accordingly extension of the muscle results in a loss of synchronisation between the inner and outer structures, and this is followed by physical damage to, and degradation of, the affected tissue. Even though this type of trauma will stimulate the body's natural repair systems, the continued lack of dystrophin eventually results in this repair–regeneration process becoming cyclical, with extensive inflammation, fibrosis and eventual loss of muscle integrity as the muscle fibre gets replaced by adipose and connective tissue. Gradually as the muscle loses structure its function is also inevitably degraded and eventually lost. Clearly this loss of function will have a major impact in any skeletal muscle, but because all muscle tissue is affected, including cardiac and respiratory muscles such as within the diaphragm, the consequences are devastating, and death is the ultimate result, usually through cardiac or respiratory failure.

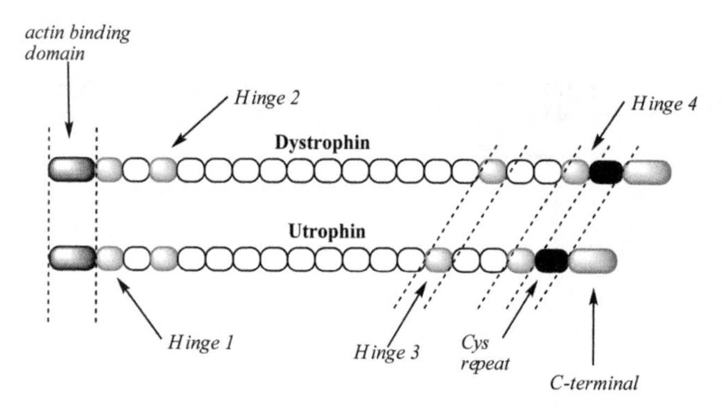

Figure 11.6 Structural homology of dystrophin and utrophin.

In early development, the structurally related protein utrophin (a contraction of 'ubiquitous dystrophin') has been shown to play a similar sarcolemmal link role in muscle structure, but after birth the production of this protein rapidly declines, to be replaced by dystrophin. Utrophin has much structural similarity to dystrophin, including up to 80% homology in the critical glycoprotein binding C-terminal region (Figure 11.6).[5]

Interestingly, the presence of utrophin can still be detected in regenerating muscle, such as that found in DMD patients, although only at low levels. Consequently, pharmacological strategies to treat DMD which are predicated on upregulation of the production of utrophin, and aiming to functionally compensate for lack of dystrophin, have attracted much attention, and will be described in more detail later.

11.2.2 Screening Models for DMD

As with any disease area, the availability of physiologically relevant assay systems with appropriate throughput is critical to enable effective drug discovery. Of particular relevance when considering downstream compound development and IND enabling studies is access to suitable predictive animal models which can both model the key attributes of the disease pathology and progression, yet at the same time be robust and cost-effective enough for use within normal animal laboratories. A range of appropriate *in vitro* assays, which have enabled much progress in the identification of new drugs for DMD to be made are described in more detail later, and often are representative of a phenotypic whole cell screening approach.

While it is generally accepted that an *in vivo* animal efficacy test such as the golden retriever disease model provides unparalleled physiological relevance to the human DMD condition, it is impractical to use in a higher throughout fashion, because the animals are difficult to breed, and have a relatively short lifespan.[17] The *mdx* mouse, by contrast, although having a similar genetic deficiency to that in humans, presents a much milder disease phenotype. Importantly, it is easy to breed, is readily transferred between different animal laboratories, thereby allowing the majority of DMD researchers ready access. As would be expected, quantities of test compounds required for *in vivo* studies are also considerably lower compared to the dog model, this being an important consideration from a medicinal chemistry perspective. For these reasons it has become adopted as the mainstay animal model for DMD, and is the yardstick by which efficacy (or lack thereof) is judged for new therapeutic approaches.[18]

Other DMD disease models, such as the zebrafish (*Danio rerio*), and new mouse models which are intended to incorporate other aspects of the disease (such as cardiomyopathy), are emerging as alternative testing systems,[19] although it is not yet clear how much of a role they will play in drug screening, or how predictive or cost-effective they are to run. A recent review has summarised the various animal models available for a range of other orphan diseases.[20]

11.2.3 Current Symptomatic Therapies

There is a real paucity of treatment options for DMD at the current time. A range of drugs are used to manage the disease in sufferers, although at best these only help to alleviate the symptoms, while providing limited therapeutic benefit.[21] The most extensively used maintenance therapy is based on corticosteroids, with these being used to manage the inflammatory component of the disease pathology.[22] According to recommendations from parent support groups, treatment with steroids such as prednisone **11.1** and deflazacort **11.2** (Figure 11.7) provide most functional benefit when treatment is started before loss of ambulation is seen in the patient. Less functional benefit is seen in patients if treatment is delayed.[9]

Chronic use of steroids is well known to produce a range of side effects in DMD patients; these include weight gain, cataracts and growth retardation.[23] Given the debilitating nature of DMD, as well as the paediatric patient population, those who are being treated with this class of agents are usually carefully monitored.

A number of other classes of drug are used to manage the symptoms of DMD. Of particular note is the use of ACE inhibitors and beta blockers.

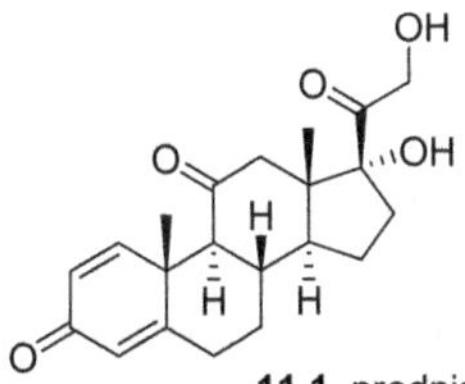

11.1, prednisone

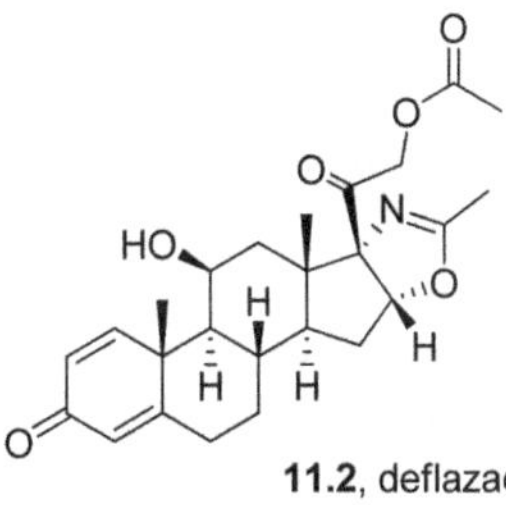

11.2, deflazacort

Figure 11.7 Current steroids used to treat DMD.

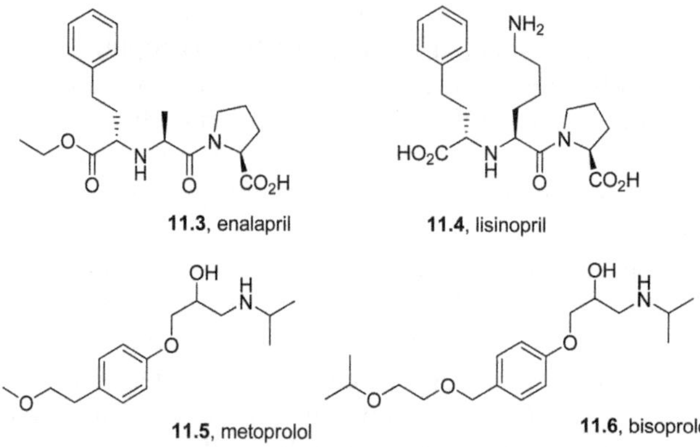

11.3, enalapril

11.4, lisinopril

11.5, metoprolol

11.6, bisoprolol

Figure 11.8 ACE inhibitors and beta blockers used to treat DMD.

Examples from both classes have been highlighted as potentially providing benefit for DMD patients who are suffering from cardiomyopathy. These include enalapril **11.3** or lisinopril **11.4** and metoprolol **11.5** or bisoprolol **11.6**, respectively (Figure 11.8).[24]

A recent study by Otaga *et al.* investigated combination therapy with both classes of drug, and concluded that this treatment paradigm gave improved survival for DMD patients suffering from cardiac dysfunction.[25]

11.2.4 Managing the Condition: New Symptomatic Approaches

A range of therapeutic strategies for DMD have been investigated, which could be classified as being intended to provide improved symptomatic relief rather than correcting the underlying primary genetic defect, with a recent review providing an up-to-date summary.[26] Examples of agents that address one or more of the consequences of disease progression, such as oxidative stress, fibrosis and inflammation, are described below. Approaches based on nutritional supplements and the like have been reviewed recently,[27] and are beyond the scope of this review.

11.2.4.1 Inflammation

New glucocorticoids are under investigation as improved therapeutics for DMD, with the novel derivative VBP15 (**11.6a**, Figure 11.9) having shown encouraging activity in murine models of inflammation.[28] Synthesis and structure–activity studies have recently been described, along with a comprehensive range of the pre-clinical ADME and safety data which led to its selection as a development candidate.[29]

Resveratrol has attracted a great deal of attention in recent years, both in the scientific literature and in the popular press, as an apparent panacea with

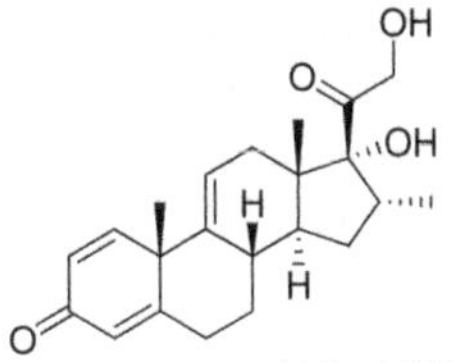

Figure 11.9 Compounds targeting the inflammation component of DMD.

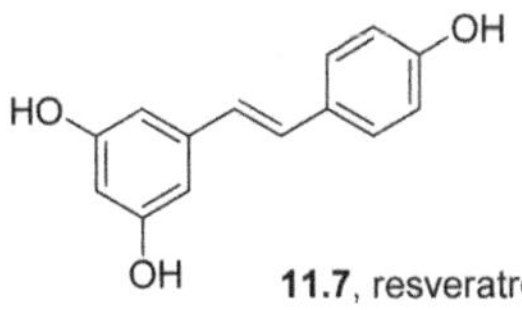

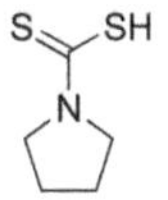

application to the relief of many common diseases,[30] although the amount and quality of evidence supporting a number of these claims varies considerably.

Very recently the activity of resveratrol (**11.7**, Figure 11.9) has been studied in *mdx* mice, the hypothesis being that it has been documented as having anti-inflammatory activity, coupled with the ability to activate the utrophin A promoter, and is therefore of interest as a potential treatment for DMD.[31] As with other compounds of this nature, whose activity has been discovered through phenotypic screening methods, the mechanism by which both of these activities are mediated is currently the subject of much debate, although activation of the sirtuins (particularly Sirt1) has been reported. The dosing regime in this case was relatively short term, with the primary objective being to establish evidence for a histological effect on muscle (reduced inflammation, *etc.*) and to demonstrate increased levels of utrophin expression. Following 10 days of oral dosing at three different dose levels (10, 100, 500 mg kg^{-1}), the animals were sacrificed and muscle samples taken for histological examination and gene expression analysis. An optimal dose of 100 mg kg^{-1} was identified, with RT-PCR analysis of samples from the *mdx* mice showing an increase in both Sirt1 and utrophin mRNA (approx. 1.5-fold), although this did not translate to significant increases in production of the respective proteins. The reason for this was suggested as being due to insufficient dosing time. From a histological perspective, immune cell infiltration and inflammation were also reduced. Although more data, such as more detailed histological examination, would clearly be instructive, other questions remain, including establishing an appropriate duration of, and level of dosing for the compound, as well as identifying an underlying mechanism of action. Nonetheless the authors concluded that these results could support the use of resveratrol as a possible treatment for DMD, although this has not been followed up to date.

Inflammation occurring *via* the activation of NF-κB has also been postulated as a significant underlying mechanism for the degradative muscle pathology seen in DMD. Animal studies using *mdx* mice have been published which utilised the pharmacological tool PDTC (pyrrolidine dithiocarbamate) (**11.8**, Figure 11.9), a compound reported to act *via* inhibition of NF-κB.[32] Given the small polar nature and functionality within the compound there is potential for pleiotropic effects, thereby confounding interpretation of the results.

Although this approach appears to offer intriguing therapeutic possibilities,[33] no follow-up experiments using more drug-like compounds or other NF-κB chemotypes have been described to date, nor has there been confirmed medicinal chemistry programmes focused on identifying specific modulators of this biological mechanism.

11.2.4.2 Antioxidants

Originally developed by Takeda as a potential therapy for Alzheimer's disease,[34] idebenone (trade names: Catena, Sovrima) (**11.9**, Figure 11.10) is a synthetic analogue of coenzyme Q10 (an important enzyme in the electron transport chain), and hence can function as an antioxidant.

As well as Alzheimer's, its use has also been explored for various other neurodegenerative diseases, particularly Friedreich's ataxia,[35] and more recently for the therapy of DMD. Given the antioxidant activity of idebenone, it seems likely that the mode of action of idebenone is connected with the compound's ability to scavenge reactive oxygen entities, thereby helping to ameliorate the pathological damage to muscle, including cardiac muscle caused by such species in DMD.

Preliminary data showing the efficacy of idebenone in an *mdx* mouse model of DMD was published a number of years ago.[36] In this study long-term treatment of mice from an early age with idebenone (3–4 weeks old to 10–12 months; 200 mg kg^{-1} once daily) showed marked improvement relative to vehicle-treated *mdx* mice. Cardiac structure and function improved, as did resistance to stress-associated cardiac failure, both of these being critical readouts when considering human trials. Furthermore, clear functional benefit was noted in terms of enhanced exercise performance (voluntary wheel running model). While the data were statistically significant, the authors urged caution because it was not clear at that time whether the improvements seen were due to the cardiac effects previously described, or direct effects on skeletal muscle. Nonetheless, the data was clearly supportive of further study, and progression of the compound to clinical trials followed shortly thereafter.

Results of a Phase 2a study (the DELPHI trial, NCT00654784) of idebenone in 21 DMD patients were published during 2011.[37] Patients were aged between 8 and 16, and received daily treatment of either 450 mg of drug ($n = 13$), or placebo ($n = 8$) for 52 weeks. The compound was found to be safe and well tolerated, with no drug-associated adverse events, but due in part to

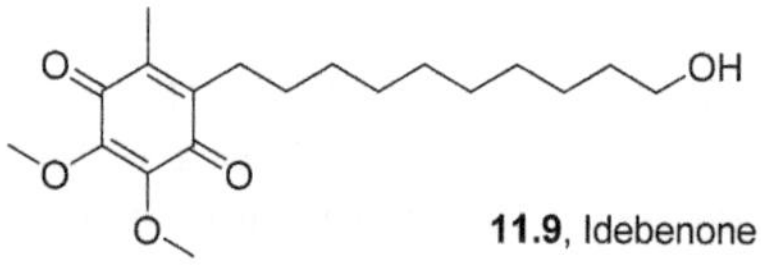

O OH O O O

11.9, Idebenone

Figure 11.10 Idebenone, an antioxidant compound.

both the small trial cohort sizes, the results did not show a statistically significant improvement after treatment. Furthermore, there was a significant age disparity, with the drug treatment group being of notably older age than the placebo cohort (13.4 ± 2.1 years *vs.* 10.8 ± 1.9 years). For progressive disease such as DMD, this latter factor in particular is likely to significantly complicate interpretation of trial results. Notwithstanding these complications, results from the trial were generally viewed as being encouraging. While the primary trial end point, an improvement in cardiac function (as assessed by changes in peak systolic radial strain) was not statistically significant, a trend to improvement was noted. More encouragingly, significant improvement in one of the secondary end points, peak expiratory flow, was recorded. From a skeletal muscle perspective, no significant improvement in upper limb muscle strength was seen. A larger Phase 2b study (NCT00758225) was completed during 2011, and although the detailed findings have not yet been published, approval has been given to progress the drug towards pivotal trials. Accordingly, Santhera has recently undertaken a Phase 3 clinical trial (the DELOS trial, NCT 01027884), with this having been completed in January 2014. No results are yet available.

11.2.4.3 *Antifibrotics*

Natural products have served as a rich source of both leads and drugs in virtually all therapeutic areas since the earliest days of medicine. Neuromuscular diseases are no exception to this, and a recent example of a novel, natural product inspired treatment directed towards DMD serves to illustrate this.

Febrifugine (**11.10**, Figure 11.11) is a natural product found in the roots of the *Dichroa febrifuga* plant ('Chang Shan', one of the 50 fundamental herbs in Chinese medicine), which has been used for many years as an antimalarial agent in China. Unsurprisingly, as with many other readily available naturally derived compounds, alternative uses have been explored. Febrifugine and its derivatives have been investigated in the treatment of a variety of other diseases, including oncology and inflammatory diseases, although the parent compound itself was found to exhibit gastrointestinal toxicity, and so was considered to be unsuitable for further development. In an effort to mitigate this undesirable side effect a range of analogues were synthesised. One derivative that emerged as being of particular interest was halofuginone

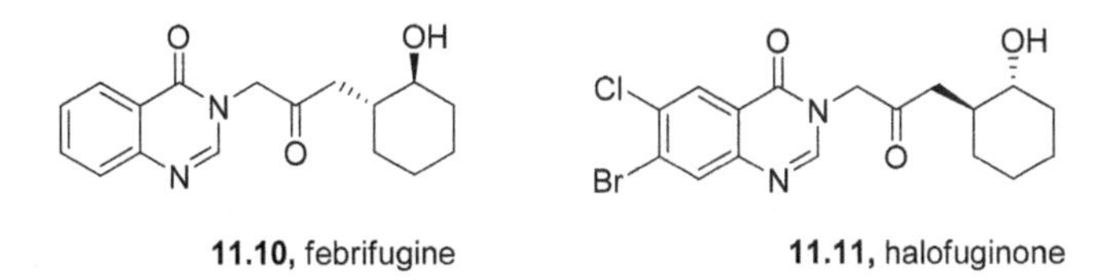

Figure 11.11 Anti-fibrotic compounds for DMD.

11.11, which in recent years has been established as having a range of pharmacological activities. From a medicinal chemistry perspective, halofuginone's heritage is rather less clear, although some limited SAR studies have described antimalarial/antiparasitic agents,[38] as well as synthesis procedures for a number of analogues.[39] Various therapeutic possibilities have been investigated for the compound, not only as a possible anticancer agent,[40] but also due to its inhibitory action on collagen synthesis, as a possible treatment for diseases involving scarring and fibrosis (the repair process which causes deposition of an excessive amount of fibrous tissue, usually in response to some form of injury).[41]

This latter opportunity was stimulated by a serendipitous observation from halofuginone's therapeutic application as an agent in veterinary medicine, where it has been used extensively in the poultry industry (and more widely) as an anticoccidiosis agent and broad-spectrum antiparasitic.[42] It was from this utility that a concomitant effect on skin integrity in birds treated with the drug was noted. The skin tearing observed suggested a deleterious effect on connective tissue, in particular collagen and the extracellular matrix (ECM), and in ensuing studies halofuginone was established as having an inhibitory effect on collagen $\alpha(1)$ gene expression, and ultimately on collagen levels in tissue.[43]

With the observation of an inhibitory effect on fibrosis and fibrotic tissue, halofuginone's potential as an antifibrotic agent was then explored on a range of tissues with relevance to multiple human diseases, including the skin, liver and lungs. The Israeli biotechnology company Collguard Pharmaceuticals pursued halofuginone for various fibrotic indications, although these never progressed beyond early clinical trials.

Results from a Phase 1 trial have been published as part of an oncology drug development programme,[44] with the intention of investigating pharmacokinetics and safety of the compound, as have various sets of animal pharmacokinetic data. Pharmacokinetic evaluation showed halofuginone to have a long plasma half-life of ~30 hours, with plasma levels of ~0.5 ng mL^{-1} seen at an oral dose level of 0.5 mg. Escalating the amount of compound gave an approximately dose-proportional increase to both compound AUC (area under the curve) and peak plasma concentration. Repeat-dose studies showed compound accumulation, a not unexpected consequence given the long half-life of the compound.

From a toxicological point of view, single ascending dose (SAD) evaluation with doses ranging from 0.07 to 0.5 mg recorded no adverse events; increasing this to 1.5 mg and above gave incidents of mild to moderate nausea and vomiting, although these were transient and reversible on cessation of drug treatment. Following dosing to refractory cancer patients, an acute maximum tolerated dose (MTD) of 3.5 mg per day was established, with the major dose-limiting toxicity being nausea, vomiting and fatigue. Chronic dosing was possible, but at the lower dose of 0.5 mg per day. Even in this study toxicity was seen, and co-administration of anti-emetic agents was recommended to control the GI toxicity, which again resulted in nausea and vomiting. Of rather more concern were a number of unexplained bleeding events recorded during the trial, some of which proved fatal. The underlying reason for these, and how (or if) they were connected with the drug treatment or the disease itself, were not established. During any future clinical evaluation of halofuginone, this would clearly represent an area that would require very careful monitoring, particularly in a paediatric indication like DMD where it would be anticipated that chronic life-long dosing would be required.

More recently the potential use of halofuginone as a new treatment for DMD has emerged, probably due to the well-established inflammatory component and fibrotic pathology observed in dystrophic muscle tissue. The effect of halofuginone on fibrosis in animal models of muscular dystrophy has been studied in some detail, and therapeutically relevant levels of efficacy seen.[45] In *mdx* mice, dosing of halofuginone was shown to downregulate expression of the collagen $\alpha(1)$ gene in the diaphragm in a dose-dependent manner, and in a different study a similar effect was noted in cardiac muscle following drug treatment. Notable improvements in muscle function were also seen in these studies. Investigation of the nature of the dosing therapy was also carried out, and it was established that if dosing of halofuginone was discontinued, then collagen content of muscle was elevated relative to the corresponding muscle in animals where continued dosing was maintained. This implies that chronic treatment using this type of antifibrotic treatment would be required in patients in order to have continued therapeutic benefit.[46,47]

Following these promising *in vivo* results in *mdx* mice, the compound came to the attention of a commercial organisation, Halo Therapeutics, formed during 2011 by two DMD-focused organisations, The Nash Avery Foundation and Charley's Fund, and who had the financial backing of several not-for-profit organisations. Halo Therapeutics licensed halofuginone from Collguard Pharmaceuticals (at which point the compound was denoted as HT-100), and have been progressing the compound towards clinical trials in DMD boys. In January 2012 Halo announced that it had received orphan designation for HT-100 from the FDA, and this was followed a few months later in May 2012 by the exciting news that it had raised $1.1 million to undertake a Phase 2 clinical trial in DMD boys.[48] The first clinical study ('Safety, Tolerability, and Pharmacokinetics of Single and Multiple Doses of HT-100 in Duchenne Muscular Dystrophy'; ClinicalTrials.Gov ID# NCT01847573) is currently recruiting patients.

It is only in recent years that more specific details of the mode of action of halofuginone in disease have started to be unravelled, although it is important to note that the parent compound itself is known to exist in equilibrium with a cyclic hemi-acetal derivative,[38] and it is possible that the compound has activity on multiple pharmacological targets and pathways.

Data published in 2004 by Xavier *et al.* showed that halofuginone inhibits TGF-β signalling, reputedly through downregulation of the TGF-β type V receptor,[49] and this in part has led to its continued study as an anticancer agent.[50] TGF-β is a critical component of the inflammatory response, and therefore probably accounts at least in part for the antifibrotic activity. More recently, an inhibitory activity of halofuginone on prolyl-tRNA synthetase has been established, leading to activation of the amino acid response pathway in autoimmunity,[51] although to what extent, if any, cross-talk exists between these two mechanisms is unclear to date. Given the relatively low molecular weight and multiple potential pharmacophores present within the compound though, it seems likely that other biological targets of halofuginone will be discovered, which may contribute to its pharmacological effects. Furthermore, this knowledge will allow further possible toxicological effects to be anticipated and tested proactively.

11.2.4.4 Drug Reprofiling Approaches

Reprofiling (or repurposing) is a recently popularised strategy in drug discovery, and can be considered to be a largely opportunistically driven approach, predicated on maximising value from past drug candidate molecules, at both the marketed and clinical development stages.[52–55] In essence, the approach seeks to establish new therapeutic opportunities for existing drugs, and is described elsewhere in this book.

A critical consideration with any reprofiling approach is availability of a defined target hypothesis and accompanying assay system to test the compounds. Alignment of both the optimal compound set with a suitable test system with disease relevance is therefore essential. The more established approaches to drug discovery undertaken within the pharmaceutical industry are as valid with neuromuscular disease as with the study of any other disease class, and indeed, a distinct paradigm shift has occurred in recent years, effectively amounting to a rediscovery (or reinvention) of phenotypic screening as an effective means to both validate disease targets and identify novel compounds,[56] either through *de novo* screening or reprofiling strategies.[57]

The following examples are illustrative of drugs that formally represent reprofiling approaches, and have been investigated for DMD (Figure 11.12).

Agents that improve blood flow, such as NO pathway modulators, were described as having therapeutic benefit in DMD a number of years ago, and have more recently been explored in the context of their ability to upregulate utrophin expression (see Section 11.2.5.2.1). The use of antihypertensives such as ACE inhibitors and beta blockers has already been described. More recently, PDE5 inhibitors such as sildenafil **11.12** and tadalafil **11.13** are poised to enter clinical evaluation for DMD (ClinicalTrials.Gov ID# NCT01168908; NCT01359670; NCT01865084), with both either in the planning stage or currently recruiting patients. A therapeutic approach using PDE5 inhibitors is also predicated on improving blood flow in muscle and/or the heart. Given the encouraging results seen using the *mdx* mouse, results in patients will be of great interest to the scientific community.[58] Neither of these agents are disease modifying, however, so they would serve only to slow progression, although they would offer an attractive alternative to steroids as

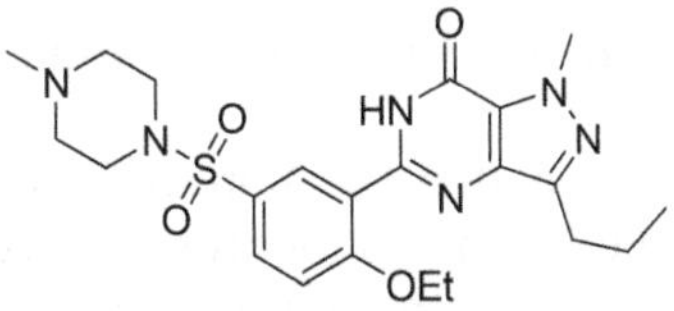

11.12, sildenafil

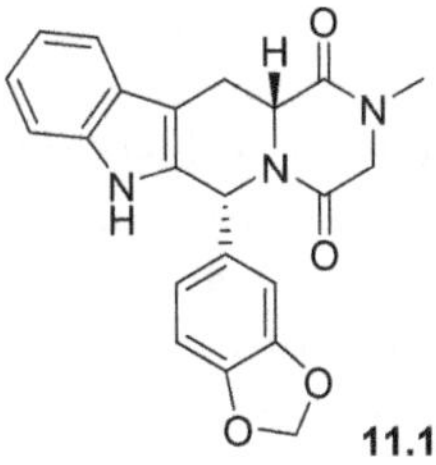

11.13, tadalafil

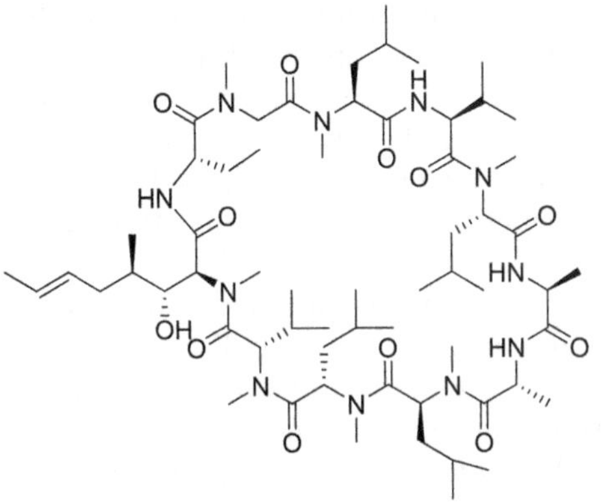

11.14, cyclosporin A

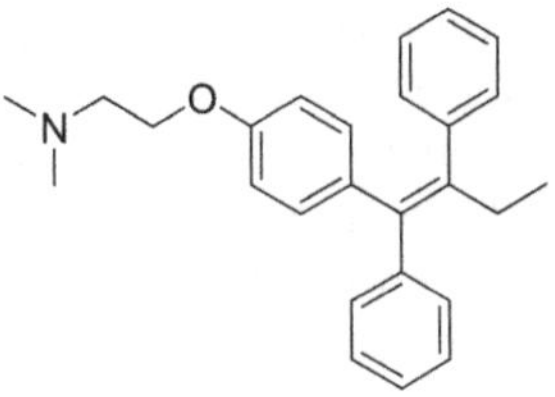

11.15, tamoxifen

Figure 11.12 Marketed drugs currently under 'reprofiling' investigation for DMD.

a first-line maintenance therapy given their cleaner toxicological profile, and should also be viable for use in combination with other agents.

Immunosuppressive agents have attracted attention as potential therapeutics for DMD, with the hypothesis being that they may contribute to amelioration of the inflammatory component associated with the disease, and represent an alternative, or combination option with steroids. Cyclosporin A **11.14** was evaluated in the *mdx* mouse based on this premise, and following multi-week dosing was found to significantly reduce the drop in exercise-induced forelimb grip strength.[59] This functional efficacy was also associated with improvement of muscle structure (as assessed by histological evaluation) and a drop in the level of the DMD-associated biomarker, creatine kinase (CK).

More recently, and following positive reports of improvements in muscle strength in patients following dosing with cyclosporin A, a larger clinical trial was undertaken in Germany (DRKS reference DRKS00000445).[60] This was designed to evaluate the effectiveness of either cyclosporin A monotherapy, or in combination with steroids, both compared to placebo as a therapy in 146 DMD boys. Following completion of the study, there was no evidence of efficacy with either the cyclosporin as a monotherapy, nor in combination with prednisone in providing improvement in the muscle strength of trial participants. Whether or not any further investigation of this agent is undertaken remains to be seen. Studies in worms and fish, very recently reported by Giacomotto *et al.*, point towards a possible mode of action involving an effect of cyclosporin A on mitochondrial fragmentation which could contribute to muscle protection.[61]

The *in vivo* effects of chronically dosed tamoxifen **11.15** on the dystrophic phenotype have recently been published. Following 15 months of once daily oral dosing to *mdx* mice, striking effects on both muscle structure and function were seen, although the mechanism through which it acted was not clear. Importantly, significant improvements were also seen in both the diaphragm and cardiac muscles, which often prove refractory to experimental therapeutics.[62] Given the established nature of tamoxifen as a relatively safe anticancer therapy which has been dosed in thousands of patients, once these results have been confirmed, it could be anticipated that clinical evaluation to establish if similar benefits are observed in DMD patients would be seen as a priority.

Very recently, the activity of a novel androgen receptor modulator GLPG0492 was described.[63] Although no structural details have been disclosed, the non-steroidal compound was described as providing functional benefit in the exercised *mdx* mouse model, as well as giving more modest effects on other muscle and biochemical markers relevant to DMD disease progression.

11.2.5 Correcting the Primary Defect: New Therapeutic Approaches

In recent years, as a greater understanding of DMD has been gleaned from continued research, as well as continued efforts for improved symptomatic therapies, there have been a range of therapeutic approaches to the pharmacological treatment of DMD by correction of the primary lesion, namely the lack of functional dystrophin.[64] Because this dystrophin deficiency arises from a genetic mutation on the DMD gene, a series of points to intervene from the transcription of DNA through to production of functional protein can be envisaged (Figure 11.13).[65]

Direct replacement of the faulty gene is perhaps the most obvious approach, while pharmacological targeting of the mechanisms influencing translational and other aspects of downstream gene processing are perhaps less obvious from a small-molecule perspective.[66] Nonetheless, these and others have all enjoyed success to a greater or lesser extent, and the following text serves to provide an overview of the various strategies adopted, involving both biological and small molecules, their scope, limitations, key results and possible future potential.[21,67]

11.2.5.1 Reintroduction of Dystrophin Expression

11.2.5.1.1 Gene Therapy. Alternative therapeutic modalities other than small molecules have been the subject of much investigation. Gene therapy, whereby the missing or damaged genetic material is introduced using an

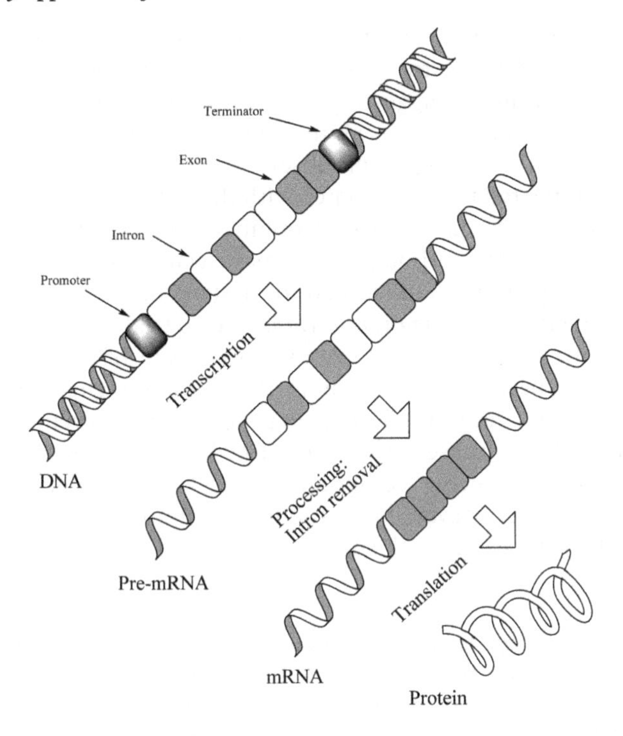

Figure 11.13 Schematic showing the basic process for protein production from DNA.

external delivery system such as the adeno-associated virus (AAV), is an important therapeutic approach being explored for DMD, and one that was originally thought to show real promise. Although encouraging progress, including testing in DMD boys, has been demonstrated, the strategy has stalled in recent years following unexpected problems associated with a variable and unpredictable immune response during early human trials. Results from the Phase 1 clinical evaluation of a dystrophin mini-gene have been published,[68,69] and the prospects for the approach recently reviewed.[70,71]

A mini-gene approach is necessary for dystrophin, this restriction arising from the limit of the requisite AAV cloning vector to incorporate genes of around a few thousand base pairs or less. Because the full-length dystrophin gene has around 2.4 million base pairs (bp), and coding mRNA size of 14 kbp, an alternative strategy is required. It has been established that the mini-gene approach is viable, and even a truncated form of the dystrophin protein transcribed from a gene missing multiple exons retains partial function in muscle, resulting in a BMD-like disease phenotype.[72] Validation of this dystrophin mini-gene approach was established *in vivo* using the *mdx* mouse model,[73] and this was translated to a clinical study in March 2006, which was undertaken by the MDA (Muscular Dystrophy Association) and Asklepios Biopharmaceuticals in the USA. In this study, six DMD patients received intramuscular injections of biostrophin, a mini-dystrophin gene delivered using a modified AAV vector.[74] Although dystrophin-positive fibres were detected after treatment, suggesting successful incorporation of the transgene, an unanticipated T-cell based immune response to the foreign genetic material was seen. Although not observed in all treated patients, this phenomenon appeared to be associated with the exons downstream from the missing sections of the patient's dystrophin gene.[69] Accordingly, while the gene therapy approach is thought to remain viable, and indeed is under continued investigation, establishment of dystrophin immunity in patients is likely to be a critical inclusion criteria in future trials. Aside from the fact that this agent required direct injection to the site of action, and issues therefore remain over delivery to the heart and diaphragm muscle, it highlights an important additional advantage for small-molecule therapeutic approaches such as those described elsewhere in this section. Aside from some cases involving compounds that form covalent bonds with biological targets (*i.e.* irreversible inhibitors),[75] immunological side effects are far less likely to be seen with small-molecule drugs as compared to larger biomolecules.

11.2.5.1.2 Exon Skipping. As described earlier, mutations or deletions in one or more exons of the DMD gene is the reason for production of either truncated or unstable dystrophin protein, this being the underlying cause of BMD or DMD, respectively. Further compounding the hereditary aspects of the disease is the fact that dystrophin is the largest gene in the human genome, and therefore the chances of spontaneous mutations occurring is far more likely than in most other genes. It is also for this reason that genetic screening of adults is unlikely to completely eliminate occurrence of the disease. Morpholino, and other related compounds ('oligos'), are chemically more stable derivatives of oligonucleotides, and have been extensively investigated as agents for promoting exon skipping. Indeed, several have demonstrated efficacy using *in vivo* models such as the *mdx* mouse.[76] There are several different subtypes of drugs within this functional class,[77] all of which are thought to exert their function by modifying the splicing of RNA. This has been suggested as taking place by the drug binding to, and thereby preventing transcription of, particular target 'problem' sections of RNA, followed by moving on to the next exon (Figure 11.14). Hence they can be viewed as a development from anti-sense gene silencing strategies.[78] Encouragingly, it has been calculated that applying exon-skipping therapeutic strategies (such as anti-sense oligonucleotides) for one or two exons could allow up to around 85% of cases of DMD to be treated.[79]

To date, several examples of this class of compounds have progressed to clinical evaluation for DMD, and it is known that a number of others are under active development.[79] The advantages of these type of agents are that they would be expected to be highly selective; the oligonucleotide has to sequence match the RNA in order to suppress transcription, and because

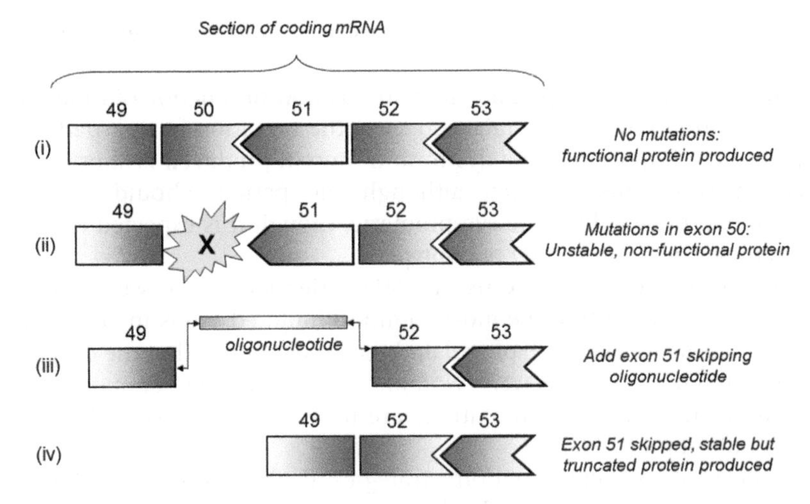

Figure 11.14 Exon skipping overview: (i) all exons present, (ii) exon 50 absent, (iii) oligonucleotide treated, (iv) exon-skipped mRNA following oligonucleotide treatment.

agents may be over 25 bases in length, and depending on the coding sequence, alternative matches occurring by chance are relatively unlikely, although this assumption would require confirmation through appropriate biochemical and toxicological studies. While progress has been encouraging, with activity of morpholinos seen in clinical trials, a number of challenging technological questions remain, and have hampered the wider applicability of the strategy. Foremost amongst them relates to how to achieve the most effective and systematically widespread delivery of drug to patients, because the compounds evaluated clinically to date are not orally available molecules and thus far have had to be injected directly to the site of the affected muscle in both pre-clinical experiments using animals,[80] as well as in clinical studies. This also suggests that there will be limited opportunity to treat cardiac muscle using this strategy, although one can imagine combination therapy with other therapeutic paradigms being explored. The oligonucleotide needs to access the cell cytosol in order to have an effect. Normally this would be problematic, but in DMD patients it does appears to adequately penetrate the muscle cytosol, and although the mechanism is not yet clear, this does allow for improved local availability of the agent. Microencapsulated derivatives have shown encouraging data, but nonetheless wider systematic bioavailability is still a problem.[81] Concerns have been raised regarding the off-target effects of exon-skipping drugs, for example the propensity to have an effect on the expression of other genes in the DMD patients, as well as in healthy volunteers (for example, during Phase 1 clinical trials). While it is theoretically possible that these agents could provoke an immune response or even *unintended* exon skipping (resulting in a faulty dystrophin protein in a healthy patient), the risk of this actually occurring is unknown at this time, and so will have to be monitored carefully during clinical evaluation.

Ultimately, exon-skipping strategies should allow production of functional dystrophin protein in individuals who carry the mutation in that particular exon, although importantly the dystrophin protein produced is shorter. The consequences of this are that although the patient should show an improvement, they will still be symptomatic, though it is expected that this will resemble the milder BMD phenotype.

The majority of damaged exons in DMD patients occur between exon 45 and exon 55. A selection of the most common mutated exons in DMD, their incidence, the agents being explored and the clinical stage that the drugs are at is illustrated in Table 11.1.[82] Notable is the fact that even the most common of these, in which mutations are present in exon 51, only has an incidence of around 13%.

Regardless of any combination strategies devised, one exon-skipping oligonucleotide will not be applicable to anything other than a subset of DMD boys, and a new drug will need to be designed *for every exon* that carries a mutation.

In an extension of this strategy, a screening approach was employed by Miceli and co-workers to search for small molecules that could enhance exon skipping. This resulted in the identification of dantrolene (**11.16**, Figure 11.15), which acts at least in part through blockade of ryanodine receptors (a type of calcium channel) and was effective in combination with exon-skipping oligonucleotides.[85]

Table 11.1 Status of the various exon-skipping oligonucleotide therapies.

Exon	*Incidence (%)*	*Drug*	*Clinical stage*	*Trial ID (no data published)*
44	6	PRO044	Phase 1/2	NCT01037309
45	8	ND	Phase 2b	NCT01826474
50	4	ND	Preclinical	
51	13	AVI-4658	Phase 2	NCT01396239 NCT01540409 NCT00159250 [83] NCT00844597 [84]
		PRO051	Phase 3	NCT01803412, NCT01480245
52	4	ND	Preclinical	N/A
53	8	ND	Preclinical	N/A
55	2	ND	Preclinical	N/A

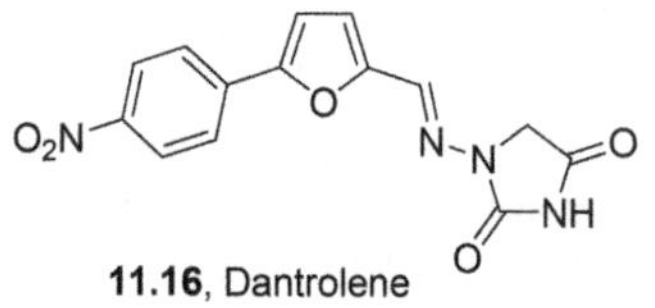

11.16, Dantrolene

Figure 11.15 Dantrolene, studied in combination with oligonucleotides.

Data from both *in vitro* assays (*mdx* and human muscle cells), and functional tests in *mdx* mice indicated the effects of anti-sense were augmented by dantrolene, and although the precise mechanisms in play are not clear at this time, the results were encouraging in their own right and provide precedent for the use of drug combinations.

Eteplirsen, or AVI-4658, is an phosphorodiamidate morpholino (PMO) exon-skipping drug intended to treat the 13% or so of DMD patients who suffer from a mutation in exon 51 (Figure 11.16).

Encouraging increases in dystrophin levels were seen in patient biopsies, and no series safety issues were noted. Two additional clinical trials have been completed or are active. A 28 week study using 30 mg kg^{-1} and 50 mg kg^{-1} doses (NCT01396239), as well as an 80 week extension to this (NCT01540409), are listed on the ClinicalTrials.gov database. Results have not yet been published, but Sarepta Therapeutics has stated that following 84 weeks of treatment the trial participants have maintained increased levels of dystrophin, as well as demonstrating stabilised walking ability, as measured by the 6 minute walk test (6MWT) (a standard clinical assessment for Duchenne boys).[86]

GlaxoSmithKline (GSK), in collaboration with Prosensa, is developing a similar exon-skipping treatment for DMD, PRO051 (drisapersen, GSK2402968). Chemically it is distinct from AVI-4658, being a 2′-*O*-methyl-phosphorothioate oligonucleotide, and is also intended to induce skipping of exon 51, thereby producing truncated but partially functional dystrophin protein (Figure 11.17).

Increases in dystrophin protein have been seen in preclinical[87] and clinical studies.[88,89] Furthermore, a modest improvement in walking distance in the 6MWT was also noted, although the authors urge caution because there was no placebo control group in the latter trial, and furthermore there were no biopsy samples taken at the end of the study. It was therefore not possible to definitively correlate changes in dystrophin levels with the functional improvements seen in the 6MWT. Overall, the results from this trial, albeit in

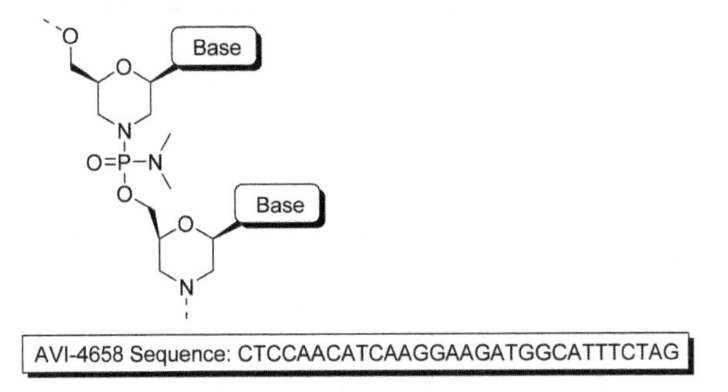

Figure 11.16 Generic structure of nucleotide backbone, and sequence for AVI-4658. Results from Phase 1 and 2 clinical trials for this agent were published recently.[83,84]

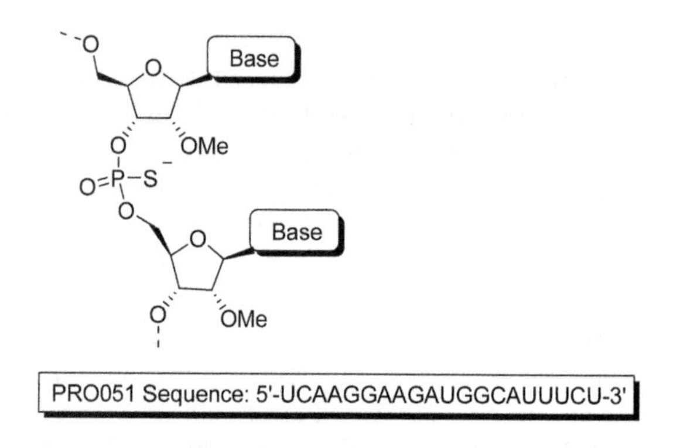

Figure 11.17 Generic structure of nucleotide backbone, and sequence for PRO051.

a small patient population, were deemed sufficient to progress the compound further, with GSK having reported that a Phase 3 clinical trial is currently being recruited.

11.2.5.1.3 Nonsense Mutation Read-Through. A nonsense mutation is a genetic mutation which gives rise to a premature stop codon in the coding region of mRNA. This causes translation to be terminated early, and results in the formation of truncated and usually non-functional proteins, which are responsible for a significant proportion of many inherited diseases, including DMD.[90] Indeed, it has been estimated that just over 10% of the 7000 known human inherited diseases can be attributed to in-frame nonsense mutations.[91] Restoration of the production of protein by only a relatively small amount (up to 5–10% of normal levels) through one of various mechanisms has been suggested as being sufficient to reduce disease severity and therefore to be of potential therapeutic value.[92] As well as muscular dystrophy, a proportion of other well-known rare genetic diseases such as CF have premature stop codons as an underlying root cause and therefore may be treatable using this type of strategy.[93] Accordingly, pharmacological agents that are able to selectively override these premature stop codons, or nonsense mutations (sometimes termed 'read-through' agents) are of great interest, and represent an attractive target for the drug discovery community.

The aminoglycoside antibiotics (Figure 11.18) as a compound class have been extensively studied as potential therapies for genetic disorders driven by nonsense mutations,[94] with the initial proof-of-concept application being demonstrated for CF in the 1990s. The advantage of the class was that the drugs had already been evaluated in humans as antibiotics (which was in fact where the ground-breaking observation of their stop codon read-through ability was originally made),[95] and therefore had a known toxicological profile. Thus, in a way they could also be seen as a very early example of the reprofiling approach paying therapeutic dividends. Specifically, Geneticin™ **11.17** and gentamicin **11.18** were shown to produce full-length CTFR protein

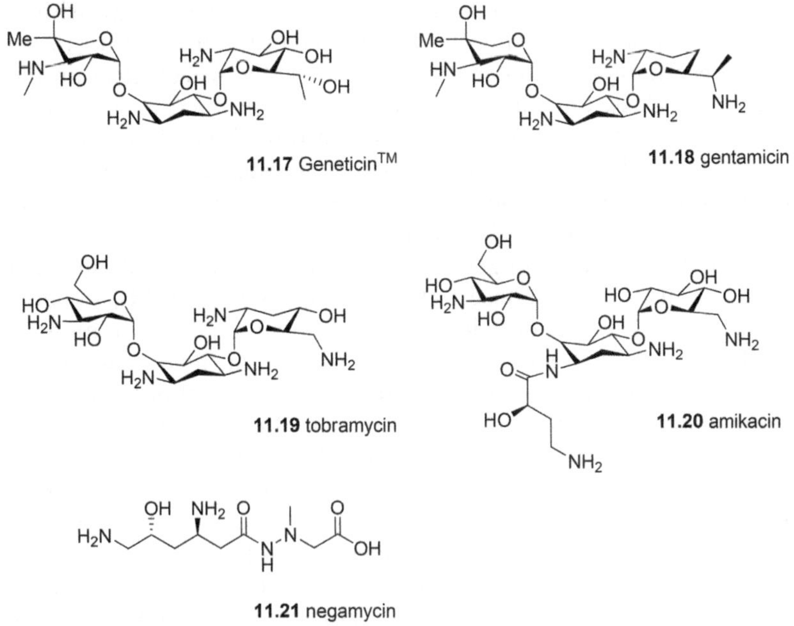

11.17 Geneticin™

11.18 gentamicin

11.19 tobramycin

11.20 amikacin

11.21 negamycin

Figure 11.18 Aminoglycoside read-through agents.

and demonstrate functional benefit in both cellular and animal models of CF,[96] with the range of efficacious aminoglycosides expanding to include tobramycin **11.19** [97] and amikacin **11.20** [98] in later work. Although the detailed mechanism of these compounds is not fully understood, the mode of action postulated to aminoglycosides is that the compounds bind to the decoding region of the ribosomal RNA, and cause a conformational change to take place. The net result is that the usual discrimination which takes place in reading of codons during translation is relaxed somewhat, and allows for incorporation of an amino acid, rather than termination of translation occurring. Translation then continues through to the 'normal' stop codon. As more details of the mechanism arise, the observed premature stop codon selectivity (UGA *vs.* UAG *vs.* UAA) seen with different members of the aminoglycoside class will be able to be more clearly rationalised on more conventional structure–activity considerations.

The extension of this approach to DMD followed only a few years later. Specifically, the expression of dystrophin in cultured *mdx* muscle cells as well as *in vivo* in the *mdx* mouse itself was initially demonstrated using gentamicin **11.18**,[99] as well as the non-aminoglycoside antibiotic negamycin **11.21**.[100]

Other *mdx* studies using gentamicin showed both increased level of dystrophin and reduced CK levels, as well as improved muscle histology.[101] This positive data eventually led to evaluation of gentamicin in patients, with a series of Phase 1 trials being run in DMD patients between 2001 and 2010.

All were undertaken using a relatively conservative trial design, with only small patient cohorts (4–10 subjects) and limited duration of dosing (2 weeks in total), presumably due to concerns regarding the toxicological profile of the compound class. Encouragingly, gentamicin appeared to be well tolerated in patients, with no signs of toxicity.[102] Although no clinically relevant improvement in physiology and histopathology was seen in any of these trials, a number of positive signs were noted, for example decreases in CK levels, as well as detection of dystrophin protein.[103]

Based on these results longer-term studies have been undertaken. For example, in a more recent Phase 1 clinical trial (NCT00451074), 16 patients were treated either weekly or twice weekly with gentamicin for 6 months.[104] No significant toxicity was observed in any patient. At the end of the study, muscle biopsies revealed that dystrophin levels in several of these patients increased significantly, from <5% to over 15%. This is particularly striking, because protein levels have reached a point that is thought to be therapeutically relevant based on the efficacy results previously noted for other readthrough agents.[92] As predicted by the earlier studies, CK levels were also reduced, although disappointingly no functional improvement was noted in any patient in terms of walking time or ability to climb steps. Of more concern was the observation that in one subject an immune system T-cell response was noted, although whether this was related to the expression of dystrophin was unclear. It does highlight, however, that careful monitoring of subjects for propensity of immunological adverse events is critical in this paediatric patient population (both pre- and post-trial), particularly due to the chronic dosing anticipated for the agents.

Aside from the well-known dose-limiting toxicological liabilities of aminoglycosides, particularly ototoxicity[105] and nephrotoxicity,[106] the necessary intravenous dosing route is not ideal for a long-term daily dosing regimen, although it is noteworthy that members of the compound class, when administered for up to 6 months in the aforementioned studies, had not demonstrated any significant issues or liabilities at the dose levels used. Furthermore the production of clinical grade (GMP) material is anticipated to be more challenging, and indeed different variants of aminoglycosides such as gentamicin **11.18** are known in which minor side chain modifications have taken place (Figure 11.19).[94]

It is possible that the subtle structural differences therein could also have unanticipated biological consequences (including altered selectivity for the

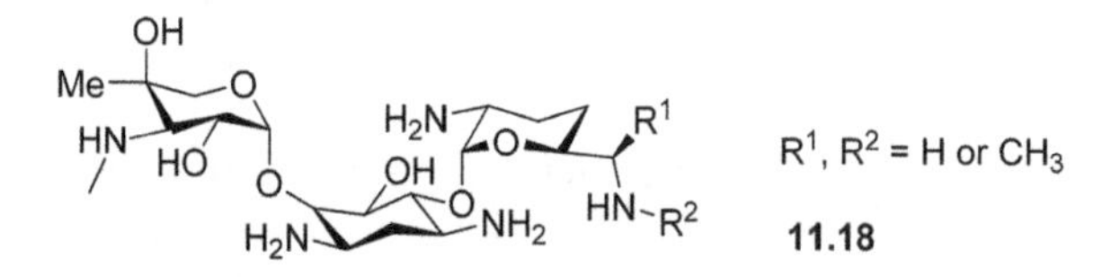

Figure 11.19 Gentamicin structural variants.

premature stop codon, *vide infra*); both of these latter concerns would be circumvented by the use of orally available small-molecule drugs.

PTC Therapeutics conducted a screening campaign in which some 800 000 small molecules were evaluated in a luciferase-based reporter assay designed to identify compounds which promoted read-through of nonsense mutations. Cellular and enzymatic assay systems were developed which used a luciferase-based reporter linked to a premature stop motif; in this case the mRNA contained a UGA codon. Compounds that were able to promote read-through of this premature stop codon would therefore result in an increased amount of luciferase expression, and more luminescence. The result of this screen was identification of a series of hit compounds based on an oxadiazole core, which were subsequently optimised further using conventional library-based medicinal chemistry synthesis techniques. All in all, following the high-throughput screen, approximately 3500 compounds were synthesised and evaluated in follow-up biological tests.[107]

The structure of PTC-124 (**11.22**, Figure 11.20) itself first appeared as 'Example 28' in a US patent application published in 2004.[108] This document disclosed a series of approximately 100 compounds which were claimed to have therapeutic utility as agents for enabling read-through of premature stop codons, also known as nonsense mutations.

Although full details of the medicinal chemistry and SAR strategy which led from the original screening hit(s) to PTC-124 itself has not been disclosed, other variants are known in the literature. For example, a closely related series of compounds that appear to be imide derivatives of the PTC-124 chemotype were disclosed in a later patent application.[109] Data published therein included efficacy studies in the *mdx* mouse, which surprisingly had not been included in the original 2004 filing.

Given the putative mode of action of the compound, it might be anticipated that any compounds discovered using this paradigm would similarly have the potential to find utility as a therapeutic agent for use in other diseases caused by premature stop codons. This was indeed the case, with the use of PTC-124 as a potential treatment for CF also being under active investigation at the current time.[110]

The information that has been disclosed for PTC-124 suggested that the compound was well tolerated and safe after oral dosing, and was also well absorbed, with no overt toxicological effects at up to 1500 mg kg^{-1} as assessed in rat and dog models using both chronic and acute dosing.[111] Furthermore, the compound was metabolically stable, showing only limited

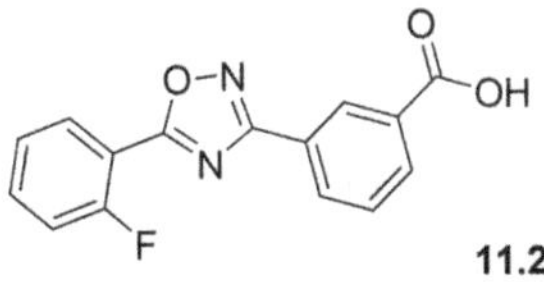

Figure 11.20 Structure of PTC-124.

degradation when incubated in liver microsomes, and was not anticipated to show cardiotoxicity problems, because it exhibited no binding activity to the hERG ion channel. Importantly, it was not genotoxic or mutagenic. In terms of target plasma concentration for efficacy, detailed information on the plasma levels required to see a therapeutic effect were not available, but the pre-clinical efficacy models used had suggested that maintaining plasma concentrations between 2 and 10 $\mu g\ mL^{-1}$ would prove efficacious. Accordingly, these were the trough levels targeted during the clinical studies for the compound.

Clinical evaluation of PTC-124 was initiated in the mid-2000s, and following a co-development deal with Genzyme the results of a clinical trial were published in 2007. The design for this Phase 1 evaluation of PTC-124 was a standard single ascending dose/multiple ascending dose (SAD/MAD) study.[111] In total, 62 healthy adult volunteers were recruited, although it is interesting to note that since the intended patient population would be paediatric, the study designers opted to select volunteers from a younger age group than usual (18–30 years). In addition, a qualitative assessment of drug taste (palatability) of the orally delivered liquid suspension was included in the trial design, again to cater for the intended paediatric patients. The SAD arm of the study was undertaken using doses of 3 to 200 mg kg^{-1}. No adverse events were recorded at the lower doses, but levels of 150 mg kg^{-1} and above produced a range of side effects including mild headaches, dizziness and GI disturbance. Based on the dose dependence of these events, the investigators concluded that the occurrences were linked to C_{max} plasma levels of drug. An MTD of 100 mg kg^{-1} was determined based on these data. In line with the results from the SAD arm of the study, the MAD evaluation was undertaken using lower doses of drug which would not be anticipated to cause adverse events, these being in the range 10–50 mg kg^{-1} twice daily for up to 14 days. In line with the investigators' expectations, no toxicological issues were encountered.

Pharmacokinetics from the SAD group showed approximately dose-proportional changes in C_{max} and AUC, with a modest food effect. Data from the MAD volunteers established that the pharmacokinetic profile largely mirrored that of the SAD group, with there being no significant dose accumulation following multiple doses, and no issues with enzyme induction which could adversely affect compound plasma levels. Critically, given the compound mode of action, analysis of blood samples for evidence of non-selective read-through of stop codons was also undertaken by looking for extended length marker proteins. None were seen, supporting the previous findings that the compound worked in a selective fashion against the dystrophin RNA.[111] Finally, and most importantly, trough plasma levels were recorded in all studies that were within the range predicted to be required in order to demonstrate efficacy. Following these encouraging results, the compound's safety and exposure profile was deemed appropriate for further evaluation, and supportive of progression into Phase 2 evaluation for diseases caused by nonsense mutations, such as CF and DMD.

Phase 2a evaluation of PTC-124 was completed during 2007,[112] with a summary of clinical results being published very recently in a review of the compound and its effects.[113] Using a thrice daily dosing schedule, three dose levels (4/4/8; 10/10/20; 20/20/40 mg kg^{-1}) were evaluated in a total of 38 DMD patients (ages 5–13 years, each having the premature stop codons UGA, UAG or UAA, and the majority of whom were already being treated with steroids). *In vitro* treatment of patient muscle biopsies with PTC-124 showed a dose-dependent increase in dystrophin expression, and qualitative increases in protein level following assessment of dystrophin levels *in vivo*, although boys who were being treated with the lower and mid-levels of compound did not achieve compound levels associated with maximal clinical efficacy based on previous studies.

A Phase 2b study was subsequently undertaken in 174 patients of 5–20 years old. At the time, it was one of the largest prospective clinical studies ever performed in DMD.

Two thrice daily doses were investigated – low (10/10/20 mg kg^{-1}) and high (20/20/40 mg kg^{-1}), with the primary clinical end point being an increase in the distance walked in the 6MWT. Results were somewhat puzzling, with a 'bell-shaped' response being noted – the mean change in distance in the 6MWT was similar in the high dose group *vs.* placebo, whereas in the lower dose group a mean change of 29 m was noted *vs.* placebo. Although the drug was well tolerated and this latter result was encouraging, it was not statistically significant, and the study was discontinued.[113]

Following these results, in September 2011 Genzyme announced that it was returning the rights to PTC-124 to PTC Therapeutics, who in July 2012 disclosed that they succeeded in raising additional funds to support the compounds continued clinical development, presumably in support of further Phase 2b or Phase 3 trials.[114]

More recently, concerns and questions have been raised regarding the mode of action of this class of substituted oxadiazoles, as well as other compounds whose primary activity readout is based on quantification of luminescence using a luciferase reporter assay. For PTC-124 itself, data have been published which demonstrate that under the screening conditions used in a firefly luciferase reporter system, the test compound is converted in an ATP-dependent fashion to a so-called 'high affinity multi-substrate adduct inhibitor' (MAI).[115] This can cause misleading readings in the reporter system and often result in false-positives.[116]

Further data has been published comparing the activity of PTC-124 to geneticin in a wide range of read-through assays,[117] with geneticin giving effective read-through, whereas no activity is noted for PTC-124. Furthermore, the confounding activity in the firefly luciferase assay was confirmed. Because other researchers have explored the use of PTC-124 and reported positive results (*vide infra*), it would seem that more detailed independent investigations are required before firm conclusions can be drawn. Given the importance of the results for genetic diseases like DMD and CF, it is critical that these questions are answered, for example by ensuring mechanistically appropriate and therapeutically predictive secondary assays are conducted to validate the proposed drug mode of action before further clinical trials are considered.

Bertoni *et al.* have also described *in vivo mdx* results from small-molecule read-through agents directed towards DMD.[118] Two compounds, termed RTC-13 **11.23** and RTC-14 **11.24** (Figure 11.21), were identified as enhancing read-through employing a HTS strategy, with the evaluation set consisting of over 30 000 compounds. In contrast to many of the other assay systems described in this chapter, the assay system used in this work was not predicated on a luciferase reporter gene readout, rather an ELISA-based protocol. The assay protocol was validated using ataxia-telangiectasia as a disease model, and in separate experiments also established that the treatment of *mdx* cells with the compounds resulted in the restoration of dystrophin expression.[119] The ELISA-based readout would be anticipated to circumvent some of the problems of luciferase interference often encountered in drug screens by workers who use these assay systems,[115] although care is needed with any assay, because all such indirect measurements come with an attendant danger of highlighting compounds which are false-positive hits.

Both compounds were confirmed as being active in cell-based assay systems (*i.e.* they increased the expression of dystrophin), but **11.24** was found to be ineffective when injected directly into the muscle of *mdx* mice, possibly as a result of the instability of the Schiff base *in vivo*, although no data to confirm or refute this supposition are presented either in the paper or supplementary data. In any event, no further work was undertaken on that compound. In contrast, **11.23** was found to induce expression of encouraging amounts of dystrophin both when injected directly into muscle, but more importantly following repeat dosing using the intraperitoneal route (once every 5 days, repeated for 4 weeks). Although a detailed evaluation of the compound's pharmacokinetics was not reported, bioanalysis of **11.23** levels in various tissues, as well as plasma, was undertaken. Interestingly no parent compound was detectable in plasma at any time point following intraperitoneal injection, although levels up to around 3 μM were detected in most muscles sampled, including the heart. This observation is of particular relevance for a muscular dystrophy therapeutic where cardiac muscle has historically proven difficult to target with drugs. Based on the *in vitro* cellular data, compound levels of between 2 and 10 μM would be predicted to increase dystrophin levels by around 1–3%, and indeed this is what was

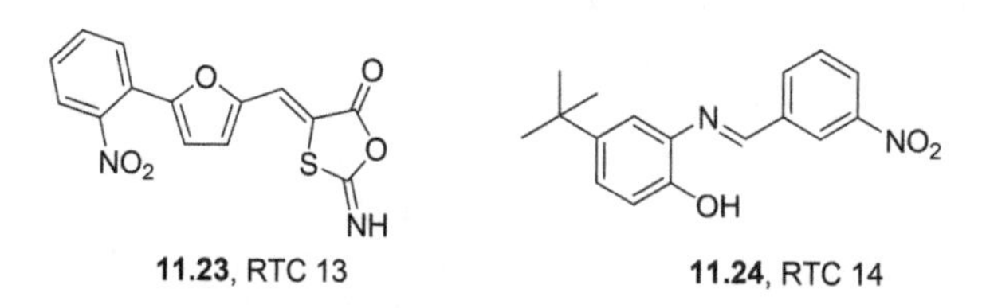

Figure 11.21 Other read-through agents.

observed when various types of muscle were analysed for dystrophin-positive fibres. CK levels were also appreciably reduced (by approximately 50%), relative to vehicle-treated control samples, this being an often used surrogate marker for muscle degeneration and physiological benefit. PTC-124 **11.22** was used as a positive control compound in these studies, and encouragingly, **11.23** proved superior in all histological readouts.

More importantly, these positive histological data translated to functional benefit following treatment with **11.23**, with improved forelimb grip strength and better performance in the four-limb hanging wire test. Again, the positive control compound PTC-124 proved to be inferior.

Although these data are clearly encouraging, it is important to place this project in context. The compound described, **11.23**, represents a very early stage drug that is likely to require significant optimisation before it can progress further down the drug development path. A more detailed analysis of the compound/class pharmacokinetics will be essential, as well as concomitant delineation of structure–activity relationships in order to translate the intraperitoneal dosing regimen into (ideally) an orally delivered agent. There are also functional groups within the compound that may raise concern, for example the nitrophenyl motif, as well as the iminothiazolone ring, because these types of functionalities have been commonly associated with both assay interference and poor drug profiles in the past, and so appropriate replacements and/or safety assessments will be critical.[120] Furthermore, in order to expedite the progress of the programme towards the clinic, exploration of the compound's mode of action would be advisable. Although contrary to widely held beliefs, a specifically defined mode of action/pharmacological target is not an FDA regulatory requirement for a new drug, in order to facilitate the regulatory review process, and to aid prediction of possible toxicological or safety issues, this information can be invaluable (see 21CFR312.23, subsection (a)(8)(i) in the notes for completing IND submission documentation).[121]

11.2.5.2 Dystrophin Replacements: The Utrophin Approach

Utrophin is a structural homologue of dystrophin, and is known to be co-expressed with dystrophin in early development. For reasons that are not yet clear, utrophin expression decreases significantly with maturity during foetal development, and is replaced almost exclusively by dystrophin. As well as having structural similarity, utrophin has been established as playing a functionally equivalent role to dystrophin, this having been conclusively demonstrated by Davies *et al.* in 1996 where delivery of a truncated utrophin transgene was shown to ameliorate the dystrophic phenotype in *mdx* mice.[122] In follow-up work, the expression of full-length utrophin was shown to prevent the development of muscular dystrophy, with approximately a two-fold increase being sufficient to completely recover the phenotype in the mouse.[123] A similar result was seen in the dog model of DMD,[124] and furthermore, the severity of muscular dystrophy (judged by the age at which patients become wheelchair bound) has been shown to inversely correlate with the amount of utrophin that is expressed.[125]

Since this seminal work, strategies predicated on the upregulation of utrophin have been postulated as a viable therapeutic approach to DMD.[126] These are viewed as being particularly attractive because the approach is independent of patient DMD mutation status, and therefore should be applicable to all DMD patients. Although the proof-of-concept murine experiments were conducted using transgenes, alternative strategies using pharmacological approaches can be envisaged, and are potentially attractive as a small-molecule drug can in principle be delivered orally, would be relatively inexpensive compared to a biologic agent, and should be systemically available, thereby having the potential for treating all muscles, including the difficult to target cardiac tissue.[127] These properties of small-molecule therapeutics are particularly important considerations for muscle diseases of this nature, because skeletal muscle is one of the most abundant and widely distributed tissues in the human body.

11.2.5.2.1 Small-Molecule Modulators of Utrophin Expression. The potential of both biologics and low molecular weight biochemicals to upregulate the production of utrophin has good precedent, with agents such as heregulin[128] and L-arginine[129] having been shown to ameliorate the dystrophic phenotype when dosed to *mdx* mice. Heregulin is thought to work by activation of the utrophin A promoter, with the mode of action of L-arginine being postulated as being through activation of the nitric oxide pathway, indirectly activating utrophin.

More recently combination approaches have been examined using a mixture of L-arginine and butyric acid, and seeking to combine the activities of two biological pathways which have individually been shown to have beneficial effects, these being NO pathway modulation combined with histone deacetylase inhibition.[130] Positive results were seen from these studies, with both histological and functional measurements showing improvement as compared to saline-treated control *mdx* mice. Data presented were consistent with an additive effect on CK levels *in vivo*, although more comparable dosing levels would be required to confirm this, and to support whether an additive effect is seen *in vitro* on histone acetylation levels.

Although providing a critical proof-of-concept for the approach, none of these agents represents a viable drug therapy at this stage, because many questions remain unanswered, particularly how an appropriate dosing regimen can be established, as well as whether or not there are any longer-term compound-associated toxicological consequences. Exploration of heregulin mimetics, or alternative modulators of the NO pathway, in combination with more recently identified HDAC inhibitor molecules, may represent interesting and fruitful avenues of research.

A number of companies, including large pharmaceutical organisations as well as biotechnology companies, are seeking to develop small-molecule upregulators of utrophin, including BioFocus and Summit plc, and the therapeutic approach has been reviewed recently by Khurana *et al.*[131] Summit plc has the most advanced drug discovery programme, having recently announced the imminent initiation of a Phase 2 trial with their compound SMT-C1100 (**11.24**, Figure 11.22), a naphthalene-substituted benzoxazole derivative.[132] This compound represents a potential first-in-class opportunity for a utrophin upregulator as a therapeutic agent for all sufferers of DMD. It was discovered during a collaborative programme with scientists from the University of Oxford's Chemistry, Physiology, Anatomy and Genetics Departments. The medicinal chemistry hit discovery and lead optimisation work for this project has been published recently.[133,134] The drug discovery process which resulted in the identification of SMT-C1100 (RMM = 337, $C \log P = 4$)[135] began following a relatively unconventional (at the time) high-throughput phenotypic screening approach using a novel engineered murine H2K muscle cell line. Following hit confirmation, a more straightforward lead optimisation approach was undertaken, based on evaluating the structure–activity relationships of a series of hit compounds.

The initial assay used for primary screening of the compound libraries was conducted in H2K cells, which had been engineered to express the utrophin A promoter linked to a luciferase reporter construct. Accordingly, any compounds that interacted with and activated the respective utrophin promoter would be easily detected and quantified using a luminescent readout. As with any screen of this nature, elimination of false-positives is a critical consideration; in this case it was undertaken using the same H2K cell line, with UTRN mRNA levels being quantified using RT-PCR. Since that time additional utrophin promoters have been identified, and therefore it is possible that this screen would not necessarily identify all compounds that are potentially able to upregulate the production of utrophin using this or a related mechanism.[136]

Following the high-throughput screen of several thousand drug-like molecules, a number of hit molecules were identified, including benzoxazole **11.25**,[133] along with the structurally related 2-aryl benzotriazole **11.26**.[137]

From a medicinal chemistry perspective, although encouraging activity was noted in the *in vitro* cell assay, both compound classes required

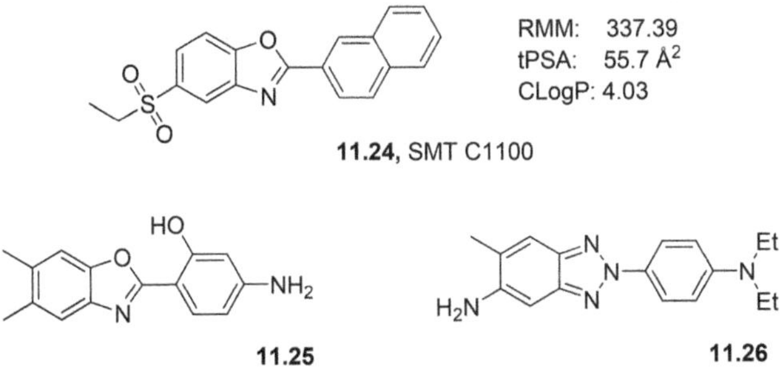

RMM: 337.39
tPSA: 55.7 Å^2
CLogP: 4.03

11.24, SMT C1100

11.25

11.26

Figure 11.22 SMT-C1100 and the original HTS screening hits.

considerable optimisation, because they were described as suffering from rapid metabolism in mouse liver microsomes and having poor physicochemical properties. Moreover, both contained functional groups that were felt to be unsuitable for progressing the compounds further, including anilines and phenols. The aniline motif contained within both examples was felt to be a particular liability, because it is known to be a potent toxicophore in some cases.[138,139] Additionally, the dimethyl substitution of **11.25** was thought to play a part in the compound's short *in vitro* half-life. The latter liability was confirmed *in vivo* when preliminary assessment of exposure levels was made by dosing lead molecules orally in mice, and plasma levels of compound were found to be very low. Optimisation of the class represented by **11.25** was felt to be the preferable proposition, and an extensive medicinal chemistry programme was undertaken. A schematic representation of the strategy used to explore the structure–activity relationships carried out is illustrated in Figure 11.23.[133]

A systematic study of substitution to replace the dimethyl groups in Region A of the molecule was carried out first. Alkyl amides were found to be active, particularly when located at the 6- and 7-positions of the benzoxazole core, and with a clear size dependence, although they were also found to suffer from poor metabolic stability, a problem that was further apparent following *in vivo* dosing. Other linking groups were investigated, including thioamides, amines and sulfonamides, and all were less active than the starting compound. Replacement of the amide with a sulfone was found to recover the biological activity, and much improved *in vitro* and *in vivo* ADME properties were noted in these examples. In particular, this structural change appeared to confer preferable pharmacokinetic properties on the compounds, as well as having improved solubility over its amide analogue.

For Region B, the benzoxazole, a range of alternative cores were explored, including the isosteric replacements benzothiazole and benzimidazole, as well as a benzofuran analogue. Of these, only the benzothiazole exhibited any appreciable activity, being approximately equipotent with the benzoxazole, but otherwise there was seen as being no advantage to a core switch, so focus was maintained on the benzoxazole.

A wide range of mono and bicyclic cycloalkyl, aryl and heteroaryl rings were examined as a replacement for the phenyl ring in Region C of the molecule. Simple acyclic alkyl derivatives were found to be inactive, as were compounds

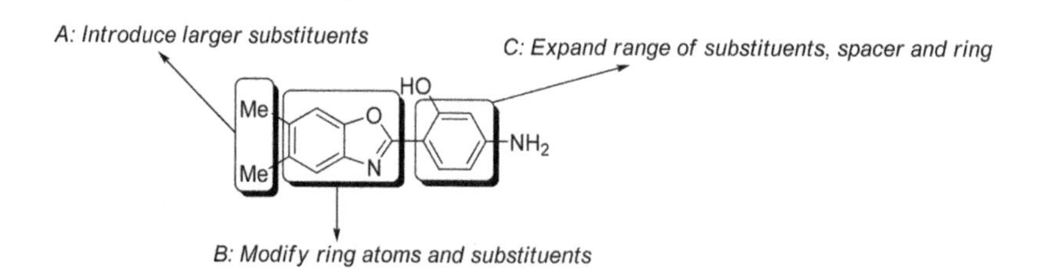

Figure 11.23 SAR strategy for benzoxazole hit compound 25.

bearing 2-aryl substituents with an *ortho* substituent. Preferable substituents on the 2-aryl ring were found to be those that were relatively lipophilic, and positioned at the 4- or 3,4-positions, with particularly favoured groups being 3,4-dichloro and 2-naphthyl.

A selection of preferred active compounds were profiled further, using standard *in vitro* tests encompassing basic ADME parameters, as well as mechanism-based assessments such as analysis of utrophin mRNA levels in H2K cells. The latter assay directly confirmed the compound's mode of action as upregulating the production of utrophin mRNA rather than an indirect effect associated with interference due to the luciferase reporter readout. On the basis of these data, compound **11.24**, latterly denoted as SMT-C1100, was identified as a promising lead compound which upregulated utrophin expressing and had suitable drug-like properties for further progression.

Following the identification of SMT-C1100 as a transcriptional upregulator of utrophin production, more detailed data on its *in vitro* pharmacology have been published, along with supporting data illustrating the compound's *in vivo* activity in a range of murine *mdx* models which are viewed as being the gold standard pre-clinical disease models of DMD before progressing a compound into human clinical trials.[140] The conclusion from the authors was that the biochemical and *in vivo* efficacy data obtained represented proof-of-principle for small-molecule utrophin upregulators, with the *in vivo* efficacy results in *mdx* mice positioning the project very strongly for translation into clinical efficacy trials.

SMT-C1100 was progressed to a Phase 1 trial in January 2010,[141–143] with a follow-up Phase 1 trial using an improved drug formulation, a nanoparticulate aqueous suspension taking place later. Results were presented at the World Muscle Congress in October 2012.[144]

This second Phase 1 trial involving 48 healthy volunteers gave encouraging results, with the compound appearing to be safe and well tolerated. Compound plasma concentrations stabilised after an initial drop, and the level being seen was felt by the authors to be above that which was anticipated to provide therapeutic benefit for at least 60% of the time.[145]

On the basis of the Phase 1 clinical results in volunteers to date, Summit have stated that they believe the data to be supportive of moving the compound into patient trials, and have recently announced their intention to progress SMT-C1100 into Phase 1b evaluation, which was expected to start recruitment during the second half of 2013.[146] Summit has also published preliminary data on an additional bicyclic series of utrophin modulators, structurally related to SMT-C1100, but based around alternative heterocyclic cores.[137] Starting from the previously described benzotriazole amine (**11.26**, Figure 11.22), a systematic investigation of the structure–activity relationships of this series was carried out. Modification of the substituents on the benzotriazole core was undertaken, with groups ranging from amides to sulfones being introduced, the latter in an attempt to determine if the SAR from SMT-C1100 crossed over to this series. While small amides (*e.g.* Et, iPr) exhibited moderate levels of activity in the H2K luciferase reporter assay

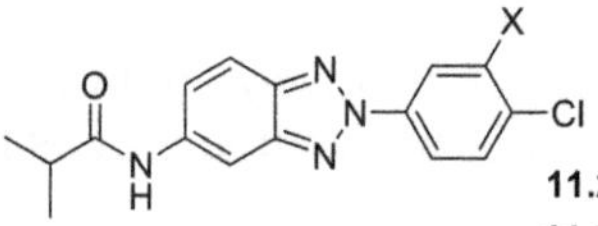

	11.27	11.29
EC_{50} / μM	0.8	1.7
KinSol / μM	1	3 – 10
HLM t½ / min	291	198

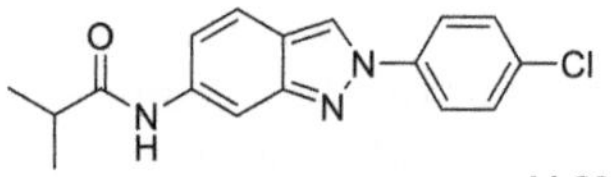

Figure 11.24 Comparison of benzotriazole and indazole classes.

(*e.g.* **11.27**, $EC_{50} = 0.76$ μM), larger heteroaryl substituents were not tolerated. Interestingly, the sulfone analogues which were directly analogous to SMT-C1100 were also only very weakly active, suggesting a fundamental change has been made in modifying the benzoxazole core to the triazole analogue.

Modification of the benzotriazole to the less polar indazole was also investigated, with the authors synthesising a number of key compounds which crossed over with the corresponding benzotriazoles. Similar structure–activity trends to those seen in the corresponding benzoxazole series were observed, with only the amide derivative showing any appreciable activity (**11.29**, $EC_{50} = 1.73$ μM).

Comparative analysis of several *in vitro* ADME properties was undertaken for compounds **11.27** and **11.29** (Figure 11.24). Both were found to have low to moderate kinetic solubility, but more encouragingly they had low metabolic turnover upon incubation with human liver microsomes. The authors conclude by stating that these data were encouraging enough to progress the compounds for further evaluation, although no *in vivo* data, such as pharmacokinetic profiling and/or efficacy testing, has been reported for either to date.

11.2.5.2.2 A Reprofiling Approach to Utrophin Upregulation. Khurana and co-workers have also recently described their efforts to identify upregulators of utrophin production, using a screen of small molecules in an assay designed to assess the ability to activate the utrophin A promoter in $C2C12_{utrn}$ cells (C2C12 cells which have been stably transfected with the utrophin A promoter linked to a luciferase reporter).[147] They screened the Prestwick Chemical Library, which at that time contained 1120 compounds. Of these, approximately 90% were drugs which were approved for use in humans, with the remainder being natural products. Importantly then, the vast majority of these compounds will have entered clinical trials at some stage. Details on dosing, efficacy, alternative potential modes of action and most importantly (given the paediatric, chronic nature of the disease) toxicology profiles should also be accessible.

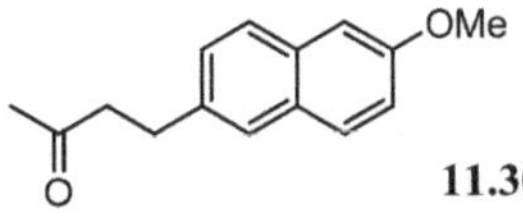

Figure 11.25 Nabumetone, identified from Khurana's UTRN screen.

Dose–response assays on all 14 confirmed hit compounds generated data showing dose-dependent responses for most, but for several examples cytotoxicity was observed at higher screening concentrations, as adjudged by a drop in luciferase response.

Confirmatory screening, using assessment of utrophin mRNA levels, as well as Western blots for utrophin protein itself, were carried out in wild-type non-transfected C2C12 cells (see final two columns in Table 11.2), although this was only carried out for one compound, nabumetone (**11.30**, Figure 11.25), a regulatory approved non-steroidal anti-inflammatory drug, a COX2 inhibitor.

The authors suggest that following the positive *in vitro* data described for nabumetone, the drug may represent a possible therapeutic option for DMD. Further, they note that follow-up experiments of a similar nature to those previously described for nabumetone are under way for several of the other non-cytotoxic hit compounds, although there is no mention of *in vivo* testing of any of the compounds in the *mdx* mouse model.

The authors acknowledge that there are other utrophin promoters, activation of which could also increase levels of the protein, as well as post-translational strategies. In an effort to address the latter deficiency in more recent work, they have described a new cell-based assay designed to identify compounds which upregulate utrophin levels through post-translational mechanisms, although no reports of compound libraries being screened using it have appeared yet.[148]

BioFocus DPI has also recently reported details of its utrophin upregulation project in poster form, the work being supported financially by Charley's Fund.[149] Biofocus scientists developed a cell-based immunoassay to measure the upregulation of utrophin protein when the cells were treated with small molecules. Following a screening programme which evaluated the complete NIH Clinical Collection (~450 compounds, all of which have entered clinical development in humans, and have therefore passed pre-clinical toxicology evaluation), and using heregulin as a positive control, nine compounds were highlighted as upregulating utrophin levels in their assay. Compound structures, and detailed information about activity levels and any follow-up confirmatory tests, have not yet been published. Whether this is directly connected with other work on utrophin modulation by the same organisation is unclear.[63]

11.2.5.2.3 Calpain Inhibition. The calpain enzymes are a family of cysteine proteases consisting of around 15 members, and which have been established as having diverse physiological functions including signal

Table 11.2 Hit compounds from Prestwick 1120 screen.[a]

			Primary assays		Secondary validation assays	
Compound	*Class*	*Description/drug MOA*	*Max UTRN upreg (fold)*	*Cytotoxicity seen*	*mRNA @25 μM*	*Western @25 μM*
Nabumetone	Approved drug	NSAID, COX2i	2.6	N	1.8×	1.2×
Chrysin	Natural product	Flavone	1.4	Y	N.T.	N.T.
Piperine	Natural product	Stilbenyl alkaloid	1.3	Y	N.T.	N.T.
Apigenin	Natural product	Flavone	1.5	Y	N.T.	N.T.
Riluzole·HCl	Approved drug	Na^+ channel blocker	2.0	Y	N.T.	N.T.
Phenazopyridine·HCl	Approved drug	Local analgesic	2.9	Y	N.T.	N.T.
Resveratrol	Natural product	Stilbenoid	2.9	Y	N.T.	N.T.
Tiabendazole	Approved drug	Fungicide/antiparasitic	2.2	N	N.T.	N.T.
Hesperetin	Natural product	Flavanone	2.8	N	N.T.	N.T.
Leflunomide	Approved drug	Dihydroorotate dehydrogenase inhibitor	1.8	Y	N.T.	N.T.
Kawain	Natural product	Stilbenoid	2.5	N	N.T.	N.T.
Clorgyline·HCl	Approved drug	Monoamine oxidase inhibitor	3.5	Y	N.T.	N.T.
Equilin	Approved drug component	Equine oestrogen	2.4	N	N.T.	N.T.

[a]N.T. = not tested.

transduction, proliferation, differentiation and apoptosis. They are calcium-dependent enzymes, with various isoforms being ubiquitously expressed, and others being more specifically localised in tissues including skeletal muscle (calpain 3) and the testis (calpains 5, 11 and 13). They are believed to play a role in the pathology of DMD, as well as a number of debilitating disorders including Alzheimer's and other neurodegenerative diseases.[150] Utrophin itself has been shown to be a direct *in vivo* substrate for calpain, therefore a calpain inhibitor would be anticipated to increase utrophin stability and therefore have value for DMD, and potentially for other muscle-wasting diseases.[151] Given the association with disease areas of much unmet need, a number of organisations and research groups have undertaken drug discovery programmes aimed at identifying selective small-molecule inhibitors of calpain. A wide range of compound classes have been described in the literature in recent years for the treatment of various diseases, including DMD (Figure 11.26).[152] Many of these have been shown to have potent enzymatic inhibition of calpain, with the most advanced having been shown to also have activity using cellular and *in vivo* systems. To date though, none appear to have progressed beyond testing in animal models.

Of particular note for DMD, Santhera has an active interest in new drugs to treat muscular dystrophy, and have published work describing a novel series of ketoamide derivatives which are both potent calpain inhibitors and have encouraging levels of cellular activity. Of particular note was one example, compound **11.31**, which gave an improvement in dystrophic muscle histology after 4 weeks intraperitoneal dosing in *mdx* mice, although no physiological parameters were reported.[153] In other studies, the same organisation has also described bifunctional molecules containing additional

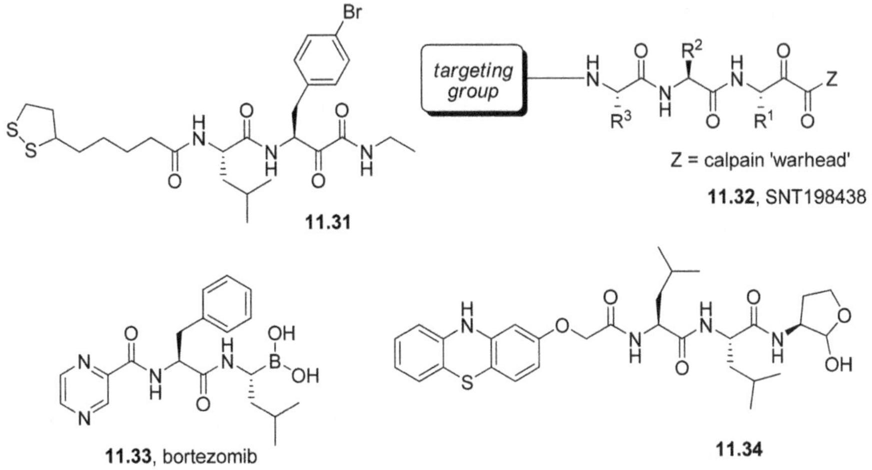

Figure 11.26 Calpain inhibitors, and related peptidomimetics.

functional groups as well as the lipoic acid derivative, all of which are intended to act as muscle-targeting motifs.[154]

The current status of the project is unclear, because more recently data was published on a further bifunctional series of calpain/proteasome inhibitors which cast doubt on the use of calpain inhibitors for DMD.[155] In these studies, compounds such as the tripeptide-based SNT198438, **11.32** (full structural details of which were not revealed) were found to have inhibitory activity in the low nanomolar range against both calpain and the 20S proteasome, along with a profile that was amenable to *in vivo* evaluation. Following efficacy testing in *mdx* mice, **11.32** itself showed encouraging amelioration of the dystrophic muscle histology using intraperitoneal dosing at 20 mg kg^{-1} every second day. However, when studies were undertaken using a transgenic mouse overexpressing the endogenous calpain inhibitor calpastatin, crossed with the *mdx* mouse, no histopathological improvement was seen. While it is clearly important that further detailed studies are undertaken, the suggestion from these results is that the observed benefit gained from treatment with these bifunctional molecules was seen solely due to inhibition of the proteasome activity. Furthermore, the data also suggest a potentially productive line of research would be a detailed evaluation of monofunctional proteasome inhibitors, because these represent a class of drugs including bortezomib **11.33**, which has been shown to be tolerated in a clinical setting, and indeed, has also shown encouraging preliminary histopathological data in the *mdx* mouse.[156]

Ipsen has also published details of a series of bifunctional molecules, these being novel peptidomimetic compounds such as **11.34** containing a calpain inhibitor warhead motif coupled through an amino acid linkage to a reactive oxygen scavenger (ROS) unit. Although the target indication of interest to the project team was neurodegenerative disease, given the therapeutic possibilities associated with modulation of both functional motifs, wider application of these compounds could be reasonably anticipated.[157,158] Specifically, because reactive oxygen species have also been implicated in the disease pathology of various forms of muscular dystrophy and other disorders, application of these types of combination agents may be beneficial.

As well as the structure–activity relationships described in the original medicinal chemistry papers, the compound series advanced further, with examples also having undergone *in vivo* testing.[159] Notable therapeutic benefit was seen in exercised *mdx* mice, with a significant reduction in forelimb grip strength decline following drug treatment, as well as a reduction in serum CK levels. As with the previously described ketoamide dual inhibitors,[153] it is not clear to what extent any improvement seen is attributable to calpain inhibition alone.

11.2.5.2.4 PPAR Modulators. Peroxisome proliferator-activated receptors (PPAR) are nuclear receptors that can modulate gene expression, and therefore have the potential to provide therapeutic benefit in a number of diseases. Three subtypes of receptor have been described: α, β/γ and δ, with

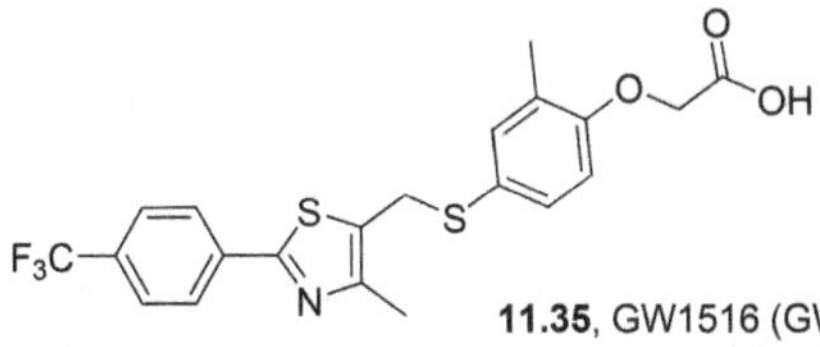
11.35, GW1516 (GW501516)

Figure 11.27 PPAR modulator GW1516.

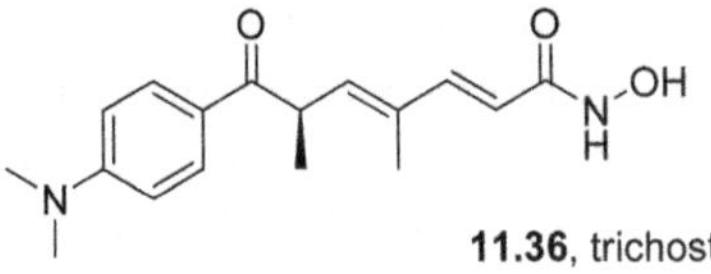
11.36, trichostatin A

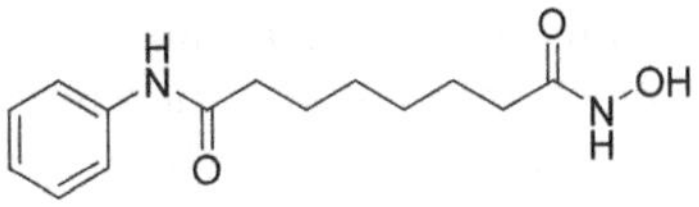
11.37, vorinostat

Figure 11.28 HDAC inhibitors which have been studied for application to DMD.

activation of the β/γ isoform in particular having been shown to play a role in muscle structure and function. This appears to be at least in part through activation of PGC-1α and ensuing modulation of utrophin expression. A small-molecule agonist of PPARβ/γ, GW1516 (**11.35**, Figure 11.27) has been shown to increase PGC-1α and utrophin A mRNA levels and provide functional benefit and improvements to disease pathology in *mdx* mice.[160]

To date there have been no further data published on this compound, so the current status of the approach is unclear.

11.2.5.3 Other Discovery Strategies/Mechanisms of Action

11.2.5.3.1 Epigenetic Approaches. Histone deacetylase inhibitors fall into the class of agents known as epigenetic modulators.[161] This class of drugs was originally identified in the late 1970s,[162] and exert their action by modifying gene expression patterns through inhibition of the enzymes responsible for deacetylation of lysine residues on histones. In recent years a number of HDAC inhibitors have been approved for treatment of various diseases, including cancer, and given their well-studied ability to affect gene expression levels, it is unsurprising that evaluation in rare genetic diseases has been the subject of intense study.[163]

There is growing evidence in the literature that HDAC inhibitors may be promising therapeutic agents for the treatment of DMD and other genetic diseases.[164]

Studies in *mdx* mice with compounds including trichostatin A **11.36** and vorinastat **11.37** (Figure 11.28) have yielded promising results including improved histological and functional readouts. Although the precise mechanisms in play are not clear, inhibitor treatment was shown to increase levels of the myostatin antagonist follistatin in muscle satellite cells, which was suggested to contribute to the functional improvements.[165] None of the existing HDAC inhibitors that are approved for use in man have yet been investigated clinically in DMD patients, though it can only be a matter of time before such studies take place.

11.2.5.3.2 AMPK Modulators. As well as defective muscle function, abnormalities in various metabolic processes and pathways are commonly seen in dystrophin-deficient DMD muscle. AMP kinase (AMPK) is a key metabolic signalling kinase that responds to changes in cellular energy levels. If deficiencies in cellular energetics are sensed, AMPK is activated and switches on mechanisms which generate ATP, thereby rectifying the deficiency.[166] Activators of AMPK would therefore be anticipated to demonstrate a beneficial effect in DMD muscle. This was demonstrated recently in the *mdx* mouse model where the dystrophic phenotype was significantly improved following 500 mg kg^{-1} treatment (5 days on, 2 days off) for 4 weeks with AICAR (5-aminoimidazole-4-carboxamide 1-β-D-ribofuranoside, **11.38**, Figure 11.29).[167] Because the agent was dosed using the intraperitoneal route it is not viable as a drug in its current form, but may well form the start point for future drug discovery programmes.

Combinations of AMPK and PPAR modulators have also been suggested as a viable therapeutic strategy; recently the dual activity of both agents was found to elicit functional improvements using various measurements of muscle function.[168]

11.2.5.3.3 Carbonic Anhydrase Inhibitors. Following a novel *in vivo* screen of 1000 'already approved' compounds using a *Caenorhabditis elegans* disease model,[169] two small-molecule inhibitors of carbonic anhydrase (CA), methazolamide **11.39** and dichlorphenamide **11.40** (Figure 11.30) were identified as partially rescuing the disease phenotype. What was particularly encouraging was that this activity translated into efficacy in the *mdx* mouse

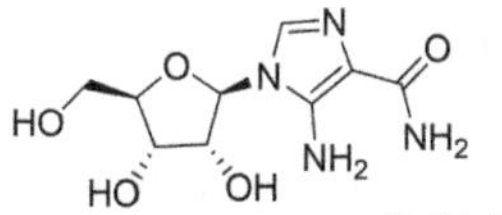

11.38, AICAR

Figure 11.29 5-Aminoimidazole-4-carboxamide 1-β-D-ribofuranoside (AICAR).

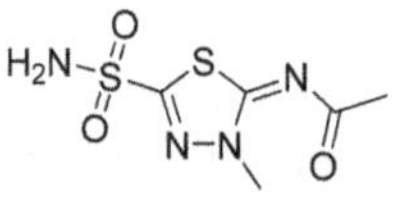

11.39, methazolamide

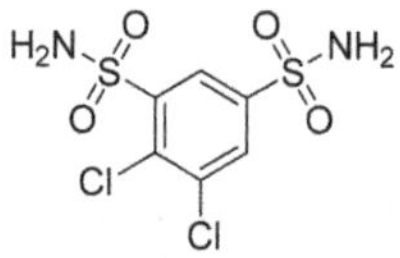

11.40, dichlorphenamide

Figure 11.30 Carbonic anhydrase inhibitors with activity in the *mdx* mouse model.

following dosing for 120 days, with improvement in histological and physiological readouts of dystrophic phenotype, including reduction in centrally nucleated fibres (CNF), and increases in muscle strength.

Of note in this latter section of the experiment was that the *mdx* mice used were 10 weeks old when dosing was initiated. This is unusual in experiments intended to assess the effect of new drugs on the *mdx* phenotype, because by that stage there has already been a considerable amount of muscle degeneration and regeneration taking place; dosing from around the 3 week postnatal period is more usual. Furthermore, although the compounds were dosed orally, this was not undertaken using oral gavage, but by mixing compound with the food. Although there appeared to be a reasonably consistent amount of food intake between the various animals, gavage dosing might be expected to give more consistent dosing results.

The authors speculated that the mechanism of action could involve calcium trafficking. Altering pH and hence transmembrane potential in turn influences specific ion channel activities, particularly Na^+ and Ca^{2+}. Because the progression of the DMD phenotype is known to be affected by calcium channel activity, this hypothesis seems plausible, although it warrants further study.[170] Investigation of the effect of CA isoform selective compounds would be of particular interest, and aid in the selection of the most appropriate analogues to progress further.

Further compound screening using the *C. elegans* assay system would also be valuable. Although there are clear limitations to the screening platform, such as clarity/consistency on compound dose levels, the value of using an *in vivo* disease model with a dystrophin-like gene is clear. For any future screening programme, as well as identifying new lead molecules it would be important to establish the profiles of previously described compounds which work through the full range of mechanisms described herein. In this manner it would be possible to assess the scope and limitations of the assay system, particularly for evaluating compound modes of action which are independent of dystrophin.

11.2.5.3.4 Myostatin Inhibition. Myostatin is an important member of the TGF-β superfamily of proteins, and has been established as playing a critical role in muscle myogenesis, with a 1997 study showing that inactivating mutations in the gene coding for myostatin resulted in an increase in muscle mass.[171] The hypothesis that inhibitors of myostatin might be useful therapies for muscle-wasting diseases such as DMD followed shortly thereafter, with validation studies describing the approach using myostatin antibodies in the *mdx* mouse being published in 2002,[172] and more recently in the canine model.[173] A neutralising antibody to myostatin, MYO-029, even advanced into human trials for other muscular dystrophies, and although it was found to be clinically safe, no efficacy was observed.[174] While work utilising this approach continues to be described,[175] there have not as yet been any successful attempts to identify and develop small-molecule myostatin inhibitors, although a screening approach has been reported.[176] This is presumably due to the significant medicinal chemistry challenges facing those seeking to design and reduce to practice examples of this type of protein–protein interaction inhibitor compound, although as the field advances it may become more viable to do so.

A dual approach, combining myostatin knockdown with myostatin inhibition, has been investigated by several groups, and shown to be beneficial.[177,178] Although the authors acknowledged that further work is needed, preliminary results of the dual anti-sense strategy appeared very encouraging as a new therapeutic paradigm for DMD.

11.2.5.3.5 Utrophin Stabilisation. A recombinant version of an ECM protein biglycan, rhBGN has recently been described by Fallon *et al.* as a new approach to the treatment of DMD.[179] The endogenous protein itself is expressed in, and plays an important role in, developing muscle, where it is a key structural component of the DAPC complex. The therapeutic potential of the protein was further illustrated in studies using biglycan null mice, which were shown to exhibit reduced levels of utrophin expression, along with reduced muscle function. Conversely, *mdx* mice treated with the recombinant protein were shown to have enhanced levels of various components of the DAPC complex as well as utrophin itself. Histological improvement in muscle structure and functional benefit were also seen.

Based on the comprehensive studies described by the Fallon group, biglycan's mode of action seems to be by stabilisation of endogenous utrophin, rather than modulation and enhancing its transcription, because no change in utrophin transcript level was seen on dosing rhBGN in *mdx* mice.[179] This raises a number of important questions, including whether a strategy of this nature will provide anything other than an additional option for combination therapy along with other drugs, rather than a monotherapy in its own right due to low basal levels of endogenous utrophin protein. Furthermore, efficacy has yet to be demonstrated using *in vivo* systems other than the *mdx* mouse. An important advantage of the approach relative to gene therapy is that the protein can be delivered systemically using intraperitoneal injection. Fallon also demonstrated that the agent is well tolerated following chronic dosing and appears to be physiologically stable for sufficient time to provide sustained functional benefit. Given the adverse immunological findings with biostrophin,[69] development of any biologically based agent for DMD is likely to focus at an early stage on immunogenic potential, although RhBGN would not be anticipated to invoke this type of response on dosing, due to its highly conserved nature and expression in both wild-type and dystrophic muscle. It is under development by Tirvorsan, a spin-out company from Brown University co-founded by Fallon.[180]

11.2.5.3.6 Cell-Based Therapies. Although not the subject of this review, cell-based therapeutics are having a significant impact on the development of new therapies for DMD.[181] Of particular note is the use of tissue-derived stem cell therapeutics, an area of intense current interest following encouraging results from several groups using *mdx* mice,[182,183] and for which clinical trials such as NCT01834066 (Bone Marrow Derived Autologous Cells) and NCT01834040 (Umbilical Cord Derived Cell) have been initiated.

11.2.5.3.7 Phenotypic In Vivo Screening. Kunkel and co-workers have recently described an *in vivo* phenotypic screen using two dystrophin-deficient zebrafish models (*sapje* and *sapje*-like), for compounds which may be potential therapeutics for DMD.[184] Homozygous dystrophin null zebrafish of this type present abnormal muscle pathology, as assessed using a birefringence readout, and usually do not survive beyond 9dpf (days post-fertilisation). Using this assay the Prestwick Chemical Library (1120 compounds) was screened, and seven compounds found to give a birefringement readout equivalent to wild-type, although further analysis using an antibody to dystrophin established that this effect was not due to restoration of dystrophin production.

Interestingly, when 4dpf dystrophin null zebrafish were treated with the seven compounds of interest, one, the non-selective PDE inhibitor aminophylline (**11.41**, Figure 11.31) restored normal muscle pathology and approximately 60% survived to 30dpf. Furthermore, increases in PKA activation were seen in aminophylline-treated fish, presumably due to the effect of a non-selective PDE inhibitor on cyclic AMP levels. Testing of a range of other PDE inhibitors then followed, with the PDE5 inhibitor sildenafil **11.21** giving similar results to aminophylline **11.41**. These results confirm the published work on sildenafil in the *mdx* mouse (see Section 11.2.4.4),[58] and illustrate the potential for the use of phenotypic assay systems for rare muscle diseases using model organisms such as zebrafish.

Taken together, the significant amount of novel research now being undertaken for new supportive care drugs and non-symptomatic treatments, such as utrophin upregulation therapies for DMD, clearly highlights the paradigm shift towards the development of therapies for orphan diseases that has taken place in drug discovery organisations in recent years. For the first time this offers real hope for all DMD patients that this devastating disease will finally be genuinely treatable, with a range of therapeutic options in the foreseeable future.

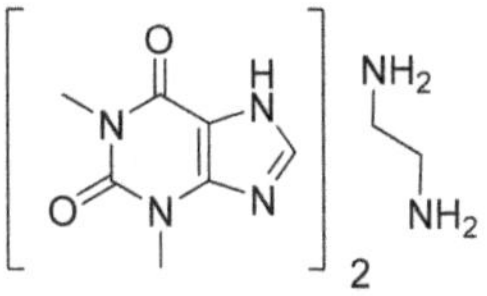

11.41, aminophylline

Figure 11.31 Aminophylline, a PDE inhibitor hit from a phenotypic zebrafish screen.

11.3 Spinal Muscular Atrophy

11.3.1 Introduction

Spinal muscular atrophy (SMA) is an autosomal recessive disorder, caused by a deficiency in the survival motor neuron (SMN) protein.[185] It is less common than DMD, occurring in approximately 1 in 6–10 000 live births, although a sobering fact is that SMA is thought to be the leading genetic cause of infant death worldwide. The gene is estimated to be carried by around 1 in every 40–50 adults, and in the UK alone there may be as many as 1 million carriers.[186] As with DMD it is a devastating, extremely debilitating and progressive disease with no known cure.

The SMN protein is the product of the *SMN* gene, with defects in the gene being identified as the causative effect of SMA in 1995.[6] Humans, unlike most organisms, have two copies of the *SMN* gene, *SMN1* and *SMN2*, both of which can code for the protein. The two genes are almost identical in sequence, although while *SMN1* produces full-length, functional SMN protein, the majority (~85%) of the protein produced from *SMN2* is truncated, less stable and non-functional. This is due to a deletion of exon 7 (denoted as SMNΔ7) in *SMN2* which results from a C to T transition, the end result being a modification in splicing. Individuals suffering from SMA have deletions or mutations in these genes, with loss of protein produced from expression of the *SMN1* gene being the critical factor in occurrence of SMA. The small amount of full-length protein produced from the *SMN2* gene (*i.e.* the 15% or so which is correctly spliced and includes exon 7) can compensate to a small degree for the loss of *SMN1*, but is insufficient to compensate fully. Disease severity is known to correlate with copy number of (and therefore the quantity of protein produced from) the *SMN2* gene.

SMN is a 294 amino acid, ubiquitously expressed protein with a mass of approximately 38 kDa. It is known to have in excess of 40 protein binding partners, and is found at varying levels in most tissues, with particularly high concentrations being localised in the brain and spinal cord. Although SMA in its various forms presents as a muscle-wasting disease, characterised by muscle weakness, formally it is a neurodegenerative disorder arising due to degeneration of motor neurons (particularly those of the spinal cord and CNS).

Protein produced from *SMN1* appears to be essential for survival of motor neurons, with absence of the SMN protein resulting in motor neuron cell loss. The reasons for this are not clear, although it is thought to play a role in mRNA splicing. At a cellular level, SMN is found both in the cytoplasm and nucleus. In the latter location it is concentrated in so-called 'gem' ('Gemini of the coiled bodies') structures.[187] Quantification of these 'gems' is often used as a surrogate readout in screens for compounds which increase levels of SMN protein.

The physiological characteristics of SMA are muscle wasting and impairment of mobility, the severity of which inversely correlates with the copy number of the *SMN2* gene,[188] and amount of SMN protein that is produced.[189] Sufferers of the disease are categorised as having one of four subtypes of SMA (Table 11.3). Type 1 SMA is the most serious, with around 50% of sufferers falling into this category.

Unanswered questions relating to the development of therapeutics to treat SMA are many, and primarily include challenges relating to the underlying pathology. For example, it will be critical to investigate and clarify pharmacodynamic considerations such as the optimum patient age and duration for dosing drugs, as well as how much SMN protein will be sufficient to provide therapeutic benefit. As with DMD, problems associated with the progressive nature of the disease and a heterogeneous patient population coupled with selection of an appropriate duration for testing of putative drugs has caused significant complications in both study design and interpretation of data from *in vitro* cellular systems all the way through to clinical trials.[190,191] In a number of animal studies it has been demonstrated that the timing of the treatment has a critical effect, and while the use of other model organisms is increasing,[192,193] and suggestions have been made for standardised animal testing protocols in order to attempt to homogenise the various discovery efforts throughout the scientific community,[194] no one paradigm currently appears to predominate.

11.3.2 Therapeutic Approaches

Current treatment options for the most severe forms (types 1 and 2) are limited to supportive care, and although there has been much work in recent years aimed at finding an effective therapy,[195–198] to date no curative small-molecule or other treatments have received regulatory approval.[199]

As with DMD, therapeutic agents aimed at treating SMA can be divided into two categories: biological therapies, principally cell- and gene-based, are under intense investigation, with small-molecule approaches also receiving a great deal of attention. Gene therapies are intended to replace the missing *SMN1* gene. Because SMA only manifests when both alleles of *SMN1* are missing, replacement of only one of the alleles would be expected to provide therapeutic benefit and be curative. Although intravenous delivery of *SMN1* transgenes into SMAΔ7 mice has been demonstrated,[200] and indeed resulted in a significant increase in lifespan (27 days in untreated to over 340 days for treated mice), this paradigm has not yet been translated into human studies.

Stem cell therapies, particularly those intended to replace the missing cells, have been described.[201] Studies using direct delivery of embryonic neural stem cells *via* intrathecal injection have been shown to provide benefit in murine models.[202] Studies in which direct injection of motor neuron clusters, generated from embryonic stem cells, has taken place have also been pursued, although there is a clinical hold on this study from the FDA at present.[203] In all stem cell-based therapeutic approaches the production of

Table 11.3 Comparison of SMA types.

Classification	*Type 1*	*Type 2*	*Type 3*	*Type 4*
Pseudonym	Werdnig–Hoffman disease	—	Kugelberg–Welander disease	Adult onset
Severity	Severe	Intermediate	Mild	Very mild
Age of onset	<6 months	7–18 months	Childhood to adulthood	20–30 years
Age of death	<2 years	Adolescence	~Normal	~Normal
Major symptoms	Extensive muscle weakness, including respiratory and skeletal	Difficulty standing, poor breathing, swallowing, digit tremors	Variable, ranging from difficulty walking to only minor muscle problems	Mild motor impairment

appropriate grade cells is a continued technical and economic challenge. The potential for teratoma formation is also often highlighted as a concern for embryonic stem cell (ESC) derived therapies; furthermore SMA cell-based therapies would typically require spinal (intrathecal) injection of the stem cells. These and any other allogenic therapies also carry the attendant risk of immune rejection, or the requirement for immunosuppression, which may cause further problems in the targeted patient population.

Notwithstanding these challenges, other stem cell-based approaches are expected to have a significant impact on SMA as well as a multitude of other diseases.[204,205] Of particular note is the use of patient-derived induced pluripotent stem cells (iPSCs), which are anticipated to provide a fully disease-relevant assay system,[206,207] thereby negating the caveats associated with the use of reporter readouts or cells from less disease-relevant tissue types.[208] It could reasonably be anticipated that the significant technological advances arising from the use of iPSCs may have most impact in historically challenging neurodegenerative disorders such as SMA.[209] In the case of SMA these would be expected to overcome many of the major limitations in existing disease model systems. For instance, many model organisms such as mice and flies lack the *SMN2* gene and therefore therapies based on modulating SMN expression cannot be evaluated in these systems without prior genetic manipulation/knock-ins, *etc.* Moreover, the development of defined differentiation protocols allows for the production of specific cell types of interest in the disease pathology (*e.g.* motor neurons), negating the reliance on more readily accessible, but less disease-relevant patient cell types (*e.g.* fibroblasts).[208]

In 2009, in the first study of its kind, Ebert *et al.* described the generation of iPSCs from skin fibroblasts from a SMA patient and compared them to iPSCs derived from his non-SMA mother. They went on to differentiate these iPSCs into motor neurons without loss of genomic integrity and were able to show characteristic deficiencies in the patient-derived cells, most notably decreased neuron production and/or increased degeneration after culturing for 6 weeks. Importantly they were also able to demonstrate that compounds already reported to increase SMN levels, such as valproic acid[210,211] and tobramycin,[212] were able to selectively increase SMN levels and nuclear gem number in SMA-iPSC-derived motor neurons, effectively validating the approach as a screening strategy for SMA.[208] Providing the not insignificant technological challenges involved in assay development are overcome, this would in principle allow for the screening of large compound collections in order to identify new therapeutics for SMA. Most of the current assay systems used, while allowing for HTS campaigns to take place, have attendant disadvantages, particularly limited disease relevance, so the use of iPSC derived systems would represent a significant step forward for the drug discovery community.[213]

Small-molecule therapeutics for SMA are being evaluated with increased frequency, and summaries of the various approaches, along with potential pitfalls, have been published in several review articles.[214–216]

11.3.2.1 Neuroprotective Agents

Repurposing of known drugs has emerged as a new drug discovery paradigm and can be defined as identifying an alternative, unanticipated therapeutic use for a drug which may be connected either with its originally intended pharmacology, or a previously undiscovered off-target (beneficial) biological activity.[52] Reprofiling is now a common strategy for companies seeking to develop therapies in many disease areas, with SMA being no exception. This approach could be envisaged as being particularly advantageous because toxicological studies in man will have already taken place for the compounds/agents in question. In the main, repurposing efforts for SMA have been focused on drugs already known to have an effect in other degenerative disorders (Figure 11.32).

For example, riluzole **11.42**, a benzothiazole derivative marketed by Sanofi-Aventis for the treatment of ALS (amyotrophic lateral sclerosis) has shown activity in mouse models of SMA, giving around a 15% increase in lifespan (38 days, as opposed to 33 days for SMNΔ7 mutants),[217] and as a result was taken into clinical trials. Results from a small Phase 1 clinical trial were reported in 2003, and although the study was not powered sufficiently to establish significant efficacy, encouraging results were noted, in that several patients were still alive many months after dosing started.[218] Although no data have yet been published, at least one additional trial appears to be under way (type 2/3 SMA, Clinicaltrials.gov NCT00774423).

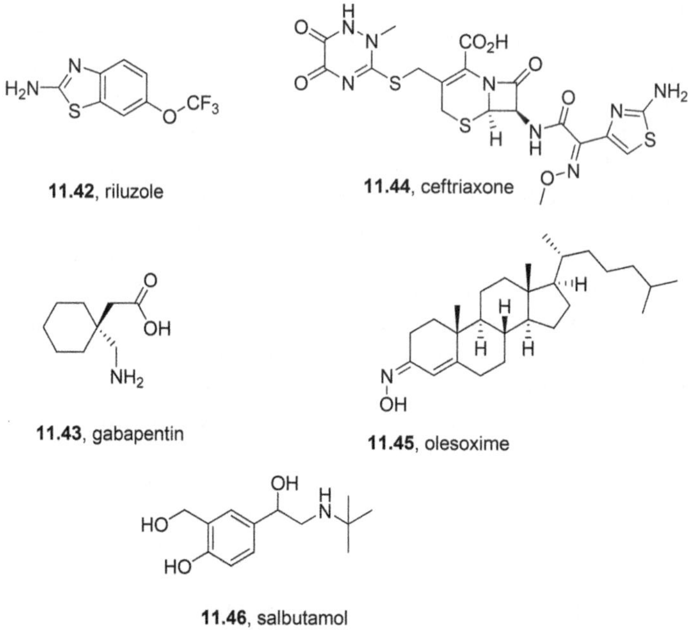

11.42, riluzole

11.44, ceftriaxone

11.43, gabapentin

11.45, olesoxime

11.46, salbutamol

Figure 11.32 Neuroprotective agents.

Riluzole is a relatively promiscuous small molecule, having multiple pharmacological activities associated with it, including acting as an ion channel blocker and disrupter of glutamate signalling. Through these modes of action it is thought to exert its action as a neurotrophic factor, promoting the growth, survival and maintenance of motor neurons.[219]

Gabapentin **11.43**, a well-known GABA analogue primarily used to treat neuropathic pain, has been investigated in early stage clinical trials for SMA in recent years,[220] and was found to have some functional benefit in various tests of muscle strength and dexterity. The mode of action of the compound remains to be elucidated fully, but was suggested as being at least in part due to a neuroprotective effect.[221]

Beta lactam antibiotics have also been reported as providing therapeutic benefit in SMA models.[222] Multiple mechanisms of action are thought to be involved, with the overall effect provided being thought to be one of neuroprotection. For example, the beta lactam ceftriaxone **11.44**, when dosed to SMAΔ7 mice, was found to provide a 25% increase in lifespan. This encouraging level of efficacy was supported by increased levels of both SMN protein and mRNA. Despite these positive data, and somewhat surprisingly given the extensive body of data on the dosing of beta lactams in man, no corresponding investigations in human trials for SMA have yet been reported with this class of drug, although encouraging safety and exposure data was recently published for use of the drug as a neuroprotective to treat ALS.[223]

Olesoxime (**11.45**, Trophos, TRO19622) is a steroid derivative which is also purported to act as a neuroprotective agent, and has been evaluated for the treatment of ALS. More recently, it has been studied as a potential treatment for other neurodegenerative disorders; of particular note was the recent disclosure that it is currently in Phase 2 clinical trials in SMA patients (approximately 150 patients with SMA type 2 and 3; Clinicaltrial.gov NCT01302600), although no results have yet been published.[224]

Finally, β2 adrenergic receptor modulators have been shown to act as neuroprotective agents. One of the more recent publications in this therapeutic compound class, involving studies using both *in vitro* and patient dosing, has described the use of the bronchodilatory drug salbutamol **11.46** as a possible treatment for SMA. Treatment of SMA patient fibroblasts with the compound was found to increase the amount of full-length SMN transcripts (*i.e.* those containing exon 7), although the mechanism through which this occurs is unclear.[225] A subsequent clinical trial in type 2 and 3 SMA patients in which subjects took daily oral doses of salbutamol over 6 months confirmed these results *in vivo*, although no efficacy data were included.[226]

11.3.2.2 Approaches that aim to Restore SMN Protein Production

11.3.2.2.1 Read-Through and Exon-Skipping Agents. The use of read-through agents has also been studied in order to identify possible therapeutics for SMA, with the first publication in 2005 of increases in SMN protein level in SMA patient fibroblasts following treatment with the aminoglycosides tobramycin **11.19** and amikacin **11.20**.[212] Following this work, a selection of other aminoglycoside antibiotics including geneticin **11.17** [227] and TC007 [228,229] have also been evaluated and found to have effects on SMN protein levels using *in vitro* and *in vivo* assays. Efficacy was noted both when the compounds were administered directly into the CNS (if the blood–brain barrier penetration of the compound was unknown), or *via* intraperitoneal injection. As with DMD, the continued development of aminoglycosides for SMA remains associated with a number of unanswered questions relating to the potential for toxicological problems following chronic dosing, of particular concern in a paediatric patient population. Whether this will indeed prove to be the case remains to be seen following appropriately designed long-term studies.

The use of anti-sense oligonucleotides that can restore the production of full-length SMN transcript from the *SMN2* gene is under active investigation for SMA. This approach aims to alleviate the usual propensity of *SMN2* to produce only around 15% of functional SMN protein, due to the C to T mutation in exon 7. Encouraging results have been reported in various animal studies, although no clinical evaluation has yet taken place in human patients.[230]

11.3.2.2.2 Modifiers of Gene Splicing. While a significant number of compound classes have been identified which increase the level of *SMN2* RNA and full-length SMN protein, many have been identified using cell-based assays, and accordingly it has often proven difficult to attribute the compound's activity to a specific molecular target, or even to define more details surrounding its potential mode of action. Recent work from Krainer *et al.* has described a series of tetracycline derivatives (**11.47**, Figure 11.33) which were shown, using a cell-free assay system, to promote exon 7 splicing

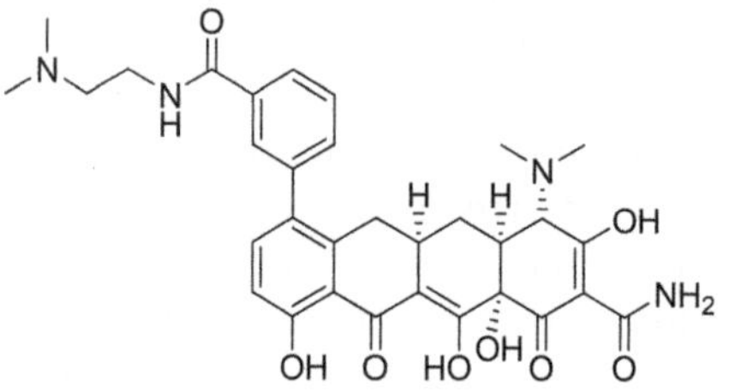

11.47, PTK-SMA1

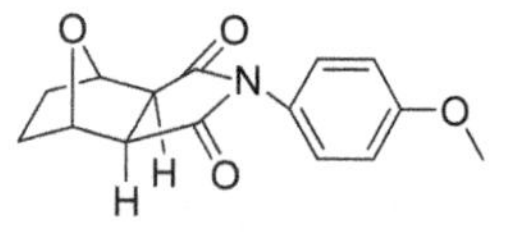

11.48, Pseudocantharidin A

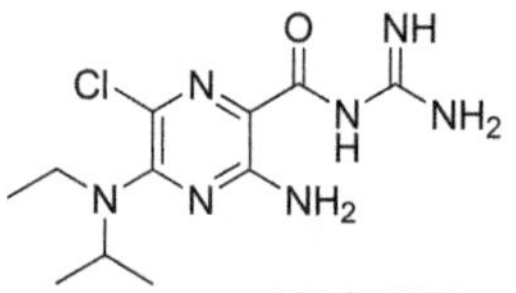

11.49, EIPA

Figure 11.33 Compounds reported to modify *SMN2* gene splicing.

of *SMN2*, and thence to increase levels of full-length SMN protein in cell-based assays. Encouraging levels of efficacy in two different mouse models of SMA (type 1 and type 3) were also described, in which increases in exon 7 spliced RNA, as well as full-length SMN protein, should be following intraperitoneal dosing.[231]

The key differentiator with this work was that the compounds were established to have a defined mode of action, although the specific molecular target within the splicing mechanism was not identified.[232]

Other compound classes including phosphatase modulators **11.48** [233] and Na^+/H^+ membrane ion exchange inhibitors such as EIPA **11.49** [234] (Figure 11.33) have been described, which are purported to increase SMN protein levels by acting on the *SMN2* splicing mechanism, although the possibility of differing mechanisms in some cases was acknowledged, and whether the effects were direct or indirect was also unclear.

11.3.2.2.3 Histone Deacetylase Inhibitors and other Epigenetic Regulators. It could reasonably be anticipated that HDAC inhibitors could have either a direct effect on expression levels of SMN protein itself, or alternatively by modifying the expression of other genes which indirectly affect its expression. Although studies have been undertaken which have linked levels of histone acetylation, and particular HDAC enzymes (for example, HDAC2) to expression of the SMN protein,[235,236] these findings have yet to be translated into clinical efficacy. More recently, HDAC enzymes have been shown to regulate protein stability,[237] which may have some bearing on the use of inhibitors as therapeutics for diseases such as SMA and DMD.

In vitro studies using SMA patient-derived cell lines with older, clinically approved agents such as the short chain carboxylic acid derivatives sodium butyrate **11.50**,[238] phenyl butyrate **11.51** [239] and valproic acid **11.52** (Figure 11.34)[210,211] have been reported, and increases in SMN protein expression noted. These results attracted a great deal of interest, and led to the proposal that HDAC inhibitors may offer a new therapeutic approach to SMA because the compounds have already been evaluated in clinical trials.[240] Of these 'first generation' histone deacetylase inhibitors, valproic acid **11.52**

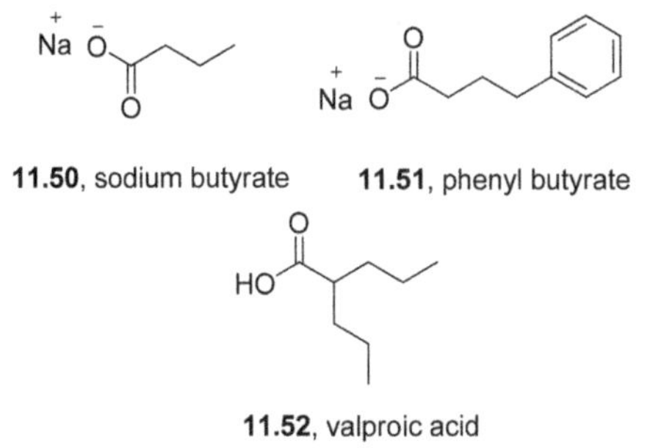

11.50, sodium butyrate **11.51**, phenyl butyrate

11.52, valproic acid

Figure 11.34 First-generation histone deacetylase inhibitors.

appears to have attracted most attention as a possible treatment for SMA.[241] Preliminary indications of efficacy were seen in two small Phase 1 trials,[242,243] and a recent review has summarised the scope, limitations and possible side effects associated with its use.[244] Following a recent Phase 2 evaluation of **11.52** in a small group ($n = 42$) of SMA patients of mixed type 1/2/3, the compound appeared to be well tolerated, although side effects including weight gain and carnitine depletion were noted.[245] While there was evidence of improved motor function, the study clearly demonstrated that use of a more heterogeneous patient population caused complications when analysing the results. Accordingly, a number of recommendations were made based on the aforementioned points, including close monitoring of carnitine levels (and possible supplemental dosing during the trial to correct the deficit), along with greater patient stratification relating to SMA disease subtype. While only limited data has been published to date,[246,247] Phase 3 clinical trials with valproic acid and carnitine co-dosing at least appear to be planned.[248]

More recently, newer second-generation HDAC inhibitors have also been investigated in murine and cellular models of SMA and shown a positive effect (*e.g.* increases in SMN RNA and/or protein levels).[249] These are compounds which are already either approved or in the clinic for other indications, and include members of both the hydroxamic and bis-aniline classes, for example vorinostat (SAHA) **11.37**,[250] M344 **11.53**,[251] panobinostat (LBH589) **11.54**,[237] trichostatin A **11.36**,[252] as well as entinostat (MS-275) **11.55** (Figure 11.35). As yet, however, none have been progressed to clinical evaluation for SMA, possibly due to the fact that adequate CNS penetration remains to be demonstrated, this being a critical factor for a drug to treat

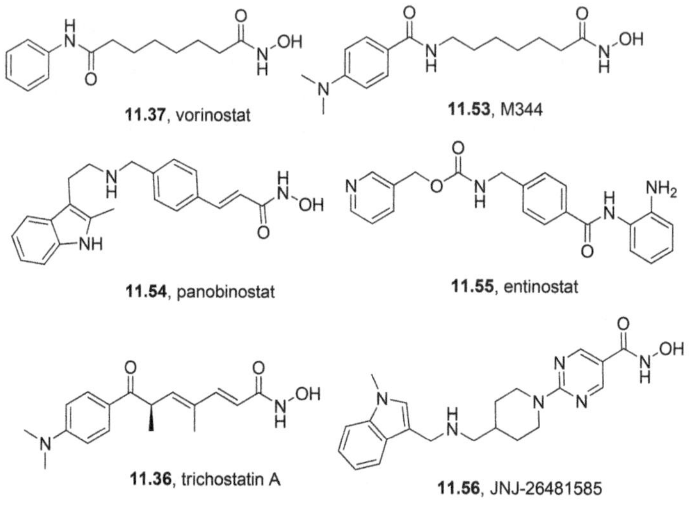

11.37, vorinostat

11.53, M344

11.54, panobinostat

11.55, entinostat

11.36, trichostatin A

11.56, JNJ-26481585

Figure 11.35 Second-generation histone deacetylase inhibitors.

SMA. Furthermore, recently published data for the pan-HDAC inhibitor JNJ-26481585 **11.56** demonstrated that for this compound, while increases in levels of SMN protein were seen *in vitro*, this effect did not translate to *in vivo* efficacy in animal models.[253] Whether or not this was due to a pharmacokinetic/pharmacodynamic effect, HDAC isoform selectivity, or related to the timing of dosing was not clear, and would require further investigation.

Given the pan-HDAC inhibitory activity of the majority of these compounds, as well as numerous possible off-target effects for the older agents, it seems likely that more detailed studies of isoform selectivity for the various HDAC classes will be needed, along with evaluation of the impact of this, in particular the downstream effect on production of SMN protein in order to determine if this class of agents represents a viable therapeutic option for SMA patients. Furthermore, the use of agents originally designed for oncology indications in a potentially chronically dosed paediatric indication seems optimistic, and will in all likelihood require more selective compounds with a significantly cleaner toxicological profile than that seen with compounds thus far. Furthermore, the issue of blood–brain barrier (BBB) penetration remains an important issue to be addressed.

Other compounds which have been established as being able to modulate gene expression have been investigated as potential therapies for SMA. For example, hydroxyurea (**11.57**, Figure 11.36) has been reported to increase both the amount of exon 7 containing SMN protein,[254] as well as 'gem' number in SMA cells, although this positive *in vitro* data did not translate into efficacy when evaluated in Phase 2 clinical studies.[255]

As well as its primary reported mode of action as a ribonucleotide reductase inhibitor, it is possible that it may have a number of additional pharmacological modes of action; accordingly a more detailed investigation would be instructive.

Interestingly, despite their known effects on gene expression, other epigenetic modulators such as bromodomain binding compounds have not yet been studied in genetic diseases such as SMA. Given the encouraging results seen so far with HDAC inhibitors, coupled with the increasing availability of information on each of the various compound classes,[256] evaluation of the effects of these and alternative epigenetic modulators[257] in cellular and animal models of SMA would be justified.

11.3.2.2.4 Proteasome Inhibitors and Stabilisation of SMN protein. Proteasome inhibitors such as MG-132 **11.58** (Figure 11.37) have been evaluated as a potential therapeutic for SMA, the rationale being that they should

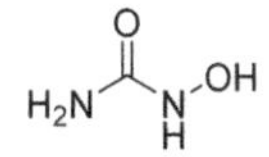

11.57, hydroxyurea

Figure 11.36 Hydroxyurea, a putative gene expression modulator.

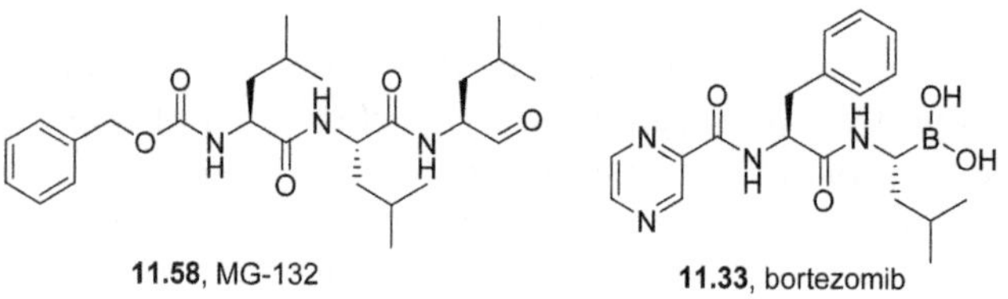

11.58, MG-132

11.33, bortezomib

Figure 11.37 Proteasome inhibitors studied for SMA.

stabilise any SMN protein that is produced, with the result of negating the disease pathology.[258]

The approved drug bortezomib **11.33** has also been studied.[259] While *in vivo* evaluation was found to improve the disease pathology and deliver functional improvement in terms of motor function, no increase in survival was noted in SMA model mice. Interestingly, when this agent was combined with a histone deacetylase inhibitor in a murine model, an additive effect was seen on SMN expression levels, along with significantly enhanced survival.

11.3.2.2.5 Growth Factor Signalling. Growth factors and their modulators have also been reported as having an effect on SMN protein levels (Figure 11.38). For example, platelet derived growth factor (PDGF) has been shown to increase SMN expression levels, and conversely, PDGFR inhibitors (*i.e.* compounds which bind to and inhibit autophosphorylation mediated by the intracellular kinase domain) have been shown to decrease SMN protein levels. Other well-studied inhibitors of important intracellular signalling pathways, including LY29002 (LY294002) **11.59**,[260] PI-103 **11.60**, alsterpaullone **11.61**, Y-27632 **11.62** [261] and fasudil **11.63** [262] have also been shown to modulate the expression of SMN protein, with the up- or downregulation being dependent on which pathways are being inhibited. These compounds will be discussed in more detail later.[215]

As with HDAC inhibitors, however, wider biological profiling of the aforementioned compounds has shown them to be relatively non-specific in their inhibitory activity. Therefore, definitive conclusions about the effect of particular pathways on SMN levels or the cross-talk between them cannot be drawn at this stage, although alternative chemotypes regulating these pathways should be explored further. Furthermore, if increasing the activity of these signalling pathways is shown unambiguously to increase levels of SMN protein, this would be anticipated as providing a significantly more challenging target, because few small-molecule, non-peptidic analogues have been described as demonstrating an agonist effect on extracellular growth factor receptors.

In an interesting recent development, celecoxib **11.64** has been shown to increase SMN levels and survival in a mouse model of SMA.[263] The study was predicated on the demonstrated ability of celecoxib to activate p38,[264] and given both the favourable pharmacokinetic profile of the drug (particularly its blood–brain penetrative ability) and the effect of stimulation of the p38

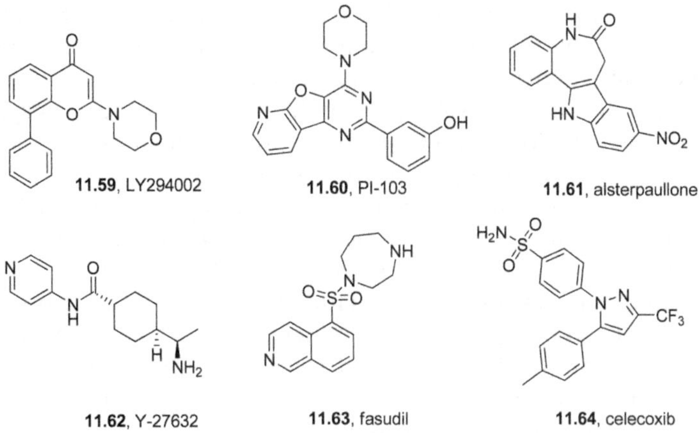

11.59, LY294002 **11.60**, PI-103 **11.61**, alsterpaullone

11.62, Y-27632 **11.63**, fasudil **11.64**, celecoxib

Figure 11.38 Growth factor signalling pathway modulators.

pathway on SMN levels, further development of this agent for SMA will be worth watching.

11.3.2.3 HTS Approaches to Increasing SMN Levels

Based on the known link of increased copy number of *SMN2* to increased levels of SMN protein and reduced severity of the SMA phenotype, a number of assay systems have been described to allow for screening of compounds which activate the *SMN2* promoter, particularly to promote the expression of full-length SMN protein, and are also amenable to high-throughput strategies. In 2001 Zhang *et al.* described details of an assay system which, for the first time, could stimulate exon 7 inclusion into the otherwise truncated *SMN2* mRNAs.[265] Others have subsequently used this assay system to screen a range of compounds as potential agents to treat SMA, including Sakla and co-workers, who identified a number of polyphenolic natural products (**11.7, 11.65, 11.66,** Figure 11.39) which increased the amount of *SMN2* mRNA transcripts with an intact exon 7.[266]

Due to the well-characterised pleiotropic effects of compounds with this type of extensively hydroxylated arene structure, it seems unlikely that these and related compounds will progress beyond use as biological tools, but they may be suitable for exploitation as mechanistic probes, or even represent start points for chemical optimisation. Interestingly, resveratrol **11.7** has been identified in other studies as having possible therapeutic potential for SMA, albeit through different proposed mechanisms.[267] For example, its previously described antioxidant properties may be responsible, at least in part, for any therapeutic benefit.[30]

Although details of the original screen were not described, Lo and co-workers have investigated the activity of the unusual polyepoxide containing natural product triptolide (**11.67**, Figure 11.39) in murine and SMA patient fibroblast cells.[268] Following intraperitoneal administration of the compound to SMA-like mice, indications of mechanistic and functional activity were seen through increases in SMN protein levels. Interestingly, despite the presence of reactive electrophilic functional groups such as the triepoxide and butenolide motifs, no apparent toxicity was noted.

An alternative assay using an SMN–luciferase reporter system has also been described.[269] The protocol is amenable to high-throughput optimisation, and consequently following screening of 47 000 compounds from various chemical classes, indoprofen (**11.68**, Figure 11.40) was subsequently validated as a hit, showing a very modest increase (~10%) in SMN protein in SMA fibroblasts. Although there was a trend to improved survival noted following *in vivo* dosing in SMA mice, this was not significant. Interestingly, no effects on mRNA levels were noted, leading the authors to the conclusion that the effects were post- or co-translational, rather than at the transcriptional

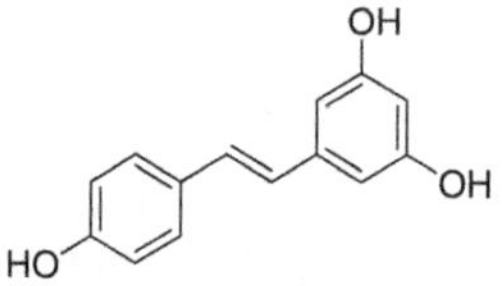

11.7, resveratrol

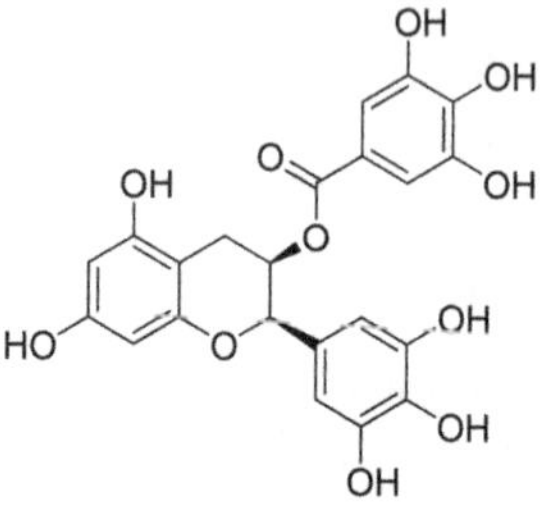

11.65, EGCG

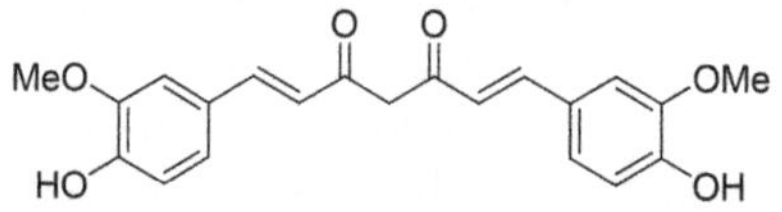

11.66, curcumin

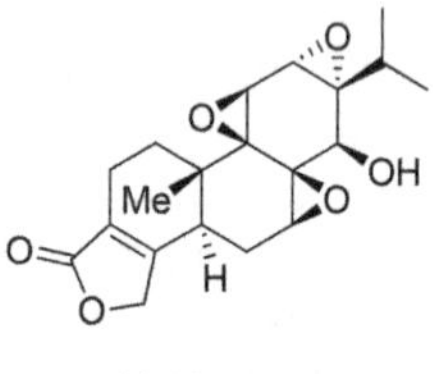

11.67, triptolide

Figure 11.39 Natural product hits from SMA drug screens.

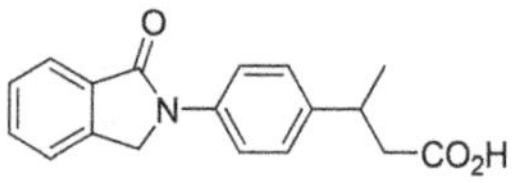

11.68, indoprofen

Figure 11.40 Indoprofen, screening hit for SMA.

level. Because the effects on SMN protein level were relatively small, further studies would be required before drawing firm conclusions on the underlying mechanism(s).

No other non-steroidal anti-inflammatory drugs or COX inhibitors tested showed this effect, suggesting the mechanism of action is independent of cyclooxygenase inhibition. Extensive delineation of compound SAR was not disclosed in this paper, although some limited data has been presented at a scientific meeting.[270]

The most advanced SMA drug discovery project to date, in which a potential new therapeutic agent has been identified from chemical library screening and subsequently optimised, is likely to be that originating from a collaborative venture between workers at Aurora Biosciences and Ohio State University. Initially, researchers designed a reporter-based screening assay to identify compounds such as the anthracycline aclarubicin (**11.69**, Figure 11.41), which could increase levels of *SMN2* mRNA in murine NSC-34 motor neuron cells.[271]

Approximately 558 000 unique compounds were then screened using an HTS strategy, and following deconvolution and dose–response analysis, 17 compounds were confirmed as hits.[272] The hit compounds were assigned to nine distinct structural classes, although only six (consisting of 10 compounds) were deemed 'drug-like' enough to be considered for follow-up evaluation. No details of the criteria which qualified these six classes as preferable for further evaluation were described.

Next, secondary screening assays designed to directly assess production of both full-length *SMN2* mRNA and SMN protein were conducted using cells from type 1 SMA patients, after which only two compound series remained – an indole analogue **11.70**, and a small series of quinazoline derivatives exemplified by **11.71** (Figure 11.41).

Although both series were felt to represent viable start points for medicinal chemistry optimisation, only work on the latter series appears to have been undertaken. The project represents a collaborative drug discovery programme between deCODE Genetics Inc. and The Families of SMA (FSMA), and is arguably one of the most advanced '*de novo*' drug discovery programmes directed towards discovering a cure for SMA.

Optimisation of the 2,4-diaminoquinazoline analogues has been published recently, and described the structure–activity studies which led to the discovery of lead candidate D156844 (**11.72**, Figure 11.41). This compound, one of approximately 1000 made during the project,[273,274] was highly active in the β-lactamase reporter assay used ($EC_{50} = 4$ μM), giving a 2.3-fold increase in signal, and increasing SMN levels in SMA patient fibroblasts in a dose-dependent fashion. Most notably, and (so far) uniquely for this disease, other key compound data were described including pharmacokinetic parameters, and systemic and central plasma and tissue compound levels, all of which were found to be extremely encouraging. Illustrating a typical challenge faced during a drug discovery programme, the off-target pharmacology of compounds was also assessed, and examples of the 2,4-diaminoquinazoline

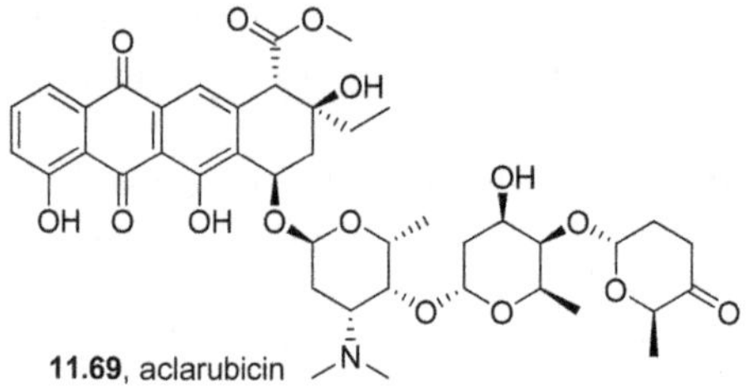

11.69, aclarubicin

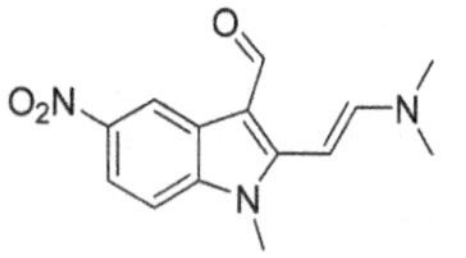

11.70, indole hit

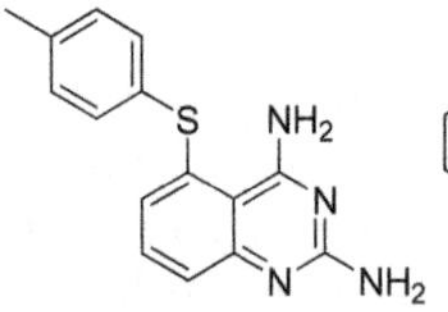

11.71, quinazoline lead

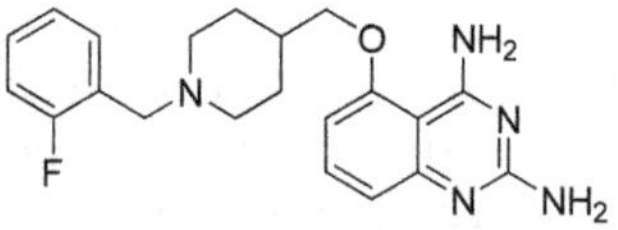

11.72, D156844

Figure 11.41 Evolution of the first series of SMA HTS leads to progress to clinical trials.

series were established as having potent inhibition of DHFR, which led to ATP depletion. Accordingly, this had to be designed out of subsequent analogues. Of particular note was the additional publication of exposure and efficacy data for several analogues in the series.[275]

Following further lead optimisation efforts a compound denoted as Quinazoline495 was identified and nominated for progression into human clinical trials. This compound has been described as having a similar chemical structure to quinazoline **11.72**, but exact details have not yet been published. As well as giving high brain levels of SMN, it demonstrated significant efficacy in two mouse models of SMA.[276] Protein microarray analysis was carried out using a ^{125}I-radiolabelled molecular probe from the quinazoline class in order to investigate putative molecular targets for the compound.[277] In-depth follow-up studies, including X-ray crystallography, were conducted, and from this it was established that the compound was binding to DcpS, which is an RNA decapping enzyme involved in RNA metabolism. Following licensing of this compound to Repligen Corp., FDA approval for Phase 1 clinical trials of the drug (now called RG3039) was granted in 2011, and the first cohorts of this double-blind, single-dose ascending study were anticipated to be undertaken during the first half of 2013. Repligen has very recently announced that it has signed a licensing agreement with Pfizer to further develop the programme.[278] According to company documentation, Pfizer is expected to assume responsibility for the continued clinical development of RG3039 and back-ups following completion of the first Phase 1 cohorts. Because a molecular target for the compound has now been identified and crystal structures are available, it could reasonably be anticipated that development of next-generation compounds using both screening and computational drug discovery platforms will follow.

Details of an alternative firefly luciferase-based *SMN2* reporter cell assay have also recently been published.[279] The primary difference between this and the previously described test system[269] is that the new cell line was designed to detect compounds that are able to modulate SMN expression *via* activation of the *SMN2* promoter. In contrast, the groups previously reported assay measured increases based on splicing alone. Using this novel assay a collection of 115 000 compounds were screened, which resulted in the identification of 462 hits. Confirmatory screening removed a significant proportion of hit molecules, leaving 294 compounds which fell into 19 structural classes. Dose–response analysis on authentic solid material obtained from commercial sources then further trimmed the number of hits to three, all of which had acceptable activity; pyrazole LDN-72939 (EC_{50} = 1.1 μM), benzofuran LDN-79199 (EC_{50} = 0.75 μM) and benzo[*cd*]indol-2(1*H*)-one LDN-109657 (EC_{50} = 2.4 μM) (**11.73–11.75** respectively, Figure 11.42).

Following this confirmatory screen, the compounds were evaluated for their ability to upregulate the expression of SMN protein in patient-derived 3813 fibroblasts. Of the three series, only **11.74** and **11.75** showed an increase in protein levels, with the results from **11.73** suggesting that it may have interfered in the readout of the primary reporter-based assay (*i.e.* been a false-positive).[280] Further development of the remaining two compounds has very recently been reported, along with detailed evaluation of further examples LDN-75654 **11.76** and LDN-76070 **11.77** (Figure 11.42).[281] The latter

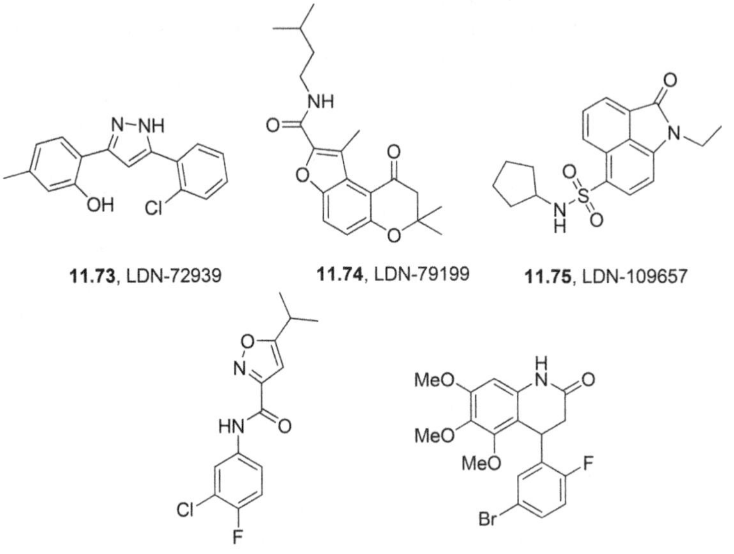

11.73, LDN-72939 **11.74**, LDN-79199 **11.75**, LDN-109657

11.76, LDN-75654 **11.77**, LDN-76070

Figure 11.42 Hit compounds from a splicing/SMN2 promoter-based HTS.

compound was of particular interest, because efficacy was seen following dosing in the *SMAΔ7* mouse model. Hits giving an EC_{50} in the range of 1–10 μM were identified using the same luciferase reporter assay, this being designed to identify compounds which upregulated SMN protein expression from the *SMN2* gene. In order to assess the specificity of the mode of action of the various compounds (*i.e.* that they were not just non-specific gene activators), selectivity assays were conducted using two additional cell lines with the *SMN1*–luciferase reporter and a non-specific SV40–luciferase reporter, respectively. Although all were found to exhibit some non-specific activation of luciferase using these screens, activity in the *SMN2*–luciferase reporter systems was generally more pronounced. Confirmation of increased levels of SMN protein production was obtained by Western blot analysis (because the assay produced an SMN–luciferase fusion protein, antibodies to luciferase were used as the detection system). The EC_{50} values obtained were found to be broadly in line with those observed in the primary luciferase reporter assay. In order to investigate the mechanism of action in more detail, PCR quantification of *SMN*–luciferase mRNA was carried out. A dose-dependent increase in both total and exon 7 containing transcripts, as well as luciferase mRNA was noted from cells treated with **11.77**, whereas only an increase in luciferase was seen with **11.76** – no increase in SMN transcript level was seen. The authors proposed that these results indicated differing mechanisms of action were in play for the different structural classes, with **11.77** acting at the transcriptional level, and **11.76** having activity at the post-transcriptional level. For any future studies it would be useful to investigate these in more detail, particularly whether they translate to non-genetically modified cells such as patient-derived fibroblasts as there are important implications for future progression of compounds, for example they may act in an additive manner and offer possible co-treatment options.

The effect of **11.76** and **11.77** was next investigated using SMA patient-derived cells, 3813 primary fibroblasts, where the ability of the compounds to modulate two important characteristics was evaluated. Initially, the effect on total SMN protein level was assessed following compound dosing for 3 days. Both compounds showed reasonable levels of activity, increasing the amount of SMN protein relative to control experiments by around 1.5-fold, although some sensitivity of the primary cells to compound was noted compared to the immortalised SMN–luciferase line. Secondly the effect on 'gem' formation (localisation of protein to nuclear bodies) at lower compound concentration was assessed. 'Gem' number is a common measurement used to quantify activity of compounds because it has been shown to both positively correlate with total SMN level in cells, and negatively with disease severity in a range of patient-derived fibroblasts. Importantly it is a 'per cell' based measure which, unlike the SMN–protein quantification method, is not influenced by any effects the compounds may have on cell proliferation.

Because it had been hypothesised that the compound classes may have differing modes of action, a novel experiment undertaken in these studies was an evaluation of combination effects. Evidence of improved efficacy was seen with several combinations of **11.76, 11.77** and also the potent HDAC inhibitor SAHA, although notably this work was only done in the *SMN2*–luciferase reporter system and no confirmatory studies at the protein level in the primary SMA fibroblasts was reported.

Finally, after a brief assessment of some ADME parameters using *in vitro* assays, evaluation of **11.76** and **11.77** using *SMNΔ7* mice was undertaken. Compound **11.75** was not studied, because it was found to have poor solubility and also proved to be synthetically intractable in the quantities required. Preliminary experiments showed **11.77** to give the most robust *in vivo* effect following daily intraperitoneal dosing (5 mg kg^{-1} or 20 mg kg^{-1}). A significant increase in SMN protein levels was observed, as well as a dramatic improvement in functional activity using the 'Time To Right' test, which is intended to provide an overall indication of motor function in the animals. Animal lifespan was increased relative to controls, by approximately 180%, which is also impressive. Preliminary results from **11.76** were variable and less convincing, so no functional efficacy was assessed with this compound. Given the significantly shorter *in vitro* half-life of **11.76**, the lack of efficacy may have been a result of rapid turnover *in vivo*, although surprisingly no *in vivo* pharmacokinetic data were reported to support this. Isoxazole-containing molecules such as **11.76** are known to be rapidly metabolised *in vivo*, with leflunomide being a well-characterised example.[282]

In related work Stockwell and co-workers have very recently described development and optimisation of an HTS which does not use reporter-based readouts, but instead directly measures protein levels produced in relevant cell lines – arguably a far preferable readout to a reporter-based assay. Using this screening system several compounds which upregulate the production of the SMN protein in SMA patient fibroblasts were identified.[283] This high-throughput ELISA (Enzyme Linked ImmunoSorbent Assay) based screening approach was run using a 384-well microtitre plate format, with approximately 70 000 compounds from two sub-libraries being evaluated for their ability to upregulate the production of SMN protein in 9677 SMA patient cells. Approximately 1000 compounds were identified as hits in the primary assay, following single concentration screening (5.33 μg mL^{-1} per well), a 'hit' being defined as a compound which showed a median upregulation of SMN greater than two standard deviations from the median of the untreated control screening wells. These hits were subsequently tested in dose–response analysis using three alternative SMA patient cell lines (232 and 3813 as well as the original 9677), and the 105 molecules which showed a discernable DR activity in at least one cell line were then selected for final deconvolution – this being based on those compounds which showed upregulation of SMN in two or more cell lines in the dose–response assay. Following confirmatory screening using Western blot analysis, three compounds were found to do this (**11.78–11.80**, Figure 11.43, EC_{50} values of 10–20 μM), representing an overall confirmed hit rate of 0.004%.

Of the hit compounds identified during the screening process, pyridyl ketone, denoted as 'cuspin 1' **11.80**, which upregulated the production of

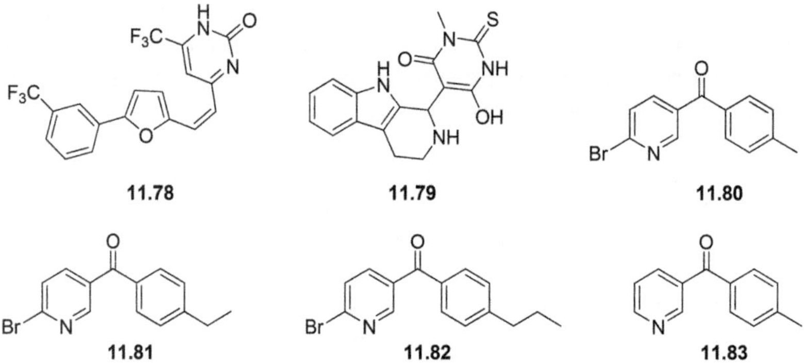

Figure 11.43 Hits obtained from an HTS campaign assessing protein levels.

SMN protein by approximately 60% above DMSO control in the primary assay after treating the cells for 48 hours with 10 μM compound, was of particular interest. A number of simple structural analogues of this compound were obtained, and screened for their ability to upregulate SMN production. While one example, **11.81**, substituted with a ethyl group rather than the methyl group present in the hit 'cuspin-1' **11.80** was active at approximately the same level, all other analogues were found to be inactive, even those with only relatively minor structural changes (examples **11.82** and **11.83**). Accordingly, the ensuing mechanistic studies were undertaken on compound **11.80**. Although two other hit compounds (**11.78** and **11.79**) had been identified in this study, it was not apparent whether the hit confirmation and subsequent dose–response analysis had taken place on authentic solid materials (either purchased or synthesised) for all examples. Further examination of their properties and mode of action would need to be carried out; this would be facilitated by the fact that multiple structural analogues appear to be commercially available. Mechanistic evaluation of the mode of action of hit compounds was next undertaken through probing the effect of compounds on common intracellular signalling pathways including Erk and PI3K. These studies indicated that they were exerting their effects through modulation of the Ras signalling pathway. The role of Ras in modulating SMN protein levels was supported by further studies using the constitutively active Ras +ve cell line BJeLR. This work also showed that Ras did not increase *SMN* mRNA levels, nor did it slow the rate of degradation of the SMN protein. A time course study revealed that rather the compound was exerting its effect through increasing the translation rate of SMN by approximately 2.5-fold relative to control cells.

Further mechanistic, structure–activity and ADME studies are clearly needed to establish more detailed information about the mode of action of this compound, and establish the full potential (or liabilities) of a mode of action involving Ras activation, given its well-known oncogenic potential.[284] Nonetheless, these early results provide encouraging precedent for the identification of more compounds which may be useful, and provide fruitful lines of research for drug discovery efforts directed towards the treatment of SMA. Furthermore, the assessment of SMN protein level changes in a direct manner using an SMA patient-derived cell line as the primary screen is likely to provide *direct* evidence for the utility of compounds in the intended application because it is a more clinically relevant assay.

Recently, Marugan and co-workers at the NIH have published the results of a screening campaign conducted using the NIH Molecular Libraries Small Molecule Repository (MLSMR).[285] This library, which contains over 210 000 compounds (all available in PubChem), was screened in a HTS assay designed to identify compounds which increased expression of full-length SMN protein, using a luciferase-based readout in HEK-293 cells. This assay construct differed from those used by other workers and utilised a novel strategy in that luciferase was only expressed when exon 7 was included in the mRNA transcript and full-length SMN protein was produced, thereby in principle allowing identification of compounds which worked using multiple mechanisms for full-length, stable SMN protein production, because if exon 7 was excluded from the transcript it would result in a frame-shifted *luciferase* mRNA and no production of luciferase protein.

A large number of hits were identified from the primary screen (6128 compounds, approximately 3%), and unsurprisingly this raised concerns amongst the co-workers about the occurrence of false-positives, particularly because this screen used a luciferase-based readout. The deconvolution process to remove putative false-positives involved a number of steps. These included sourcing solid or solution samples of selected hits and evaluating them in additional biochemical and enzymatic assays, for example using a *SMN1*-luciferase counter-screen to remove non-specific gene activators, as well as purified luciferase enzyme itself to assess luciferase inhibition/stabilisation. Furthermore, chemoinformatic triage played an important role. This latter step was intended to remove singletons, compounds with undesirable physiochemical properties or structural motifs, although further details of these steps were not provided.

Ultimately, 21 compounds were selected for screening in SMA patient-derived fibroblast cell assays which would directly measure SMN protein level, and not use a readout linked to luciferase activity. The authors have deposited full results from these screens in PubChem. Two compound classes were selected for study in more detail. These were the triazine **11.84** and aminothiazole **11.85** classes (Figure 11.44), which as well as showing encouraging levels of activity in the initial luciferase-based *SMN2* assay, also gave dose-dependent increases in the level of SMN protein in primary SMA patient fibroblasts.

However, more detailed analysis of triazine **11.84** showed that it inhibited firefly luciferase, leading to concerns of luciferase stabilisation occurring in the reporter assay. It was also found to have a very steep SAR, and was therefore not pursued further.

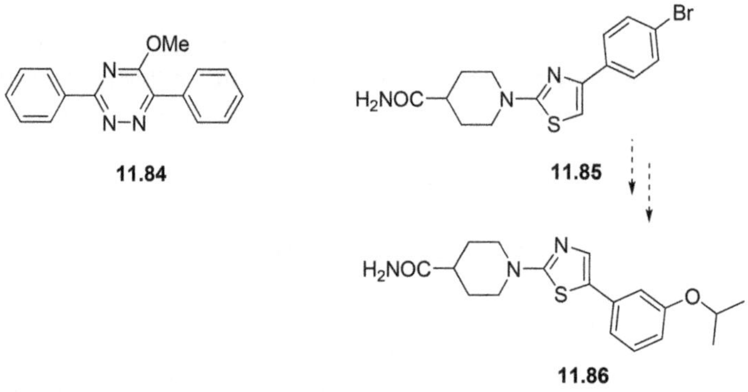

11.84

11.85

11.86

Figure 11.44 Hits and lead from a novel reporter HTS assessing exon 7 inclusion.

Thiazole derivative **11.85**, however, was the subject of a considerable amount of medicinal chemistry optimisation work. The molecule was notionally broken into three regions, the two peripheral substituents, and the thiazole core itself, and these were explored individually. Structure–activity relationships were again assessed using the luciferase readout only, with follow-up tests only being carried out on a selected subset of the most potent examples. A number of selected examples were shown to have reasonable *in vitro* ADME properties, as well as significantly improved potency (EC_{50} < 1 μM). Moreover, increases in SMN protein levels (assessed by Western blot) were seen, coupled with cellular activity consistent with increasing the amount of functional SMN protein, *i.e.* they increased the number of SMN-positive foci in SMA type 3183 cell nuclei.

Finally, the best examples, including **11.86** which had reasonable metabolic stability and permeability, were selected for *in vivo* testing, and progressed to dosing in mice. There, after dosing orally at 30 mg kg^{-1}, compound concentrations exceeding the EC_{50} were measured in the brain for over 4 hours, these levels being felt to be therapeutically relevant. Because no *in vivo* efficacy data have yet been published on these compounds, what is not clear at this point is whether (i) just exceeding EC_{50} would be sufficient, or whether higher levels would be required; and (ii) the length of time for which these levels would need to be maintained. As yet, no further work has been published, but more detail about the pharmacokinetics and pharmacodynamics (PK–PD) of the system is clearly required, and this will be driven at least in part by a deeper knowledge of transcription, translation and degradation times for SMN proteins, as well as the critical level of SMN needed to show therapeutic benefit.

More advanced high-throughput screens have also been run, for example that of Rubin and co-workers, who carried out a high-content screen in which they treated SMA patient fibroblasts with a series of 'annotated chemical libraries' comprising 3500 compounds with known biological activities.[260] High-content screening is a development from conventional HTS,[286] and is

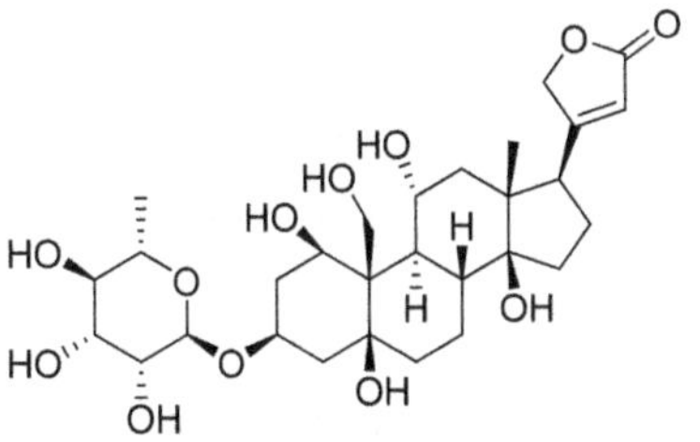

11.87, ouabain 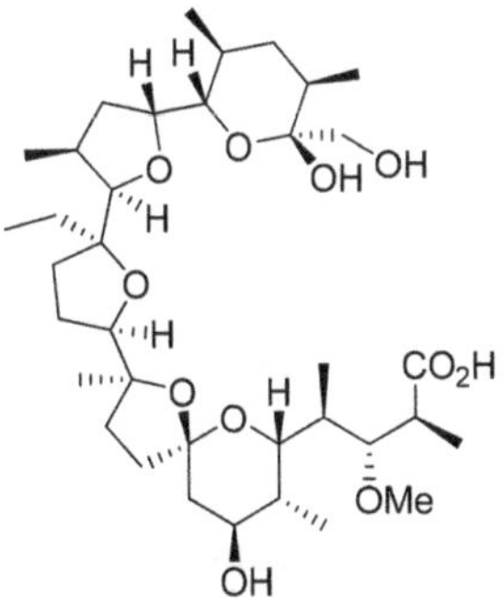 **11.88**, monensin

Figure 11.45 Hits identified using a high-content screening approach.

an emerging and potentially very powerful paradigm for phenotypic screening in which detailed measurement and analysis of multiple parameters is simultaneously undertaken.[287,288] The phenomenal advances in both imaging and computing hardware which have taken place in recent years have greatly facilitated this process. Using image-based analysis of different cellular compartments, cytoplasmic and nuclear levels of SMN protein were simultaneously assessed. After dose–response determination and elimination of false-positives, 188 compounds were confirmed as active, upregulating the amount of SMN in one or more of the cellular compartment studies, and these were clustered according to their chemical and biological properties. Compounds with a range of biological activities were found to be hits, including ion channel modulators such as ouabain **11.87** and monensin **11.88** (Figure 11.45), which were postulated to act by modulating cellular levels of calcium.

A decrease in SMN protein levels was also noted following treatment with inhibitors of the PDGF growth factor receptor. Through follow-up studies on their possible modes of action, a link was found between inhibition of GSK3β and a reduced rate of degradation of the SMN protein, with both GSK3β inhibitors from multiple chemotypes (for example alsterpaullone **11.61**, CHIR98014 **11.89** and AR-0A14418 **11.90**, Figure 11.46) and shRNA GSK knockdown studies confirming this.

Although these represent interesting leads, the selectivity of kinase inhibitors can often be a confounding factor in biological assays (both enzymatic and cellular). Inhibition of GSK3β for example has proved to be a challenging target for some time, with the identification of truly selective compounds with no appreciable off-target kinase inhibition remaining a challenge.[289] Clearly new discoveries and insights from a structural and selectivity perspective will have a pivotal effect on these and other drug programmes.

Further studies using GSK3β inhibition have recently been established by Kozikowski *et al.*[290] The staurosporine analogue BIP-135 (**11.91**, Figure 11.46)

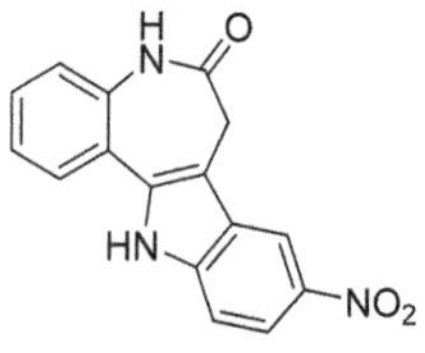

11.61, alsterpaullone

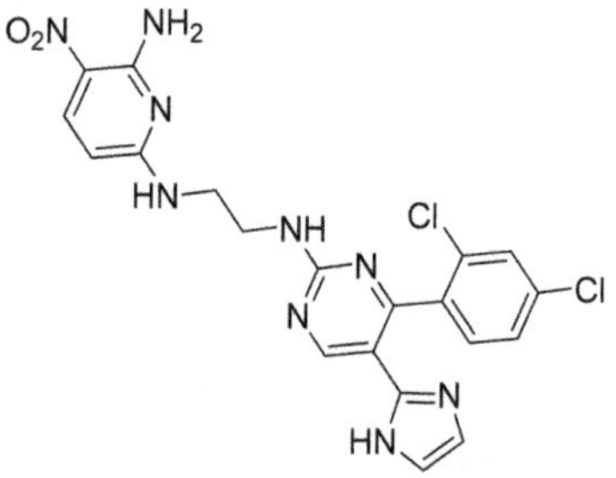

11.89, CHIR98014

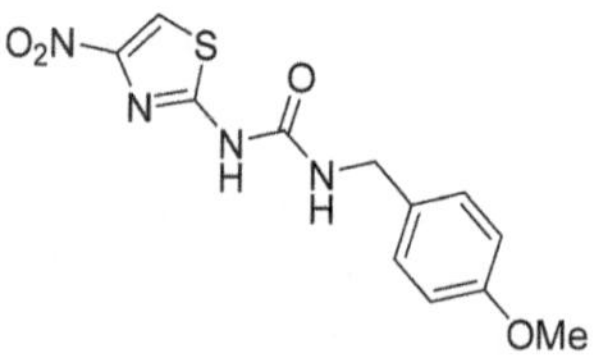

11.90, AR-0A14418

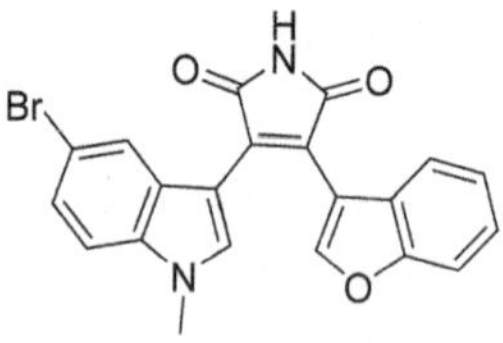

11.91, BIP-135

Figure 11.46 GSK3β and other kinase inhibitors with possible SMA utility.

was found to show activity using an *in vitro* oxidative model of neuroprotection, and furthermore when dosed in the Δ7 SMA mouse model of SMA, provided a modest median survival benefit of 2 days. The authors comment that results, while positive, were using a compound which was 'reasonably selective', and further studies with alternative GSK3β inhibitor chemotypes would need to be explored. Care needs to be taken when interpreting these data, however, because staurosporine and its structural relatives are known to be promiscuous, inhibiting a wide range of other members of the kinase superfamily of enzymes.[291]

Although the number of screening campaigns that have been run for SMA have been limited, it is clear that the factors which have often confounded HTS hit analysis are just as prevalent here. Due to the extensive use of luciferase-based readouts a large amount of follow-up and confirmatory study resource is applied to compounds that are later found to be false-positives. It is therefore critical to either eliminate these classes of compounds from reporter-based screens at as early a stage as possible, using either physical or chemoinformatic methods, and crucially to move compounds into luciferase-free confirmatory assays as soon as possible in order to establish whether the apparent hits have a genuine effect on the desired mode of action. For example, with SMA, this would typically mean screening more compounds faster in a Western blot based assay assessing SMN levels in SMA-derived patient cells, or perhaps an induced pluripotent stem cell derived model.

11.4 Summary and Future Outlook

From the foregoing text it is clear that a considerable amount of effort has taken place to date on both assay and HTS development, as part of screening for the rare neuromuscular diseases DMD and SMA. Because only limited examples of clinical trials have taken place on any of these agents, it is not surprising that relatively few of the examples described have followed the traditional 'screen, optimise, nominate clinical candidate' approach typically followed for non-orphan, target-based drug discovery. The majority of screening exercises have been opportunistic, and evaluated pre-existing commercial compound sets, and while these typically provide valuable pharmacological tools for future researchers, there are attendant risks, including the effects of off-target reactivity and the need to recognise that further structure–activity optimisation will be necessary. Even for drug reprofiling based approaches, the likelihood that any compounds identified would represent anything more than an opportunity for a fairly speculative clinical study is low. Despite these caveats, studies to date have provided a variety of valuable probe compounds, several of which have demonstrated activity in industry-accepted disease models, and allowed the identification of a range of points for possible therapeutic intervention. As long as the data is placed in the appropriate context there now exists a multitude of molecular and biological start points for projects which could accelerate drug discovery for these and other rare diseases.

New screening technologies are likely to continue to play a critical role in the development of new therapeutic agents to treat neuromuscular and other genetic diseases such as those reviewed here. As is evident from the case studies presented, much reliance has been placed on reporter assays, particularly luciferase-based systems, rather than assays in which direct readout of either a mechanistic or pharmacological endpoint is measured. Much critique has been presented in the literature on luciferase assays, and potential confounding factors.[280] While they certainly have value in drug discovery projects involving reporter-based screening approaches, care needs to be taken with assay selection and interpretation of the results. It is also vital that appropriate deconvolution tests are carried out to rule out false-positives associated with compounds having a direct effect on luciferase such as inhibition or stabilisation. Assuming these precautionary measures are adequately accounted for, these along with (re) emergent technologies such as phenotypic and high-content screening[57,288] and newer drug discovery platforms which comprise more physiological/pathologically relevant systems such as patient-derived stem cell models are anticipated to be critical in providing more disease- and patient-relevant models.

Whatever the assays chosen within projects, it is critical that appropriate validation occurs to determine (for example) the extent of modulation (level and duration) required of a new target in order to establish therapeutic benefit in the clinic. Of the examples described here, the compounds that have progressed to clinical studies are first generation, and so will provide valuable information on these pharmacodynamic aspects.

Coupled with the increase in disease-relevant screening systems, refinement of corporate screening sets in order to remove problem compounds must continue. While this will restrict the number of compounds screened it should also improve the quality of hits obtained, thereby reducing downstream attrition. All too frequently within drug discovery programmes, and despite the greater emphasis in modern pharmaceutical and biotechnology companies on improving compound quality, problems with molecules which are either false-positives or unsuitable for further development persist. Appropriate forward-thinking synthetic strategies within medicinal chemistry teams will widen the structural diversity of molecules tested, while often the incorporation of relatively simple cross-checks into screening cascades can help ensure rapid elimination of unsuitable molecules that would otherwise lead to project and clinical trial failures, and potentially setting back discovery efforts in rare diseases many years. Otherwise the disturbing possibility exists that the failure of an 'unsuitable' compound in clinical trials may discourage further efforts on an otherwise feasible mechanism for the treatment of a particular disease.

The two case studies described here, as well as being representative of the rapid and merciless progression of both diseases present in a paediatric population, and it is critically important to establish as soon as possible the appropriate clinical trial inclusion criteria so that the chances of seeing therapeutic benefit are maximised. As illustrated from review of a number of the agents for DMD, factors such as age matching and stage of disease are factors which are likely to be pivotal in allowing statistically significant results to be seen. Cohort size, as with any clinical trial, will also play a crucial role, as will availability of the appropriate patient groups – by definition the diseases are rare and so the patient numbers will be limited.

What is clear at this stage is that there are two clear emergent paradigms for curative treatment of rare neuromuscular disease, as opposed to the development of improved symptomatic treatments. The first of these is predicated on inventing a therapy to treat the disease's underlying cause, in these cases this being a genetic mutation. Approaches using oligonucleotides to enable exon skipping, or employing small-molecule read-through agents, have made fantastic progress, and are starting to deliver encouraging results in later stage clinical trials. However, the possibility of the disease encompassing a more heterogeneous group of sufferers with multiple mutations limits the applicability of each specific therapy to a smaller subset of patients. The alternative is, through a detailed knowledge of the disease in question, to identify a therapeutic approach which is independent of the primary lesion. While this may be more technically challenging, and relies on the existence of an appropriate redundant/compensatory mechanism to target, the advantages are hugely significant, in that the opportunity for treatment of *all* patients becomes potentially viable.

There is of course a middle ground, in which a combination of drugs, each addressing a specific point in progression of the disease is used, or simply one in which an established symptomatic treatment is partnered with an emerging disease-modifying drug; examples of both of these paradigms having been summarised in the preceding text. In reality, this latter approach is likely to be the first to be reduced to clinical practice and receive regulatory approval, with combinations of disease-modifying agents coming next, subject of course to the appropriate combination clinical trials taking place first. This pathway parallels established development pathways, which have taken place in other therapeutic areas such as the oncology and anti-infective fields.

Over the past decades pioneering work has taken place to elucidate the underlying pathological mechanisms of many rare neuromuscular diseases. This in turn has inspired the development of several truly innovative therapeutic strategies aimed at correcting the underlying pathology. We are now at a hugely exciting juncture in this translational research effort, particularly for DMD and SMA, and during the next decade it can be anticipated that there should be an explosion in the number of agents evaluated in patients, hopefully followed by the first approvals of disease-modifying agents that show true benefit for sufferers of these devastating and debilitating disorders.

Acknowledgements

The authors wish to thank Professor Dame Kay Davies, Professor Steve Davies and Dr Robert Westwood for helpful advice and comments, and for proof-reading this manuscript.

References

1. N. Marpillat, *Prevalence of rare diseases*, http://www.orpha.net/orphacom/cahiers/docs/GB/Prevalence_of_rare_diseases_by_alphabetical_list.pdf, accessed 8 August 2013.
2. http://www.muscular-dystrophy.org/about_muscular_dystrophy/conditions, accessed 8 August 2013.
3. Databases: Chemical Abstracts and PubMed; searched using the search terms 'Duchenne Muscular Dystrophy' and 'Spinal Muscular Atrophy' respectively.
4. E. P. Hoffman, R. H. Brown and L. M. Kunkel, *Cell,* 1987, **51**, 919–928.
5. J. M. Tinsley, D. J. Blake, A. Roche, U. Fairbrother, J. Riss, B. C. Byth, A. E. Knight, J. Kendrick-Jones, G. K. Suthers and D. R. Love, *Nature,* 1992, **360**, 591–593.
6. S. Lefebvre, L. Bürglen, S. Reboullet, O. Clermont, P. Burlet, L. Viollet, B. Benichou, C. Cruaud, P. Millasseau and M. Zeviani, *Cell,* 1995, **80**, 155–165.

7. K. N. Meekings, C. S. Williams and J. E. Arrowsmith, *Drug Discovery Today,* 2012, **17**, 660–664.
8. J. Bivona, *Duchenne Muscular Dystrophy Drug could Unlock Huge Potential for this Pharmaceutical*, http://beta.fool.com/jordobivona/2013/01/31/duchenne-muscular-dystrophy-drug-could-unlock-huge/22741/, accessed 8 August 2013.
9. K. Bushby, R. Finkel, D. J. Birnkrant, L. E. Case, P. R. Clemens, L. Cripe, A. Kaul, K. Kinnett, C. McDonald, S. Pandya, J. Poysky, F. Shapiro, J. Tomezsko, C. Constantin and D. C. C. W. Group, *Lancet Neurol.,* 2010, **9**, 77–93.
10. K. Bushby, R. Finkel, D. J. Birnkrant, L. E. Case, P. R. Clemens, L. Cripe, A. Kaul, K. Kinnett, C. McDonald, S. Pandya, J. Poysky, F. Shapiro, J. Tomezsko, C. Constantin and D. C. C. W. Group, *Lancet Neurol.,* 2010, **9**, 177–189.
11. D. E. McNeil, C. Davis, D. Jillapalli, S. Targum, A. Durmowicz and T. R. Coté, *Muscle Nerve,* 2010, **41**, 740–745.
12. *Duchenne muscular dystrophy*, http://www.muscular-dystrophy.org/about_muscular_dystrophy/conditions/97_duchenne_muscular_dystrophy, accessed 8 August 2013.
13. M. Koenig, E. P. Hoffman, C. J. Bertelson, A. P. Monaco, C. Feener and L. M. Kunkel, *Cell,* 1987, **50**, 509–517.
14. A. H. Beggs, E. P. Hoffman, J. R. Snyder, K. Arahata, L. Specht, F. Shapiro, C. Angelini, H. Sugita and L. M. Kunkel, *Am. J. Hum. Genet.,* 1991, **49**, 54–67.
15. K. S. Saladin, in *Human anatomy*, McGraw-Hill Higher Education, New York, 2010, pp. 235–262, 264.
16. K. E. Davies and K. J. Nowak, *Nat. Rev. Mol. Cell Biol.,* 2006, 7, 762–773.
17. C. A. Collins and J. E. Morgan, *Int. J. Exp. Pathol.,* 2003, **84**, 165–172.
18. M. D. Grounds, H. G. Radley, G. S. Lynch, K. Nagaraju and A. De Luca, *Neurobiol. Dis.,* 2008, **31**, 1–19.
19. F. Mourkioti, J. Kustan, P. Kraft, J. W. Day, M. M. Zhao, M. Kost-Alimova, A. Protopopov, R. A. Depinho, D. Bernstein, A. K. Meeker and H. M. Blau, *Nat. Cell Biol.,* 2013, **15**, 895–904.
20. G. Vaquer, F. Rivière, M. Mavris, F. Bignami, J. Llinares-Garcia, K. Westermark and B. Sepodes, *Nat. Rev. Drug Discovery,* 2013, **12**, 287–305.
21. R. J. Fairclough, A. Bareja and K. E. Davies, *Exp. Physiol.,* 2011, **96**, 1101–1113.
22. C. Angelini and E. Peterle, *Acta Myol.,* 2012, **31**, 9–15.
23. C. Angelini, *Muscle Nerve,* 2007, **36**, 424–435.
24. Y. Ishikawa, J. R. Bach and R. Minami, *Am. Heart J.,* 1999, **137**, 895–902.
25. H. Ogata, Y. Ishikawa and R. Minami, *J. Cardiol.,* 2009, **53**, 72–78.
26. V. Malik, L. R. Rodino-Klapac and J. R. Mendell, *Expert Opin. Emerging Drugs,* 2012, **17**, 261–277.
27. Z. E. Davidson and H. Truby, *J. Hum. Nutr. Diet.,* 2009, **22**, 383–393.
28. J. M. Damsker, B. C. Dillingham, M. C. Rose, M. A. Balsley, C. R. Heier, A. M. Watson, E. J. Stemmy, R. A. Jurjus, T. Huynh, K. Tatem,

K. Uaesoontrachoon, D. M. Berry, A. S. Benton, R. J. Freishtat, E. P. Hoffman, J. M. McCall, H. Gordish-Dressman, S. L. Constant, E. K. Reeves and K. Nagaraju, *PLoS One,* 2013, **8**, e63871.
29. E. K. Reeves, E. P. Hoffman, K. Nagaraju, J. M. Damsker and J. M. McCall, *Bioorg. Med. Chem.,* 2013, **21**, 2241–2249.
30. J. A. Baur and D. A. Sinclair, *Nat. Rev. Drug Discovery,* 2006, **5**, 493–506.
31. B. S. Gordon, D. C. Delgado Díaz and M. C. Kostek, *Clin. Nutr.,* 2013, **32**, 104–111.
32. S. Messina, A. Bitto, M. Aguennouz, L. Minutoli, M. C. Monici, D. Altavilla, F. Squadrito and G. Vita, *Exp. Neurol.,* 2006, **198**, 234–241.
33. S. Messina, G. L. Vita, M. Aguennouz, M. Sframeli, S. Romeo, C. Rodolico and G. Vita, *Acta Myol.,* 2011, **30**, 16–23.
34. H. Gutzmann and D. Hadler, *J. Neural Transm., Suppl.,* 1998, **54**, 301–310.
35. C. Tonon and R. Lodi, *Expert Opin. Pharmacother.,* 2008, **9**, 2327–2337.
36. G. M. Buyse, G. Van der Mieren, M. Erb, J. D'hooge, P. Herijgers, E. Verbeken, A. Jara, A. Van Den Bergh, L. Mertens, I. Courdier-Fruh, P. Barzaghi and T. Meier, *Eur. Heart J.,* 2009, **30**, 116–124.
37. G. M. Buyse, N. Goemans, M. van den Hauwe, D. Thijs, I. J. de Groot, U. Schara, B. Ceulemans, T. Meier and L. Mertens, *Neuromuscular Disord.,* 2011, **21**, 396–405.
38. Y. Takaya, H. Tasaka, T. Chiba, K. Uwai, M. Tanitsu, H. S. Kim, Y. Wataya, M. Miura, M. Takeshita and Y. Oshima, *J. Med. Chem.,* 1999, **42**, 3163–3166.
39. B. R. Baker, J. P. Joseph, R. E. Schaub, F. J. McEvoy and J. H. Williams, *J. Org. Chem.,* 1952, **17**, 157–163.
40. M. Pines, I. Vlodavsky and A. Nagler, *Drug Dev. Res.,* 2000, **50**, 371–378.
41. M. Pines and A. Nagler, *Gen. Pharmacol.,* 1998, **30**, 445–450.
42. M. F. Manuel, E. Morales and E. Trovela, Philipp. *J. Vet. Med.,* 1977, **16**, 20–30.
43. O. Halevy, A. Nagler, F. Levi-Schaffer, O. Genina and M. Pines, *Biochem. Pharmacol.,* 1996, **52**, 1057–1063.
44. M. J. de Jonge, H. Dumez, J. Verweij, S. Yarkoni, D. Snyder, D. Lacombe, S. Marréaud, T. Yamaguchi, C. J. Punt, A. van Oosterom and E. N. D. D. G. (NDDG), *Eur. J. Cancer,* 2006, **42**, 1768–1774.
45. M. Pines and O. Halevy, *Histol. Histopathol.,* 2011, **26**, 135–146.
46. K. D. Huebner, D. S. Jassal, O. Halevy, M. Pines and J. E. Anderson, *Am. J. Physiol.: Heart Circ. Physiol.,* 2008, **294**, H1550–H1561.
47. L. K. McLoon, *Am. J. Physiol.: Heart Circ. Physiol.,* 2008, **294**, H1505–H1507.
48. *Halo Therapeutics Raises $1.1 Million to Expedite Phase 2 Study of HT-100,* http://www.halotherapeutics.com/category/press/, accessed 21 August 2013.
49. S. Xavier, E. Piek, M. Fujii, D. Javelaud, A. Mauviel, K. C. Flanders, A. M. Samuni, A. Felici, M. Reiss, S. Yarkoni, A. Sowers, J. B. Mitchell, A. B. Roberts and A. Russo, *J. Biol. Chem.,* 2004, **279**, 15167–15176.
50. P. Juárez, K. S. Mohammad, J. J. Yin, P. G. Fournier, R. C. McKenna, H. W. Davis, X. H. Peng, M. Niewolna, D. Javelaud, J. M. Chirgwin, A. Mauviel and T. A. Guise, *Cancer Res.,* 2012, **72**, 6247–6256.

51. T. L. Keller, D. Zocco, M. S. Sundrud, M. Hendrick, M. Edenius, J. Yum, Y. J. Kim, H. K. Lee, J. F. Cortese, D. F. Wirth, J. D. Dignam, A. Rao, C. Y. Yeo, R. Mazitschek and M. Whitman, *Nat. Chem. Biol.,* 2012, **8**, 311–317.
52. E. L. Tobinick, *Drug News Perspect.,* 2009, **22**, 119–125.
53. E. Dorey, *Chem. Ind.,* 2012, 2.
54. C. Southan, A. J. Williams and S. Ekins, *Drug Discovery Today,* 2013, **18**, 58–70.
55. *Alzheimer's, cancer and rare disease research to benefit from landmark MRC-AstraZeneca compound collaboration,* http://www.mrc.ac.uk/news-events/news/alzheimere28099s-cancer-and-rare-disease-research-to-benefit-from-landmark-mrc-astrazeneca-compound-collaboration/.
56. Y. Feng, T. J. Mitchison, A. Bender, D. W. Young and J. A. Tallarico, *Nat. Rev. Drug Discovery,* 2009, **8**, 567–578.
57. D. C. Swinney and J. Anthony, *Nat. Rev. Drug Discovery,* 2011, **10**, 507–519.
58. J. M. Percival, N. P. Whitehead, M. E. Adams, C. M. Adamo, J. A. Beavo and S. C. Froehner, *J. Pathol.,* 2012, **228**, 77–87.
59. A. De Luca, B. Nico, A. Liantonio, M. P. Didonna, B. Fraysse, S. Pierno, R. Burdi, D. Mangieri, J. F. Rolland, C. Camerino, A. Zallone, P. Confalonieri, F. Andreetta, E. Arnoldi, I. Courdier-Fruh, J. P. Magyar, A. Frigeri, M. Pisoni, M. Svelto and D. Conte Camerino, *Am. J. Pathol.,* 2005, **166**, 477–489.
60. J. Kirschner, J. Schessl, U. Schara, B. Reitter, G. M. Stettner, E. Hobbiebrunken, E. Wilichowski, G. Bernert, S. Weiss, F. Stehling, G. Wiegand, W. Müller-Felber, S. Thiele, U. Grieben, M. von der Hagen, J. Lütschg, C. Schmoor, G. Ihorst and R. Korinthenberg, *Lancet Neurol.,* 2010, **9**, 1053–1059.
61. J. Giacomotto, N. Brouilly, L. Walter, M. C. Mariol, J. Berger, L. Ségalat, T. S. Becker, P. D. Currie and K. Gieseler, *Hum. Mol. Genet.,* 2013, **22**, 4562–4578.
62. O. M. Dorchies, J. Reutenauer-Patte, E. Dahmane, H. M. Ismail, O. Petermann, O. Patthey-Vuadens, S. A. Comyn, E. Gayi, T. Piacenza, R. J. Handa, L. A. Décosterd and U. T. Ruegg, *Am. J. Pathol.,* 2013, **182**, 485–504.
63. A. Cozzoli, R. F. Capogrosso, V. T. Sblendorio, M. M. Dinardo, C. Jagerschmidt, F. Namour, G. M. Camerino and A. De Luca, *Pharmacol. Res.,* 2013, **72**, 9–24.
64. R. J. Fairclough, K. J. Perkins and K. E. Davies, *Curr. Gene Ther.,* 2012, **12**, 206–244.
65. F. Le Roy, K. Charton, C. L. Lorson and I. Richard, *Trends Mol. Med.,* 2009, **15**, 580–591.
66. G. Cossu and M. Sampaolesi, *Trends Mol. Med.,* 2007, **13**, 520–526.
67. C. Pichavant, A. Aartsma-Rus, P. R. Clemens, K. E. Davies, G. Dickson, S. Takeda, S. D. Wilton, J. A. Wolff, C. I. Wooddell, X. Xiao and J. P. Tremblay, *Mol. Ther.,* 2011, **19**, 830–840.

68. D. E. Bowles, S. W. McPhee, C. Li, S. J. Gray, J. J. Samulski, A. S. Camp, J. Li, B. Wang, P. E. Monahan, J. E. Rabinowitz, J. C. Grieger, L. Govindasamy, M. Agbandje-McKenna, X. Xiao and R. J. Samulski, *Mol. Ther.,* 2012, **20**, 443–455.
69. J. R. Mendell, K. Campbell, L. Rodino-Klapac, Z. Sahenk, C. Shilling, S. Lewis, D. Bowles, S. Gray, C. Li, G. Galloway, V. Malik, B. Coley, K. R. Clark, J. Li, X. Xiao, J. Samulski, S. W. McPhee, R. J. Samulski and C. M. Walker, *N. Engl. J. Med.,* 2010, **363**, 1429–1437.
70. J. R. Mendell, L. Rodino-Klapac, Z. Sahenk, V. Malik, B. K. Kaspar, C. M. Walker and K. R. Clark, *Neurosci. Lett.,* 2012, **527**, 90–99.
71. R. J. Fairclough, M. J. Wood and K. E. Davies, *Nat. Rev. Genet.,* 2013, **14**, 373–378.
72. S. B. England, L. V. Nicholson, M. A. Johnson, S. M. Forrest, D. R. Love, E. E. Zubrzycka-Gaarn, D. E. Bulman, J. B. Harris and K. E. Davies, *Nature,* 1990, **343**, 180–182.
73. B. Wang, J. Li and X. Xiao, *Proc. Natl. Acad. Sci. U. S. A.,* 2000, **97**, 13714–13719.
74. *First U.S. Trial of DMD Gene Therapy Under Way*, http://static.mda.org/research/060329dmd_gene_therapy.html#, accessed 8 August 2013.
75. J. Singh, R. C. Petter, T. A. Baillie and A. Whitty, *Nat. Rev. Drug Discovery,* 2011, **10**, 307–317.
76. S. Fletcher, K. Honeyman, A. M. Fall, P. L. Harding, R. D. Johnsen and S. D. Wilton, *J. Gene Med.,* 2006, **8**, 207–216.
77. H. A. Heemskerk, C. L. de Winter, S. J. de Kimpe, P. van Kuik-Romeijn, N. Heuvelmans, G. J. Platenburg, G. J. van Ommen, J. C. van Deutekom and A. Aartsma-Rus, *J. Gene Med.,* 2009, **11**, 257–266.
78. *What is exon skipping and how does it work?*, http://www.muscular-dystrophy.org/about_muscular_dystrophy/research_faqs/612_what_is_exon_skipping_and_how_does_it_work, accessed 8 August 2013.
79. A. G. Douglas and M. J. Wood, *Mol. Cell. Neurosci.,* 2013, **56C**, 169–185.
80. J. Alter, F. Lou, A. Rabinowitz, H. Yin, J. Rosenfeld, S. D. Wilton, T. A. Partridge and Q. L. Lu, *Nat. Med.,* 2006, **12**, 175–177.
81. M. N. Uddin, N. J. Patel, T. Bhowmik, B. D'Souza, A. Akalkotkar, F. Etzlar, C. W. Oettinger and M. D'Souza, *J. Drug Targeting,* 2013, **21**, 450–457.
82. A. Aartsma-Rus, I. Fokkema, J. Verschuuren, I. Ginjaar, J. van Deutekom, G. J. van Ommen and J. T. den Dunnen, *Hum. Mutat.,* 2009, **30**, 293–299.
83. M. Kinali, V. Arechavala-Gomeza, L. Feng, S. Cirak, D. Hunt, C. Adkin, M. Guglieri, E. Ashton, S. Abbs, P. Nihoyannopoulos, M. E. Garralda, M. Rutherford, C. McCulley, L. Popplewell, I. R. Graham, G. Dickson, M. J. Wood, D. J. Wells, S. D. Wilton, R. Kole, V. Straub, K. Bushby, C. Sewry, J. E. Morgan and F. Muntoni, *Lancet Neurol.,* 2009, **8**, 918–928.
84. S. Cirak, V. Arechavala-Gomeza, M. Guglieri, L. Feng, S. Torelli, K. Anthony, S. Abbs, M. E. Garralda, J. Bourke, D. J. Wells, G. Dickson, M. J. Wood, S. D. Wilton, V. Straub, R. Kole, S. B. Shrewsbury, C. Sewry, J. E. Morgan, K. Bushby and F. Muntoni, *Lancet,* 2011, **378**, 595–605.

85. G. C. Kendall, E. I. Mokhonova, M. Moran, N. E. Sejbuk, D. W. Wang, O. Silva, R. T. Wang, L. Martinez, Q. L. Lu, R. Damoiseaux, M. J. Spencer, S. F. Nelson and M. C. Miceli, *Sci. Transl. Med.*, 2012, **4**, 164ra160.
86. C. M. McDonald, E. K. Henricson, J. J. Han, R. T. Abresch, A. Nicorici, G. L. Elfring, L. Atkinson, A. Reha, S. Hirawat and L. L. Miller, *Muscle Nerve*, 2010, **41**, 500–510.
87. A. Aartsma-Rus, A. A. Janson, W. E. Kaman, M. Bremmer-Bout, J. T. den Dunnen, F. Baas, G. J. van Ommen and J. C. van Deutekom, *Hum. Mol. Genet.*, 2003, **12**, 907–914.
88. J. C. van Deutekom, A. A. Janson, I. B. Ginjaar, W. S. Frankhuizen, A. Aartsma-Rus, M. Bremmer-Bout, J. T. den Dunnen, K. Koop, A. J. van der Kooi, N. M. Goemans, S. J. de Kimpe, P. F. Ekhart, E. H. Venneker, G. J. Platenburg, J. J. Verschuuren and G. J. van Ommen, *N. Engl. J. Med.*, 2007, **357**, 2677–2686.
89. N. M. Goemans, M. Tulinius, J. T. van den Akker, B. E. Burm, P. F. Ekhart, N. Heuvelmans, T. Holling, A. A. Janson, G. J. Platenburg, J. A. Sipkens, J. M. Sitsen, A. Aartsma-Rus, G. J. van Ommen, G. Buyse, N. Darin, J. J. Verschuuren, G. V. Campion, S. J. de Kimpe and J. C. van Deutekom, *N. Engl. J. Med.*, 2011, **364**, 1513–1522.
90. J. T. Mendell and H. C. Dietz, *Cell*, 2001, **107**, 411–414.
91. K. M. Keeling and D. M. Bedwell, *Wiley Interdiscip. Rev.: RNA*, 2011, **2**, 837–852.
92. E. Kerem, *Curr. Opin. Pulm. Med.*, 2004, **10**, 547–552.
93. H. L. Lee and J. P. Dougherty, *Pharmacol. Ther.*, 2012, **136**, 227–266.
94. V. Malik, L. R. Rodino-Klapac, L. Viollet and J. R. Mendell, *Ther. Adv. Neurol. Disord.*, 2010, **3**, 379–389.
95. J. Davies, W. Gilbert and L. Gorini, *Proc. Natl. Acad. Sci. U. S. A.*, 1964, **51**, 883–890.
96. M. Howard, R. A. Frizzell and D. M. Bedwell, *Nat. Med.*, 1996, **2**, 467–469.
97. M. Du, J. R. Jones, J. Lanier, K. M. Keeling, J. R. Lindsey, A. Tousson, Z. Bebök, J. A. Whitsett, C. R. Dey, W. H. Colledge, M. J. Evans, E. J. Sorscher and D. M. Bedwell, *J. Mol. Med.*, 2002, **80**, 595–604.
98. M. Du, K. M. Keeling, L. Fan, X. Liu, T. Kovaçs, E. Sorscher and D. M. Bedwell, *J. Mol. Med.*, 2006, **84**, 573–582.
99. E. R. Barton-Davis, L. Cordier, D. I. Shoturma, S. E. Leland and H. L. Sweeney, *J. Clin. Invest.*, 1999, **104**, 375–381.
100. M. Arakawa, M. Shiozuka, Y. Nakayama, T. Hara, M. Hamada, S. Kondo, D. Ikeda, Y. Takahashi, R. Sawa, Y. Nonomura, K. Sheykholeslami, K. Kondo, K. Kaga, T. Kitamura, Y. Suzuki-Miyagoe, S. Takeda and R. Matsuda, *J. Biochem.*, 2003, **134**, 751–758.
101. A. De Luca, B. Nico, J. F. Rolland, A. Cozzoli, R. Burdi, D. Mangieri, V. Giannuzzi, A. Liantonio, V. Cippone, M. De Bellis, G. P. Nicchia, G. M. Camerino, A. Frigeri, M. Svelto and D. C. Camerino, *Neurobiol. Dis.*, 2008, **32**, 243–253.

102. K. R. Wagner, S. Hamed, D. W. Hadley, A. L. Gropman, A. H. Burstein, D. M. Escolar, E. P. Hoffman and K. H. Fischbeck, *Ann. Neurol.,* 2001, **49**, 706–711.
103. L. Politano, G. Nigro, V. Nigro, G. Piluso, S. Papparella, O. Paciello and L. I. Comi, *Acta Myol.,* 2003, **22**, 15–21.
104. V. Malik, L. R. Rodino-Klapac, L. Viollet, C. Wall, W. King, R. Al-Dahhak, S. Lewis, C. J. Shilling, J. Kota, C. Serrano-Munuera, J. Hayes, J. D. Mahan, K. J. Campbell, B. Banwell, M. Dasouki, V. Watts, K. Sivakumar, R. Bien-Willner, K. M. Flanigan, Z. Sahenk, R. J. Barohn, C. M. Walker and J. R. Mendell, *Ann. Neurol.,* 2010, **67**, 771–780.
105. E. Selimoglu, *Curr. Pharm. Des.,* 2007, **13**, 119–126.
106. M. P. Mingeot-Leclercq and P. M. Tulkens, *Antimicrob. Agents Chemother.,* 1999, **43**, 1003–1012.
107. E. M. Welch, E. R. Barton, J. Zhuo, Y. Tomizawa, W. J. Friesen, P. Trifillis, S. Paushkin, M. Patel, C. R. Trotta, S. Hwang, R. G. Wilde, G. Karp, J. Takasugi, G. Chen, S. Jones, H. Ren, Y. C. Moon, D. Corson, A. A. Turpoff, J. A. Campbell, M. M. Conn, A. Khan, N. G. Almstead, J. Hedrick, A. Mollin, N. Risher, M. Weetall, S. Yeh, A. A. Branstrom, J. M. Colacino, J. Babiak, W. D. Ju, S. Hirawat, V. J. Northcutt, L. L. Miller, P. Spatrick, F. He, M. Kawana, H. Feng, A. Jacobson, S. W. Peltz and H. L. Sweeney, *Nature,* 2007, **447**, 87–91.
108. G. M. Karp, S. Hwang, G. Chen, N. G. Almstead, US 2004/0204461 A1 (Published 14th October 2004).
109. J. A. Campbell, A. Kahn, J. Takasugi, E. Welch, WO 2006/044682 A1 (Published 27th April 2006).
110. M. Wilschanski, L. L. Miller, D. Shoseyov, H. Blau, J. Rivlin, M. Aviram, M. Cohen, S. Armoni, Y. Yaakov, T. Pugatsch, T. Pugatch, M. Cohen-Cymberknoh, N. L. Miller, A. Reha, V. J. Northcutt, S. Hirawat, K. Donnelly, G. L. Elfring, T. Ajayi and E. Kerem, *Eur. Respir. J.,* 2011, **38**, 59–69.
111. S. Hirawat, E. M. Welch, G. L. Elfring, V. J. Northcutt, S. Paushkin, S. Hwang, E. M. Leonard, N. G. Almstead, W. Ju, S. W. Peltz and L. L. Miller, *J. Clin. Pharmacol.,* 2007, **47**, 430–444.
112. C. Bönnemann, R. Finkel, B. Wong, K. Flanigan, J. Sampson, L. Sweeney, A. Reha, G. Elfring, L. Miller and S. Hirawat, in *Neuromuscular Disorders*, 2007, vol. 17, p. 783.
113. S. W. Peltz, M. Morsy, E. M. Welch and A. Jacobson, *Annu. Rev. Med.,* 2013, **64**, 407–425.
114. *PTC Therapeutics Closes $30 Million Financing*, http://ir.ptcbio.com/releasedetail.cfm?ReleaseID=693986, accessed 21 August 2013.
115. N. Thorne, J. Inglese and D. S. Auld, *Chem. Biol.,* 2010, **17**, 646–657.
116. D. S. Auld, N. Thorne, W. F. Maguire and J. Inglese, *Proc. Natl. Acad. Sci. U. S. A.,* 2009, **106**, 3585–3590.
117. S. P. McElroy, T. Nomura, L. S. Torrie, E. Warbrick, U. Gartner, G. Wood and W. H. McLean, *PLoS Biol.,* 2013, **11**, e1001593.

118. R. Kayali, J. M. Ku, G. Khitrov, M. E. Jung, O. Prikhodko and C. Bertoni, *Hum. Mol. Genet.*, 2012, **21**, 4007–4020.
119. L. Du, R. Damoiseaux, S. Nahas, K. Gao, H. Hu, J. M. Pollard, J. Goldstine, M. E. Jung, S. M. Henning, C. Bertoni and R. A. Gatti, *J. Exp. Med.*, 2009, **206**, 2285–2297.
120. J. B. Baell and G. A. Holloway, *J. Med. Chem.*, 2010, **53**, 2719–2740.
121. *21CFR312.23 Investigational New Drug Application (IND)*, http://www.accessdata.fda.gov/scripts/cdrh/cfdocs/cfcfr/cfrsearch.cfm?fr=312.23, accessed 8 August 2013.
122. J. M. Tinsley, A. C. Potter, S. R. Phelps, R. Fisher, J. I. Trickett and K. E. Davies, *Nature*, 1996, **384**, 349–353.
123. J. Tinsley, N. Deconinck, R. Fisher, D. Kahn, S. Phelps, J. M. Gillis and K. Davies, *Nat. Med.*, 1998, **4**, 1441–1444.
124. M. Cerletti, T. Negri, F. Cozzi, R. Colpo, F. Andreetta, D. Croci, K. E. Davies, F. Cornelio, O. Pozza, G. Karpati, R. Gilbert and M. Mora, *Gene Ther.*, 2003, **10**, 750–757.
125. K. A. Kleopa, A. Drousiotou, E. Mavrikiou, A. Ormiston and T. Kyriakides, *Hum. Mol. Genet.*, 2006, **15**, 1623–1628.
126. P. Miura and B. J. Jasmin, *Trends Mol. Med.*, 2006, **12**, 122–129.
127. T. S. Khurana and K. E. Davies, *Nat. Rev. Drug Discovery*, 2003, **2**, 379–390.
128. T. O. Krag, S. Bogdanovich, C. J. Jensen, M. D. Fischer, J. Hansen-Schwartz, E. H. Javazon, A. W. Flake, L. Edvinsson and T. S. Khurana, *Proc. Natl. Acad. Sci. U. S. A.*, 2004, **101**, 13856–13860.
129. V. Voisin, C. Sébrié, S. Matecki, H. Yu, B. Gillet, M. Ramonatxo, M. Israël and S. De la Porte, *Neurobiol. Dis.*, 2005, **20**, 123–130.
130. S. Vianello, H. Yu, V. Voisin, H. Haddad, X. He, A. S. Foutz, C. Sebrié, B. Gillet, M. Roulot, F. Fougerousse, C. Perronnet, C. Vaillend, S. Matecki, D. Escolar, L. Bossi, M. Israël and S. de la Porte, *FASEB J.*, 2013, **27**, 2256–2269.
131. C. Moorwood and T. S. Khurana, *Expert Opin. Drug Discovery*, 2013, **8**, 569–581.
132. *Summit Outlines Clinical Development Plans For Utrophin Modulator Programme For Duchenne Muscular Dystrophy*, http://www.summitplc.com/userfiles/file/2013_RNS_07%20Clinical%20plans%20for%20DMD%20programme%20FINAL.pdf, accessed 21 August 2013.
133. D. R. Chancellor, K. E. Davies, O. De Moor, C. R. Dorgan, P. D. Johnson, A. G. Lambert, D. Lawrence, C. Lecci, C. Maillol, P. J. Middleton, G. Nugent, S. D. Poignant, A. C. Potter, P. D. Price, R. J. Pye, R. Storer, J. M. Tinsley, R. van Well, R. Vickers, J. Vile, F. J. Wilkes, F. X. Wilson, S. P. Wren and G. M. Wynne, *J. Med. Chem.*, 2011, **54**, 3241–3250.
134. See for example, G. M. Wynne, S. P. Wren, P. D. Johnson, P. D. Price, O. De Moor, G. Nugent, R. Storer, R. J. Pye, C. R. Dorgan, US 8, 518, 980 B2 (published 27th August 2013).
135. Calculated using ChemBioOffice v11 (CambridgeSoft, 2008).
136. K. J. Perkins and K. E. Davies, *FEBS Lett.*, 2003, **538**, 168–172.

137. O. De Moor, C. R. Dorgan, P. D. Johnson, A. G. Lambert, C. Lecci, C. Maillol, G. Nugent, S. D. Poignant, P. D. Price, R. J. Pye, R. Storer, J. M. Tinsley, R. Vickers, R. Well, F. J. Wilkes, F. X. Wilson, S. P. Wren and G. M. Wynne, *Bioorg. Med. Chem. Lett.*, 2011, **21**, 4828–4831.
138. F. P. Guengerich and J. S. MacDonald, *Chem. Res. Toxicol.*, 2007, **20**, 344–369.
139. A.-C. Macherey and P. M. Dansette, in *The Practice of Medicinal Chemistry*, 3rd edition Academic Press, 2008, pp. 674–696.
140. J. M. Tinsley, R. J. Fairclough, R. Storer, F. J. Wilkes, A. C. Potter, S. E. Squire, D. S. Powell, A. Cozzoli, R. F. Capogrosso, A. Lambert, F. X. Wilson, S. P. Wren, A. De Luca and K. E. Davies, *PLoS One*, 2011, **6**, e19189.
141. *Biomarin Initiates Phase 1 Clinical Study of SMT C1100 for Duchenne Muscular Dystrophy*, http://www.summitplc.com/userfiles/file/10_RNS_01%20DMD%20Phase%20I%20clinical%20trial%20start%20FINAL.pdf, accessed 21 August 2013.
142. *DMD Phase I Clinical Trial Results & Future Plans*, http://www.summitplc.com/userfiles/file/10_RNS_13%20BMRN%20FINAL.pdf, accessed 21 August 2013.
143. *Summit Secures $1.5M Agreement with us DMD Organisations*, http://www.summitplc.com/userfiles/file/11_RNS_17%20SMT%20C1100%20research%20agreement%20funding%20FINAL.pdf, accessed 21 August, 2013.
144. *SMT C1100: A Potential Breakthrough Treatment for Duchenne Muscular Dystrophy (DMD)*, http://www.summitplc.com/userfiles/file/SMT%20C1100%20WMS%20poster%20FINAL%202.pdf, accessed 21 August 2013.
145. T. John, *Utrophin Upregulation Programme A Year of Progress for SMT C1100*, http://www.summitplc.com/userfiles/file/2012%20Action%20Duchenne_Summit%20FINAL.pdf, accessed 21 August 2013.
146. *Summit Outlines Clinical Development Plans for Utrophin Modulator Programme for Duchenne Muscular Dystrophy*, http://www.summitplc.com/userfiles/file/2013_RNS_07%20Clinical%20plans%20for%20DMD%20programme%20FINAL.pdf, accessed 21 August 2013.
147. C. Moorwood, O. Lozynska, N. Suri, A. D. Napper, S. L. Diamond and T. S. Khurana, *PLoS One*, 2011, **6**, e26169.
148. C. Moorwood, N. Soni, G. Patel, S. D. Wilton and T. S. Khurana, *J. Biomol. Screening*, 2013, **18**, 400–406.
149. S. Griffioen, B. Mille-Baker, G. Vella, D. F. Fischer and R. A. J. Janssen, *Identification of Compounds Enhancing Utrophin Expression in Primary Human Skeletal Muscle Cells*, http://www.biofocus.com/_downloads/posters/2009/identification-of-compounds.pdf, accessed 21 August 2013.
150. M. E. Saez, R. Ramirez-Lorca, F. J. Moron and A. Ruiz, *Drug Discovery Today*, 2006, **11**, 917–923.
151. I. Courdier-Fruh and A. Briguet, *Muscle Nerve*, 2006, **33**, 753–759.

152. I. O. Donkor, *Expert Opin. Ther. Pat.*, 2011, **21**, 601–636.
153. C. Lescop, H. Herzner, H. Siendt, R. Bolliger, M. Henneböhle, P. Weyermann, A. Briguet, I. Courdier-Fruh, M. Erb, M. Foster, T. Meier, J. P. Magyar and A. von Sprecher, *Bioorg. Med. Chem. Lett.*, 2005, **15**, 5176–5181.
154. P. Weyermann, H. Herzner, C. Lescop, H. Siendt, R. Bolliger, M. Hennebohle, C. Rummey, A. Briguet, I. Courdier-Fruh, M. Erb, M. Foster, J. P. Magyar, A. von Sprecher and T. Meier, *Lett. Drug Des. Discovery*, 2006, 7, 152–158.
155. A. Briguet, M. Erb, I. Courdier-Fruh, P. Barzaghi, G. Santos, H. Herzner, C. Lescop, H. Siendt, M. Henneboehle, P. Weyermann, J. P. Magyar, J. Dubach-Powell, G. Metz and T. Meier, *FASEB J.*, 2008, **22**, 4190–4200.
156. E. Gazzerro, S. Assereto, A. Bonetto, F. Sotgia, S. Scarfì, A. Pistorio, G. Bonuccelli, M. Cilli, C. Bruno, F. Zara, M. P. Lisanti and C. Minetti, *Am. J. Pathol.*, 2010, **176**, 1863–1877.
157. S. Auvin, B. Pignol, E. Navet, D. Pons, J. G. Marin, D. Bigg and P. E. Chabrier, *Bioorg. Med. Chem. Lett.*, 2004, **14**, 3825–3828.
158. S. Auvin, B. Pignol, E. Navet, M. Troadec, D. Carré, J. Camara, D. Bigg and P. E. Chabrier, *Bioorg. Med. Chem. Lett.*, 2006, **16**, 1586–1589.
159. R. Burdi, M. P. Didonna, B. Pignol, B. Nico, D. Mangieri, J. F. Rolland, C. Camerino, A. Zallone, P. Ferro, F. Andreetta, P. Confalonieri and A. De Luca, *Neuromuscular Disord.*, 2006, **16**, 237–248.
160. P. Miura, J. V. Chakkalakal, L. Boudreault, G. Bélanger, R. L. Hébert, J. M. Renaud and B. J. Jasmin, *Hum. Mol. Genet.*, 2009, **18**, 4640–4649.
161. C. H. Arrowsmith, C. Bountra, P. V. Fish, K. Lee and M. Schapira, *Nat. Rev. Drug Discovery*, 2012, **11**, 384–400.
162. L. Sealy and R. Chalkley, *Cell*, 1978, **14**, 115–121.
163. M. Haberland, R. L. Montgomery and E. N. Olson, *Nat. Rev. Genet.*, 2009, **10**, 32–42.
164. S. Consalvi, V. Saccone, L. Giordani, G. Minetti, C. Mozzetta and P. L. Puri, *Mol. Med.*, 2011, **17**, 457–465.
165. G. C. Minetti, C. Colussi, R. Adami, C. Serra, C. Mozzetta, V. Parente, S. Fortuni, S. Straino, M. Sampaolesi, M. Di Padova, B. Illi, P. Gallinari, C. Steinkühler, M. C. Capogrossi, V. Sartorelli, R. Bottinelli, C. Gaetano and P. L. Puri, *Nat. Med.*, 2006, **12**, 1147–1150.
166. V. Ljubicic, P. Miura, M. Burt, L. Boudreault, S. Khogali, J. A. Lunde, J. M. Renaud and B. J. Jasmin, *Hum. Mol. Genet.*, 2011, **20**, 3478–3493.
167. M. Pauly, F. Daussin, Y. Burelle, T. Li, R. Godin, J. Fauconnier, C. Koechlin-Ramonatxo, G. Hugon, A. Lacampagne, M. Coisy-Quivy, F. Liang, S. Hussain, S. Matecki and B. J. Petrof, *Am. J. Pathol.*, 2012, **181**, 583–592.
168. C. R. Bueno Júnior, L. C. Pantaleão, V. A. Voltarelli, L. H. Bozi, P. C. Brum and M. Zatz, *PLoS One*, 2012, 7, e45699.
169. J. Giacomotto, C. Pertl, C. Borrel, M. C. Walter, S. Bulst, B. Johnsen, D. L. Baillie, H. Lochmüller, C. Thirion and L. Ségalat, *Hum. Mol. Genet.*, 2009, **18**, 4089–4101.

170. M. C. Mariol and L. Ségalat, *Curr. Biol.,* 2001, **11**, 1691–1694.
171. A. C. McPherron, A. M. Lawler and S. J. Lee, *Nature,* 1997, **387**, 83–90.
172. S. Bogdanovich, T. O. Krag, E. R. Barton, L. D. Morris, L. A. Whittemore, R. S. Ahima and T. S. Khurana, *Nature,* 2002, **420**, 418–421.
173. L. T. Bish, M. M. Sleeper, S. C. Forbes, K. J. Morine, C. Reynolds, G. E. Singletary, D. Trafny, J. Pham, J. Bogan, J. N. Kornegay, K. Vandenborne, G. A. Walter and H. L. Sweeney, *Hum. Gene Ther.,* 2011, **22**, 1499–1509.
174. K. R. Wagner, J. L. Fleckenstein, A. A. Amato, R. J. Barohn, K. Bushby, D. M. Escolar, K. M. Flanigan, A. Pestronk, R. Tawil, G. I. Wolfe, M. Eagle, J. M. Florence, W. M. King, S. Pandya, V. Straub, P. Juneau, K. Meyers, C. Csimma, T. Araujo, R. Allen, S. A. Parsons, J. M. Wozney, E. R. Lavallie and J. R. Mendell, *Ann. Neurol.,* 2008, **63**, 561–571.
175. S. Bogdanovich, K. J. Perkins, T. O. Krag, L. A. Whittemore and T. S. Khurana, *FASEB J.,* 2005, **19**, 543–549.
176. J. N. Cash, E. B. Angerman, R. J. Kirby, L. Merck, W. L. Seibel, M. D. Wortman, R. Papoian, S. Nelson and T. B. Thompson, *J. Biomol. Screening,* 2013, **18**, 837–844.
177. D. U. Kemaladewi, W. M. Hoogaars, S. H. van Heiningen, S. Terlouw, D. J. de Gorter, J. T. den Dunnen, G. J. van Ommen, A. Aartsma-Rus, P. ten Dijke and P. A. 't Hoen, *BMC Med. Genomics,* 2011, **4**, 36.
178. A. Malerba, J. K. Kang, G. McClorey, A. F. Saleh, L. Popplewell, M. J. Gait, M. J. Wood and G. Dickson, *Mol. Ther.–Nucleic Acids,* 2012, **1**, e62.
179. A. R. Amenta, A. Yilmaz, S. Bogdanovich, B. A. McKechnie, M. Abedi, T. S. Khurana and J. R. Fallon, *Proc. Natl. Acad. Sci. U. S. A.,* 2011, **108**, 762–767.
180. *Tirvorsan Pharmaceuticals Company Profile,* http://www.tivorsan.com/company-profile/, accessed 9 August 2013.
181. B. Péault, M. Rudnicki, Y. Torrente, G. Cossu, J. P. Tremblay, T. Partridge, E. Gussoni, L. M. Kunkel and J. Huard, *Mol. Ther.,* 2007, **15**, 867–877.
182. R. Benchaouir, M. Meregalli, A. Farini, G. D'Antona, M. Belicchi, A. Goyenvalle, M. Battistelli, N. Bresolin, R. Bottinelli, L. Garcia and Y. Torrente, *Cell Stem Cell,* 2007, **1**, 646–657.
183. J. L. Chun, R. O'Brien, M. H. Song, B. F. Wondrasch and S. E. Berry, *Stem Cells Transl. Med.,* 2013, **2**, 68–80.
184. G. Kawahara, J. A. Karpf, J. A. Myers, M. S. Alexander, J. R. Guyon and L. M. Kunkel, *Proc. Natl. Acad. Sci. U. S. A.,* 2011, **108**, 5331–5336.
185. M. R. Lunn and C. H. Wang, *Lancet,* 2008, **371**, 2120–2133.
186. *Population Statistics for SMA,* http://www.jtsma.org.uk/population_statistics_for_sma_in_the_uk.html, accessed 9 August 2013.
187. D. D. Coovert, T. T. Le, P. E. McAndrew, J. Strasswimmer, T. O. Crawford, J. R. Mendell, S. E. Coulson, E. J. Androphy, T. W. Prior and A. H. Burghes, *Hum. Mol. Genet.,* 1997, **6**, 1205–1214.
188. K. J. Swoboda, T. W. Prior, C. B. Scott, T. P. McNaught, M. C. Wride, S. P. Reyna and M. B. Bromberg, *Ann. Neurol.,* 2005, **57**, 704–712.

189. E. Zanoteli, J. R. Maximino, U. Conti Reed and G. Chadi, *Funct. Neurol.*, 2010, **25**, 73–79.
190. T. O. Crawford, *Neuromuscular Disord.*, 2004, **14**, 456–460.
191. E. Also-Rallo, L. Alías, R. Martínez-Hernández, L. Caselles, M. J. Barceló, M. Baiget, S. Bernal and E. F. Tizzano, *Eur. J. Hum. Genet.*, 2011, **19**, 1059–1065.
192. M. E. R. Butchbach and A. H. M. Burghes, *Drug Discovery Today: Dis. Models*, 2004, **1**, 151–156.
193. l. T. Hao, A. H. Burghes and C. E. Beattie, *Mol. Neurodegener.*, 2011, **6**, 24.
194. L. K. Tsai, M. S. Tsai, T. B. Lin, W. L. Hwu and H. Li, *Neurobiol. Dis.*, 2006, **24**, 286–295.
195. K. J. Swoboda, J. T. Kissel, T. O. Crawford, M. B. Bromberg, G. Acsadi, G. D'Anjou, K. J. Krosschell, S. P. Reyna, M. K. Schroth, C. B. Scott and L. R. Simard, *J. Child Neurol.*, 2007, **22**, 957–966.
196. D. M. Sproule and P. Kaufmann, *Ther. Adv. Neurol. Disord.*, 2010, **3**, 173–185.
197. M. Sendtner, *Nat. Neurosci.*, 2010, **13**, 795–799.
198. R. M. Pruss, *Expert Opin. Drug Discovery*, 2011, **6**, 827–837.
199. A. Lewelt, T. M. Newcomb and K. J. Swoboda, *Curr. Neurol. Neurosci. Rep.*, 2012, **12**, 42–53.
200. E. Dominguez, T. Marais, N. Chatauret, S. Benkhelifa-Ziyyat, S. Duque, P. Ravassard, R. Carcenac, S. Astord, A. Pereira de Moura, T. Voit and M. Barkats, *Hum. Mol. Genet.*, 2011, **20**, 681–693.
201. E. O'Hare and P. J. Young, *Mol. Med. Rep.*, 2009, **2**, 3–5.
202. S. Corti, M. Nizzardo, M. Nardini, C. Donadoni, S. Salani, D. Ronchi, C. Simone, M. Falcone, D. Papadimitriou, F. Locatelli, N. Mezzina, F. Gianni, N. Bresolin and G. P. Comi, *Brain*, 2010, **133**, 465–481.
203. *MotorGraft Clinical Trial for SMA Placed on FDA Clinical Hold*, http://smaheadlines.com/2011/02/08/motorgraft-clinical-trial-for-sma-placed-on-fda-clinical-hold/, accessed 21 August 2013.
204. L. L. Rubin and K. M. Haston, *BMC Biol.*, 2011, **9**, 42.
205. E. J. Culme-Seymour, N. L. Davie, D. A. Brindley, S. Edwards-Parton and C. Mason, *Regener. Med.*, 2012, 7, 455–462.
206. J. J. Unternaehrer and G. Q. Daley, *Philos. Trans. R. Soc., B*, 2011, **366**, 2274–2285.
207. S. Corti, M. Nizzardo, C. Simone, M. Falcone, M. Nardini, D. Ronchi, C. Donadoni, S. Salani, G. Riboldi, F. Magri, G. Menozzi, C. Bonaglia, F. Rizzo, N. Bresolin and G. P. Comi, *Sci. Transl. Med.*, 2012, **4**, 165ra162.
208. A. D. Ebert, J. Yu, F. F. Rose, V. B. Mattis, C. L. Lorson, J. A. Thomson and C. N. Svendsen, *Nature*, 2009, **457**, 277–280.
209. J. Sandoe and K. Eggan, *Nat. Neurosci.*, 2013, **16**, 780–789.
210. C. J. Sumner, T. N. Huynh, J. A. Markowitz, J. S. Perhac, B. Hill, D. D. Coovert, K. Schussler, X. Chen, J. Jarecki, A. H. Burghes, J. P. Taylor and K. H. Fischbeck, *Ann. Neurol.*, 2003, **54**, 647–654.

211. L. Brichta, Y. Hofmann, E. Hahnen, F. A. Siebzehnrubl, H. Raschke, I. Blumcke, I. Y. Eyupoglu and B. Wirth, *Hum. Mol. Genet.,* 2003, **12**, 2481–2489.
212. E. C. Wolstencroft, V. Mattis, A. A. Bajer, P. J. Young and C. L. Lorson, *Hum. Mol. Genet.,* 2005, **14**, 1199–1210.
213. M. A. Saporta, M. Grskovic and J. T. Dimos, *Stem Cell Res. Ther.,* 2011, **2**, 37.
214. R. M. Pruss, M. Giraudon-Paoli, S. Morozova, P. Berna, J. L. Abitbol and T. Bordet, *Future Med. Chem.,* 2010, **2**, 1429–1440.
215. J. J. Cherry and E. J. Androphy, *Future Med. Chem.,* 2012, **4**, 1733–1750.
216. M. A. Lorson and C. L. Lorson, *Future Med. Chem.,* 2012, **4**, 2067–2084.
217. H. Haddad, C. Cifuentes-Diaz, A. Miroglio, N. Roblot, V. Joshi and J. Melki, *Muscle Nerve,* 2003, **28**, 432–437.
218. B. S. Russman, S. T. Iannaccone and F. J. Samaha, *Arch. Neurol.,* 2003, **60**, 1601–1603.
219. M. Dimitriadi, M. J. Kye, G. Kalloo, J. M. Yersak, M. Sahin and A. C. Hart, *J. Neurosci.,* 2013, **33**, 6557–6562.
220. R. G. Miller, D. H. Moore, V. Dronsky, W. Bradley, R. Barohn, W. Bryan, T. W. Prior, D. F. Gelinas, S. Iannaccone, J. Kissel, R. Leshner, J. Mendell, M. Mendoza, B. Russman, F. Samaha, S. Smith and S. S. Group, *J. Neurol. Sci.,* 2001, **191**, 127–131.
221. L. Merlini, A. Solari, G. Vita, E. Bertini, C. Minetti, T. Mongini, E. Mazzoni, C. Angelini and L. Morandi, *J. Child. Neurol.,* 2003, **18**, 537–541.
222. M. Nizzardo, M. Nardini, D. Ronchi, S. Salani, C. Donadoni, F. Fortunato, G. Colciago, M. Falcone, C. Simone, G. Riboldi, A. Govoni, N. Bresolin, G. P. Comi and S. Corti, *Exp. Neurol.,* 2011, **229**, 214–225.
223. J. D. Berry, J. M. Shefner, R. Conwit, D. Schoenfeld, M. Keroack, D. Felsenstein, L. Krivickas, W. S. David, F. Vriesendorp, A. Pestronk, J. B. Caress, J. Katz, E. Simpson, J. Rosenfeld, R. Pascuzzi, J. Glass, K. Rezania, J. D. Rothstein, D. J. Greenblatt, M. E. Cudkowicz and N. A. Consortium, *PLoS One,* 2013, **8**, e61177.
224. T. Bordet, P. Berna, J.-L. Abitbol and R. M. Pruss, *Pharmaceuticals,* 2010, **23**, 345–368.
225. C. Angelozzi, F. Borgo, F. D. Tiziano, A. Martella, G. Neri and C. Brahe, *J. Med. Genet.,* 2008, **45**, 29–31.
226. F. D. Tiziano, R. Lomastro, A. M. Pinto, S. Messina, A. D'Amico, S. Fiori, C. Angelozzi, M. Pane, E. Mercuri, E. Bertini, G. Neri and C. Brahe, *J. Med. Genet.,* 2010, **47**, 856–858.
227. C. R. Heier and C. J. DiDonato, *Hum. Mol. Genet.,* 2009, **18**, 1310–1322.
228. C. W. Chang, Y. Hui, B. Elchert, J. Wang, J. Li and R. Rai, *Org. Lett.,* 2002, **4**, 4603–4606.
229. V. Mattis, C. Chang and C. Lorson, *Neurosci. Lett.,* 2012, **525**, 72–75.
230. A. H. Burghes and V. L. McGovern, *Genes Dev.,* 2010, **24**, 1574–1579.

231. M. L. Hastings, J. Berniac, Y. H. Liu, P. Abato, F. M. Jodelka, L. Barthel, S. Kumar, C. Dudley, M. Nelson, K. Larson, J. Edmonds, T. Bowser, M. Draper, P. Higgins and A. R. Krainer, *Sci. Transl. Med.,* 2009, **1**, 5ra12.
232. B. Khoo and A. R. Krainer, *Curr. Opin. Mol. Ther.,* 2009, **11**, 108–115.
233. Z. Zhang, O. Kelemen, M. A. van Santen, S. M. Yelton, A. E. Wendlandt, V. M. Sviripa, M. Bollen, M. Beullens, H. Urlaub, R. Lührmann, D. S. Watt and S. Stamm, *J. Biol. Chem.,* 2011, **286**, 10126–10136.
234. C. Y. Yuo, H. H. Lin, Y. S. Chang, W. K. Yang and J. G. Chang, *Ann. Neurol.,* 2008, **63**, 26–34.
235. L. E. Kernochan, M. L. Russo, N. S. Woodling, T. N. Huynh, A. M. Avila, K. H. Fischbeck and C. J. Sumner, *Hum. Mol. Genet.,* 2005, **14**, 1171–1182.
236. M. C. Evans, J. J. Cherry and E. J. Androphy, *Biochem. Biophys. Res. Commun.,* 2011, **414**, 25–30.
237. L. Garbes, M. Riessland, I. Hölker, R. Heller, J. Hauke, C. Tränkle, R. Coras, I. Blümcke, E. Hahnen and B. Wirth, *Hum. Mol. Genet.,* 2009, **18**, 3645–3658.
238. J. G. Chang, H. M. Hsieh-Li, Y. J. Jong, N. M. Wang, C. H. Tsai and H. Li, *Proc. Natl. Acad. Sci. U. S. A.,* 2001, **98**, 9808–9813.
239. C. Brahe, T. Vitali, F. D. Tiziano, C. Angelozzi, A. M. Pinto, F. Borgo, U. Moscato, E. Bertini, E. Mercuri and G. Neri, *Eur. J. Hum. Genet.,* 2005, **13**, 256–259.
240. R. Butler and G. P. Bates, *Nat. Rev. Neurosci.,* 2006, **7**, 784–796.
241. L. K. Tsai, M. S. Tsai, C. H. Ting and H. Li, *J. Mol. Med.,* 2008, **86**, 1243–1254.
242. C. C. Weihl, A. M. Connolly and A. Pestronk, *Neurology,* 2006, **67**, 500–501.
243. L. K. Tsai, C. C. Yang, W. L. Hwu and H. Li, *Eur. J. Neurol.,* 2007, **14**, e8–e9.
244. G. Natasha, K. G. Brandom, E. C. Young and P. J. Young, *Mol. Med. Rep.,* 2008, **1**, 161–165.
245. K. J. Swoboda, C. B. Scott, S. P. Reyna, T. W. Prior, B. LaSalle, S. L. Sorenson, J. Wood, G. Acsadi, T. O. Crawford, J. T. Kissel, K. J. Krosschell, G. D'Anjou, M. B. Bromberg, M. K. Schroth, G. M. Chan, B. Elsheikh and L. R. Simard, *PLoS One,* 2009, **4**, e5268.
246. K. J. Swoboda, C. B. Scott, T. O. Crawford, L. R. Simard, S. P. Reyna, K. J. Krosschell, G. Acsadi, B. Elsheik, M. K. Schroth, G. D'Anjou, B. LaSalle, T. W. Prior, S. L. Sorenson, J. A. Maczulski, M. B. Bromberg, G. M. Chan, J. T. Kissel and P. C. S. M. A. I. Network, *PLoS One,* 2010, **5**, e12140.
247. J. T. Kissel, C. B. Scott, S. P. Reyna, T. O. Crawford, L. R. Simard, K. J. Krosschell, G. Acsadi, B. Elsheik, M. K. Schroth, G. D'Anjou, B. LaSalle, T. W. Prior, S. Sorenson, J. A. Maczulski, M. B. Bromberg, G. M. Chan, K. J. Swoboda and P. C. S. M. A. I. Network, *PLoS One,* 2011, **6**, e21296.

248. *ClinicalTrials.gov Identifier: NCT01671384 and NCT00061607*, http://www.clinicaltrials.gov, accessed 14 August 2013.
249. E. Hahnen, I. Y. Eyüpoglu, L. Brichta, K. Haastert, C. Tränkle, F. A. Siebzehnrübl, M. Riessland, I. Hölker, P. Claus, J. Romstöck, R. Buslei, B. Wirth and I. Blümcke, *J. Neurochem.*, 2006, **98**, 193–202.
250. M. Riessland, B. Ackermann, A. Förster, M. Jakubik, J. Hauke, L. Garbes, I. Fritzsche, Y. Mende, I. Blumcke, E. Hahnen and B. Wirth, *Hum. Mol. Genet.*, 2010, **19**, 1492–1506.
251. M. Riessland, L. Brichta, E. Hahnen and B. Wirth, *Hum. Genet.*, 2006, **120**, 101–110.
252. A. M. Avila, B. G. Burnett, A. A. Taye, F. Gabanella, M. A. Knight, P. Hartenstein, Z. Cizman, N. A. Di Prospero, L. Pellizzoni, K. H. Fischbeck and C. J. Sumner, *J. Clin. Invest.*, 2007, **117**, 659–671.
253. J. Schreml, M. Riessland, M. Paterno, L. Garbes, K. Roßbach, B. Ackermann, J. Krämer, E. Somers, S. H. Parson, R. Heller, A. Berkessel, A. Sterner-Kock and B. Wirth, *Eur. J. Hum. Genet.*, 2013, **21**, 643–652.
254. W. C. Liang, C. Y. Yuo, J. G. Chang, Y. C. Chen, Y. F. Chang, H. Y. Wang, Y. H. Ju, S. S. Chiou and Y. J. Jong, *J. Neurol. Sci.*, 2008, **268**, 87–94.
255. T. H. Chen, J. G. Chang, Y. H. Yang, H. H. Mai, W. C. Liang, Y. C. Wu, H. Y. Wang, Y. B. Huang, S. M. Wu, Y. C. Chen, S. N. Yang and Y. J. Jong, *Neurology*, 2010, 75, 2190–2197.
256. T. A. Miller, D. J. Witter and S. Belvedere, *J. Med. Chem.*, 2003, **46**, 5097–5116.
257. D. S. Hewings, T. P. Rooney, L. E. Jennings, D. A. Hay, C. J. Schofield, P. E. Brennan, S. Knapp and S. J. Conway, *J. Med. Chem.*, 2012, **55**, 9393–9413.
258. B. G. Burnett, E. Muñoz, A. Tandon, D. Y. Kwon, C. J. Sumner and K. H. Fischbeck, *Mol. Cell. Biol.*, 2009, **29**, 1107–1115.
259. D. Y. Kwon, W. W. Motley, K. H. Fischbeck and B. G. Burnett, *Hum. Mol. Genet.*, 2011, **20**, 3667–3677.
260. N. R. Makhortova, M. Hayhurst, A. Cerqueira, A. D. Sinor-Anderson, W. N. Zhao, P. W. Heiser, A. C. Arvanites, L. S. Davidow, Z. O. Waldon, J. A. Steen, K. Lam, H. D. Ngo and L. L. Rubin, *Nat. Chem. Biol.*, 2011, 7, 544–552.
261. M. Bowerman, A. Beauvais, C. L. Anderson and R. Kothary, *Hum. Mol. Genet.*, 2010, **19**, 1468–1478.
262. M. Bowerman, L. M. Murray, J. G. Boyer, C. L. Anderson and R. Kothary, *BMC Med.*, 2012, **10**, 24.
263. F. Farooq, F. Abadía-Molina, D. Mackenzie, J. Hadwen, F. Shamim, S. O'Reilly, M. Holcik and A. Mackenzie, *Hum. Mol. Genet.*, 2013, **22**, 3415–3424.
264. P. W. Hsiao, C. C. Chang, H. F. Liu, C. M. Tsai, T. H. Chiu and J. I. Chao, *Toxicol. Appl. Pharmacol.*, 2007, **222**, 97–104.
265. M. L. Zhang, C. L. Lorson, E. J. Androphy and J. Zhou, *Gene Ther.*, 2001, **8**, 1532–1538.

266. M. S. Sakla and C. L. Lorson, *Hum. Genet.*, 2008, **122**, 635–643.
267. D. Dayangaç-Erden, G. Bora, P. Ayhan, C. Kocaefe, S. Dalkara, K. Yelekçi, A. S. Demir and H. Erdem-Yurter, *Chem. Biol. Drug Des.*, 2009, **73**, 355–364.
268. Y. Y. Hsu, Y. J. Jong, H. H. Tsai, Y. T. Tseng, L. M. An and Y. C. Lo, *Br. J. Pharmacol.*, 2012, **166**, 1114–1126.
269. M. R. Lunn, D. E. Root, A. M. Martino, S. P. Flaherty, B. P. Kelley, D. D. Coovert, A. H. Burghes, N. T. Man, G. E. Morris, J. Zhou, E. J. Androphy, C. J. Sumner and B. R. Stockwell, *Chem. Biol.*, 2004, **11**, 1489–1493.
270. *SMA Summit on Drug Development*, http://www.fsma.org/UploadedFiles/Research/Clinical/SMADrugSummit/SMADrugSummitSlides/session7slidesweb.pdf, accessed 9 August 2013.
271. C. Andreassi, J. Jarecki, J. Zhou, D. D. Coovert, U. R. Monani, X. Chen, M. Whitney, B. Pollok, M. Zhang, E. Androphy and A. H. Burghes, *Hum. Mol. Genet.*, 2001, **10**, 2841–2849.
272. J. Jarecki, X. Chen, A. Bernardino, D. D. Coovert, M. Whitney, A. Burghes, J. Stack and B. A. Pollok, *Hum. Mol. Genet.*, 2005, **14**, 2003–2018.
273. J. Thurmond, M. E. Butchbach, M. Palomo, B. Pease, M. Rao, L. Bedell, M. Keyvaņ, G. Pai, R. Mishra, M. Haraldsson, T. Andresson, G. Bragason, M. Thosteinsdottir, J. M. Bjornsson, D. D. Coovert, A. H. Burghes, M. E. Gurney and J. Singh, *J. Med. Chem.*, 2008, **51**, 449–469.
274. *Quinazoline Program Details*, http://www.fsma.org/Research/DrugDiscovery/QuinazolineProgram/, accessed 9 August 2013.
275. M. E. Butchbach, J. Singh, M. Thorsteinsdóttir, L. Saieva, E. Slominski, J. Thurmond, T. Andrésson, J. Zhang, J. D. Edwards, L. R. Simard, L. Pellizzoni, J. Jarecki, A. H. Burghes and M. E. Gurney, *Hum. Mol. Genet.*, 2010, **19**, 454–467.
276. J. P. Van Meerbeke, R. M. Gibbs, H. L. Plasterer, W. Miao, Z. Feng, M. Y. Lin, A. A. Rucki, C. D. Wee, B. Xia, S. Sharma, V. Jacques, D. K. Li, L. Pellizzoni, J. R. Rusche, C. P. Ko and C. J. Sumner, *Hum. Mol. Genet.*, 2013, **22**, 4074–4083.
277. J. Singh, M. Salcius, S. W. Liu, B. L. Staker, R. Mishra, J. Thurmond, G. Michaud, D. R. Mattoon, J. Printen, J. Christensen, J. M. Bjornsson, B. A. Pollok, M. Kiledjian, L. Stewart, J. Jarecki and M. E. Gurney, *ACS Chem. Biol.*, 2008, **3**, 711–722.
278. *Pfizer Licenses Families of Spinal Muscular Atrophy Quinazoline Drug Program from Repligen*, http://www.fsma.org/LatestNews/index.cfm?ID=7490&TYPE=1150, accessed 9 August 2013.
279. J. J. Cherry, M. C. Evans, J. Ni, G. D. Cuny, M. A. Glicksman and E. J. Androphy, *J. Biomol. Screening*, 2012, **17**, 481–495.
280. N. Thorne, M. Shen, W. A. Lea, A. Simeonov, S. Lovell, D. S. Auld and J. Inglese, *Chem. Biol.*, 2012, **19**, 1060–1072.
281. J. J. Cherry, E. Y. Osman, M. C. Evans, S. Choi, X. Xing, G. D. Cuny, M. A. Glicksman, C. L. Lorson and E. J. Androphy, *EMBO Mol. Med.*, 2013, **5**, 1103–1118.

282. E. A. Kuo, P. T. Hambleton, D. P. Kay, P. L. Evans, S. S. Matharu, E. Little, N. McDowall, C. B. Jones, C. J. Hedgecock, C. M. Yea, A. W. Chan, P. W. Hairsine, I. R. Ager, W. R. Tully, R. A. Williamson and R. Westwood, *J. Med. Chem.,* 1996, **39**, 4608–4621.
283. R. R. Letso, A. J. Bauer, M. R. Lunn, W. S. Yang and B. R. Stockwell, *ACS Chem. Biol.,* 2013, **8**, 914–922.
284. M. Malumbres and M. Barbacid, *Nat. Rev. Cancer,* 2003, **3**, 459–465.
285. J. Xiao, J. J. Marugan, W. Zheng, S. Titus, N. Southall, J. J. Cherry, M. Evans, E. J. Androphy and C. P. Austin, *J. Med. Chem.,* 2011, **54**, 6215–6233.
286. K. H. Bleicher, H. J. Böhm, K. Müller and A. I. Alanine, *Nat. Rev. Drug Discovery,* 2003, **2**, 369–378.
287. X. Xia and S. T. Wong, *Stem Cells,* 2012, **30**, 1800–1807.
288. S. Heynen-Genel, L. Pache, S. K. Chanda and J. Rosen, *Expert Opin. Drug Discovery,* 2012, **7**, 955–968.
289. M. A. Fabian, W. H. Biggs, D. K. Treiber, C. E. Atteridge, M. D. Azimioara, M. G. Benedetti, T. A. Carter, P. Ciceri, P. T. Edeen, M. Floyd, J. M. Ford, M. Galvin, J. L. Gerlach, R. M. Grotzfeld, S. Herrgard, D. E. Insko, M. A. Insko, A. G. Lai, J. M. Lélias, S. A. Mehta, Z. V. Milanov, A. M. Velasco, L. M. Wodicka, H. K. Patel, P. P. Zarrinkar and D. J. Lockhart, *Nat. Biotechnol.,* 2005, **23**, 329–336.
290. P. C. Chen, I. N. Gaisina, B. F. El-Khodor, S. Ramboz, N. R. Makhortova, L. L. Rubin and A. P. Kozikowski, *ACS Chem. Neurosci.,* 2012, **3**, 5–11.
291. M. W. Karaman, S. Herrgard, D. K. Treiber, P. Gallant, C. E. Atteridge, B. T. Campbell, K. W. Chan, P. Ciceri, M. I. Davis, P. T. Edeen, R. Faraoni, M. Floyd, J. P. Hunt, D. J. Lockhart, Z. V. Milanov, M. J. Morrison, G. Pallares, H. K. Patel, S. Pritchard, L. M. Wodicka and P. P. Zarrinkar, *Nat. Biotechnol.,* 2008, **26**, 127–132.

RARE CANCERS

CHAPTER 12

Unleashing the Power of Semi-Synthesis: The Discovery of Torisel®

JERAULD S. SKOTNICKI*[a] AND MAGID A. ABOU-GHARBIA[b]

[a]Chemical Sciences, Pfizer, Pearl River, New York 10965, USA; [b]Moulder Center for Drug Discovery Research, School of Pharmacy, Temple University, Philadelphia, Pennsylvania 10140, USA
*E-mail: clskot@verizon.net

12.1 Introduction

Natural products offer substantial opportunities for drug discovery in that they provide a rich source of privileged structures yielding a high degree of chemical diversity across a broad chemical space.[1] Isolated and characterised by virtue of biological activity, they often possess interesting pharmacological profiles and properties. Improved precision in isolation, purification, characterisation and production have increased the availability of these secondary metabolites to explore their inherent chemical and biological diversity. Enriched with complex, multifunctional and distinct molecular landscapes, natural products provide creative starting points for medicinal chemists to test hypotheses *via* semi-synthetic manipulation. With access to structural regions of the parent molecules for pharmacophore interrogation, semi-synthetic derivatives can be designed to examine biochemical principles, probe, distinguish and define fundamental biological processes, improve chemical, pharmacological and pharmaceutical properties over the

RSC Drug Discovery Series No. 38
Orphan Drugs and Rare Diseases
Edited by David C Pryde and Michael J Palmer

Published by the Royal Society of Chemistry, www.rsc.org

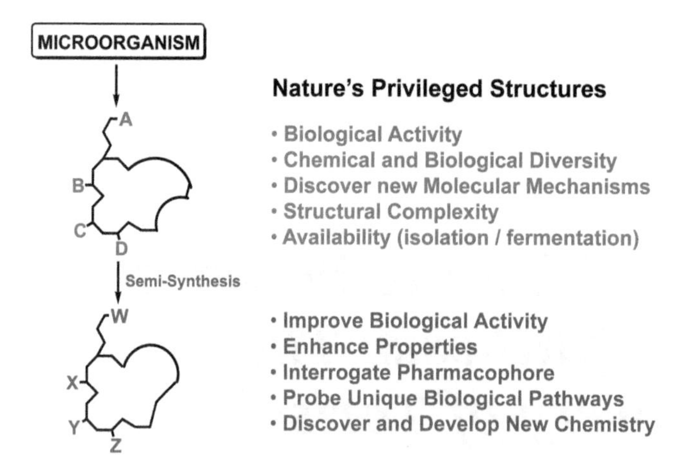
MICROORGANISM
A
B
C
D
Semi-Synthesis
W
X
Y
Z
Nature's Privileged Structures
• Biological Activity
• Chemical and Biological Diversity
• Discover new Molecular Mechanisms
• Structural Complexity
• Availability (isolation / fermentation)
• Improve Biological Activity
• Enhance Properties
• Interrogate Pharmacophore
• Probe Unique Biological Pathways
• Discover and Develop New Chemistry

Figure 12.1 Practice and rewards of semi-synthesis.

parent, and to extend the therapeutic arena of the structural class, all leading to the generation of innovative therapies. There are significant challenges to this research: availability of the natural product in sufficient quantity with suitable purity, chemical stability of the molecule, limitations of available and requisite transformations, analyses and purification methods, and notably precise structure–activity requirements. Achievement of semi-synthetic goals mandate accountability for the above-mentioned synthetic limitations coupled with synthetic efficiency afforded by judicious design of synthetic pathways. Thus, successful strategies further result in the development of new synthetic methods and reagents, optimisation of reaction conditions, and creation of enabling analytical and purification technologies (Figure 12.1). One example of a successful semi-synthetic campaign is the Wyeth Rapamycin Analogs programme, culminating in the discovery and development of Torisel® (temsirolimus, CCI-779) approved for the treatment of renal cell carcinoma.

12.2 Background

Rapamycin (Rapamune®, Wyeth, now Pfizer†) is a potent polyketide immunosuppressive agent, produced by *Streptomyces hygroscopicus* (Figure 12.2).[2] It is approved for the treatment of transplantation rejection. Rapamycin forms a complex with FKBP and mTOR which elicits its unique mechanism of action that is central in a number of biological processes. The approximate FKBP (red) and mTOR (blue) binding regions of rapamycin are depicted. mTOR is a large, polypeptide, serine/threonine specific kinase of the PI3K-related kinase family and is regulated by the availability of nutrients and

†Wyeth was acquired by Pfizer on 16 October 2009.

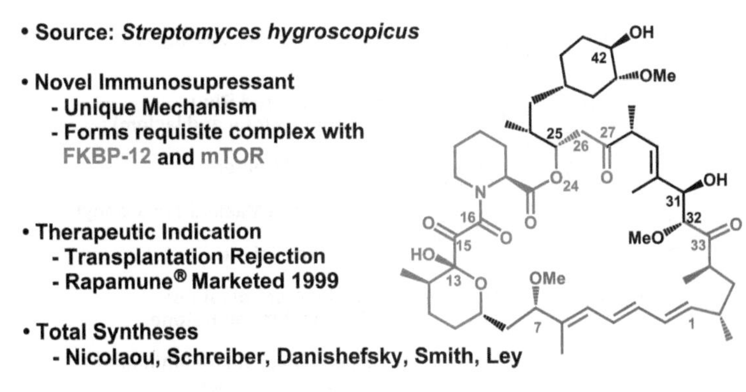
• Source: Streptomyces hygroscopicus
• Novel Immunosupressant
- Unique Mechanism
- Forms requisite complex with FKBP-12 and mTOR
• Therapeutic Indication
- Transplantation Rejection
- Rapamune® Marketed 1999
• Total Syntheses
- Nicolaou, Schreiber, Danishefsky, Smith, Ley
• Degradation Studies - Limited Semi-synthesis
OH
42
OMe
25
26
27
O24
OH
31
32
MeO
33
16
15
HO
13
OMe
7
1

Figure 12.2 Rapamycin.

growth factors in the cell environment. When mTOR is activated through growth factor receptors, it phosphorylates two proteins – p70S6K1 and 4E-BP1 – that regulate the translation of proteins that control cell division. Consequently, inhibition of mTOR activity produces cell cycle arrest, stopping progression from the G_1 to the S phase of the cell cycle. A recent article describing the salient features of the interplay of mTOR, its protein partners and allosteric inhibitors has been published.[3] Because of the critical interest in the molecule as an important therapeutic agent and its ability to serve as a biochemical tool to provide greater understanding of signalling pathways, combined with an exquisite architectural motif, several leading synthetic organic chemists have investigated the total synthesis of rapamycin. To date, ingenious and divergent solutions to this daunting problem have been achieved evinced by the five total syntheses that have been reported.[4]

To further the clinical success of Rapamune® a rapamycin analogues programme was initiated. The genesis of the programme involved two main goals. One goal was the identification of development candidates for transplantation and for other therapeutic indications. In this regard, the team needed to be cognisant of pharmacological parameters (potency, efficacy, metabolism, bioavailability), physical properties (stability, solubility, crystallinity, solid form) and synthetic processes (reactivity, selectivity, stability, reagents, purification). An equal goal was to develop rapamycin chemistry to decipher biological processes to enable enhancement of our understanding of the dynamics and the local politics of the rapamycin–FKBP–mTOR interaction.

Candidate compounds were evaluated in a number of biological assays. In the immunoinflammatory therapeutic area, the primary *in vitro* assays were the lymphocyte activity factor (LAF) assay, a functional assay for T-cell proliferation, and the peptidyl prolyl isomerase (PPIase) assay, a functional assay for immunophilin (FKBP) affinity. Primary *in vivo* models were the mouse skin graft rejection model and the rat adjuvant arthritis model. In

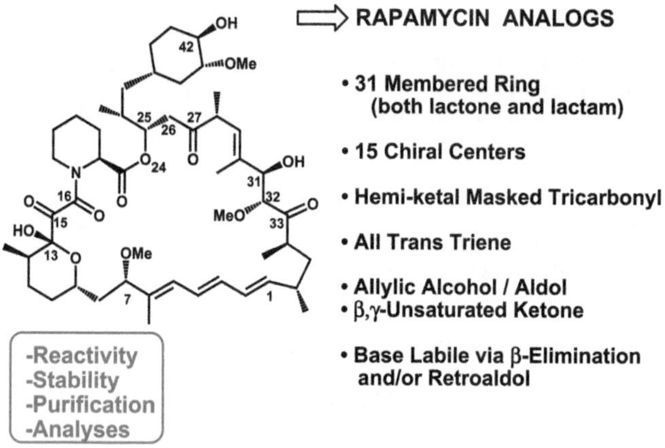

Figure 12.3 Rapamycin structural features.

oncology, compounds were subjected to a battery of cell-based assays and mouse allograft models reflective of various tumour types. An FKBP binding assay was developed later in the programme. In addition, numerous mechanistic assays and models were developed to enable advanced pharmacological assessment.

With its intricate juxtaposition of functional groups, rapamycin provides a fertile and versatile platform for semi-synthesis. Modification of rapamycin is challenging but rewarding, and must take into account a 31-membered ring containing both a lactone and a lactam, an all-trans triene unit, 15 chiral centres, a masked contiguous tricarbonyl region, an allylic alcohol, a β,γ-unsaturated ketone, and a segment susceptible to β-elimination (Figure 12.3). Rapamycin derivatives may serve as biochemical tool molecules or potential drug candidates and semi-synthetic manipulations are designed to anticipate substitution patterns or conformational changes that affect binding to either of the two protein partners.

As part of a programme aimed at the identification of novel rapamycin analogues, we have explored systematic semi-synthetic point modifications to functional groups at essential regions of the molecule. These include, amongst others, alcohol functionalisation, alteration of the triene unit, manipulation of the carbonyl groups, modification of the pipecolinate region, ring opening, contraction and expansion (Figure 12.4). Each compound was designed to probe specific properties and to expand knowledge in the rapamycin arena. Using these semi-synthetic derivatives, the Wyeth team introduced and developed creative strategies to perturb and distinguish fundamental biological processes, probe pharmacological attributes, elucidate chemical characteristics and optimise pharmaceutical properties. Structural boundaries for activity were surveyed and conformational requirements for effective interactions with FKBP and mTOR were delineated. To this end, the scope of synthetic transformations was defined,

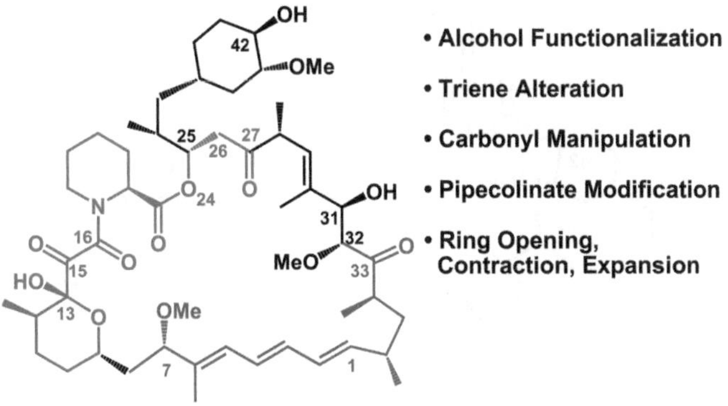

- Alcohol Functionalization
- Triene Alteration
- Carbonyl Manipulation
- Pipecolinate Modification
- Ring Opening, Contraction, Expansion

Figure 12.4 Directed rapamycin point modifications.

refined and expanded, reaction and purification conditions were optimised, enabling analytical characterisation was enhanced, and new avenues for pharmacological analyses were identified.

Through this process, Wyeth had identified and developed considerable equity in rapamycin analogues initially targeted for use in transplantation/immunoinflammatory programmes. Numerous semi-synthetic derivatives displayed potent *in vitro* activity and *in vivo* efficacy. To expand the horizon beyond these therapeutic areas, preclinical studies both at the NCI and at Wyeth were initiated to evaluate rapamycin derivatives for oncology.[5] The AKT/PTEN/mTOR pathway is mutated in a high percentage of human cancers. At that time, this mechanism of action (mTOR inhibition) was unique for an anticancer drug suggesting potential for cytostatic influence. A number of compounds from this equity displayed anti-tumour activity *in vivo*. In view of the comparable activity shown amongst a number of analogues, attention became focused on two back-up leads, WAY-129327 and WAY-130779 (Figure 12.5) from the inflammation/transplantation programmes that had shown good *in vivo* efficacy in animal models in those programmes. Both compounds exhibit potent activity against a histologically diverse panel of cell lines and displayed anti-tumour activity against a panel of human xenografts in nude mice. WAY-130779 (known as CCI-779; cell cycle inhibitor-779) was selected for advancement due to its improved solubility profile and resulting suitability for intravenous formulation that was necessary owing to the higher dose levels required to achieve an anti-tumour effect in many cancer animal models.[6] Studies in xenograft models at Wyeth and at the NCI had established that the effect of rapamycin-like drugs on tumour growth in animals often required higher doses than those needed for inflammation or transplantation models. Accordingly, the collective efforts, across a broad spectrum of functions at Wyeth, culminated in the selection of CCI-779 (temsirolimus, Torisel®). Orphan designation for the treatment of renal cell carcinoma, which accounts for approximately 3% of all cancers in the USA, was granted by the FDA in December 2004 and Torisel® was

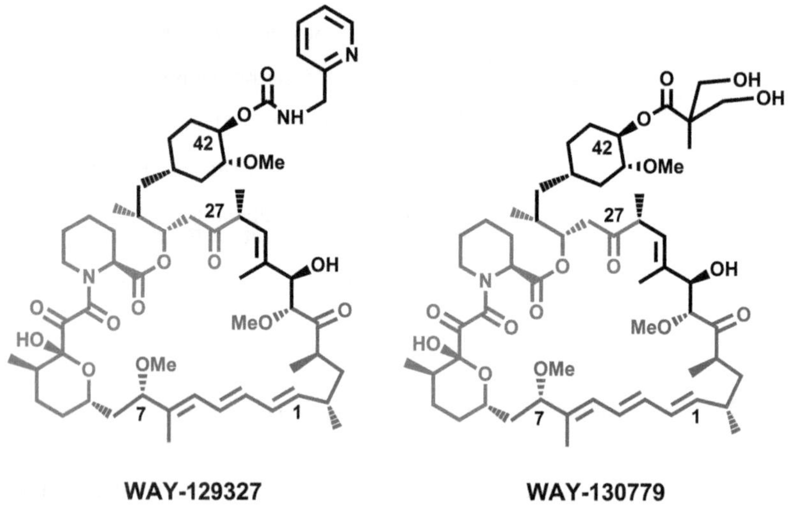

Figure 12.5 WAY-129327 and WAY-130779.

approved in the USA in May 2007 for the treatment of advanced renal cell carcinoma.[7] Torisel® is the first approved cancer therapy that specifically targets mTOR (mammalian target of rapamycin), a key protein in cells that regulates cell proliferation, cell growth and cell survival. Torisel® works along two pathways in the mTOR cascade to provide both anti-proliferative and anti-angiogenic activity.

In this chapter, the medicinal chemistry target design rationale and strategy for rapamycin analogues will be highlighted, exploring the aforementioned systematic point modification approach. The scope and limitations of synthetic transformations will be illustrated; synthetic challenges and their solutions will be described. Representative preclinical *in vitro* and *in vivo* data will be discussed. Details of the preclinical pharmacology studies of temsirolimus have been reported and will not be discussed further.[8,9] Recent publications and reviews of clinical studies of temsirolimus have appeared.[7,10,11] Capitalising on the benefits of hindsight, the authors present a retrospective summation of the salient features of these research efforts. To focus the article, not all compounds and experiments are exemplified, but an ensemble of diverse and essential chemical approaches are delineated.[12]

12.3 Alcohol Functionalisation

In addition to modifications of the core structure of macrocycle for our structure activity relationship (SAR) efforts (*vide infra*), we also examined chemistry on the cul-de-sac known as the cyclohexyl region. In this region, the initial, sustained and primary focus of these synthetic efforts was the derivatisation of the C-42 alcohol. The C-42 position is synthetically accessible, providing a fertile derivatisation platform and the alcohol functionality provides an important handle for a variety of synthetic manipulations. From our experience, we have observed that the C-42 position could accommodate chemistry to yield functional groups of various size, shape and constitution; these include, but are not limited to, preparation of ethers, acetals, ketones, carbonates, epimerised alcohols and ethers, sulfonates, esters and carbamates.[13] Potentially, these functional groups and/or their substituents could embrace chemical attributes to effect physicochemical characteristics, PK/PD parameters, oral absorption, drug-like properties, *inter alia*. In cases where a mixture of C-42, C-42/C-31 and C-31 products are formed in the reaction, these are separated chromatographically.

The C-42 esters of rapamycin have been prepared by four different methods (Scheme 12.1). Acylation of rapamycin using acid chlorides, acid anhydrides, acids (carbodiimide coupling conditions), or by the Yamaguchi[14] procedure, afforded the candidate compounds. Carbamates were prepared by direct acylation of rapamycin using appropriately substituted isocyanates or in a two-step procedure involving sequential preparation (and isolation) of the nitrophenylcarbonate, followed by reaction of appropriately substituted amines. In this manner, the complexion of the acyl group with respect to functionality, size, steric considerations or polarity was varied. The previously described lead compounds, WAY-129327 and WAY-130779 (see Figure 12.5) have resulted from these efforts. Additionally, specifically designed esters have been deployed in pursuit of gaining additional fundamental insights: as photolabile analogues for use in affinity matrices to isolate the putative effector protein,[15] in circular dichroism spectroscopy experiments to probe conformational changes in complexation with FKBP,[16]

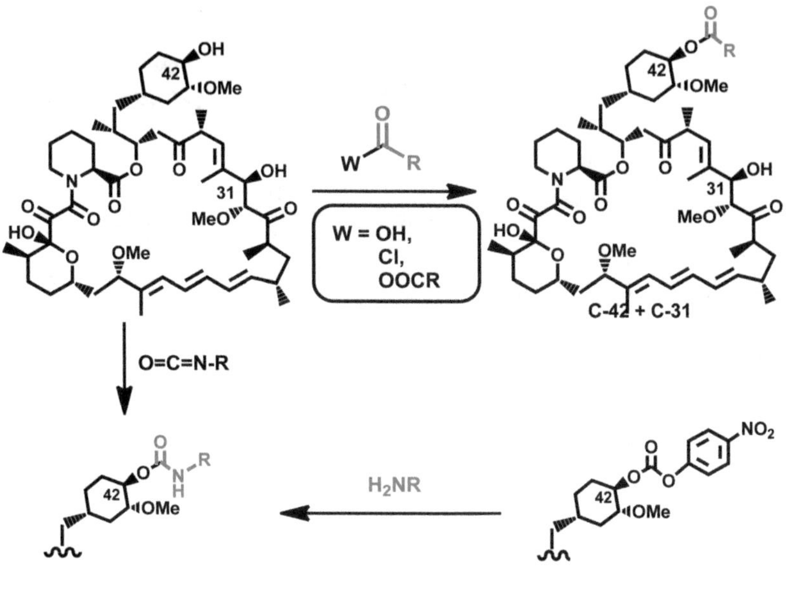

Scheme 12.1 C-42 acylation methods.

and in ion spray mass spectrometry experiments to examine and analyse the energetics of gas-phase analogue–protein complexes.[17]

12.4 Alteration of the Triene Unit

In seminal work in this area, the X-ray structure of the rapamycin/FKBP complex determined by Clardy and Schreiber shows the triene extends into solution.[18] The triene is believed to be an important contributor in the relationship of mTOR and the rapamycin/FKBP complex. Alterations in this region have provided a number of interesting compounds. Representative triene modifications have been designed to probe the boundaries of the effector region (Figure 12.6).

Catalytic hydrogenation of the triene has provided partial (both mono and di) or full reduction products, by alteration of reagents and reaction conditions. In a limited assessment, biological activity tracked with the extent of hydrogenation (C-1–C-7: rapamycin > diene > alkene > alkane).[13] Reaction of rapamycin with Lewis acids ($SnCl_4$, BF_3–OEt_2, *etc.*) in the presence of nucleophiles yielded a mixture of C-1 and C-7 ethers (or thioethers with RSH).[19] In the absence of a strong nucleophile, elimination occurred to furnish a mixture of tetraenes. The authors postulate the intermediacy of a stabilised carbocation at C-7 *via* attack at the C-7 methoxy group.

Another significant alteration in this region involved the generation of a Diels–Alder adduct.[20] This analogue had modest T-cell proliferation activity as measured in the standard LAF assay with retention of PPIase (peptidylprolyl isomerase) activity. Furthermore, this adduct antagonises the effects of rapamycin on thymocyte proliferation. These results underscore

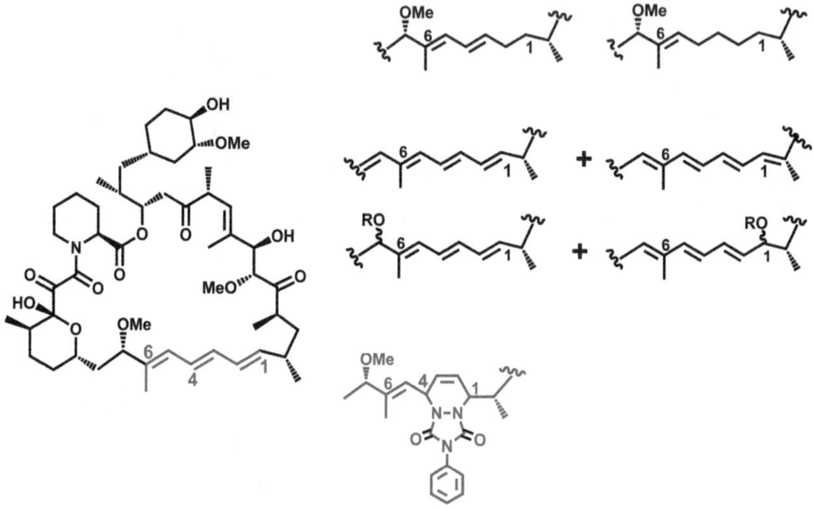

Figure 12.6 Representative triene modifications.

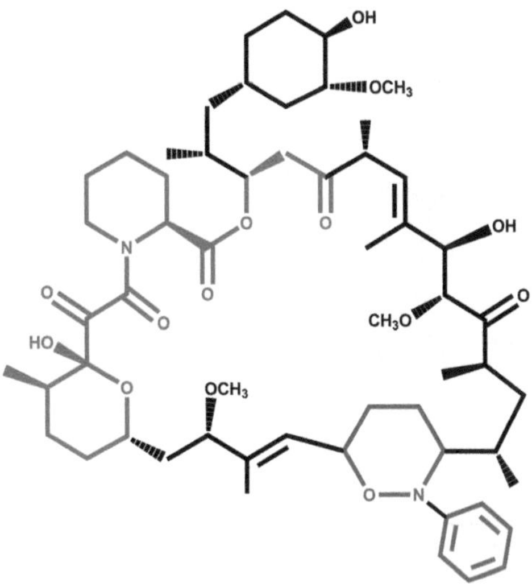

Figure 12.7 Diels–Alder adduct.

the importance of the triene region to mediate anti-proliferative effects and perturbation of the triene does not fully compromise binding to FKBP.

More recently, a novel Diels–Alder adduct (ILS-920; Figure 12.7) was designed in an attempt to disrupt immunosuppressive activity with and without significant disturbance of FKBP binding.[21] Reaction of rapamycin with nitrobenzene, followed by selective reduction of the olefin proximal to the point of addition, provided the target compound which displays potent neurotrophic profile and reduced immunosuppressive activity. With a higher affinity for FKBP-52 (*versus* FKBP-12), the compound was also used in ligand affinity purification studies.

12.5 Manipulation of the Carbonyl Groups

The primary route of rapamycin metabolism involves ring opening to provide secorapamycins.[22] The process may involve hydrolysis of the lactone and/or beta elimination in the central portion of the molecule (C-26 and C-25). To enhance the metabolic stability of rapamycin while retaining biological activity, the Wyeth group investigated manipulation of the C-27 ketone (Figure 12.8). A series of oximes[23] (as mixture of *syn* and *anti*) and hydrazones[24] have been designed to alter the electronic and steric environments at this position. In another approach, regio- and stereospecific reduction of the ketone was effected by protection of the C-42 and C-31 alcohols as triethylsilyl ethers, reaction with L-selectride, and deprotection.[25] Additionally, regioselective reductions of rapamycin at the C-15 and C-33 ketones were surveyed. Good to excellent regiocontrol was observed with proper choice of reagents and conditions.[26]

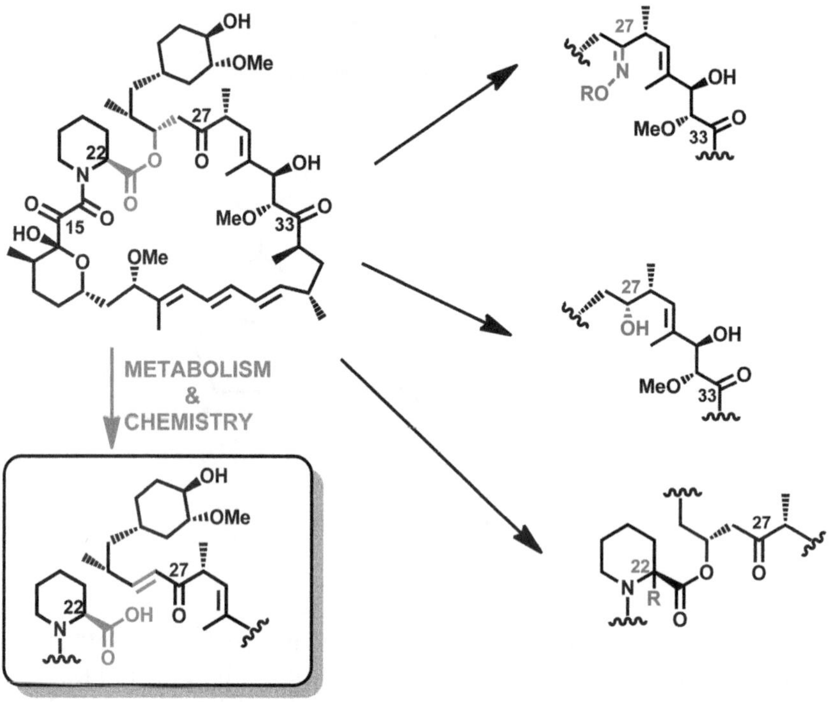

Figure 12.8 C-22/C-27 modifications.

Regioselective oxidation at C-31 was achieved using Dess–Martin periodinane (no C-42 ketone) was obtained bar small amounts of bis-oxidation (C-31, C-42). Reaction of the C-31, C-33 dicarbonyl system with hydrazine furnished the corresponding pyrazole as a mixture of tautomers.[27]

12.6 Modification of the Pipecolinate Region

Alternative efforts to effect metabolic stability involved consideration of the pipecolinate moiety. Furthermore, modifications therein could have the added value to enable probing of steric boundaries in the interactions with FKBP with its ligands. Reaction of C-42, C-31 bis-protected alcohol (TES ethers), with LDA at −78 °C, followed by the addition of methyl triflate, and deprotection (HF–pyridine) provided the C-22 methyl congener.[28] Analogous alkylation was accomplished using ethyl triflate (Scheme 12.2). The authors hypothesised that alkylation at this position provides rigidity with resultant restricted rotation of the amide bond. The methyl derivative has slightly reduced activity (LAF and PPIase), but significantly improved $T_{1/2}$, whereas the ethyl derivative displayed significantly reduced activity suggesting limited steric latitude in this area. Interestingly, reduction of the C-27 ketone (L-selectride) of the C-22 methyl congener provided an alcohol with

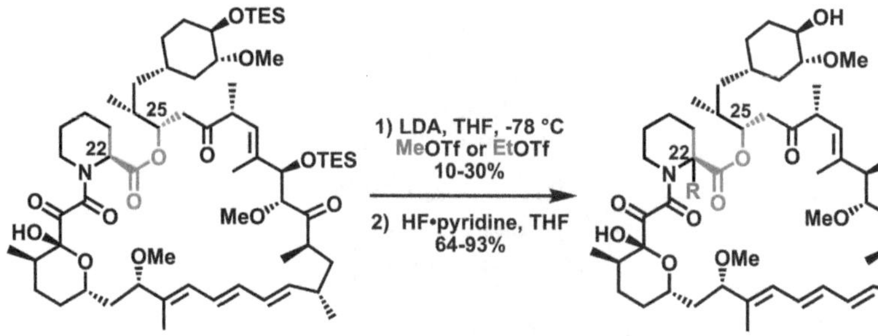

Scheme 12.2 C-22 alkylation methods.

comparable biological activity to its precursor but with a diminished $T_{1/2}$.[28] Novel sulfur isosteres of the pipecolinate moiety, derived from precursor-directed biosynthesis, displayed diminished FKBP binding affinity.[29]

12.7 Ring Opening, Expansion and Contraction

During the course of our synthetic studies we re-examined base-catalysed ring opening of rapamycin.[30] The resulting secorapamycin is significantly less active than the parent and accordingly ring-opened analogues were designed in an effort to improve this biological profile. A variety of esters and amides, reminiscent of the cyclohexyl region, were prepared by condensation with the liberated acid of the pipecolinate.[31] Conjugate adducts[32] derived from the enone C-25/C-26/C-27, were also explored (Scheme 12.3). In each case, no significant improvement in activity was observed.

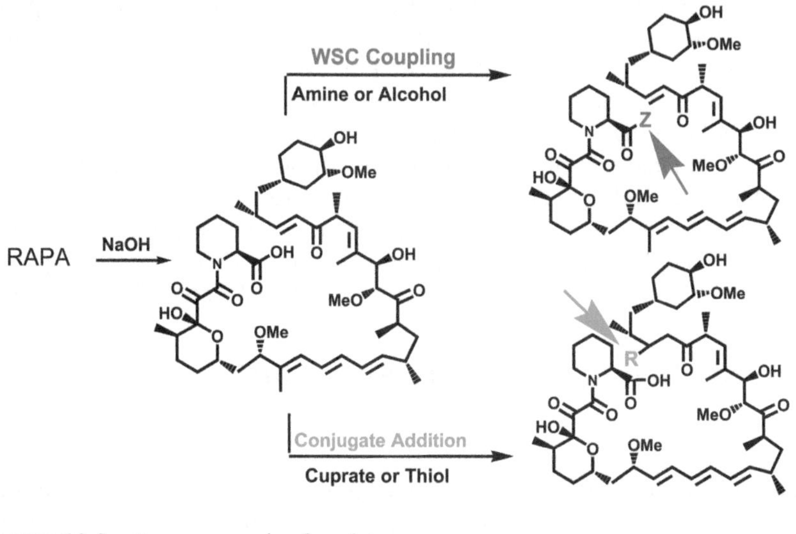

Scheme 12.3 Secorapamycin chemistry.

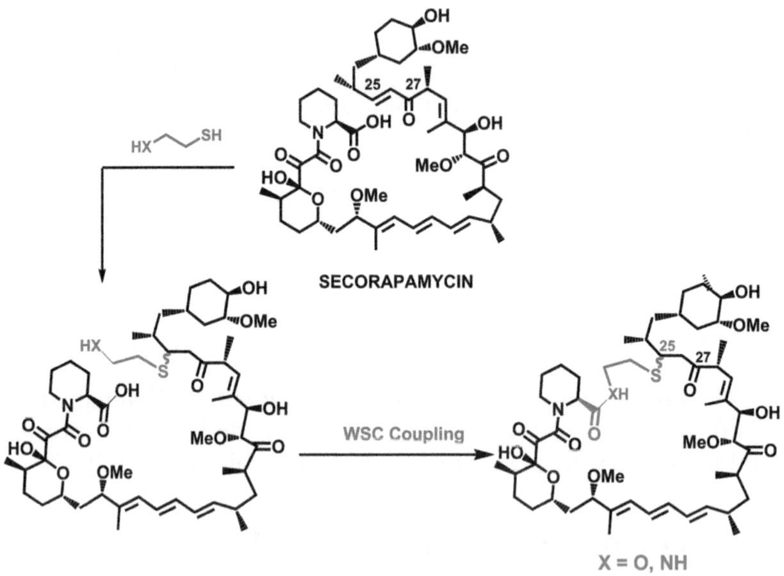

Scheme 12.4 Rapamycin ring expansion.

Nonetheless, an ulterior motive for these efforts was to serve as a model system for a tandem conjugate addition-acid functionalisation strategy to yield ring expanded rapamycin analogues.[33] In the event, treatment of secorapamycin with 2-mercaptoethanol in the presence of triethylamine furnished the conjugate adduct in high yield, which was converted to the lactone using the water-soluble carbodiimide procedure under high dilution to give the thia bis-homo rapamycin analogue. The lactam was prepared similarly, starting with 2-mercaptoethylamine (Scheme 12.4). Lactam derivatives showed reasonable activity in the LAF assay.

Rapamycin ring contraction[25] was also examined, with a goal of disabling one metabolic pathway and to probe FKBP binding events. The precursor for the ring-contracted rapamycin is the previously described C-27 alcohol (C-42, C-31 TES protected, *cf.* Figure 12.8). Translactonisation of C-27 alcohol (potassium carbonate in methanol) and removal of the protecting groups (acetic acid/THF) gave the target compound (Scheme 12.5),[25] which displayed improved $T_{1/2}$, but with diminished activity in the LAF and PPIase assays.[28]

12.8 Medicinal Chemistry Summary

In the previous sections we have described an ambitious semi-synthetic effort directed toward the identification of novel rapamycin analogues by systematic point group modifications. Our initial goal was to identify potential clinical candidates for transplantation and autoimmune diseases. Another goal of the rapamycin analogues programme was to

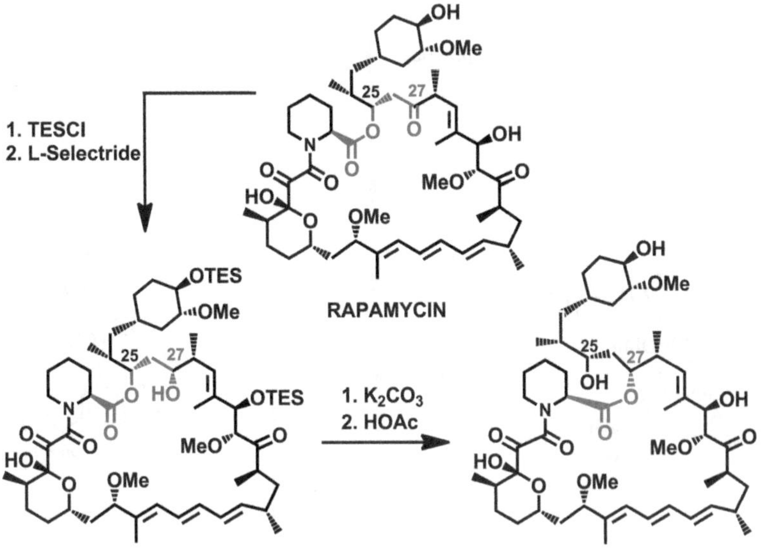

Scheme 12.5 Rapamycin ring contraction.

identify lead compounds for alternate therapeutic indications. Because the AKT/PTEN/mTOR pathway is mutated in a high percentage of human cancers, select analogues were evaluated in cell-based tumour assays as well as animal models. Additionally, we wished to develop the necessary chemistry to yield compounds to enable the deciphering of biological processes in this area.

In the event, by harnessing the power of semi-synthesis, a broad array of rapamycin analogues were designed and synthesised for biological evaluation. Biological activities and chemical properties were optimised. Fundamental principles were probed and our knowledge base concerning rapamycin chemistry and biology was expanded. New synthetic strategies and mechanistic studies to further elucidate biology were developed. The composite of these activities, which involved an intimate interplay between chemistry (and chemists) and biology (and biologists) was generation and evaluation of an expanding compound estate and the identification of CCI-779.

Although many analogues retained *in vitro* potency in the primary assays (LAF or cell-based oncology), only a limited number of compounds displayed the requisite level of activity in animal models (mouse skin graft, rat adjuvant arthritis, tumour xenograft). To discriminate from this pool of potent, efficacious molecules, other parameters including solubility, stability and crystallinity were considered. Following careful scrutiny and assessment (Figure 12.9), lead candidates were pared down to 31 to 14 to 3, to the two preproject compounds WAY-129327 and WAY-130779 (CCI-779; temsirolimus;

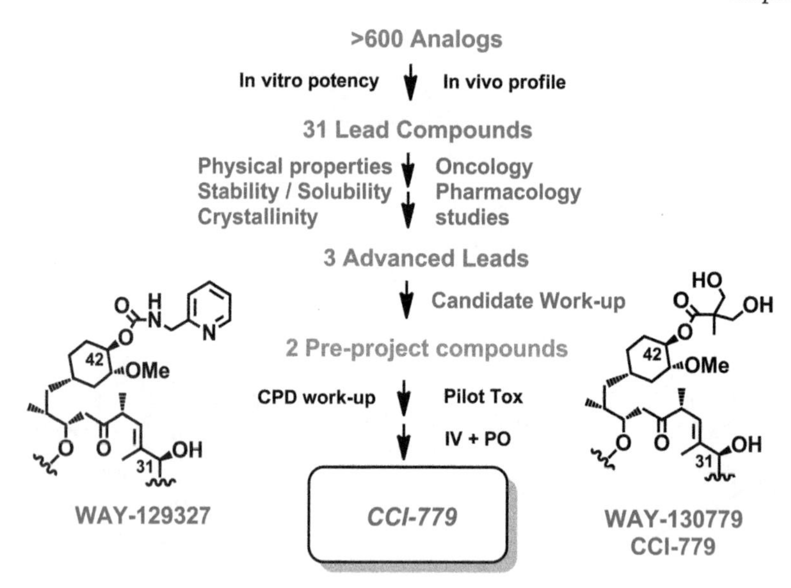

Figure 12.9 The selection process: CCI-779.

Torisel®). As was described above (Section 12.2), WAY-130779 was selected for advancement due to its superior solubility profile and resulting suitability for intravenous formulation for clinical evaluation in oncology. WAY-129327 serves as a back-up compound, but importantly, because of its biological profile, is also considered as a critical tool compound for advanced pharmacological and biochemical studies.[34]

12.9 Temsirolimus (CCI-779; Torisel®)

Temsirolimus (CCI-779) is a rapamycin ester derived from 2,2-bis-(hydroxymethyl)propionic acid and bears distinctive structural characteristics (Figure 12.10). It contains a branched and sterically hindered diol for increased hydrophilicity and to retard hydrolysis. The ester moiety is symmetrical and thus no additional chirality is introduced. There are no hydrogens α to the carbonyl so β elimination to give an acrylate is impossible. The ester is amenable to synthesis, scale-up and optimisation.

Temsirolimus is a potent inhibitor of T-cell proliferation and binds to FKBP with high affinity. It is active both ip and po in animal models of transplantation (mouse skin graft rejection model) and arthritis (rat adjuvant arthritis model).[8c] Temsirolimus is active against a histologically diverse panel of cell lines. *In vivo*, it has been shown to inhibit human tumour xenografts in nude mice of diverse histological type (glioma, breast, prostate, pancreatic, colon, renal cell carcinoma, melanoma, amongst others). *In vivo* efficacy has been achieved by both oral and iv routes of administration and

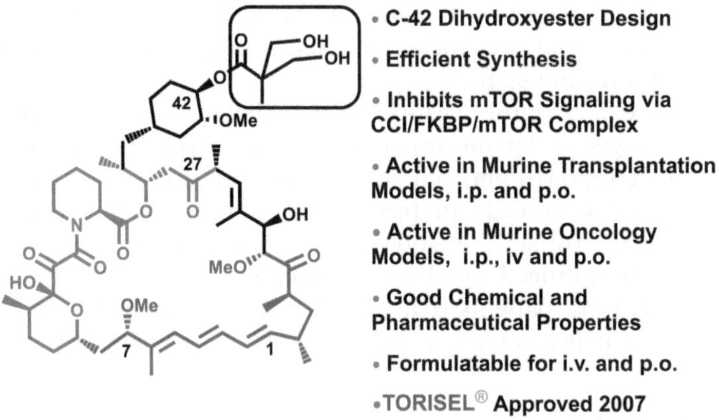
• C-42 Dihydroxyester Design
• Efficient Synthesis
• Inhibits mTOR Signaling via CCI/FKBP/mTOR Complex
• Active in Murine Transplantation Models, i.p. and p.o.
• Active in Murine Oncology Models, i.p., iv and p.o.
• Good Chemical and Pharmaceutical Properties
• Formulatable for i.v. and p.o.
•TORISEL® Approved 2007
42
27
7
1
OH
OMe
MeO
HO

Figure 12.10 Temsirolimus (CCI-779).

intermittent dosing regimens are effective. PTEN-deficient tumours are hypersensitive to mTOR inhibition.[5,8,9]

The discovery synthesis of CCI-779 involves acylation of rapamycin with acetonide protected diol acid under Yamaguchi[14] conditions. Chromatographic purification yielded the requisite monoacylation product. Liberation of the diol was accomplished using aqueous HCl in THF to afford CCI-779, after chromatography (Scheme 12.6).[35] Wyeth Chemical Development discovered improved processes for the synthesis of CCI-779. One involved replacing the acetonide with a boronate protecting group; deprotective transboration was effected using a diol reagent under mild conditions.[36] In

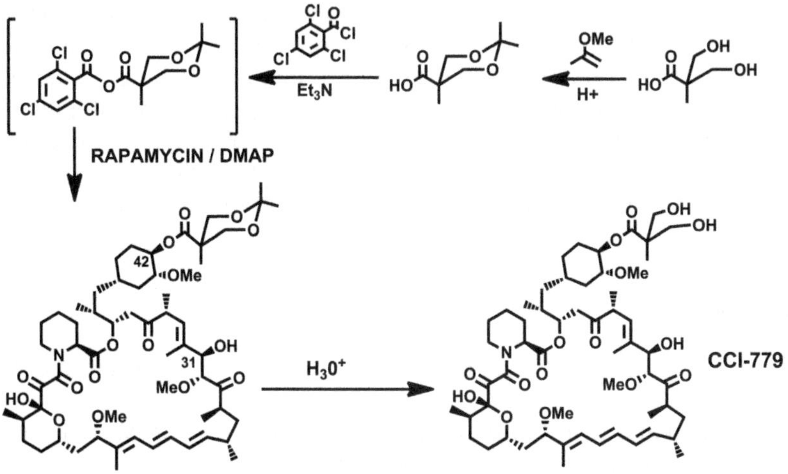
Et3N
H+
RAPAMYCIN / DMAP
H30+
CCI-779
42
31

Scheme 12.6 Discovery synthesis of CCI-779.

the second, a lipase-catalysed regioselective acylation was accomplished providing a streamlined and efficient method to prepare CCI-779 without the protection/deprotection steps.[37]

Torisel® is the first approved cancer therapy that specifically targets mTOR (mammalian target of rapamycin), a key protein in cells that regulates cell proliferation, cell growth and cell survival. Torisel® works along two pathways in the mTOR cascade to provide both anti-proliferative and anti-angiogenic activity. Torisel® has shown objective responses in investigational clinical trials in advanced renal cell carcinoma (RCC)[7,10] and mantle cell lymphoma.[11] In RCC, patients treated with Torisel® had a 49% increase in median overall survival compared with patients treated with IFN.

12.10 Conclusions

The breadth and depth of the research in the rapamycin analogues programme led to the discovery of Torisel®, inspired further activities in multiple therapeutic areas[21,38] and validated the investigation of non-rapamycin mTOR inhibitors for oncology and beyond.[39] Efforts continue directed to understanding essential details of the rapamycin biosynthetic cascade and applications are directed to bioengineered analogues not readily available by semi-synthesis.[29,40] Key experiments progress to further elucidate the membership and intricacies of this complex signal transduction pathway, and expand our knowledge of mTOR as a crucial regulator.

Acknowledgements

The authors extend their sincere appreciation to colleagues at Wyeth during the course of this research, many of which are cited in this document. These contributors to this work spanned numerous department and functional units therein, at multiple Research and Development sites. These include individuals from Chemical Sciences, Immunopharmacology, Oncology, Chemical and Pharmaceutical Development, Drug Safety and Metabolism, Clinical, Project Management, and Legal. We are also grateful to the external collaborators for their research activities. Because of a dedicated, transparent collaborative spirit, the journey was exciting, productive, successful and enjoyable.

References

1. (a) M. Abou-Gharbia, *J. Med. Chem.*, 2009, **52**, 2; (b) G. T. Carter, *Nat. Prod. Rep.*, 2011, **28**, 1783; (c) C. Bailly, *Biochem. Pharmacol.*, 2009, 77, 1447; (d) F. E. Koehn and G. T. Carter, *Nat. Rev.*, 2005, **4**, 206; (e) R. M. Wilson and S. J. Danishefsky, *J. Org. Chem.*, 2006, **71**, 8329.
2. (a) S. N. Sehgal, K. Molnar-Kimber, T. D. Ocain and B. M. Weichman, *Med. Res. Rev.*, 1994, **14**, 1; (b) S. N. Sehgal, *Clin. Biochem.*, 1998, **31**, 335; (c)

S. N. Sehgal, *Transplant. Proc.*, 2003, **35**(suppl. 3A), 7S; (d) J. Camardo, *Transplant. Proc.*, 2003, **35**(suppl. 3A), 18S.
3. H. Yang, D. G. Rudge, J. D. Koos, B. Vaidialingam, H. J. Yang and N. P. Pavletich, *Nature,* 2013, **497**, 217.
4. (a) K. C. Nicolaou, T. K. Chakraborty, A. D. Piscopio, N. Minowa and P. Bertinato, *J. Am. Chem. Soc.,* 1993, **115**, 4419; (b) D. Romo, S. D. Meyer, D. D. Johnson and S. L. Schreiber, *J. Am. Chem. Soc.,* 1993, **115**, 7906; (c) C. M. Hayward, D. Yohannes and S. J. Danishefsky, *J. Am. Chem. Soc.,* 1993, **115**, 9345; (d) A. B. Smith, III, S. M. Condon, J. A. McCauley, J. L. Leazer, Jr, J. W. Leahy and R. M. Maleczka, Jr, *J. Am. Chem. Soc.,* 1995, **117**, 5407; (e) M. L. Maddess, M. N. Tuckett, H. Watanabe, P. E. Brennan, C. D. Spilling, J. S. Scott, D. P. Osborn and S. V. Ley, *Angew. Chem., Int. Ed.,* 2007, **46**, 591.
5. (a) J. J. Gibbons, R. T. Abraham and K. Yu, *Semin. Oncol.,* 2009, **36**(6, suppl. 3), S3; (b) R. T. Abraham and J. J. Gibbons, *Clin. Cancer Res.,* 2007, **13**, 3109; (c) R. T. Abraham, J. J. Gibbons and E. I. Graziani, *Enzymes,* 2010, **27**, 329.
6. J. P. Boni, B. Hug, C. Leister and D. Sonnichsen, *Semin. Oncol.,* 2009, **36**(6, suppl. 3), S18.
7. (a) G. Hudes, M. Carducci, P. Tomczak, J. Dutcher, R. Figlin, A. Kapoor, E. Staroslawska, J. Sosman, D. McDermott, I. Bodrogi, Z. Kovacevic, V. Lesovoy, I. G. H. Schmidt-Wolf, O. Barbarash, E. Gokmen, T. O'Toole, S. Lustgarten, L. Moore and R. J. Motzer, *N. Engl. J. Med.,* 2007, **356**, 2271; (b) G. R. Hudes, A. Berkenblit, J. Feingold, M. B. Atkins, B. I. Rini and J. Dutcher, *Semin. Oncol.,* 2009, **36**(6, suppl. 3), S26.
8. (a) J. J. Gibbons, C. Discafani, R. Peterson, R. Hernandez, J. S. Skotnicki and P. Frost, *Proc. Amer. Assoc. Cancer Res.,* 1999, **40**, 2000; (b) K. Yu, W. Zhang, J. Lucas, L. Toral-Barza, R. Petersen, J. Skotnicki, P. Frost and J. Gibbons, *Proc. Amer. Assoc. Cancer Res.,* 2001, **42**, 4305; (c) J. S. Skotnicki, C. L. Leone, A. L. Smith, Y. L. Palmer, K. Yu, C. M. Discafani, J. J. Gibbons, P. Frost and M. A. Abou-Gharbia, Proc. AACR-NCI-EORTC Internat. Conf., Abstr. 477, *Clin. Cancer Res.,* 2001, **7**, S3749; (d) K. Yu, L. Toral-Barza, C. Discafani, W. G. Zhang, J. Skotnicki, P. Frost and J. J. Gibbons, *Endocr.-Relat. Cancer,* 2001, **8**, 249; (e) J. J. Gibbons, Jr, L. Toral-Barza, W.-G. Zhang, P. Frost, J. Skotnicki, C. Gaydos, J. Lucas and K. Yu, *Proc. Signal Trans. Cancer,* 2005, **16**, 78; (f) T. M. Sadler, M. Gavriil, T. Annable, P. Frost, L. M. Greenberger and Y. Zhang, *Endocr.-Relat. Cancer,* 2006, **13**, 863; (g) B. Shor, W.-G. Zhang, L. Toral-Barza, J. Lucas, R. T. Abraham, J. J. Gibbons and K. Yu, *Cancer Res.,* 2008, **68**, 2934.
9. (a) M. S. Neshat, I. K. Mellinghoff, C. Tran, B. Stiles, G. Thomas, R. Petersen, P. Frost, J. J. Gibbons, H. Wu and C. L. Sawyers, *Proc. Natl. Acad. Sci. U. S. A.,* 2001, **98**, 10314; (b) K. Podsypanina, R. T. Lee, C. Politis, I. Hennessy, A. Crane, J. Puc, M. Neshat, H. Wang, L. Yang, J. Gibbons, P. Frost, V. Dreisbach, J. Blenis, Z. Gaciong, P. Fisher,

C. Sawyers, L. Hedrick-Ellenson and R. Parsons, *Proc. Natl. Acad. Sci. U. S. A.,* 2001, **98**, 10320; (c) P. Frost, F. Moatamed, B. Hoang, Y. Shi, J. Gera, H. Yan, P. Frost, J. Gibbons and A. Lichtenstein, *Blood,* 2004, **104**, 4181.
10. (a) J. E. Dancey, R. Curiel and J. Purvis, *Semin. Oncol.,* 2009, **36**(6, suppl. 3), S46; (b) B. I. Rini, *Clin. Cancer Res.,* 2008, **14**, 1286; (c) R. A. Figlin, *Clin. Adv. Hematol. Oncol.,* 2007, **5**, 893; (d) B. J. Escudier, *Oncol. Rev.,* 2007, **1**, 73; (e) D. Simpson and M. P. Curran, *Drugs,* 2008, **68**, 631; (f) W. W. Ma and A. Jimeno, *Drugs Today,* 2007, **43**, 659.
11. (a) G. Hess, *Expert Rev. Hematol.,* 2009, **2**, 631; (b) G. Hess, S. M. Smith, A. Berkenblit and B. Coiffier, *Semin. Oncol.,* 2009, **36**(6, suppl. 3), S37; (c) S. M. Hoy and K. McKeage, *Drugs,* 2010, **70**, 1819.
12. A portion of this story has been discussed:J. S. Skotnicki, C. L. Leone, A. L. Smith, G. A. Schiehser, K. Yu, C. M. Discafani, J. J. Gibbons, P. Frost and M. A. Abou-Gharbia, *International Symposium on Advances in Synthetic and Medicinal Chemistry*, St. Petersburg, Russia, 2007.
13. C. E. Caufield, *Curr. Pharm. Des.,* 1995, **1**, 145.
14. I. Inanaga, K. Hirata, H. Saeki, T. Katsuki and M. Yamaguchi, *Bull. Chem. Soc. Jpn.,* 1979, **52**, 1989.
15. (a) Y. Chen, H. Chen, A. E. Rhoad, L. Warner, T. J. Caggiano, A. Failli, H. Zhang, C.-L. Hsiao, K. Nakanishi and K. Molnar-Kimber, *Biochem. Biophys. Res. Commun.,* 1994, **203**, 1; (b) Y. Chen, K. Nakanishi, D. Merrill, C. P. Eng, K. L. Molnar-Kimber, A. Failli and T. J. Caggiano, *Bioorg. Med. Chem. Lett.,* 1995, **5**, 1355.
16. Y. Chen, P. Zhou, N. Berova, H. Zhang, K. Nakanishi, A. Failli, R. J. Steffan, K. Molnar-Kimber and T. J. Caggiano, *J. Am. Chem. Soc.,* 1994, **116**, 2683.
17. Y.-T. Li, Y.-L. Hsieh, J. D. Henion, T. D. Ocain, G. A. Schiehser and B. Ganem, *J. Am. Chem. Soc.,* 1994, **116**, 7487.
18. (a) G. D. Van Duyne, R. F. Standaert, S. L. Schrieber and J. Clardy, *J. Am. Chem. Soc.,* 1991, **113**, 7433; (b) J. Choi, J. Chen, S. L. Schrieber and J. Clardy, *Science,* 1996, **273**, 239.
19. A. A. Grinfeld, C. E. Caufield, R. A. Schiksnis, J. F. Mattes and K. W. Chan, *Tetrahedron Lett.,* 1994, **35**, 6835.
20. T. D. Ocain, D. Longhi, R. J. Steffan, R. G. Caccese and S. N. Sehgal, *Biochem. Biophys. Res. Commun.,* 1993, **192**, 1340.
21. B. Ruan, K. Pong, F. Jow, M. Bowlby, R. A. Crozier, D. Liu, S. Liang, Y. Chen, M. L. Mercado, X. Feng, F. Bennett, D. von Schack, L. McDonald, M. M. Zaleska, A. Wood, P. H. Reinhart, R. L. Magolda, J. Skotnicki, M. N. Pangalos, F. E. Koehn, G. T. Carter, M. Abou-Gharbia and E. I. Graziani, *Proc. Natl. Acad. Sci. U. S. A.,* 2008, **105**, 33.
22. (a) C. P. Wang, H.-K. Lim, K. W. Chan, J. Scatina and S. F. Sisenwine, *J. Liq. Chromatogr.,* 1995, **18**, 2559; (b) C. P. Wang, H.-K. Lim, K. W. Chan, J. Scatina and S. F. Sisenwine, *J. Liq. Chromatogr. Relat. Technol.,* 1997, **20**, 1689.

23. (a) C. E. Caulfield and A. A. Failli, *U.S. Pat.*, 5,023,264, 1991; (b) A. A. Failli, R. J. Steffen, C. E. Caulfield, D. C. Hu and A. A. Grinfeld, *U.S. Pat.*, 5,373,014, 1994.
24. A. A. Failli and R. J. Steffen, *U.S. Pat.*, 5,120,726, 1992.
25. F. C. Nelson, S. J. Stachel and J. M. Mattes, *Tetrahedron Lett.*, 1994, **35**, 7557.
26. C. E. Caufield, F. C. Nelson, A. A. Grinfeld, D. C. Hu, S. J. Stachel, P. F. Hughes, R. Schiksnis, J. F. Mattes and K. W. Chan, *208th ACS National Meeting*, 1994, Abstr. Medi.114.
27. A. A. Failli and R. J. Steffen, *U.S. Pat.*, 5,164,399, 1992.
28. F. C. Nelson, S. J. Stachel, C. P. Eng and S. N. Sehgal, *Bioorg. Med. Chem. Lett.*, 1999, **9**, 295.
29. E. I. Graziani, F. V. Ritacco, M. Y. Summers, T. M. Zabriskie, K. Yu, V. S. Bernan, M. Greenstein and G. T. Carter, *Org. Lett.*, 2003, **5**, 2385.
30. R. J. Steffan, R. M. Kearney, D. C. Hu, A. A. Failli, J. S. Skotnicki, R. A. Schiksnis, J. F. Mattes, K. W. Chan and C. E. Caufield, *Tetrahedron Lett.*, 1993, **34**, 3699.
31. J. S. Skotnicki, R. M. Kearney and A. L. Smith, *Tetrahedron Lett.*, 1994, **35**, 197.
32. J. S. Skotnicki, R. M. Kearney, A. L. Smith and C. E. Caufield, *XVIIIth Internat. Symp. Macrocycl. Chem.*, Enschede, The Netherlands, 1993, Abstr. B71.
33. J. S. Skotnicki and R. M. Kearney, *Tetrahedron Lett.*, 1994, **35**, 201.
34. J. S. Crabtree, S. A. Jelinsky, H. A. Harris, S. E. Choe, M. M. Cotreau, M. L. Kimberland, E. Wilson, K. A. Saraf, W. Liu, A. S. McCampbell, B. Dave, R. R. Broaddus, E. L. Brown, W. Kao, J. S. Skotnicki, M. Abou-Gharbia, R. C. Winneker and C. L. Walker, *Cancer Res.*, 2009, **69**, 6171.
35. J. S. Skotnicki, C. L. Leone and G. A. Schiehser, *U.S. Pat.*, 5,362,718, 1994.
36. W. Chew and C.-C. Shaw, *20th Internat. Congr. Heterocycl. Chem.*, Palermo, Italy, 2005.
37. J. Gu, M. E. Ruppen and P. Cai, *Org. Lett.*, 2005, 7, 3945.
38. D. A. Young and C. L. Nickerson-Nutter, *Curr. Opin. Pharmacol.*, 2005, **5**, 418.
39. J. Verheijen, K. Yu and A. Zask, *Annu. Rep. Med. Chem.*, 2008, **43**, 189.
40. (a) J. N. Andexer, S. G. Kendrew, M. Nur-e-Alam, O. Lazos, T. A. Foster, A.-S. Zimmermann, T. D. Warneck, D. Suthar, N. J. Coates, F. E. Koehn, J. S. Skotnicki, G. T. Carter, M. A. Gregory, C. J. Martin, S. J. Moss, P. F. Leadlay and B. Wilkinson, *Proc. Natl. Acad. Sci. U. S. A.*, 2011, **108**, 4776; (b) S. G. Kendrew, H. Petkovic, S. Gaisser, M. A. Gregory, N. J. Coates, M. Nur-e-Alam, T. Warneck, D. Suther, T. A. Foster, L. McDonald, G. Schlingman, F. E. Koehn, J. S. Skotnicki, G. T. Carter, S. J. Moss, M.-Q. Zhang, C. J. Martin, R. M. Sheridan and B. Wilkinson, *Metab. Eng.*, 2013, **15**, 167; (c) E. I. Graziani, *Nat. Prod. Rep.*, 2009, **26**, 602; (d) B. Wilkinson and J. Micklefield, *Nat. Chem. Biol.*, 2007, **3**, 379;

(e) M.-Q. Zhang and B. Wilkinson, *Curr. Opin. Biotechnol.*, 2007, **18**, 478; (f) E. Kuscer, N. Coates, I. Challis, M. Gregory, B. Wilkinson, R. Sheridan and H. Petkovic, *J. Bacteriol.*, 2007, **189**, 4756; (g) M. A. Gregory, H. Hong, R. E. Lill, S. Gaisser, H. Petkovic, L. Lows, L. S. Sheehan, I. Carletti, S. J. Ready, M. J. Ward, A. L. Kaja, A. J. Weston, I. R. Challis, P. F. Leadlay, C. J. Martin, B. Wilkinson and R. M. Sheridan, *Org. Biomol. Chem.*, 2006, **4**, 3565; (h) R. J. M. Goss, S. E. Lanceron, N. J. Wise and S. J. Moss, *Org. Biomol. Chem.*, 2006, **4**, 4071; (i) M. A. Gregory, H. Petkovic, R. E. Lill, S. J. Moss, B. Wilkinson, S. Gaisser, P. F. Leadlay and R. M. Sheridan, *Angew. Chem., Int. Ed.*, 2005, **44**, 4747.

RESPIRATORY DISORDERS

CHAPTER 13

Treatments for Pulmonary Arterial Hypertension

MICHAEL J. PALMER

Consultant for Medicines for Malaria Venture, Discovery Park, Sandwich, Kent CT13 9NJ, UK
E-mail: mikepalmer@live.co.uk

13.1 Introduction

Pulmonary arterial hypertension (PAH) is a progressive and ultimately fatal disease that is estimated to affect approximately 12 to 50 people per million.[1,2] PAH thus qualifies as an orphan or rare disease and this chapter will set out to review the disease and current treatments, assess new treatment options in the pipeline and discuss the impact of orphan designation.

PAH comprises group 1 within the current World Health Organization (WHO) World Symposium classification of pulmonary hypertension (PH), which consists of five groups.[3] The other four groups that comprise PH are: PH owing to left heart disease, PH owing to lung diseases or hypoxia, PH owing to chronic thromboembolic disease and a miscellaneous category comprising unclear multifactorial mechanisms (Table 13.1). Although only a small category within PH, PAH has received much attention as a result of being the only WHO group that has proven patient benefit from PAH-specific therapies, while treatments for the other categories have lagged behind due to the challenges in characterising these diseases.[4]

PAH is characterised by an increase in pulmonary vascular resistance resulting in an elevation of pulmonary arterial pressure. The key pathological impact of this progressive effect is the development of right ventricular

RSC Drug Discovery Series No. 38
Orphan Drugs and Rare Diseases
Edited by David C Pryde and Michael J Palmer

Published by the Royal Society of Chemistry, www.rsc.org

Table 13.1 Updated classification of pulmonary hypertension from the 4th World Symposium on Pulmonary Hypertension[a] (Dana Point, California, 2008).

Group 1: Pulmonary arterial hypertension (PAH)	1.1 Idiopathic PAH 1.2 Heritable 1.2.1 BMPR2 1.2.2 ALK1, endoglin * 1.2.3 Unknown 1.3 Drug- and toxin-induced 1.4 Associated with 1.4.1 Connective tissue diseases 1.4.2 HIV infection 1.4.3 Portal hypertension 1.4.4 Congenital heart diseases 1.4.5 Schistosomiasis 1.4.6 Chronic hemolytic anemia 1.5 Persistent pulmonary hypertension of the newborn
Group 1′: Pulmonary veno-occlusive disease (PVOD) and/or pulmonary capillary hemangiomatosis (PCH)	
Group 2: Pulmonary hypertension owing to left heart disease	2.1 Systolic dysfunction 2.2 Diastolic dysfunction 2.3 Valvular disease
Group 3: Pulmonary hypertension owing to lung diseases and/or hypoxia	3.1 Chronic obstructive pulmonary disease 3.2 Interstitial lung disease 3.3 Other pulmonary diseases with mixed restrictive and obstructive pattern 3.4 Sleep-disordered breathing 3.5 Alveolar hypoventilation disorders 3.6 Chronic exposure to high altitude 3.7 Developmental abnormalities
Group 4: Chronic thromboembolic pulmonary hypertension (CTEPH)	
Group 5: Pulmonary hypertension with unclear multifactorial mechanisms	5.1 Hematologic disorders: myeloproliferative disorders, splenectomy 5.2 Systemic disorders: sarcoidosis, pulmonary Langerhans cell histiocytosis, lymphangioleiomyomatosis, neurofibromatosis, vasculitis 5.3 Metabolic disorders: glycogen storage disease, Gaucher disease, thyroid disorders 5.4 Others: tumoral obstruction, fibrosing mediastinitis, chronic renal failure on dialysis

[a]Reproduced from Simonneau *et al.*,[3] used with permission from Elsevier. ALK1, activin receptor-like kinase type 1; *with or without hereditary hemorrhagic telangiectasia; BMPR2, bone morphogenetic protein receptor type 2; HIV, human immunodeficiency virus.

hypertrophy, leading eventually to right ventricular failure.[5] PAH group 1 divides into five sub-categories comprising idiopathic, heritable, drug- and toxin-induced, associated, and persistent PH of the newborn derived forms of PAH (Table 13.1).[3] Of these five categories, idiopathic and associated PAH account for approximately 97% of cases, in which for associated PAH, connective tissue disease, chronic haemolytic anaemia, and portal hypertension are the most common forms.[6]

Treatment of PAH has advanced significantly in the past two decades,[7] with considerable improvement in survival rates. However, average survival after diagnosis in adults is still only estimated to be in the 5 to 7 year range.[8] Thus, with the advent of new therapies, making tailored decisions on the most appropriate therapies for an individual patient becomes an important challenge. In this respect, the Registry to Evaluate Early and Long-term Pulmonary Arterial Hypertension Disease Management (REVEAL) provides a valuable data source and had enrolled some 3500 PAH WHO group 1 patients by 2009.[6,9] The identification of multiple independent prognostic factors[10] and the development of a validated prognostic score calculator[11] provide further options for enabling patient-specific therapy. These developments further enable patients of high risk to be identified and a tailored treatment regimen of appropriate aggressiveness to be instigated. Recent reviews of therapeutic advances and the therapeutic treatment options for PAH provide more detail.[12,13]

13.2 Pathobiology

While the pathobiology of PAH is far from fully understood, proven features that affect the vascular wall are vasoconstriction, remodelling of the pulmonary vascular wall and thrombosis.[14,15] In a normal pulmonary vasculature, prostacyclin (prostaglandin I_2), nitric oxide (NO) and vasoactive intestinal peptide serve as vasodilators and mediate vasoconstriction.[16] In terms of vascular remodelling in PAH sufferers, both smooth muscle and vascular endothelial cells display cellular proliferation and abnormal growth. The net effect of these characteristics is a significant increase in pulmonary vascular resistance and a resultant increase in pulmonary arterial pressure.

An understanding in part at least of these pathobiological mechanisms has led to the current treatment options that are available for PAH.[16] For instance, prostacyclin levels are diminished in PAH patients, while NO affects vasodilation through a pathway based on cGMP production and diminished calcium supply to the vasculature. The endothelins have been shown to have an important role in the regulation of vascular smooth muscle, with key effects being direct vasoconstriction and the stimulation of cellular proliferation.[17] Therapies that target the prostacyclin, NO or endothelial pathways have been shown to offset these effects, reducing vasoconstriction and diminishing cellular proliferation. Additionally, emerging therapeutic agents are seeking to build on this understanding and target previously untapped mechanisms within the areas of vasoconstriction and excessive cell growth.

13.3 Current Treatments for PAH

The majority of approved treatments are orphan designated products. Treatments will be discussed in defined target categories that follow from the previous Pathobiology section. The aims of PAH treatments span improvement of symptoms and quality of life together with prevention of disease progression. The WHO classification of the functional status of patients with PH comprises four classes[18a,b] that can be simply described as follows:

Class I – patients with PH in whom there is no limitation of usual physical activity.

Class II – patients with PH who have mild limitation of physical activity; mild PAH symptoms.

Class III – patients with PH who have a marked limitation of physical activity; moderate PAH symptoms.

Class IV – patients with PH who are unable to perform any physical activity at rest and who may have signs of right ventricular failure; severe PAH symptoms.

Patients who respond positively to acute vasodilator testing can often be successfully treated with oral calcium channel blockers; this mature class of therapeutic agents will not be discussed here.

13.3.1 Prostanoids

Prostacyclin and related synthetic analogues were the first treatment class to emerge and remain the front-line treatment for patients with moderate to severe PAH (WHO functional classes III and IV). Mechanistically, prostacyclin-type agents act as potent vasodilators and platelet aggregation inhibitors with these effects predominantly mediated through specific G-protein coupled receptors that generate cAMP.[13] Anti-proliferative and anti-inflammatory effects have also been documented. Three therapeutic agents will be discussed.

13.3.1.1 Epoprostenol

Epoprostenol or synthetic prostacyclin **1** has orphan product designation in the USA dating back to 1984 and was initially approved by the Food and Drug Administration (FDA) in 1995 for patients with class III and IV symptoms. Approval in over 20 countries worldwide followed. Subsequent FDA label additions now extend to PAH patients with scleroderma and to all PAH patients for improvement of exercise capacity. Marketed as Flolan by GSK, epoprostenol is administered as a continuous intravenous agent and benefits exercise capacity, haemodynamic parameters and quality of life. Improved survival in patients has also been demonstrated, albeit this data is mainly from sufferers with idiopathic PAH.[19] The 6 minute walk test was used as a primary measure of efficacy in the assessment of epoprostenol with an increase of 32 metres the mean change from baseline for patients receiving epoprostenol. Subsequently the 6 minute walk test has been adopted as a key measure for all approved PAH agents.

The need for continuous infusion is because of the short (<6 minutes) half-life of prostacyclin. Intravenous infusion makes for delivery challenges with the possibility of supply interruption and venous line derived infections such as sepsis. Other adverse events include headache, nausea and diarrhoea. The advent of alternative therapies means that epoprostenol is now often reserved for severe cases that fail to respond well to oral or inhaled treatment options. The patent for Flolan expired in 2007 and a generic form is now also available. A new intravenous formulation is now marketed by Actelion. This new formulation, Veletri, is more stable at room temperature, allowing for greater convenience in respect of preparation with stability of up to 7 days at refrigerator temperature and 2 days at room temperature. Epoprostenol is not approved for paediatric use; however there is strong observational data supportive of significant benefit in children with idiopathic, heritable and congenital heart disease forms of PAH.

13.3.1.2 Treprostinil

Treprostinil **2** is a synthetic prostacyclin analogue featuring a tricyclic core. The agent is chemically stable with a half-life of 2–4 hours and can be delivered subcutaneously in addition to intravenously (Remodulin).[20] An inhaled form has subsequently become available (Tyvaso).

Remodulin was first approved in the USA by the FDA for subcutaneous use in 2002 for PAH patients in functional classes II–IV, with approval in most other major countries following. Improved exercise capacity and clinical symptoms were seen in patients with idiopathic PAH together with connective tissue and congenital heart disease associated forms.[21] In terms of 6 minute walking distance an improvement from baseline of 36.1 metres was seen for the highest dose group (>13.8 ng kg^{-1} min^{-1}). Small improvements in cardio-haemodynamic parameters were also seen. The major adverse event is discomfort at the infusion site, with approximately 80% of patients experiencing pain or erythema. Treprostinil is not approved for use in children although there is a small amount of supporting data showing benefit with paediatrics.[22]

FDA approval for intravenous use was granted in 2004. Intravenous treprostinil has shown clinically significant improvements in exercise walking capacity and pulmonary haemodynamics in clinical trials[23] and benefits from a longer 48 hour infusion reservoir change time relative to epoprostenol as a result of the increased chemical stability. One potential drawback is a higher incidence of Gram-negative infections possibly as a result of a neutral saline diluent, although this can be minimised with use of an alkaline system.[24] The intravenous form has not been as widely adopted as the subcutaneous form and key factors could be the advent of new oral therapies and the limited patient population of sufferers unable to tolerate the subcutaneous form or those being transitioned from epoprostenol.

The inhaled form of treprostinil was approved by the FDA in 2009 for functional class III patients and this latter inhaled form has been granted orphan designation in the USA, Europe and Australia. The pivotal TRIUMPH-1 clinical trial demonstrated improvement in walking distance and quality of life but no improvement in the time to clinical worsening.[25] Dosing is four times daily and the main adverse events are cough, headache and throat irritation. There has been limited use in children to date.

An oral form of treprostinil based on sustained release of the diethanolamine salt has received orphan product designation in Europe for PAH and has progressed to Phase III trials. A series of FREEDOM designated clinical trials has been undertaken by United Therapeutics. A monotherapy trial displayed a clear improvement in the 6 minute walk distance of approximately 23 metres, while combination studies with either Revatio or Tracleer failed to achieve clinical significance. These latter studies highlight the challenge of determining true treatment effects in patients already receiving PAH-specific therapy and a new FREEDOM-EV trial is currently being enrolled with a focus on longer duration and greater flexibility in end point. The aim is to seek EMA and FDA approval for use following a successful outcome of this study.[26]

13.3.1.3 Iloprost

Iloprost **3** is a synthetic prostacyclin analogue with alkyne functionality in the hydroxyl side chain. Iloprost is available as an inhaled agent and has achieved orphan product designation in the USA and Europe. The agent has a human half-life of 20–25 minutes resulting in therapy of 6 to 9 deliveries per day. Iloprost is approved for functional class III and IV and idiopathic PAH in Europe and Australia, and more widely for group I PAH patients in the USA.

Regulatory approval was based on one key clinical trial that demonstrated a significant improvement in 6 minute walking distance (36 metres overall) together with improved pulmonary haemodynamics.[27] Toleration was mostly good with cough, headache and flushing observed as the most frequent adverse events. Iloprost is not approved for use in children although the limited data available suggests beneficial acute effects and a possible role in the short-term treatment of paediatrics.[28] A new formulation that enables the same 5 μg dose to be delivered from half of the previous inhaled volume was introduced by Actelion with the aim of reducing inhalation time and increasing patient compliance.

The chemical structures of the three approved prostacyclin-based agents are shown in Figure 13.1.

13.3.2 Endothelin Antagonists

Following the proposal of an endothelial vasoconstricting factor in 1985, the field of endothelin and associated endothelin receptor antagonists became an area of intense scientific interest for some two decades.[29] Three human

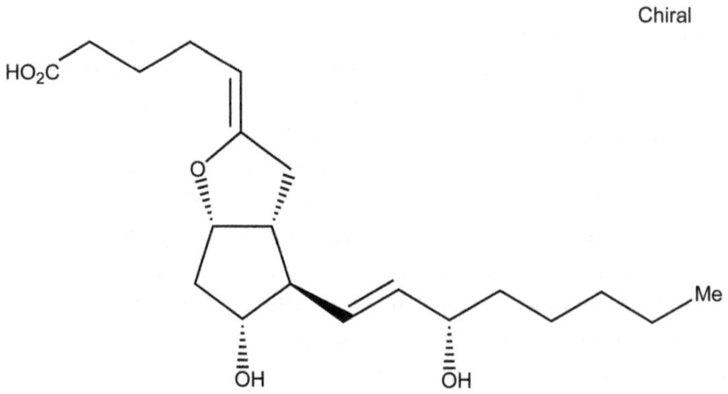

1, Epoprostenol

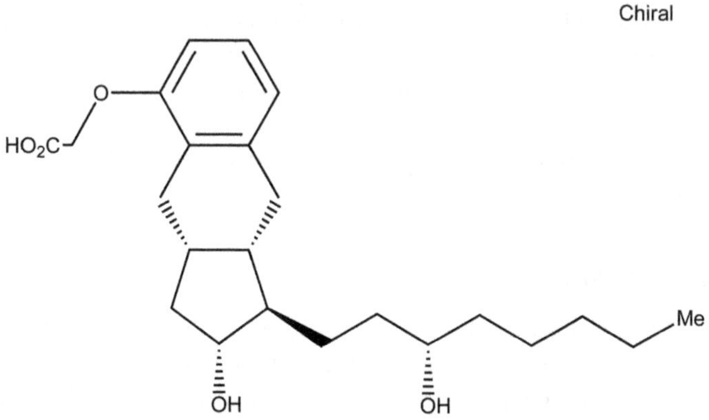

2,Treprostinil

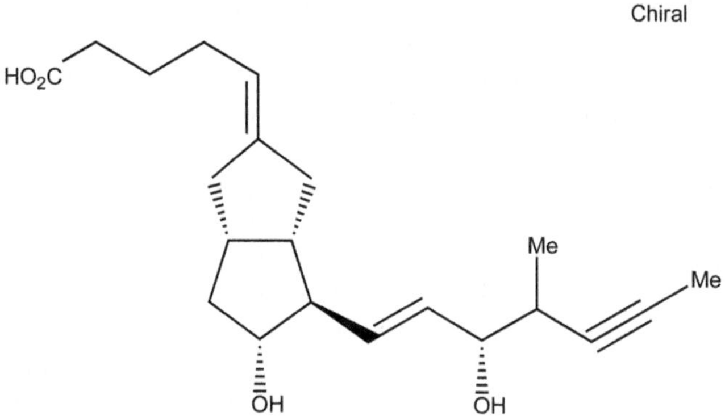

3, Iloprost

Figure 13.1 Prostacyclin and related analogues.

forms, endothelin-1, -2 and -3 were discovered and important biological roles established. Endothelin-1 is the major isoform in the human cardiovascular system and is a highly potent vasoconstrictor involved in important processes that include the regulation of vascular tone, cell proliferation and endothelial dysfunction. The biological effects of the endothelins are mediated *via* two G-protein coupled receptors (GPCR), the ET_A receptor and the ET_B receptor. While both receptors are found on vascular smooth muscle cells for mediation of vasoconstrictor effects, the ET_A receptor dominates and represents some 85% of the endothelin receptor population. However, the ET_B receptor is also found within vascular endothelial cells wherein activation results in vasodilator effects mediated through release of nitric oxide (NO) and prostacyclin. The ET_B receptor is the predominant subtype in kidney and appears to have an ET-1 clearance role that helps to regulate cardiovascular tone. With increasing knowledge of the functional role of the endothelin system, the belief arose that endothelin receptor antagonists could play an important role in mediating disease states, such as hypertension-based diseases wherein the endothelins played a key role. Additionally, the endothelin system is implicated in foetal development, appearing to play a crucial role in craniofacial and cardiovascular development. Hence all endothelin receptor antagonists are likely to be teratogenic and contraindicated in pregnancy.[29]

13.3.2.1 Bosentan

Bosentan (Tracleer) **4** was the first oral therapy to be approved for PAH and very much a front-runner in the endothelin receptor antagonist field. The field of endothelin receptor antagonists has proven a challenging one in terms of drug discovery and from some 500 preclinical agents documented as targeting the ET_A receptor,[29] only a handful have emerged as serious drug candidates. In particular, maintaining good physicochemical properties consistent with the necessary human pharmacokinetics for oral delivery while achieving sufficient efficacy and therapeutic index has proven difficult. Achieving an appropriate balance was crucial to success in the bosentan programme.

Bosentan is a mixed ET_A/ET_B receptor antagonist from the pyrimidine sulphonamide class with approximately 20-fold selectivity for the ET_A receptor (Figure 13.2). The human half-life is approximately 5 hours and oral bioavailability approximately 50%, consistent with a twice daily oral dosing profile in man, a profile well suited to the treatment of PAH. Bosentan has orphan product designation for PAH in the USA, Australia and Japan and is approved for the treatment of PAH in over 60 countries worldwide. Following initial approval by the FDA in 2001 for classes III and IV to improve exercise capacity and to decrease the rate of clinical worsening, a series of label approvals and extensions across the world have expanded scope. For instance, use for mild class II patients in the USA, EU and Canada and use for paediatrics in the EU.[13]

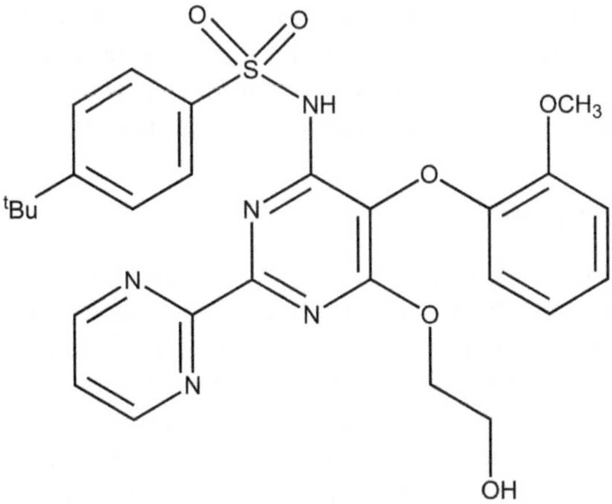

4, Bosentan

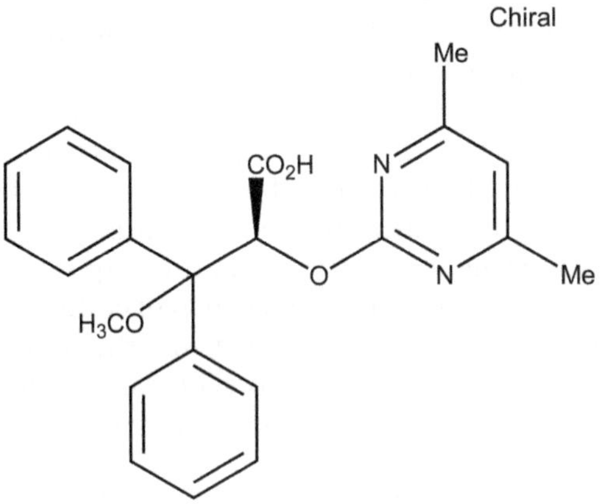

5, Ambrisentan

Figure 13.2 Endothelin receptor antagonists.

Approval was based on two key clinical trials wherein 6 minute walk distance, functional class status and time to clinical worsening were significantly improved.[30,31] Elevated liver transaminase levels were seen in these trials and although there were no reports of liver failure or jaundice, FDA approval was for the lower 125 mg dose and required patients to undertake monthly liver function tests. Headache was the most common adverse event observed and pregnancy testing is required for women of child-bearing potential. Bosentan is an inducer of the CYP3A4 P450 liver enzymes and the possibility of drug–drug interactions also has to be accounted for.

Subsequent clinical trials underpinned the label extensions. In 2009 the EMA approved a paediatric formulation based upon a clover-shaped quadrisect tablet that is dispersible in water. Bosentan was well tolerated and key haemodynamic parameters were significantly improved.

13.3.2.2 Ambrisentan

Ambrisentan **5** is a diarylpropionic acid based endothelin antagonist (Figure 13.2). This structural series is characterised by low molecular weight and good physicochemical properties relative to endothelin antagonists as a whole.[29] Ambrisentan is a potent ET_A selective antagonist and the physicochemical improvement is reflected in high oral bioavailability and a 9–15 hour human half-life that allows for once daily oral dosing.[32]

Ambrisentan (Letairis/Volibris) has orphan product designation in the USA, Europe and Australia. Ambrisentan was approved in the USA in 2007 for PAH patients with class II and III symptoms to improve exercise capacity and delay clinical worsening. Approval followed in other major countries and subsequent USA label revision extended use to PAH patients regardless of functional class. Approval was primarily based on two Phase III trials. Key data were that 6 minute walking distance improved by up to 51 metres for the higher 10 mg dose and a significant improvement in the time to clinical worsening when data from both trials were combined.[33] Ambrisentan was well tolerated, with headache the most frequent adverse event. Elevated liver serum transaminase levels were seen but none greater than three times normal levels. Monthly liver function tests were initially required by the FDA, but following the assessment of patient data this requirement was removed in 2011 leaving pregnancy monitoring in women of child-bearing potential as the sole distribution restriction. The cause of endothelin receptor antagonist induced liver toxicity is unclear, however preclinical data points to a possible inhibition of bile salt excretion mechanism, based on comparison of bosentan and ambrisentan effects.[34,35]

Ambrisentan is metabolised mainly by UDP glucuronosyltransferases and the only drug–drug interaction considered clinically relevant is with cyclosporine A. Thus ambrisentan does not interact with either of the phosphodiesterase-5 inhibitors, sildenafil or tadalafil and combination studies are ongoing.

The question as to whether selective ET_A receptor antagonism may be more beneficial than antagonism of both ET_A and ET_B receptors in PAH has been studied. For instance, an ET_A selective agent might antagonise the vasoconstrictive and mitogenic effects at the ET_A receptor, while maintaining the vasodilator and clearance responses induced *via* ET_B receptors. A review of PAH clinical data through to 2007 for endothelin receptor antagonist based agents concluded that there was no clinically meaningful difference between selective and mixed agents.[36] Other factors such as dosing flexibility, the potential for drug–drug interactions and limiting side effects are more likely to be relevant.

13.3.3 Phosphodiesterase-5 Inhibitors

Phosphodiesterase (PDE) enzymes play an important regulatory role in both cyclic adenosine monophosphate (cAMP) and cyclic guanosine monophosphate (cGMP) mediated signal transduction processes, through degradation of these second messenger nucleotides to the corresponding mononucleotides. cGMP is now recognised as the second messenger for atrial naturetic peptide and NO and as such is intimately involved in the relaxation of blood vessels. There are 11 members in the PDE superfamily of which phosphodiesterase-5 (PDE-5) is cGMP specific and widely distributed. PDE-5 is the main PDE in lung and is also found in smooth muscle and platelets.[37,38] Specific PDE-5 inhibitors were discovered and subsequent research demonstrated that inhibition of PDE-5 effects relaxation of smooth muscle in the vasculature. This effect is achieved through an increase of NO-mediated vasodilation as a result of elevation of cGMP levels.

13.3.3.1 Sildenafil

Sildenafil **6** was the prototype PDE-5 inhibitor and has played a pivotal role in driving clinical understanding of the significance of maintaining cGMP concentration and the consequent maintained relaxation of blood vessels.[39,40] This work initially led to the regulatory approval of sildenafil (Viagra) for the treatment of male erectile dysfunction (MED). Sildenafil is a pyrazolopyrimidinone based compound (Figure 13.3) with good physicochemistry, and a profile of rapid oral absorption and a short 3–5 hour human half-life that is well suited to the 'on demand' treatment of MED.

Sildenafil (Revatio) has orphan designation in Europe and was approved in 2005 by both the FDA and EMA as an oral therapy for PAH. Dosing is three times daily. Approval in the USA was to improve exercise capability for PAH regardless of functional class, while European approval was restricted to the idiopathic and connective tissue disease associated forms of PAH in class II and III to reflect the clinical trial patient population. Approval was based on the SUPER-1 clinical trial in which a clear 45–50 metre improvement in 6 minute walking distance was achieved together with improvement of functional class.[41] Sildenafil was well tolerated by PAH patients, with headache the most common adverse event. Use extensions include an FDA expanded claim in 2009 for benefit in delay of clinical worsening.

In 2009 the FDA also approved an intravenous form of sildenafil given as a bolus injection for patients unable to take an oral formulation, and similar approvals have followed in the EU and more widely. In 2010 the EU approved sildenafil for the treatment of children with PAH, the first PAH-specific therapy to be granted approval for paediatrics. This approval followed a successful trial in which peak oxygen consumption, functional class and haemodynamics improved with medium and high doses *versus* placebo.[42] Sildenafil was well tolerated over the 16 week trial period. Intravenous sildenafil has also been used in children who cannot tolerate an oral

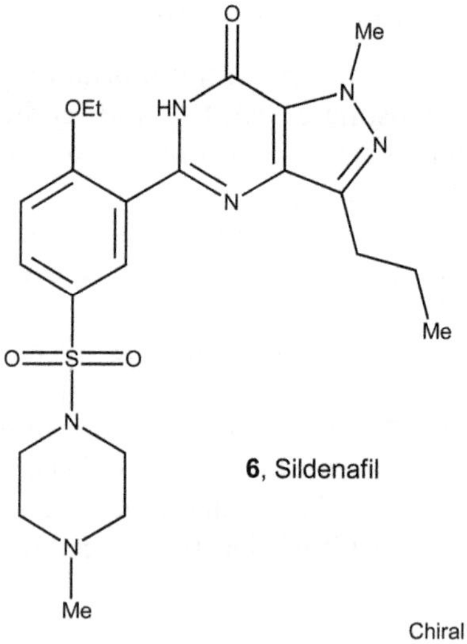

6, Sildenafil

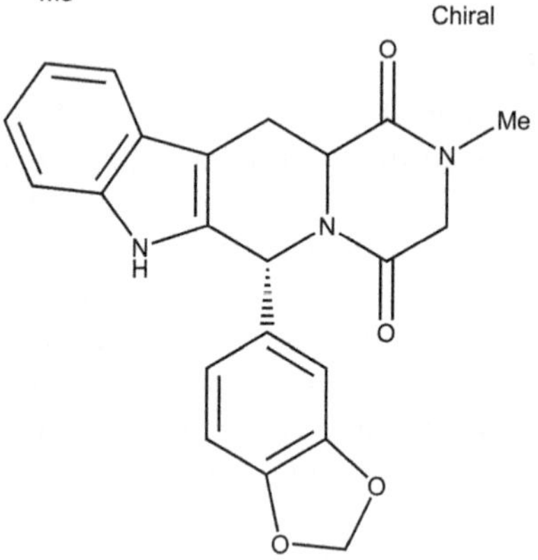

7, Tadalafil

Figure 13.3 Phosphodiesterase-5 inhibitors.

formulation and also to augment the vasodilator effect of inhaled NO in infants after cardiac surgery.[13]

13.3.3.2 Tadalafil

Tadalafil 7 is a tetracyclic PDE-5 inhibitor discovered by a joint venture between Eli Lilly and ICOS Corporation (Figure 13.3). Tadalafil has improved metabolic stability relative to sildenafil, translating into a longer ~18 hour human half-life that allows for once daily oral dosing.[20] Akin to sildenafil, tadalafil was first approved for MED with approval for PAH following in 2009 (Adcirca).

Tadalafil was first approved in 2009 in the USA and Europe as a once daily oral therapy (40 mg) for WHO group I PAH, apart from the label in Europe restricting to classes II and III, the clinical study population. Approvals in other countries followed. Approval was based upon one main 16 week trial in which tadalafil 40 mg improved 6 minute walk exercise capacity by 44 metres in treatment naïve patients, together with quality of life measures and also reduced clinical worsening.[43] Tadalafil was well tolerated, with headache the most common adverse event.

13.3.4 Nitric Oxide

As discussed previously, nitric oxide (NO) is a potent vasodilator. NO (INO_{max}) for inhalation was approved by the FDA in 1999 and the EU in 2001 for the treatment of neonates with persistent pulmonary hypertension of the newborn (PPHN).[13] In the two main clinical trials that underpinned approval, NO displayed significant benefit in reducing the need for oxygenation support and was well tolerated. Length of hospitalisation and mortality were not reduced. Later controlled studies in combination with other vasodilators have shown haemodynamic benefit in cardiac surgery patients, and additional add-on studies are in progress.

13.4 Emerging Therapies for PAH

The currently available treatments for PAH have provided significant benefits to patients in terms of improved exercise capacity, improved pulmonary haemodynamics and reduced time to clinical worsening. However, none are curative and the long-term prognosis for patients remains poor.[44] One current option for patients that doesn't appear to be optimised yet is combination treatment, drawing from the available options within the prostacyclin, endothelin and NO mechanistic pathways. Recent reports point to improved exercise capacity and reduced risk of clinical worsening with combination therapy[45,46] relative to monotherapy, and the combination option is now becoming increasingly widely used.

Beyond combination therapy, a number of new therapies are emerging that encompass both the existing vasodilation and endothelial dysfunction based therapeutic mechanisms and also several new mechanisms that target new pathways such as anti-proliferation and anti-inflammation.

13.4.1 Selexipag, a Non-Prostanoid Prostacyclin Receptor Agonist

Selexipag **8** is a non-prostanoid prostacyclin receptor agonist based on a diphenylpyrazinyl-aminobutoxy structure with a terminal *N*-methylsulphonylacetamide group (Figure 13.4). Selexipag is an orally bioavailable pro-drug and the acetamide group is readily hydrolysed to reveal a terminal acetic acid-moiety that is the active form **9**. Selexipag is selective for the prostaglandin I_2

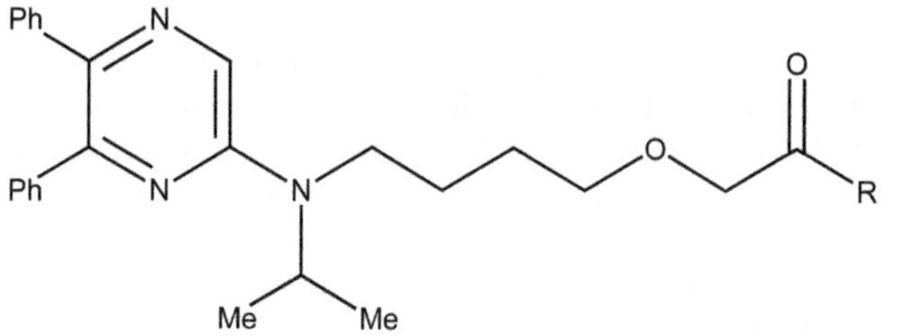

8, R = $NHSO_2Me$, Selexipag
9, R = OH, Selexipag active form

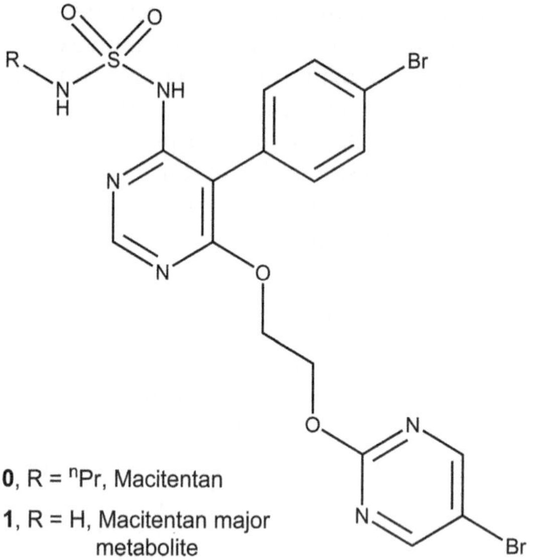

10, R = ^nPr, Macitentan
11, R = H, Macitentan major metabolite

Figure 13.4 Emerging treatments, selexipag and macitentan.

receptor and PGI_2 acts as a potent vasodilator and inhibitor of platelet aggregation *via* this receptor.[47] Selexipag has a 7.9 hour half-life which enables twice daily oral dosing. A Phase II study evaluated 43 PAH patients and the pulmonary vascular resistance and cardiac index were significantly improved after 17 weeks of treatment.[48] Selexipag was also well tolerated and a large Phase III trial is currently on-track to deliver results by mid-2014.

13.4.2 Macitentan, a Dual ET_A/ET_B Receptor Antagonist

Macitentan **10** is a new dual ET_A/ET_B receptor antagonist from Actelion that offers several improvements relative to bosentan **4** upon which the research programme was based.[49] Macitentan is a pyrimidine sulfamide based derivative (Figure 13.4) that makes for a lower propensity to ionise, and macitentan crosses lipophilic membranes effectively making for increased tissue penetration.[50] Clearance is low in both rat and dog with the primary sulfamide **11** a major metabolite and the propensity for drug–drug interactions is low. **11** is also formed in humans and evaluation reveals the compound to be a potent ET_A receptor antagonist in its own right, thus adding additional sustained pharmacological activity. Furthermore, macitentan does not increase circulating bile salts in rat and may therefore have a better liver injury profile.[51]

A Phase I study indicated a half-life in man of 17.5 hours for macitentan from a 300 mg dose, while the half-life of the metabolite **11** was 65 hours.[51] Thus macitentan is suitable as a once daily oral agent and results from a large Phase III trial indicated a significant decrease in patient morbidity and mortality.[52] Significant improvements in the 6 minute walking distance and the time to death or hospitalisation were also seen. Macitentan was generally well tolerated and elevation of liver function enzymes was no greater than placebo. Macitentan is currently under regulatory review in both the USA and the EU.

13.4.3 Riociguat, a Soluble Guanylate Cyclase Stimulator

Soluble guanylate cyclase (sGC) stimulators work in the NO–sGC–cGMP pathway akin to PDE-5 inhibitors. sGC stimulators enhance the sensitivity of sGC to the low levels of NO that are characteristic of PAH patients, thus increasing the production of cGMP and effecting vasodilation together with a reduction in smooth muscle proliferation.[53]

Riociguat **12** is a pyrazolopyridine based derivative representing a new class of agents that act as sGC stimulators (Figure 13.5) and has shown potential as a PAH treatment.[54] Riociguat directly stimulates sGC activity both independently of NO and also in synergy with endogenous NO. The compound arose from a medicinal chemistry programme that optimised the unfavourable metabolism and pharmacokinetic profiles of early pyrazolopyridine leads BAY-41-2272 and BAY-41-8443. Riociguat is an oral agent and displayed a half-life of 5–10 hours in male volunteers. The compound was well tolerated with only slight systemic hypotensive effects.[55] Following encouraging data in rodent models of PAH, a Phase II clinical trial displayed significant improvements in exercise capacity and pulmonary haemodynamics.[56] A Phase III trial for PAH functional class II and III patients followed and met the primary end point of improved exercise capacity with significant improvement in 6 minute walking distance (mean of 30 metres in 2.5 mg group), together with secondary end points that included pulmonary vascular resistance, time to clinical worsening and improved WHO functional class.[57] A FDA priority review for PAH has recently been granted.[58]

13.4.4 Tyrosine Kinase Inhibitors

Growth factors such as PDGF, EGF and bFGF have been implicated as having a role in the chronic pulmonary vascular remodelling that is inherent to PAH. This role is in addition to their recognised influence on the development and

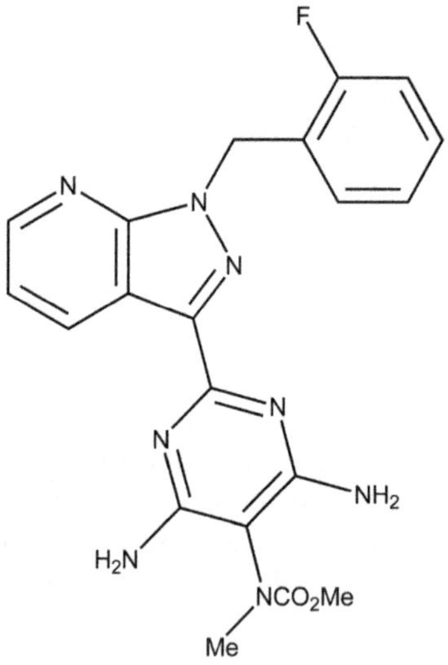

Figure 13.5 Riociguat, a sGC stimulator.

progression of human cancers and cardiopulmonary diseases. These growth factors exert their effect through transmembrane tyrosine kinase receptors, thereby activating major signal transduction pathways. Tyrosine kinases are enzymes that are implicit in this receptor mediated process as they effect the phosphorylation of tyrosine residues *via* transfer of an ATP phosphate residue to a protein.[59] PDGF is considered to be a key contributor to vascular remodelling in PAH, for example through influence on pulmonary artery smooth muscle and endothelial cells. PDGF exerts its influence through the activation of two PDGF receptors (PDGFR-α, PDGFR-β) and cellular effects include proliferation, survival and transformation.[60]

Imatinib **13** (Figure 13.6) inhibits the tyrosine kinase activity of several proteins such as the PDFGR kinases, BCR-ABL oncoprotein and the c-kit stem cell factor. The PAH beneficial effects displayed by this agent may come from inhibition of the PDGFR-α and PDGFR-β receptors. Imatinib (Gleevec) was first developed as an anti-cancer agent and is currently marketed as an oral therapy for various malignancies and most notably chronic myelogenous leukaemia (CML). Imatinib is rapidly absorbed and highly bioavailable. An initial Phase II PAH trial evaluated imatinib in patients with a poor response to established therapy, and an improvement in pulmonary haemodynamic parameters was seen, although exercise capacity did not significantly improve.[61] Post-trial analysis pointed to greater exercise and haemodynamic benefits in patients with advanced PAH. A subsequent Phase III trial in PAH patients with severe haemodynamic

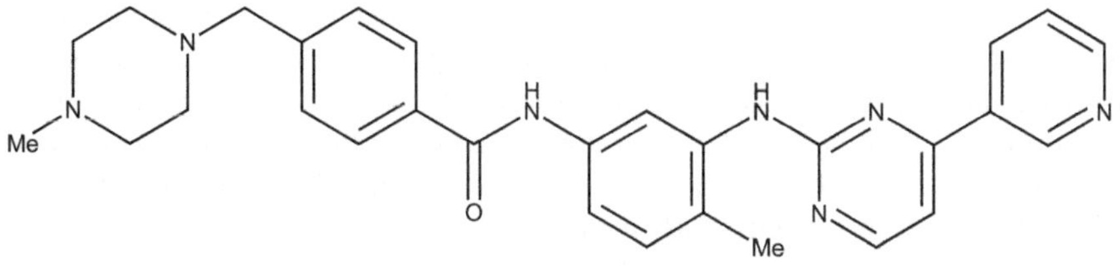

13, Imatinib

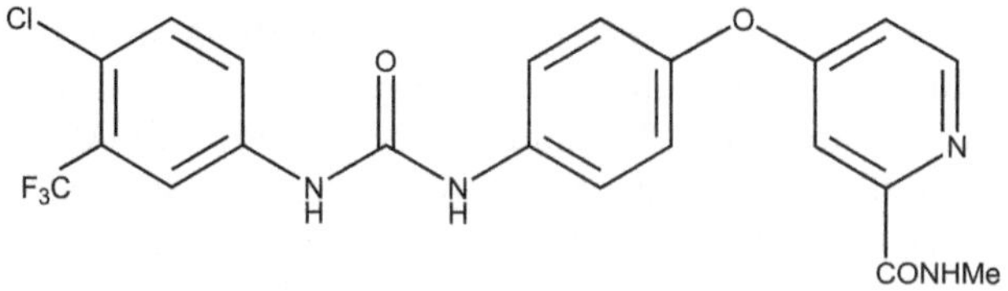

14, Sorafenib

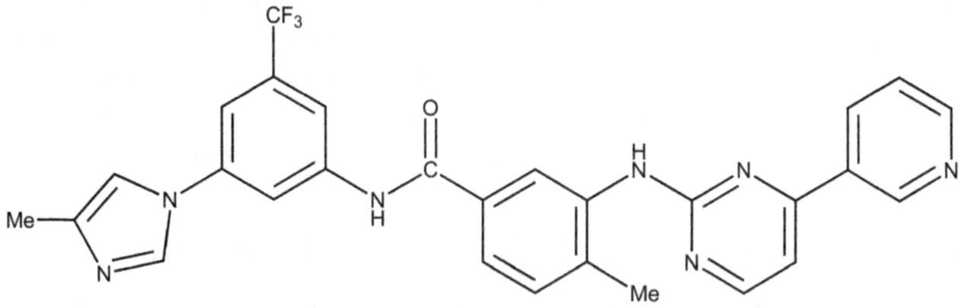

15, Nilotinib

Figure 13.6 Tyrosine kinase inhibitors.

impairment did achieve beneficial effects on exercise capacity (mean improvement 32 metres at 24 weeks) and functional class.[62] The frequency of adverse events was similar to placebo, although some concerns about a possible increase in subdural haematoma frequency were raised. Imatinib is currently undergoing regulatory review for the treatment of PAH in the USA and Europe.

Sorafenib **14** is a multikinase inhibitor that targets VEGF and Raf-1 in addition to the PDGF receptors. Although mainly used in the cancer therapy arena, where it is widely approved for treatment of liver and advanced kidney cancer, a small Phase Ib study in PAH patients on existing therapy was undertaken and a significant increase in 6 minute walking distance was achieved.[63,64] However cardiac output slightly decreased in some patients, suggesting a possible unfavourable effect on cardiac function. It is uncertain as to whether sorafenib is currently being pursued for PAH.

Nilotinib **15** is a second-generation tyrosine kinase inhibitor approved for the treatment of CML as a twice daily oral agent. Nilotinib has inhibitory activity against the PDGFR, BCR-ABL and c-kit kinases, and is more potent than imatinib against BCR-ABL.[65] In PAH, a 24 week Phase II trial to evaluate the potential of nilotinib has been undertaken and has now terminated, although no results are currently available.[66]

13.4.5 RhoA/Rho-kinase Inhibitors

The Rho-kinases (ROCK) are implicated in cellular processes such as differentiation, proliferation and endothelial cell dysfunction that are relevant to PAH. Rho-kinase interacts with the G-protein RhoA and this signalling pathway inhibits myosin phosphatase leading to a Ca^{2+} driven sensitisation of smooth muscle contraction.[67,68] Fasudil **16** is a selective RhoA/ROCK inhibitor that effects vasodilation and is used to treat cerebral vasospasm. Fasudil has also been shown to suppress cellular proliferation in a rat model of PAH.[69] An initial patient study with fasudil indicated beneficial effects on pulmonary resistance and pulmonary arterial pressure.[70] However systemic vascular resistance also decreased following systemic administration. Inhaled fasudil has also been studied in a small patient group and led to a reduced pulmonary vascular resistance.[71] A long-term clinical trial with fasudil in severe PAH patients appears to be needed to give a more conclusive outcome on the potential efficacy and safety of Rho-kinase inhibitors (Figure 13.7).[72]

13.4.6 Serotonin (5-HT) Pathway Based Treatments

Serotonin is a pulmonary vasoconstrictor and a smooth muscle cell mitogen and the serotonin pathway has been implicated as a factor in the pathogenesis of PAH.[73,74] There is not currently a full mechanistic understanding for serotonin involvement and idiopathic PAH has been associated with both increased and normal free plasma serotonin levels.[75] Escitalopram **17** is a selective serotonin reuptake inhibitor that would increase plasma serotonin levels and has been approved in the USA as an oral agent for major depressive disorder and other generalised anxiety disorder indications. A Phase III trial has evaluated escitalopram in patients with mild to moderate PAH, with the prime focus on improved exercise capacity. The trial is complete but results are not yet available.[76]

Terguride **18** is a potent serotonin 5-HT_{2B} receptor antagonist with activity documented against other 5-HT receptor subtypes,[77] and additionally partial dopamine agonist activity. An oral agent, terguride, is approved in Japan for various neurological diseases. Subsequent animal model experiments with terguride have indicated potential for the treatment of PAH.[78] For instance, pulmonary vasoconstriction and proliferation of pulmonary artery smooth muscle cells were both reduced in animal models of PAH. A Phase II trial for terguride in PAH was undertaken in 2008,[79] but no results are publicly available.

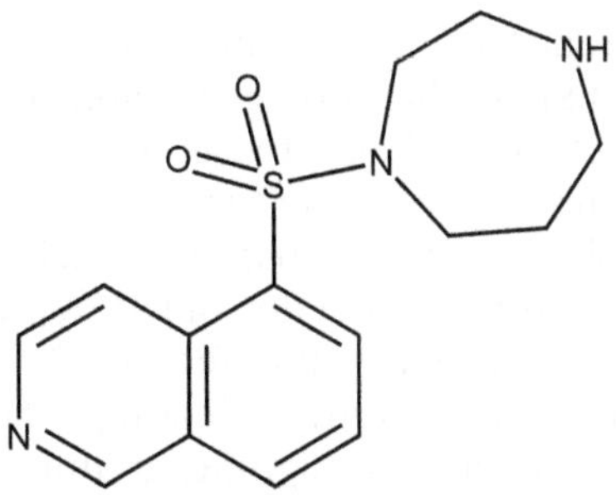

16, Fasudil

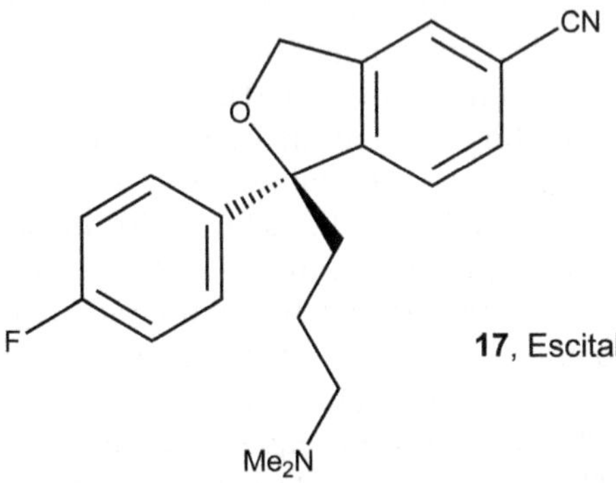

17, Escitalopram

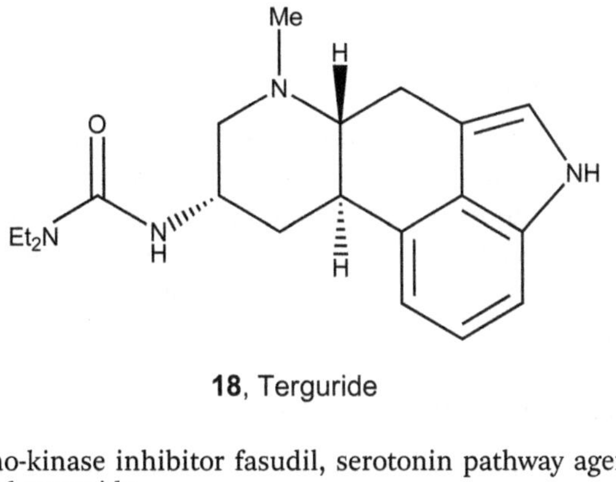

18, Terguride

Figure 13.7 Rho-kinase inhibitor fasudil, serotonin pathway agents escitalopram and terguride.

13.4.7 Dehydroepiandrosterone

Dehydroepiandrosterone (DHEA) **19** is the most abundant circulating steroid in humans (Figure 13.8) and has a range of biological effects in addition to being an intermediate in the biosynthesis of androgen and oestrogen sex steroids. DHEA is involved in a number of pathways that are involved in the regulation of cell proliferation and apoptosis, such as Akt/GSK-3/NFAT and Src/STAT3, in which the inhibitory action of DHEA prevents induced vascular remodelling and decreases cellular proliferation.[80,81] In a pilot PAH-chronic obstructive pulmonary disease study evaluating a daily 200 mg oral dose of DHEA in adults over 3 months, significant improvements in 6 minute walking distance (57 metres, $p < 0.05$), pulmonary haemodynamics and lung carbon monoxide diffusing capacity were seen. DHEA was also well tolerated.[82] The status of ongoing clinical trial study is not clear.

13.4.8 Beraprost-Modified Release

Beraprost **20** is an orally active synthetic prostacyclin analogue with improved chemical stability that enables a human half-life of 35–40 minutes (Figure 13.8). The agent has been approved in Japan for PAH,[83] and in a subsequent larger clinical trial beraprost has shown benefit in 3 and 6 month walking distance tests. However, this benefit was not sustained at 9 or 12 months,[84] thereby limiting wider regulatory approval. A modified form with an increased duration of action has been developed, beraprost-modified release, but a subsequent Phase II trial failed to meet its primary haemodynamic end point.[85] The results of the study suggested that the efficacy of beraprost may be improved by increasing therapeutic exposure through more stable and consistent plasma concentrations. United Therapeutics subsequently developed a reformulated, single isomer version of beraprost. A Phase I safety trial completed in July 2012, with the preliminary data suggesting that four times daily dosing is safe, and a Phase III study to evaluate the clinical benefit of this new formulation as an add-on therapy to inhaled treprostinil is being designed.[86]

13.4.9 Cicletanine

Cicletanine **21** is a furo-pyridine derivative (Figure 13.8) approved for the treatment of hypertension and is a once daily oral agent. Preclinical data suggest that cicletanine may act through enhanced coupling of endothelial nitric oxide synthase (eNOS) to increase vascular NO availability.[87] A Phase II clinical trial in PAH patients commenced in 2009, however results have not yet been made public.[88]

13.4.10 Other Therapeutic Options

Vasoactive intestinal peptide (VIP) is a 28-amino acid peptide hormone that has a role as a pulmonary and systemic vasodilator.[89] In PAH patients, VIP pulmonary serum levels are decreased and the expression of VIP

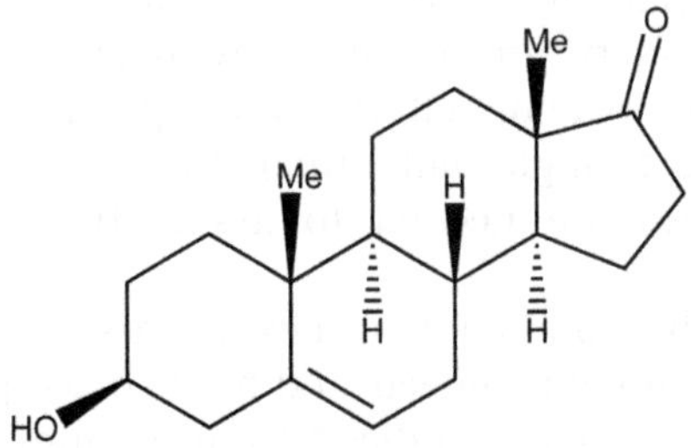

19, Dehydroepiandrosterone (DHEA)

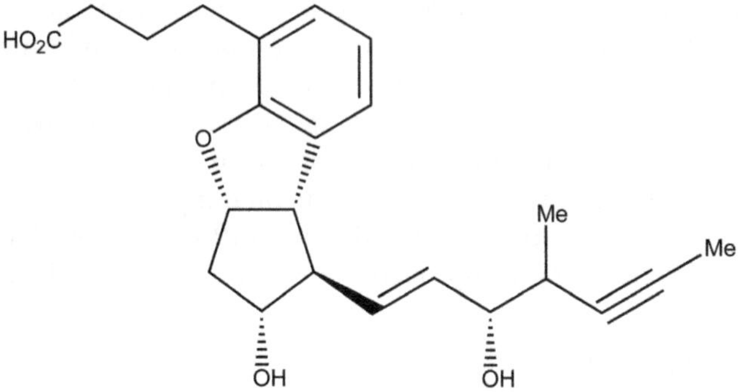

20, Beraprost

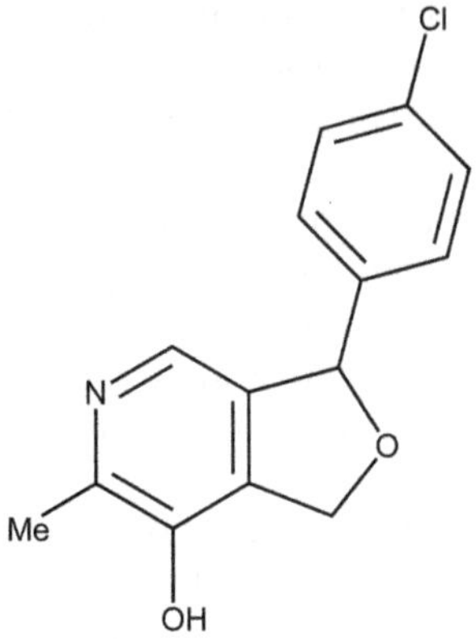

21, Cicletanine

Figure 13.8 Dehydroepiandrosterone (DHEA), beraprost and cicletanine.

mediating receptors is also increased in pulmonary artery smooth muscle cells. An inhaled form of VIP has been studied clinically in idiopathic PAH patients. In a 3 month trial with eight patients taking inhaled VIP four times daily, mean pulmonary arterial pressure decreased and cardiac output improved, while no systemic effects were reported.[90] A subsequent trial in 20 patients giving a single daily inhaled dose achieved a similar outcome.[91] Further studies are needed to assess the full therapeutic potential of VIP.

HMG-CoA reductase inhibitors (statins) have been shown to exert anti-proliferative and anti-inflammatory effects and have shown potential in the field of PAH based on animal model studies.[92] Trials in humans have shown small transient reduction in right ventricle mass but didn't significantly improve exercise capacity.[93,94] Thus the potential of statins in PAH remains uncertain currently.

Endothelial progenitor cells (EPCs) are a population of rare cells that circulate in the blood with the ability to differentiate into endothelial cells. Intravenous infusion of autologous EPCs in idiopathic PAH patients has resulted in a significant improvement in 6 minute walking distance (48 metres) and reduced pulmonary artery pressure.[95] The design of further clinical studies to assess safety and efficacy is ongoing.[96]

Additionally, therapy options based on miRNA, anti-inflammatory, metabolism and gene therapy approaches are also being investigated.[97]

13.5 Discussion

The treatment options for PAH patients have advanced significantly in the past two decades. More stable and inhaled forms of prostacyclin-based treatment have emerged for severe sufferers together with oral treatment options of increasing convenience from two new vasodilatory mechanisms (endothelin receptor antagonism and PDE-5 inhibition) that are increasingly used across all functional classes. These innovations have helped to improve the quality of life for patients in terms of treatment convenience, increased exercise capacity, improved pulmonary haemodynamics and increased time to clinical worsening. The numerous regulatory label expansions amongst these new agents, such as approval across all functional classes, and approval of a PAH specific drug for paediatrics (as with sildenafil), bear further testimony to the advancements made.

Most of the approved products and many of the emerging therapies have an orphan designation and/or approval and the field of PAH is probably one of the most active within the category of orphan diseases. Thus, a good number of PAH targeted agents have been able to take advantage of the benefits (outlined in earlier chapters) that are offered to orphan designated products, such as a period of market exclusivity and development cost benefits.

There have been challenges to the way that the orphan designation and reimbursement process works.[96] Where a product has been previously approved for another indication, but used to treat an orphan disease through a pharmacy compounded or off-label basis, the price once approved as an orphan product has become more than the previous cost. This has led to some health authorities refusing to pay reimbursement costs. This has also, in part, resulted in a call for a modified approval process in which additional factors such as the previous drug history and development costs are taken into account, together with the advantages that are offered relative to existing treatments.[97] Given that some of the current approved and emerging orphan products were originally marketed for other indications, these factors are relevant to the PAH field.

Another factor is that many of the existing PAH treatments are either off-patent or will be off-patent within the next few years, probably leading to generic products that will drive down price. Thus, for the field of PAH, from both available treatment and orphan cost-effectiveness arguments, there will be a clear need for new products to demonstrate advantage over existing treatments. These factors can be a positive for both the PAH patient and research community, as while the current treatment options have improved quality of life, they are far from perfect. The long-term prognosis for a PAH sufferer is still poor and there is a clear need for improved products that further prolong quality of life and lifespan, and ideally cure.

In this respect, there is some encouragement. The emerging agents within the existing vasodilation mechanistic fields offer hope for further patient improvements. In particular, macitentan with improved tissue penetration and prolonged duration of action offers hope of improved morbidity and mortality. If successful, riociguat from the sGC–NO–cGMP pathway would bring another new mechanism into play, while selexipag would offer a more convenient long-acting oral option within the prostacyclin arena.

In terms of mechanisms that could be anti-proliferative, pro-apoptotic or anti-inflammatory there are a number of emerging options. For example, imatinib, fasudil, escitalopram, terguride and DHEA offer hope of new first-in-class mechanisms that could supplement the predominantly vasodilatory action of the current agents. New agents of these mechanistic categories would expand the treatment paradigm available to clinicians.

Regardless of what new agents do emerge, the ability to combine with other agents from a different mechanistic class will be crucial moving forward. Combination therapy will probably become standard. Thus good physicochemical properties in terms of pharmacokinetics, metabolism and lack of drug–drug interactions will be a key requirement. More combination therapy options will offer the prospect of a multi-mechanistic attack on PAH and improved disease control. The design of clinical trials will also be crucial moving forward in terms of selection of the appropriate patient population, the trial length and the primary end points. Greater flexibility in trial design would also be aided by emerging agents having good profiles that allowed for dose variation, *etc.*

In summary, the significant advances of the past two decades have laid a firm platform upon which promising new emerging therapies can potentially take PAH treatment forward afresh and further minimise the deterioration that is still characteristic of many current patients.

Acknowledgements

The author would like to thank Gary Burgess of Conatus Pharmaceuticals for helpful advice and discussion, and for proofreading this manuscript.

References

1. A. E. Frost, D. B. Badesch, R. J. Barst, R. L. Benza, C. G. Elliott, H. W. Farber, A. Krichman, T. G. Liou, G. E. Raskob, P. Wason, K. Feldkircher, M. Turner and M. D. McGoon, *Chest,* 2011, **139**, 128–137.
2. A. J. Peacock, N. F. Murphy, J. J. McMurray, L. Caballero and S. Stewart, *Eur. Respir. J.,* 2007, **30**, 104–109.
3. G. Simonneau, I. M. Robbins, M. Beghetti, R. N. Channick, M. Delcroix, C. P. Denton, C. G. Elliott, S. P. Gaine, M. T. Gladwin, Z. C. Jing, M. J. Krowka, D. Langleben, N. Nakanishi and R. Souza, *J. Am. Coll. Cardiol.,* 2009, **54**, S43–S54.
4. S. A. van Wolferen, K. Grunberg and A. Vonk-Noordegraaf, *Respir. Med.,* 2007, **101**, 389–398.
5. N. Galie, M. M. Hoeper, M. Humbert, A. Torbicki, J.-L. Vachiery, J. A. Barbera, M. Beghetti, P. Corris, S. Gaine, J. S. Gibbs, M. A. Gomez-Sanchez, G. Jondeau, W. Klepetko, C. Opitz, A. Peacock, L. Rubin and M. Zellweger, *Eur. Heart J.,* 2009, **30**, 2493–2537.
6. D. B. Badesch, G. E. Raskob, C. G. Elliott, A. M. Krichman, H. W. Farber, A. E. Frost, R. J. Barst, R. L. Benza, T. G. Liou, M. Turner, S. Giles, K. Feldkircher, D. P. Miller and M. D. McGoon, *Chest,* 2010, **137**, 376–387.
7. R. J. Barst, J. S. Gibbs, H. A. Ghofrani, M. M. Hoeper, V. V. McLaughlin, L. J. Rubin, O. Sitbon, V. F. Tapson and N. Galie, *J. Am. Coll. Cardiol.,* 2009, **54**, S78–S84.
8. R. L. Benza, D. P. Miller, R. J. Barst, D. B. Badesch, A. E. Frost and M. D. McGoon, *Chest,* 2012, **142**, 448–456.
9. D. B. Badesch, H. W. Farber, M. D. McGoon, A. E. Frost, C. G. Elliott, R. L. Benza, A. Poms, D. P. Miller and R. J. Barst, *Chest,* 2011, **140**, 737A, meeting abstracts.
10. V. V. McLaughlin and M. D. McGoon, *Circulation,* 2006, **114**, 1417–1431.
11. R. L. Benza, M. Gomberg-Maitland, D. P. Miller, A. E. Frost, R. P. Frantz, A. J. Foreman, D. B. Badesch and M. D. McGoon, *Chest,* 2012, **141**, 354–362.
12. W. H. Fares and T. K. Trow, *Ther. Adv. Respir. Dis.,* 2012, **6**, 147–159.
13. L. R. Frumkin, *Pharma Rev.,* 2012, **64**, 583–620.
14. A. Raiesdana and J. Loscalzo, *Ann. Med.,* 2006, **38**, 95–110.
15. M. Rabinovitch, *Annu. Rev. Phytopathol.,* 2007, **2**, 369–399.
16. M. D. McGoon and G. C. Kane, *Mayo Clin. Proc.,* 2009, **84**, 191–207.

17. A. P. Davenport and J. J. Maguire, Endothelin, *Handb. Exp. Pharmacol.,* 2006, **176**(Pt1), 295–329.
18. (a) S. Rich, *Primary Pulmonary Hypertension: Executive Summary from the World Symposium on Primary Pulmonary Hypertension*, World Health Organization, Evian, France, 1998; (b) R. J. Barst, M. D. McGoon, A. Torbicki, O. Sitbon, M. J. Krowka, H. Olschewski and S. Gaine, *J. Am. Coll. Cardiol.,* 2004, **43**, 40S–47S.
19. R. J. Barst, L. J. Rubin, W. A. Long, M. D. McGoon, S. Rich, D. B. Badesch, B. M. Groves, V. F. Tapson, R. C. Bourge, B. H. Brundage, S. K. Koerner, D. Langleben, C. A. Keller, S. Murali, B. F. Uretsky, L. M. Clayton, M. M. Jobsis, S. D. Blackburn, D. Shortino and J. W. Crow, *N. Engl. J. Med.,* 1996, **334**, 296–301.
20. E. B. Rosenzweig, *Expert Opin. Emerging Drugs,* 2006, **11**, 609–619.
21. G. Simonneau, R. J. Barst, N. Galie, R. Naeije, S. Rich, R. C. Bourge, A. Keogh, R. Oudiz, A. E. Frost, S. D. Blackburn, J. W. Crow and L. J. Rubin, *Am. J. Respir. Crit. Care Med.,* 2002, **165**, 800–804.
22. M. Levy, D. S. Celermajer, E. Bourges-Petit, M. J. Del Cerro, F. Bajolle and D. Bonnet, *J. Pediatr.,* 2011, **158**, 584–588.
23. J. Hiremath, S. Thanikachalam, K. Parikh, S. Shanmugasundaram, S. Bangera, L. Shapiro, G. B. Pott, C. L. Vnencak-Jones, C. Arneson, M. Wade and R. J. White, *J. Heart Lung Transplant.,* 2010, **29b**, 137–149.
24. J. D. Rich, C. Glassner, M. Wade, S. Coslet, C. Arneson, A. Doran and M. Gomberg-Maitland, *Chest,* 2012, **141**, 36–42.
25. V. V. McLaughlin, R. L. Benza, L. J. Rubin, R. N. Channick, R. Voswinckel, V. F. Tapson, I. M. Robbins, H. Olschewski, M. Rubenfire and W. Seeger, *J. Am. Coll. Cardiol.,* 2010, **55**, 1915–1922.
26. United Therapeutics, http://www.unither.com/oral-treprostinil-for-pah, accessed 2 August 2013.
27. H. Olschewski, G. Simonneau, N. Galie, T. Higenbottam, R. Naeije, L. J. Rubin, S. Nikkho, R. Speich, M. M. Hoeper, J. Behr, J. Winkler, O. Sitbon, W. Popov, H. A. Ghofrani, A. Manes, D. G. Kiely, R. Ewert, A. Meyer, P. A. Corris, M. Delroix, M. Gomez-Sanchez, H. Siedentop and W. Seeger, *N. Engl. J. Med.,* 2002, **347**, 322–329.
28. C. Mulligan and M. Beghetti, *Pediatr. Crit. Care Med.,* 2012, **13**, 472–480.
29. M. J. Palmer, *Prog. Med. Chem.,* 2009, **47**, 203–237.
30. R. N. Channick, G. Simonneau, O. Sitbon, I. M. Robbins, A. E. Frost, V. F. Tapson, D. B. Badesch, S. Roux, M. Rainisio, F. Bodin and L. J. Rubin, *Lancet,* 2001, **358**, 1119–1123.
31. L. J. Rubin, D. B. Badesch, R. J. Barst, N. Galie, C. M. Black, A. Keogh, T. Pulido, A. E. Frost, S. Roux, I. Leconte, M. Landzberg and G. Simonneau, *N. Engl. J. Med.,* 2002, **346**, 896–903.
32. H. Vatter and V. Seifert, *Cardiovasc. Drug Rev.,* 2006, **24**, 63–76.
33. N. Galie, H. Olschewski, R. J. Oudiz, F. Torres, A. E. Frost, H. A. Ghofrani, D. B. Badesch, M. D. McGoon, V. V. McLaughlin, E. B. Roecker, M. J. Gerber, C. Dufton, B. L. Wiens and L. J. Rubin, *Circulation,* 2008, **117**, 3010–3019.

34. K. Fattinger, C. Funk, M. Pantze, C. Weber, J. Reichen, B. Stieger and P. J. Meyer, *Clin. Pharmacol. Ther.*, 2001, **69**, 223–231.
35. J. E. Frampton, *Am. J. Cardiovasc. Drugs*, 2011, **11**, 215–226.
36. C. F. Opitz, R. Ewert, W. Kirch and D. Pittrow, *Eur. Heart J.*, 2008, **29**, 1936–1948.
37. J. A. Beavo and L. L. Brunton, *Nat. Rev. Mol. Cell Biol.*, 2002, **3**, 710–718.
38. A. L. Burnett, *Am. J. Cardiol.*, 2005, **96**(12B), 29M–31M.
39. J. D. Corbin and S. H. Francis, *Int. J. Clin. Pract.*, 2002, **56**, 453–459.
40. D. P. Rotella, *Nat. Rev. Drug Discovery*, 2002, **1**, 674–682.
41. N. Galie, H. A. Ghofrani, A. Torbicki, R. J. Barst, L. J. Rubin, D. Badesch, T. Fleming, T. Parpia, G. Burgess, A. Branzi, F. Grimminger, M. Kurzyna and G. Simonneau, *N. Engl. J. Med.*, 2005, **353**, 2148–2157.
42. R. J. Barst, D. D. Ivy, G. Gaitan, A. Szatmari, A. Rudzinski, A. E. Garcia, B. K. Sastry, T. Pulido, G. R. Layton, M. Serdarevic-Pehar and D. L. Wessel, *Circulation*, 2012, **125**, 324–334.
43. N. Galie, B. H. Brundage, H. A. Ghofrani, R. J. Oudiz, G. Simonneau, Z. Safdar, S. Shapiro, R. J. White, M. Chan, A. Beardsworth, L. Frumkin and R. J. Barst, *Circulation*, 2009, **119**, 2894–2903.
44. M. Humbert, O. Sitbon, A. Chaouat, M. Bertocchi, G. Habib, V. Gressin, A. Yaici, E. Weitzenblum, J.-F. Cordier, F. Chabot, C. Dromer, C. Pison, M. Reynard-Gaubert, A. Haloun, M. Laurent, E. Hachulla, V. Cottin, B. Degano, X. Jais, D. Montani, R. Souza and G. Simonneau, *Circulation*, 2010, **122**, 156–163.
45. B. Zhu, L. Wang, L. Sun and R. Cao, *J. Cardiovasc. Pharmacol.*, 2012, **60**, 342–346.
46. B. D. Fox, A. Shimony and D. Langleben, *Am. J. Cardiol.*, 2011, **108**, 1177–1182.
47. K. Kuwano, A. Hashino, T. Asaki, T. Hamamoto, T. Yamada, K. Okubo and K. Kuwabara, *J. Pharmacol. Exp. Ther.*, 2007, **322**, 1779–1783.
48. G. Simonneau, A. Torbicki, M. M. Hoeper, M. Delcroix, K. Karlocai, N. Galie, B. Degano, D. Bonderman, M. Kurzyna, M. Efficace, R. Giorgino and I. M. Lang, *Eur. Respir. J.*, 2012, **40**, 874–880.
49. M. H. Bolli, C. Boss, C. Binkert, S. Buchmann, D. Bur, P. Hess, M. Iglarz, S. Meyer, J. Rein, M. Rey, A. Treiber, M. Clozel, W. Fischli and T. Weller, *J. Med. Chem.*, 2012, **55**, 7849–7861.
50. M. Iglarz, C. Binkert, K. Morrison, W. Fischli, J. Gatfield, A. Treiber, T. Weller, M. H. Bolli, C. Boss, S. Buchmann, B. Capeleto, P. Hess, C. Qiu and M. Clozel, *J. Pharmacol. Exp. Ther.*, 2008, **327**, 736–745.
51. P. N. Sidharta, P. L. van Gisbergen, A. Halabi and J. Dingemanse, *Eur. J. Clin. Pharmacol.*, 2011, **67**, 977–984.
52. L. Rubin, T. Pulido, R. Channick, M. Delcroix, N. Galie, H.-A. Ghofrani, P. Jansa, F.-O. Le Brun, S. Mehta, C. Mittelholzer, L. Perchenet, K. S. Sastry, O. Sitbon, R. Souza, A. Torbicki and G. Simonneau, *Chest*, 2012, **142**, 1026A.
53. O. V. Evgenov, P. Pacher, P. M. Schmidt, G. Hasko, H. H. H. W. Schmidt and J.-P. Stasch, *Nat. Rev. Drug Discovery*, 2006, **5**, 755–768.

54. J. Mittendorf, S. Weigand, C. Alonso-Acilia, E. Bischoff, A. Feurer, M. Gerisch, A. Kern, A. Knorr, D. Lang, K. Muenter, M. Radtke, H. Schirok, K.-H. Schlemmer, E. Stahl and A. Straub, *ChemMedChem*, 2009, **4**, 853–865.
55. R. Frey, W. Mück, S. Unger, U. Artmeier-Brandt, G. Weimann and G. Wensing, *J. Clin. Pharmacol.*, 2008, **48**, 926–934.
56. H.-A. Ghofrani, M. M. Hoeper, M. Halank, F. J. Meyer, G. Staehler, J. Behr, R. Ewert, G. Weimann and F. Grimminger, *Eur. Respir. J.*, 2010, **36**, 792–799.
57. H.-A. Ghofrani, N. Galie, F. Grimminger, E. Grunig, M. Humbert, Z.-C. Jing, A. M. Keogh, D. Langleben, M. O. Kilama, A. Fritsch, D. Neuser and L. J. Rubin, *N. Engl. J. Med.*, 2013, **369**, 330–340.
58. http://www.pharma.bayer.com/html/pdf/Final-US-vers-Riociguat-Priority-Review-press-release.pdf, accessed 5 August 2013.
59. M. Klein, R. T. Schermuly, P. Ellinghaus, H. Milting, B. Riedl, S. Nikolova, S. S. Pullamsetti, N. Weissmann, E. Dony, R. Savai, H. A. Ghofrani, F. Grimminger, A. E. Busch and S. Schafer, *Circulation*, 2008, **118**, 2081–2090.
60. D. Dumitrescu, C. Seck, H. ten Freyhaus, F. Gerhardt, E. Erdmann and S. Rosenkranz, *Eur. Respir. J.*, 2011, **38**, 218–220.
61. H. A. Ghofrani, N. W. Morrell, M. M. Hoeper, H. Olschewski, A. J. Peacock, R. J. Barst, S. Shapiro, H. Golpon, M. Toshner, F. Grimminger and S. Pascoe, *Am. J. Respir. Crit. Care Med.*, 2010, **182**, 1171–1177.
62. M. M. Hoeper, R. J. Barst, R. C. Bourge, J. Feldman, A. E. Frost, N. Galie, M. A. Gomez-Sanchez, F. Grimminger, E. Grunig, P. M. Hassoun, N. W. Morrell, A. J. Peacock, T. Satoh, G. Simonneau, V. F. Tapson, F. Torres, D. Lawrence, D. A. Quinn and H.-A. Ghofrani, *Circulation*, 2013, **127**, 1128–1138.
63. G. Ranieri, G. Gadaleta-Caldarola, V. Goffredo, R. Patruno, A. Mangia, A. L. Rizzo, R. D. Sciorsci and C. Gadaleta, *Curr. Med. Chem.*, 2012, **17**, 938–944.
64. M. Gomberg-Maitland, M. L. Maitland, R. J. Barst, L. Sugeng, S. Coslet, T. J. Perino, L. Bond, M. E. LaCouture, S. L. Archer and M. J. Ratain, *Clin. Pharmacol. Ther.*, 2010, **87**, 303–310.
65. H. Kantarjian, F. Giles, L. Wunderle, K. Bhalla, S. O'Brien, B. Wassmann, C. Tanaka, P. Manley, P. Rae, W. Mietlowski, K. Bochinski, A. Hochhaus, J. D. Griffin, D. Hoelzer, M. Albitar, M. Dugan, J. Cortes, L. Alland and O. G. Ottmann, *N. Engl. J. Med.*, 2006, **354**, 2542–2551.
66. Clinicaltrials.gov, Novartis, http://clinicaltrials.gov/ct2/show/results/NCT01179737, accessed 6 August 2013.
67. G. Loirand, P. Guerin and P. Pacaud, *Circ. Res.*, 2006, **98**, 322–334.
68. N. W. Morrell, S. Adnor, S. L. Archer, J. Dupuis, P. Lloyd Jones, M. R. MacLean, I. F. McMurty, K. R. Stenmark, P. A. Thistlethwaite, N. Weissmann, J. X.-J. Yuan and K. Weir, *J. Am. Coll. Cardiol.*, 2009, **54**, S20–S31.

69. K. Abe, H. Shimokowa, K. Morikawa, T. Uwatoku, K. Oi, Y. Matsumoto, T. Hattori, Y. Nakashima, K. Kaibuchi, K. Sueishi and A. Takeshita, *Circ. Res.,* 2004, **94**, 385–393.
70. Y. Fukumoto, T. Matoba, A. Ito, H. Tanaka, T. Kishi, S. Hayashidani, K. Abe, A. Takeshita and H. Shimokawa, *Heart,* 2005, **91**, 391–392.
71. H. Fujita, Y. Fukumoto, K. Saji, K. Sugimura, J. Demachi, J. Nawata and H. Shimokawa, *Heart Vessels,* 2010, **25**, 144–149.
72. P. Pacaud and G. Loirand, *Eur. Respir. J.,* 2010, **36**, 709–711.
73. J. M. Launay, P. Herve, K. Peoc'h, C. Tournois, J. Callebert, C. G. Nebigil, N. Etienne, L. Drouet, M. Humbert, G. Simonneau and L. Maroteaux, *Nat. Med.,* 2002, **8**, 1129–1135.
74. L. Long, M. R. MacLean, T. K. Jeffery, I. Morecroft, X. Yang, N. Rudarakanchana, M. Southwood, V. James, R. C. Trembath and N. W. Morrell, *Circ. Res.,* 2006, **98**, 818–827.
75. S. J. Shah, M. Gomberg-Maitland, T. Thenappan and S. Rich, *Chest,* 2009, **136**, 694–700.
76. Clinicaltrials.gov, http://www.clinicaltrials.gov/ct2/show/NCT00190333, accessed 6 August 2013.
77. A. Newman-Tancredi, D. Cussac, Y. Quentric, M. Touzard, L. Verriele, N. Carpentier and M. J. Millan, *J. Pharmacol. Exp. Ther.,* 2002, **303**, 815–822.
78. R. Dumitrascu, C. Kulcke, M. Konigshoff, F. Kouri, X. Yang, N. Morrell, H. A. Ghofrani, N. Weissmann, R. Reiter, W. Seeger, F. Grimminger, O. Eickelberg, R. T. Schermuly and S. S. Pullamsetti, *Eur. Respir. J.,* 2011, **37**, 1104–1118.
79. R. Agarwal and M. Gomberg-Maitland, *Am. Heart J.,* 2011, **162**, 201–213.
80. S. Bonnet, R. Paulin, G. Sutendra, P. Dromparis, M. Roy, K. O. Watson, J. Nagendran, A. Haromy, J. R. B. Dyck and E. D. Michelakis, *Circulation,* 2009, **120**, 1231–1240.
81. R. Paulin, J. Meloche, M. H. Jacob, M. Bisserier, A. Courboulin and S. Bonner, *Am. J. Physiol.: Heart Circ. Physiol.,* 2011, **301**, H1798–H1809.
82. E. Dumas De La Roque, J.-P. Savineau, A.-C. Metivier, M.-A. Billes, J.-P. Kraemer, S. Doutreleau, J. Jougon, R. Marthan, N. Moore, M. Fayon, E.-E. Baulieu and C. Dromer, *Annee Endocrinol.,* 2012, **73**, 20–25.
83. Y. Okano, T. Yoshioka, A. Shimouchi, T. Satoh and T. Kunieda, *Lancet,* 1997, **349**, 1365.
84. R. J. Barst, M. McGoon, V. McLaughlin, V. Tapson, R. Oudiz, S. Shapiro, I. M. Robbins, R. Channick, D. Badesch, B. K. Rayburn, R. Flinchbaugh, J. Sigman, C. Arneson and R. Jeffs, *J. Am. Coll. Cardiol.,* 2002, **41**, 2119–2125.
85. http://clinicaltrials.gov/show/NCT00989963, accessed 10 August 2013.
86. http://www.unither.com/bps-for-pah, accessed 10 August 2013.
87. L. Kalinowski, I. T. Dobrucki and T. Malinski, *J. Cardiovasc. Pharmacol.,* 2001, **37**, 713–724.
88. http://clinicaltrials.gov/show/NCT00581087, accessed 19 August 2013.

89. R. J. Henning and D. R. Sawmiller, *Cardiovasc. Res.*, 2001, **49**, 27–37.
90. V. Petkov, W. Mosgoeller, R. Ziesche, M. Raderer, L. Stiebellehner, K. Vonbank, G.-C. Funk, G. Hamilton, C. Novotny, B. Burian and L.-H. Block, *J. Clin. Invest.*, 2003, **111**, 1339–1346.
91. H. H. Leuchte, C. Baezner, R. A. Baumgartner, D. Bevec, G. Bacher, C. Neurohr and J. Behr, *Eur. Respir. J.*, 2008, **32**, 1289–1294.
92. X. Sun and D. D. Ku, *Am. J. Physiol.: Heart Circ. Physiol.*, 2008, **294**, H801–H809.
93. M. R. Wilkins, O. Ali, W. Bradlow, J. Wharton, A. Taegtmeyer, C. J. Rhodes, H. A. Ghofrani, L. Howard, P. Nihoyannopoulos, R. H. Mohiaddin and J. S. R. Gibbs, *Am. J. Respir. Crit. Care Med.*, 2010, **181**, 1106–1113.
94. S. M. Kawut, E. Bagiella, D. J. Lederer, D. Shimbo, E. M. Horn, K. E. Roberts, N. S. Hill, R. G. Barr, E. B. Rosenzweig, W. Post, R. P. Tracy, H. I. Palevsky, P. M. Hassoun and R. E. Girgis, *Circulation*, 2011, **123**, 2985–2993.
95. X. X. Wang, F. R. Zhang, Y. P. Sharp, J.-H. Zhu, X.-D. Xie, Q.-M. Tao, J.-H. Zhu and J.-Z. Chen, *J. Am. Coll. Cardiol.*, 2007, **49**, 1566–1571.
96. P. Jurasz, D. Courtman, S. Babiae and D. J. Stewart, *Pharmacol. Ther.*, 2010, **126**, 1–8.
97. S. Malenfant, G. Margaillan, J. E. Loehr, S. Bonnet and S. Provencher, *Expert Rev. Respir. Med.*, 2013, 7, 43–55.

90. [illegible] M. V. [illegible], A. M. [illegible], M. Kaderon, L. [illegible], K. Vonborke, G. G. [illegible], C. Hamilton, [illegible] and [illegible] Block, [illegible], 2003, [illegible]
91. H. J. [illegible]
92. [illegible], 2008, 294, [illegible]
93. M. [illegible] 110, [illegible]
94. S. [illegible], [illegible], [illegible], [illegible], M. [illegible], A. E. Roberts, [illegible] Hill, [illegible] Rosenwerk, W. [illegible], K. B. Tracy, [illegible], [illegible] and [illegible], 2011, 124, [illegible]
95. [illegible] Wang, [illegible], [illegible] Chu and [illegible] Chen, [illegible], 2007, 44, [illegible]
96. [illegible], 2008, 196, [illegible]
97. [illegible]

BLOOD DISORDERS

CHAPTER 14

Soliris (Eculizumab): Discovery and Development

MATTHEW A. LAMBERT AND WILLIAM J. J. FINLAY*

Pfizer, Global Biotherapeutics Technologies, Grange Castle Business Park, Clondalkin, Dublin 22, Ireland
*E-mail: william.finlay@pfizer.com

14.1 Paroxysmal Nocturnal Haemoglobinuria (PNH)

Paroxysmal nocturnal haemoglobinuria (PNH), sometimes referred to as Marchiafava–Micheli syndrome, is a potentially fatal disease characterised by intravascular haemolytic anaemia, thrombosis and eventual bone marrow failure.[1] The thrombotic events that occur in patients with this disease are the principal cause of its associated morbidity and mortality. Indeed, the risk of such an event is high for PNH patients, with their relative risk of venous thromboembolism being 60-fold greater than that of the general population. Subclinical thromboses have been observed in ~60% of PNH patients, irrespective of anticoagulant or transfusion treatments.[2] Recent improvements in supportive care have improved the median survival for sufferers, which was ~10–15 years from diagnosis in the late 1990s.[3]

The median age for PNH diagnosis is in the early 30s, but it affects people of all ages. As PNH is a heterogeneous disease with multiple manifestations its prevalence is not clear, but under its classical definition it is thought to affect <1 in 200 000 in the population.[4] As a result, it is placed in the 'ultra-orphan' disease category. Classical symptoms of PNH include haemoglobinuria, shortness of breath, extreme lethargy, abdominal pain, dysphagia and erectile failure in men.[5,6]

RSC Drug Discovery Series No. 38
Orphan Drugs and Rare Diseases
Edited by David C Pryde and Michael J Palmer

Published by the Royal Society of Chemistry, www.rsc.org

14.1.1 Molecular Basis of PNH

PNH is primarily caused by somatic mutation of the phosphatidylinositol glycan complementation class A gene (PIGA), which is found on the X-chromosome, meaning that only one active copy of the gene for PIGA is present in each cell (females silence one copy through X-inactivation).[7,8] The PIGA gene is important in the biosynthetic pathway used by the cell to make glycosyl phosphatidylinositol (GPI), an essential anchoring molecule which anchors the presentation of a variety of important proteins on the cell surface. If this mutation occurs in a bone marrow stem cell, followed by non-malignant clonal expansion, then all resulting haematopoietic cells will be deficient in GPI-anchored cell surface proteins.[8]

In PNH patients, the loss of GPI-anchored proteins in general leads to blood cells that are essentially normal in their function, but causes the absence of the important proteins CD55 (*i.e.* decay-accelerating factor, DAF) and CD59 (*i.e.* protectin) on erythrocytes.[9] Both of these proteins are negative regulators of the complement cascade and act to prevent erythrocytes from activating complement fixation and its associated downstream pro-inflammatory and cytolytic effects. In the absence of both CD55 and CD59, the enzymatic C3 and C5 convertases are not restricted, eventually leading to the generation of the terminal complement complex at the erythrocyte surface and resulting in cell lysis.[10,11]

Chronic haemolysis is therefore the primary driver of the pathology in PNH, particularly in the case of haemoglobinuria and anaemia, but secondary symptoms are believed to be the result of smooth muscle dystonia and the dysregulation of endothelial function.[12] Both of these issues are believed to be caused by haemolysis leading to high circulating levels of haemoglobin in the blood, which irreversibly binds and depletes nitric oxide. Nitric oxide depletion is also a potential trigger factor for venous thrombosis in PNH, as it plays a role in adhesion and aggregation of platelets.[6,13] Haemolysis also leads to the release of many other cell contents into the intravascular space, such as lactate dehydrogenase (LDH).[10]

Importantly, the PIGA gene mutation is not enough to cause disease in all patients and additional autoimmune mechanisms may be involved.[14,15] Additionally, clinical manifestations for PNH relate to the severity of haemolysis in a given patient, which may be driven by how many PIGA-mutant derived erythrocytes are generated and whether they are fully or only partially deficient in CD55 and CD59.[10] Common triggers of clinical symptoms in PNH patients include infections, surgery, strenuous exercise, blood transfusions and excessive alcohol intake.[16]

Flow cytometry is commonly used in PNH diagnosis, with staining for CD55 and CD59 on white and red blood cells. Depending on the relative levels of expression observed for these cell proteins, erythrocytes can be classified as Type I, II or III PNH cells. Normal levels of CD55 and CD59 are graded as Type I; cells with reduced levels are Type II; if both markers are absent, Type III.[9] Increasingly, an alternative method known as the fluorescein-labelled proaerolysin (FLAER) test is also being used to diagnose PNH. The bacterial aerolysin used in FLAER binds selectively to all cell surface proteins containing the GPI anchor and is more accurate in demonstrating a general deficit in GPI-linked proteins than flow cytometry for CD59 or CD55.[17]

Laboratory testing for PNH also includes examination of the urine for breakdown products of red blood cells (haemoglobin and hemosiderin). Haematological tests may show changes consistent with haemolytic anaemia, such as: elevations in the levels of LDH, reticulocytes, bilirubins, and decreased levels of haemoglobin and haptoglobin. The direct antiglobulin test (DAT) is negative, as the haemolysis in PNH is not caused by autoantibodies.[9,17]

14.1.2 The Complement System in Disease and C5 as a Drug Target

The complement system cascade is classically divided into two sections, proximal and terminal (Figure 14.1). The proximal cascade may proceed through different pathways: classical, alternative and lectin.[18] Each proximal pathway also has a different initiation trigger: (i) classical: antibody–antigen binding and interaction with C1q; (ii) alternative: binding of C3b onto many membrane-presented structures, including those of bacterial pathogens and erythrocytes; (iii) lectin: activation *via* mannose-specific lectins binding to any of a series of mannose-containing polysaccharides. All of these pathways ultimately end with the generation of C3 convertases that cleave C3 into C3a and C3b. C3a is a potent anaphylatoxin and C3b is critical in the progression of the complement cascade in its immunoprotective role. Genetic deficiencies in these proximal complement components are associated with high risk for potentially lethal infections from bacterial pathogens that have polysaccharide coats such as *Streptococcus pneumoniae*, *Haemophilus influenzae* and *Neisseria meningitidis*.[19]

The terminal complement system begins where all proximal pathways converge, at the cleavage of C5 into C5a and C5b by the C5 convertases.[20] Like C3a, C5a is an anaphylatoxin that is a chemotaxin, induces secondary inflammatory mediators, increases vascular permeability and alters smooth muscle tone, through the CD88 and C5L2 receptors.[21,22] The action of C5b is particularly important in PNH, as it recruits C6, C7, C8 and C9 to form the terminal complement complex (TCC) on cell membranes, leading to cell lysis. Genetic deficiencies of any members of the TCC lead to increased susceptibility to infection by *Neisseria meningitidis*, although these infections are usually of low severity.[16]

Complement factor C5 is therefore an attractive target in inflammatory and haemolytic diseases, as it is a lynchpin in the terminal cascade. The effective neutralisation of C5, stopping formation of C5a and C5b, should jointly reduce inflammation and halt the production of the cell-lytic TCC.

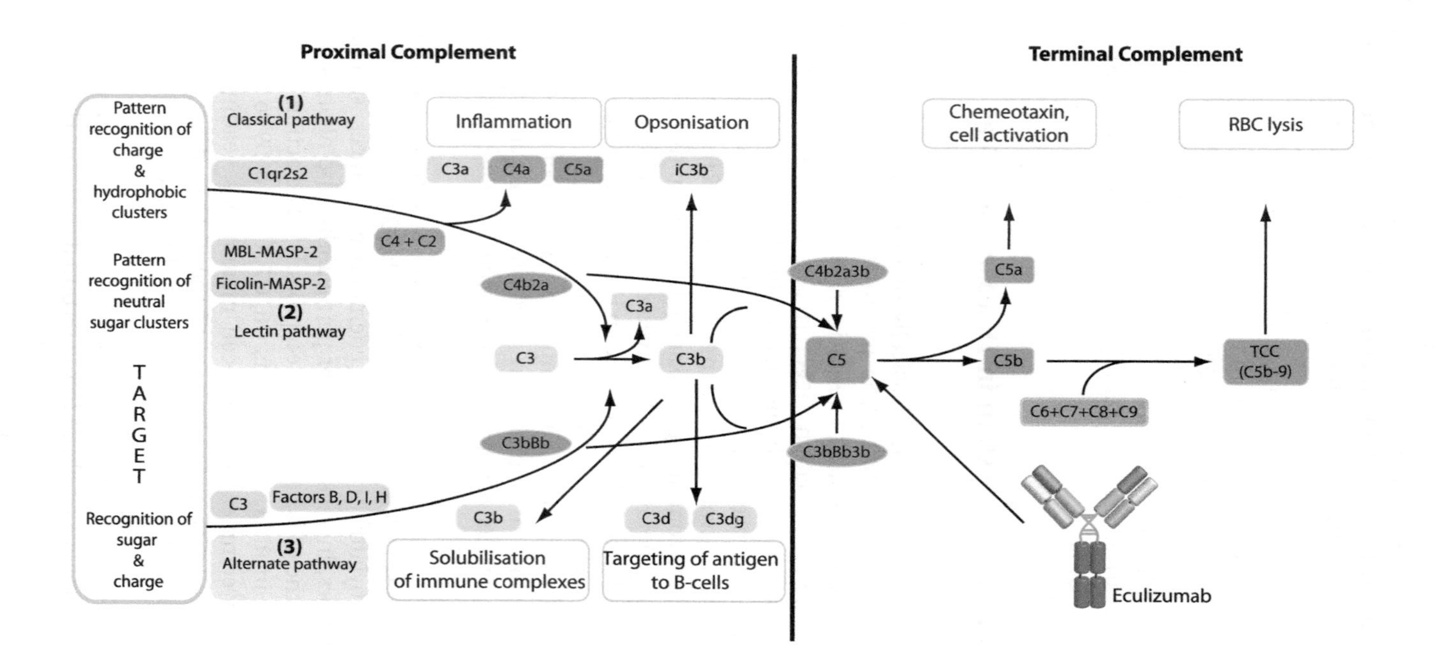

Figure 14.1 Eculizumab: rationale for blocking the activity of C5 in PNH. The complement cascade can be induced through classical (1), lectin (2) and alternate (3) pathways found in the proximal system (left). While the proximal complement system is essential for immune complex clearance and the opsonisation of microbes, all pathways ultimately converge at C5, primary factor of the terminal complement system. Cleavage of C5 generates molecules with inflammatory (C5a) and cell lysis characteristics, as C5b recruits C6–C9 to form the terminal complement complex (TCC), which lyses PNH erythrocytes.

Importantly, this should be without suppressing the critical immune effects of the upstream proximal cascade. To achieve proof of concept that this hypothesis was supported, a monoclonal antibody to murine C5 (BB5.1) was developed.[23] This antibody was a potent neutraliser of complement activity in mouse serum, allowing it to be tested in a series of *in vivo* mouse models of inflammatory disease. As no mouse models of PNH have ever been successfully developed, the BB5.1 antibody was not tested for such a function in mice. However, the antibody did show potent efficacy in murine models of arthritis (collagen induced arthritis, CIA), renal disease and a lupus-like autoimmunity model. In the CIA model, the BB5.1 antibody was initially shown to be efficacious in preventing the development of arthritis symptoms. When collagen-immunised mice were dosed biweekly with BB5.1 *versus* control, the BB5.1-treated animals demonstrated an approximately 60% reduction in haemolytic activity in their serum and complete absence of clinical signs of arthritis. The control animals developed severe disease symptoms. Additionally, the histological analysis of tissues from treated animals showed that the control animals had distinct indications of arthritic joint damage, whereas animals that received BB5.1 had normal joint architecture.[23]

In a model of established arthritis disease, dosing of BB5.1 was only begun after clear clinical signs could be seen in one or more limbs of each animal.[23] Treatment with the anti-C5 antibody led to a dramatic attenuation of disease progression in comparison to controls, as measured in reduced limb swelling and lowered arthritis index score. Again, post-study histological analyses showed significantly lower joint damage in the BB5.1-treated animals. This provided strong supporting evidence for the role of C5 in not only establishing, but also maintaining, disease progression.

Further evidence for the broad potential efficacy of C5 neutralisation was then generated in the NZB/WF1 mouse model of lupus-like autoimmune disease.[24] In this model, disease symptoms were fully suppressed in the BB5.1 antibody treatment group *versus* controls. Importantly, in all studies in mice with the BB5.1 antibody, no compound-related toxicities, abnormalities or mortalities were observed. These data established a clear rationale for the development of a new anti-C5 antibody that could mimic BB5.1 by potently neutralising human C5, rather than mouse.

14.2 Discovery and Characterisation of Eculizumab (Soliris)

In an effort to generate a potently neutralising anti-C5 antibody, a large immunisation and hybridoma screening campaign was carried out.[25] After immunising mice, hybridoma fusions were performed and approximately 30 000 hybridomas were screened for C5 neutralisation and potential leads were cloned. Despite this major investment of resource, only one mouse monoclonal antibody (m5G1.1) was identified which had the capacity to block both TCC-mediated haemolysis and C5a generation, at a molar ratio of antibody to C5 of 0.5 : 1. Epitope mapping of the m5G1.1 antibody showed that it mediates its action by binding the N-terminal region of the C5 alpha chain.

To begin engineering the m5G1.1 for use as a potential therapeutic, the V_H and V_L gene sequences were amplified from the hybridoma line by PCR and cloned for DNA sequencing.[25] The V_H gene of the m5G1.1 was found to be closely related to the murine V_H1/J558 family and J_H1, while the V_L chain was related to the murine NYC V_κ and $J_\kappa 5$ genes. For humanisation, the CDR grafting method was used, onto closest homology human v-gene sequences. This process involves performing alignments of the amino acid sequence of each murine v-domain against cloned v-gene sequences from human origin.[26] When highly homologous human v-genes have been identified, the sequences for the human CDR regions are transposed into the human framework regions by gene synthesis, replacing the human CDR loops. This process hopes to retain the potency and specificity of the parental murine antibody, while significantly reducing the murine sequence content and the potential for immunogenicity in man.

In the case of the m5G1.1 V_H domain, CDR grafting was performed into the V_H gene sequence H20C3H, which was derived from the human V_H1 family germ line gene IGHV1-46, plus the germ line J-segment J_H5.[25] For the V_L domain, the murine CDRs were grafted into the human $V_\kappa 1$ family gene I.23, which is derived from the IGKV1-39 germ line gene, plus the genomic $J_\kappa 1$ J-segment. The CDR grafting method often leads to slight structural perturbation of the murine CDR loops that have been placed in a human framework, with a resulting loss of affinity.[27] These perturbations are known to be frequently caused by the amino acid variation between the donor mouse v-gene sequence and the human acceptor sequence, at key CDR-supporting residues. With this is mind, *in silico* structural modelling of CDR-grafted m5G1.1 (h5G1.1) was used to predict any such difficulties. At one site in the V_κ and two sites in the V_H, potential liabilities were sampled by mutating those framework residues back to the murine residue found at the corresponding position in the murine donor m5G1.1. These mutations are known as 'back-mutations'.

To test the function of the designed humanised antibody sequences, Fab and scFv fragment-encoding plasmid expression vectors were constructed for the murine, chimeric, grafted and grafted + back-mutated versions.[25] The Fab and scFv versions of the proteins were expressed, purified and tested for competition with the parental IgG in an ELISA assay and for neutralisation in a chicken erythrocyte lysis assay using human serum. In both assays it was shown that the simple CDR-grafted Fab was equivalent to the murine and chimeric Fabs, showing the framework back-mutations to be unnecessary. The straight CDR-grafted humanised clone, now designated h5G1.1, was compared to the parental m5G1.1 and both were found to have similar affinities for human C5 (K_D = 120 pM). Antibody h5G1.1 was therefore deemed ready for conversion to full-length IgG and pre-clinical testing.

The conversion of the h5G1.1 v-regions to full-length IgG involved further rational engineering of the final molecule structure, in the constant regions (Figure 14.2). As PNH is an inflammatory disease, it was highly desirable that

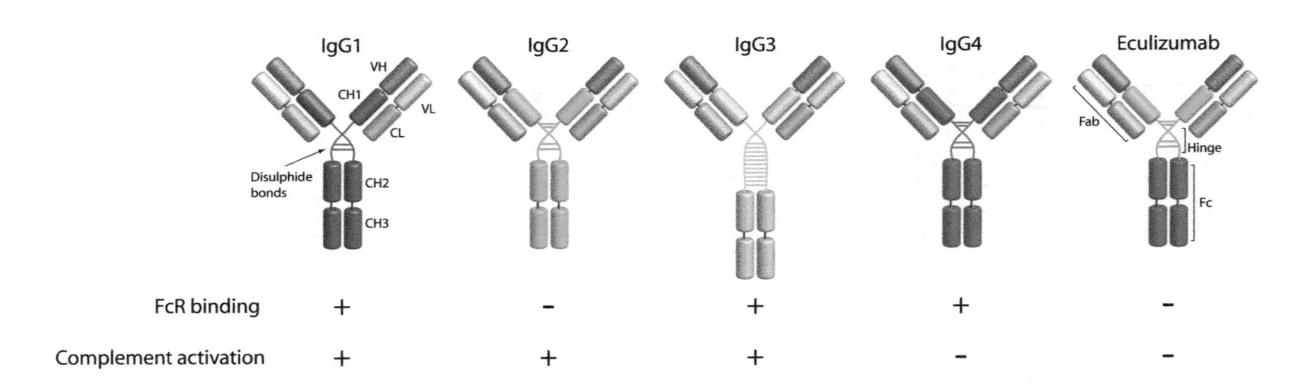

Figure 14.2 Human IgG subclasses and the design of the Fc-engineered eculizumab. Human IgG is found in four related subclasses which have distinct structural and functional differences. All four subclasses are comprised of four disulfide-linked polypeptide chains: two heavy and two light. The light chains are comprised of the V_L and C_L domains, while the heavy chains contain V_H, C_H1, hinge, C_H2 and C_H3 domains. The light chain, plus V_H and C_H1, make up the 'Fab' fragments, while the hinge, C_H2 and C_H3 domains collectively form the 'Fc' fragment. Differences in structure and sequence of the four different Fc subclasses lead to differences in FcR binding and complement activation (scored above as +/– to denote strong or weak activity). As IgG1 and IgG3 both have strong dual effector functions, they were dismissed as potential scaffolds for eculizumab. To minimise the effector function in the final molecule, eculizumab was built on an engineered scaffold that includes the C_H1-hinge region of IgG2 (removing FcR binding), fused to the C_H2–C_H3 domains of IgG4 (removing complement activation activity).

the final IgG should have no ability to induce pro-inflammatory responses. This is complicated by the intrinsic effector functions in the different isotypes of human IgG that aid the activation of complement and/or engage pro-inflammatory Fc receptors. To ameliorate this problem, the h5G1.1 VH domain was cloned in frame with a 'hybrid' Fc containing portions of IgG2 and IgG4.[25] This construct style had been used previously with humanised porcine antibodies[28] and contains the CH1 and hinge domains of IgG2 (no Fc-receptor binding), fused to the CH3 and CH4 domains of IgG4 (no complement activation).

14.3 Clinical Development of Eculizumab

As the activity of complement C5 is believed to be influential in a wide array of inflammatory diseases, eculizumab was investigated as a potential treatment for several chronic inflammatory diseases, including rheumatoid arthritis (RA), systemic lupus erythematosus (SLE), nephritis and in patients with myocardial infarction (MI) or coronary artery bypass surgery.[16] Initial eculizumab studies were designed to determine the safety of C5 inhibition in humans and also to determine the appropriate dosing schedule that should be used in full clinical trials. Dose-ranging was performed in healthy subjects between 0.1 and 8 mg kg^{-1} and all doses were observed to be safe and well tolerated. At the 8 mg kg^{-1} dose level, full blockade of terminal complement activity was observed for as long as 7–14 days.[29] In chronic dosing regimen testing in patients with RA and idiopathic membranous nephropathy, eculizumab was also shown to be safe and to have evidence of efficacy.[16] Pexelizumab (scFv version of h5G1.1) was also tested in a number of acute clinical indications, such as percutaneous coronary intervention in MI.

14.3.1 Development of Eculizumab for PNH

As the absence of CD59 on erythrocytes is a defining cause of PNH pathology, early attempts to repair this functional defect in PNH cells aimed to target this issue specifically. Groups used gene therapy to replace the missing CD59 on the membrane of PNH cells. This led to the generation of a recombinant, transmembrane form of CD59 which proved to be protective in a membrane damage assay using a complement-sensitive b-cell line derived from a PNH patient.[30] The introduction of a functional PIGA gene into PNH erythroid cells *via* retroviral transduction can also restore GPI anchor protein expression on these cells and thereby protect them from complement damage.[31] Unfortunately, however, gene therapy approaches have not yet matured to the point where effective targeting of early erythroid progenitor cells is possible, which would be essential to create a renewable source of protected erythrocytes in PNH patients. As a result, investigators turned to eculizumab as a potential disease-modifying agent in PNH.

14.3.2 Pharmacokinetic Properties of Eculizumab in Patients with PNH

The above studies allowed determination of the pharmacokinetic (PK) properties of eculizumab. These values were analysed using a one-compartment model and data derived from 40 adult PNH patients after multiple doses of the appropriate clinical regimen: 600 mg eculizumab given every 7 days by intravenous infusion for 4 weeks, followed by 900 mg at week 5, then maintenance at 900 mg every 2 weeks. Using this model, eculizumab exhibited good PK with a mean half-life of 272 ± 82 hours. In a patient weighing 70 kg, mean clearance was approximately 22 mL per hour with a mean volume distribution of 7.7 L. Treatment for 26 weeks with eculizumab led to an observed peak concentration in the serum of 194 μg mL^{-1}, with a trough of 97 μg mL^{-1}. Serum concentrations appear to reach steady state after ~57 days. As pharmacodynamic activity of eculizumab correlates directly with its serum concentration, trough levels above 35 μg mL^{-1} were found to fully block the haemolytic activity of complement *in vivo* in the majority of patients.[32]

The PK of eculizumab has not yet been evaluated in patient sub-populations on the basis of age, gender, genetic background or in the case of renal or hepatic disease, but there are no indications in the current data set to suggest that these factors are influential. A small study in pregnant women with PNH (n = 7) suggested that eculizumab did not appear to undergo placental transfer to the foetus and it was not found in breast milk, despite the ability of human IgGs to cross the placenta *via* FcRn transport.[33] Efficacy evidence in this small cohort was positive, but requires further investigation. All babies born on the study were found to be healthy.

14.3.3 Clinical Trials of Eculizumab in Patients with PNH

With basic efficacy measurements and PK characteristics having been established, eculizumab was thereafter tested in three separate clinical studies: one Phase 2 pilot study and two Phase 3 studies, which both built upon the prior experience in the clinic in other indications. The Phase 2 pilot study was small, involving only 11 participants, all of whom had haemolytic PNH.[34] This study was 3 months long and open label, evaluating terminal complement inhibition. All participants in this study had received a minimum of four blood transfusions in the previous year and received the recommended regimen, as outlined above (600 mg per week × 4, then 900 mg repeatedly up to 12 weeks). All participants were vaccinated against *N. meningitidis* a minimum of 14 days before receiving their first dose in the study, as an important precautionary measure.

To track the response in patients, multiple clinical and biochemical measurements of haemolysis were taken throughout the trial.[34] These analyses showed that eculizumab significantly reduced haemolytic activity in comparison to baseline levels, with mean LDH levels decreasing from 3111

IU L^{-1} to 594 IU L^{-1} ($p = 0.002$) after the full 12 weeks of treatment. Eculizumab also led to a significant decrease in the number of required transfusions of packed red blood cells, which are given to patients when they exhibit symptoms of anaemia. Transfusion rates were measured in units of transfusions/patient/month for the year preceding treatment and during eculizumab therapy. After eculizumab therapy, the transfusion rates dropped from 1.8 U to 0 U (median rates comparison $p = 0.003$). Mean incidence of haemoglobinuria was radically reduced, by 96% ($p < 0.001$), and the mean percentage of PNH erythrocytes in circulation increased from 37% to 59% of the total erythrocyte population. Importantly, health-related quality of life scores in eculizumab-treated patients showed statistically significant improvements in a number of physical and emotional areas, as established by the European Organization for Research and Treatment of Cancer Quality of Life Questionnaire (EORTC LQ) C30 instrument.[34]

This study provided strong evidence that eculizumab was both safe and well tolerated in patients with PNH, effectively ameliorated physical symptoms and improved quality of life in this small cohort of patients. As a result, the patients were offered an extension to the study for a further 52 weeks, which was accepted and completed by all 11 participants.[32] This study extension further corroborated the remarkable reductions in haemolysis and transfusion rates observed in the initial pilot study and also confirmed durable improvements in quality of life scores and associated physical symptoms such as haemoglobinuria and smooth muscle dystonias (abdominal pain, dysphagia, erectile dysfunction).[5] Interestingly, two participants exhibited insufficient levels of eculizumab and subsequently had clinical breakthrough events and the return of symptoms. For these patients, an increase in the total dose rapidly suppressed their symptoms again and reinstated the suppression of terminal complement activation with a resulting abrogation of haemolysis. This observation of the reversibility of the effect of eculizumab provided confirmation of the importance of its mechanism of action. In 2007, it was reported that 10 of the original 11 participants in this extended Phase 2 trial had continued on eculizumab therapy for at least 5 years.[16]

The results in the Phase 2 pilot trial outlined above inspired the progression of eculizumab into two separate Phase 3 clinical trials in PNH patients. These studies were registered and performed in the USA and Europe, after 'orphan' designation was given for eculizumab in PNH in both regions. The first Phase 3 trial was named 'TRIUMPH' and this was the pivotal trial designed to establish the safety and efficacy of eculizumab in haemolytic PNH patients who (as in the Phase 2 study), had required a minimum of four transfusions in the preceding year.[35] The second Phase 3 trial was named 'SHEPHERD' and aimed to establish the safety and efficacy of eculizumab in a broader, more heterogeneous population of PNH patients who had a history of transfusions and suffering from haemolysis.

The TRIUMPH trial was designed as a randomised, double-blind, multicentre study in adults and was conducted over a total of 26 weeks, beginning in 2004. Primary end points for this trial were defined as: stabilisation of haemoglobin levels and the reduction in number of packed red cells transfused. Biochemical indicators of haemolysis were evaluated throughout the study, as were quality of life scores. Potential participants in the trial were monitored for 13 weeks and defined as ineligible if they did not have an infusion requirement during that period. A total of 87 participants were recruited to the trial and underwent randomisation, over 34 separate international sites. Of these 87 patients, 43 received eculizumab and 44 received a placebo treatment. Both were dosed intravenously, as in the Phase 2 study, being given the regime of 600 mg weekly for 4 weeks, followed 1 week later by 900 mg and 900 mg subsequently, every 14 days (on average), up to the end of week 26 of the trial.[35]

After the study was completed and de-blinding had been performed, the data showed that 100% of participants who had received eculizumab achieved an objective response, while there was little or no improvement in symptoms for the patients in the placebo group. The eculizumab-treated patients demonstrated a significant decrease in haemolysis indicators such as LDH. This reduction in LDH was found to have a very rapid onset 1 week after the first dose of eculizumab and these reductions were durable, lasting to the end of the 26 week study. No significant reductions in LDH were observed in the placebo group. Indeed, when the area under the curve (AUC) *versus* time was calculated for LDH levels (U per L per day), the eculizumab-treated group showed an 86% lower median than the placebo group. This translated to a mean AUC of 58 587 U L^{-1} in the eculizumab group, but 411 822 U L^{-1} in the placebo group.

For 21 of 43 (49%) eculizumab-treated participants, *versus* 0 of 44 in the placebo group, stabilisation of haemoglobin levels was observed, in the absence of transfusions ($p < 0.000001$). This translated to a mean packed red blood cell administration rate of 0 units in the eculizumab group, with the placebo group having a mean of 10 units ($p < 0.000001$). A further indicator of efficacy was that 51% of participants in the eculizumab group remained transfusion independent for the full 26 weeks of the study. In the placebo group, every patient required at least one infusion during this period. For the eculizumab-treated group, amongst the 22 individuals who did not achieve full independence from infusions during the study, infusion rates were still reduced by 44%. Those individuals also demonstrated a significant reduction in haemolysis scores. While it is not clear what differentiates these patients from the transfusion-independent individuals, it is believed that they may have entered the trial in a state of more severe bone marrow aplasia, or may have a generally higher incidence of low-level extravascular haemolysis.[35]

In assessment of quality of life indicators, both the Functional Assessment of Chronic Illness Therapy-Fatigue (FACIT-fatigue) and EORTC QLQ-C30 instruments were used. In both cases, statistically (both $p < 0.00001$) and clinically significant improvements in fatigue were observed in the eculizumab group, including in those patients who did not become transfusion independent. Importantly, after completion of the 26 day study, placebo patients were transitioned to eculizumab treatment. While the LDH levels of the placebo group had remained high across the initial study, after being transitioned to eculizumab treatment, all patients exhibited a rapid decline in LDH level, ending just above a 'normal' baseline level of 103–223 U L^{-1}. When kept on the eculizumab dosing regimen, all patients maintained this low level of LDH (and, therefore, haemolysis) up to 52 weeks, when monitoring ended.

Serious adverse events were reported for four individuals in the eculizumab group, as opposed to nine in the placebo group. None of these events appeared to be treatment-related and all 13 patients recovered fully, with no observed sequelae. Common low-severity adverse events included back pain, headache, nausea and nasopharyngitis. Adverse events did not appear to be the result of anti-drug antibody (ADA) responses in the eculizumab-treated patients.[35] When anti-eculizumab antibody responses were analysed, all titres were low, transient and found across both the eculizumab and placebo groups. The TRIUMPH study therefore provided clear demonstration that eculizumab could be an efficacious and well-tolerated therapy for PNH.

In canon with the TRIUMPH trial, a second trial, named SHEPHERD, was also performed.[16,36] In contrast to the TRIUMPH trial, the SHEPHERD trial was an open-label, non-placebo-controlled Phase 3 study examining safety and efficacy in a clinically diverse population of PNH patients over the age of 18, including those with thrombocytopenia and little or no prior history of transfusions. Exclusion criteria included: patients who had received any other investigational drug in the preceding 30 days; those suffering from complement deficiency or active bacterial infection; an absolute neutrophil count $< 0.5 \times 10^9$ L^{-1}; prior history of meningococcal disease or bone marrow transplant. Beginning in 2005, 97 patients were enrolled at a total of 33 international sites. All patients received the same eculizumab dosing regimen as used in the TRIUMPH study, over a period of 52 weeks.

In this study, haemolysis levels were again significantly reduced, with improvements in haemolysis (mean LDH reduced by 87%), anaemia and either improvement or elimination of the requirement for transfusions. The patients also reported improvements in fatigue and health-related quality of life scores. Eculizumab treatment led to complete inhibition of haemolytic activity in the serum of 92% of patients receiving a maintenance dose every 14 days, but eight patients exhibited a return of haemolytic activity in the last 2 days of the dosing interval. For six of the eight patients, this problem was successfully overcome by reducing the dosing interval to 12 days. Importantly, the adverse event reporting level was consistent with that observed for the placebo group in the TRIUMPH trial.[36]

Taken together, the TRIUMPH and SHEPHERD trials had a very high total completion rate of 96% (187 out of 195), with all completing participants electing to join a Phase 3b open-label extension study, lasting a further 104 weeks. Across both studies, there was no statistically significant increase in infection rates amongst patients receiving eculizumab, in comparison to rates observed in the placebo group. There were, however, two patients who developed meningococcal sepsis, despite being vaccinated against *N. meningitidis*. Both patients were successfully treated for the infection and recovered with no clinical sequelae being reported. In September 2006, these data were considered compelling enough to submit applications for marketing authorisation to both the EMEA and FDA. After accelerated assessment, in March and June 2007, respectively, after almost a decade of clinical analyses in multiple indications, the FDA and EMEA both approved the use of eculizumab for the treatment of PNH.[16]

14.4 Post-Approval Outcomes

A comprehensive analysis has now been performed of the 195 participants of the first clinical investigation, and the subsequent TRIUMPH and SHEPHERD trials, as well as the long-term extension study showing the beneficial and long-term (over 66 months) effects of persistent eculizumab administration.[37] All three parent trials had the same dosing regimen: 600 mg intravenous infusions weekly for 4 weeks, followed by a 900 mg infusion after a further week. A 900-mg dose was then given every 14 days indefinitely. All patients showed a reduction in LDH levels (an indicator of haemolysis), a decrease in the number of thrombotic events, an improvement of renal function and a decrease in the dependency on blood transfusions, with the number of units required decreasing for those still dependent. Overall, eculizumab had a substantial impact on PNH sufferers and resulted in major improvements in patient survival and standard of life.[35,36,38]

Results from the study indicated that eculizimab was well tolerated by patients with PNH. Serious adverse events (AEs) such as haemolysis, abdominal pain and anaemia were reported in 38.5% of patients. However these generally coincided with infection or other biological stresses. Other serious AEs reported by at least four patients were pyrexia (4.6%), viral infection (3.1%) and cholecystectomies (3.1%). Two women who became pregnant during the trial period received eculizumab for the first 4 and 5 weeks of pregnancy, with both babies being delivered without complication.[33] Indeed since then, evidence from seven women exposed to eculizumab during pregnancy would suggest that the drug is safe to use and is likely to prevent many of the complications usually observed with pregnant PNH patients.[33]

Forty patients (20.5%) developed 67 serious infection-related AEs during treatment. These events were mainly manifested as pyrexia and viral infections, none of which were fatal. Of note is the fact that eculizumab administration increases the risk of *Neisseria meningitidis* infection. Such is the risk that a boxed warning is included in the US prescribing information. As such, the use of eculizumab is contraindicated in patients not vaccinated against *N. meningitidis*. Indeed all patients within the trials were vaccinated with a tetravalent meningococcal vaccine against subgroup A, C, W and Y (no vaccine is currently available to serogroup B). In the UK, serogroup B is the most prevalent serotype isolated from meningococcal sepsis. In January 2010 policy was altered to include, as well as vaccine treatment, the administration of antibiotic prophylaxis to all patients in an attempt to prevent serogroup B infection. Two cases of meningococcal sepsis were reported during the 66-week treatment period. These infections were cleared with antibiotic treatment. It is noteworthy that neither of the patients was vaccinated against the specific strain of their infection.[37]

The longest study to date was conducted in Leeds, UK, where 79 patients, treated between 2002 and 2010, as well as a further 45 treated since then, were compared to 30 patients who were cared for at the centre during the 7 years prior to the availability of eculizumab.[39] Results from these studies agreed with the trial results, confirming that eculizumab is of great benefit to PNH sufferers. Importantly, there was no significant difference between the rates of survival of patients receiving eculizumab as compared to age- and sex-matched control averages from the UK. Those patients that were eligible for eculizumab treatment, but not receiving it, between 1997 and 2004 had a significantly higher risk of death than when subsequently enlisted in treatment. The 5-year survival rate of these patients was 66.8% as compared to 95.5% for those treated with eculizumab.

During the 8 years of the study, the number of thrombotic episodes (TEs) declined in the overall patient cohort. Of the 79 patients, 21 suffered TEs of which 17 patients were not anti-coagulated. Despite frequent prophylactic warfarin use (the recommended treatment for PNH patients before eculizumab), management of TEs is difficult in PNH patients.[39–41] However, several patients with history of a previous TE have now been able to stop anti-coagulation therapy with no further reoccurrence while on eculizumab alone.[39,42] The rate of TEs per 100 patient-years for those patients in care before the launch of eculizumab was 5.6 as compared to 0.8 for those receiving treatment.[39]

Transfusion-dependent haemolysis (four or more per year) is one of the UK's nationally commissioned indications for treatment. Prevention of C5 cleavage by eculizumab had a dramatic effect on both the level of haemolysis, as measured by LDH levels, and the number of transfusions required. The median LDH value was reduced from 2872 IU L^{-1} (pre-treatment) to 477 IU L^{-1}, where 430 IU L^{-1} is the upper limit of normal, while the number of transfusions required decreased from 19.3 units to 5.0 units when comparing the last year of treatment with the year immediately prior to treatment. Of the 75 patients in the study, 61 patients had required transfusions prior to eculizumab; of these, 40 became transfusion independent after continuous treatment. Encouragingly, amongst the remaining 21 patients, there was a significant reduction in the number of transfusions needed.

It has previously been reported that PNH patients suffering from thrombocytopenia may experience an increase in platelet counts due to the use of eculizumab,[43] however this was not found to be the case in the Kelly *et al.* study. There was no difference seen in the number of platelets present in 61 patients before and after eculizumab use. Of these 61 patients, 12 suffered from thrombocytopenia and again there was no increase in platelet production upon commencement of eculizumab treatment.

A prominent feature of the Kelly *et al.* long-term efficacy study was the relationship between eculizumab use and the prevalence of aplastic anaemia (AA).[39] Eleven patients entered the trial receiving cyclosporine therapy; however three additional patients developed AA while on eculizumab. All three patients were treated with cyclosporine and continued to receive eculizumab as they all continued to have PNH clones. Of note is the fact that three patients have undergone a clonal change in their disease. Two of these developed myelodysplasia while the third developed myeloid leukaemia. All three patients' conditions developed from GPI-negative cells, and as of July 2010 they all remained on eculizumab.

Of great significance is the fact that only four people have died while receiving or having received eculizumab.[37,39] Three of these deaths were caused by concomitant conditions such as chronic myelomonocytic leukaemia, metastases of pre-existing adenocarcinoma and cerebral herniation, while the fourth death was due to sepsis and happened 19 days after the patient had stopped eculizumab treatment.[44]

The main studies so far have been performed within the UK and specifically concentrated in Leeds. As a result, the patient cohorts have been predominantly non-Asian. However, small pilot studies in Japan[45] and Korea[46] would suggest that eculizumab is well tolerated and effective in the treatment of PNH. In Korean patients the standard dosing regimen achieved a 69–88% reduction in LDH levels.[46] This was echoed by results from a 12-week Japanese study where both the number of transfusions required, and the level of LDH, fell significantly.[45]

In conclusion, long-term treatment with eculizumab improved the standard of living and patient outcome by significantly reducing the level of haemolysis experienced, and the need for transfusion. Patients experienced a decline in life-threatening morbidities and an overall improvement in survival, showing conclusively that eculizimab has become a very effective treatment for a previously unmet need.

References

1. C. J. Parker, *Exp. Hematol.*, 2007, **35**, 523–533.
2. C. J. Parker, *Hematology/the Education Program of the American Society of Hematology*, American Society of Hematology, Education Program, 2008, pp. 93–103.
3. P. Hillmen, S. M. Lewis, M. Bessler, L. Luzzatto and J. V. Dacie, *N. Engl. J. Med.*, 1995, **333**, 1253–1258.
4. D. J. Araten, K. Nafa, K. Pakdeesuwan and L. Luzzatto, *Proc. Natl. Acad. Sci. U. S. A.*, 1999, **96**, 5209–5214.

5. A. Hill, R. P. Rother and P. Hillmen, *Haematologica,* 2005, **90**, ECR40.
6. R. P. Rother, L. Bell, P. Hillmen and M. T. Gladwin, *J. Am. Med. Assoc.,* 2005, **293**, 1653–1662.
7. C. J. Parker, *Stem Cells,* 1996, **14**, 396–411.
8. J. Takeda, T. Miyata, K. Kawagoe, Y. Iida, Y. Endo, T. Fujita, M. Takahashi, T. Kitani and T. Kinoshita, *Cell,* 1993, **73**, 703–711.
9. C. Parker, M. Omine, S. Richards, J. Nishimura, M. Bessler, R. Ware, P. Hillmen, L. Luzzatto, N. Young, T. Kinoshita, W. Rosse, G. Socie and P. N. H. I. G. International, *Blood,* 2005, **106**, 3699–3709.
10. C. Parker, *Lancet,* 2009, **373**, 759–767.
11. C. J. Parker, S. Kar and P. Kirkpatrick, *Nat. Rev. Drug Discovery,* 2007, **6**, 515–516.
12. R. Kelly, S. Richards, P. Hillmen and A. Hill, *Ther. Clin. Risk Manage.,* 2009, **5**, 911–921.
13. A. Hill, S. J. Richards and P. Hillmen, *Br. J. Haematol.,* 2007, **137**, 181–192.
14. L. Gargiulo, S. Lastraioli, G. Cerruti, M. Serra, F. Loiacono, S. Zupo, L. Luzzatto and R. Notaro, *Blood,* 2007, **109**, 5036–5042.
15. L. Gargiulo, M. Papaioannou, M. Sica, G. Talini, A. Chaidos, B. Richichi, A. V. Nikolaev, C. Nativi, M. Layton, J. de la Fuente, I. Roberts, L. Luzzatto, R. Notaro and A. Karadimitris, *Blood,* 2013, **121**, 2753–2761.
16. R. P. Rother, S. A. Rollins, C. F. Mojcik, R. A. Brodsky and L. Bell, *Nat. Biotechnol.,* 2007, **25**, 1256–1264.
17. R. A. Brodsky, *Blood,* 2009, **113**, 6522–6527.
18. L. A. Matis and S. A. Rollins, *Nat. Med.,* 1995, **1**, 839–842.
19. G. D. Overturf, *Clin. Infect. Dis.,* 2003, **36**, 189–194.
20. H. J. Muller-Eberhard, *Annu. Rev. Biochem.,* 1988, **57**, 321–347.
21. M. J. Rabiet, E. Huet and F. Boulay, *Biochimie,* 2007, **89**, 1089–1106.
22. N. J. Chen, C. Mirtsos, D. Suh, Y. C. Lu, W. J. Lin, C. McKerlie, T. Lee, H. Baribault, H. Tian and W. C. Yeh, *Nature,* 2007, **446**, 203–207.
23. Y. Wang, S. A. Rollins, J. A. Madri and L. A. Matis, *Proc. Natl. Acad. Sci. U. S. A.,* 1995, **92**, 8955–8959.
24. Y. Wang, Q. Hu, J. A. Madri, S. A. Rollins, A. Chodera and L. A. Matis, *Proc. Natl. Acad. Sci. U. S. A.,* 1996, **93**, 8563–8568.
25. T. C. Thomas, S. A. Rollins, R. P. Rother, M. A. Giannoni, S. L. Hartman, E. A. Elliott, S. H. Nye, L. A. Matis, S. P. Squinto and M. J. Evans, *Mol. Immunol.,* 1996, **33**, 1389–1401.
26. L. Riechmann, M. Clark, H. Waldmann and G. Winter, *Nature,* 1988, **332**, 323–327.
27. W. J. Finlay, O. Cunningham, M. A. Lambert, A. Darmanin-Sheehan, X. Liu, B. J. Fennell, C. M. Mahon, E. Cummins, J. M. Wade, C. M. O'Sullivan, X. Y. Tan, N. Piche, D. D. Pittman, J. Paulsen, L. Tchistiakova, S. Kodangattil, D. Gill and S. E. Hufton, *J. Mol. Biol.,* 2009, **388**, 541–558.
28. J. P. Mueller, M. A. Giannoni, S. L. Hartman, E. A. Elliott, S. P. Squinto, L. A. Matis and M. J. Evans, *Mol. Immunol.,* 1997, **34**, 441–452.

29. R. P. Rother, C. F. Mojcik and E. W. McCroskery, *Lupus,* 2004, **13**, 328–334.
30. R. P. Rother, S. A. Rollins, J. Mennone, A. Chodera, S. A. Fidel, M. Bessler, P. Hillmen and S. P. Squinto, *Blood,* 1994, **84**, 2604–2611.
31. J. Nishimura, K. L. Phillips, R. E. Ware, S. Hall, L. Wilson, T. L. Gentry, T. A. Howard, Y. Murakami, M. Shibano, T. Machii, E. Gilboa, Y. Kanakura, J. Takeda, T. Kinoshita, W. F. Rosse and C. A. Smith, *Blood,* 2001, **97**, 3004–3010.
32. A. Hill, P. Hillmen, S. J. Richards, D. Elebute, J. C. Marsh, J. Chan, C. F. Mojcik and R. P. Rother, *Blood,* 2005, **106**, 2559–2565.
33. R. Kelly, L. Arnold, S. Richards, A. Hill, C. Bomken, J. Hanley, A. Loughney, J. Beauchamp, G. Khursigara, R. P. Rother, E. Chalmers, A. Fyfe, E. Fitzsimons, R. Nakamura, A. Gaya, A. M. Risitano, J. Schubert, D. Norfolk, N. Simpson and P. Hillmen, *Br. J. Haematol.,* 2010, **149**, 446–450.
34. P. Hillmen, C. Hall, J. C. Marsh, M. Elebute, M. P. Bombara, B. E. Petro, M. J. Cullen, S. J. Richards, S. A. Rollins, C. F. Mojcik and R. P. Rother, *N. Engl. J. Med.,* 2004, **350**, 552–559.
35. P. Hillmen, N. S. Young, J. Schubert, R. A. Brodsky, G. Socie, P. Muus, A. Roth, J. Szer, M. O. Elebute, R. Nakamura, P. Browne, A. M. Risitano, A. Hill, H. Schrezenmeier, C. L. Fu, J. Maciejewski, S. A. Rollins, C. F. Mojcik, R. P. Rother and L. Luzzatto, *N. Engl. J. Med.,* 2006, **355**, 1233–1243.
36. R. A. Brodsky, N. S. Young, E. Antonioli, A. M. Risitano, H. Schrezenmeier, J. Schubert, A. Gaya, L. Coyle, C. de Castro, C. L. Fu, J. P. Maciejewski, M. Bessler, H. A. Kroon, R. P. Rother and P. Hillmen, *Blood,* 2008, **111**, 1840–1847.
37. P. Hillmen, P. Muus, A. Roth, M. O. Elebute, A. M. Risitano, H. Schrezenmeier, J. Szer, P. Browne, J. P. Maciejewski, J. Schubert, A. Urbano-Ispizua, C. de Castro, G. Socie and R. A. Brodsky, *Br. J. Haematol.,* 2013, **162**, 62–73.
38. P. Hillmen, P. Muus, U. Duhrsen, A. M. Risitano, J. Schubert, L. Luzzatto, H. Schrezenmeier, J. Szer, R. A. Brodsky, A. Hill, G. Socie, M. Bessler, S. A. Rollins, L. Bell, R. P. Rother and N. S. Young, *Blood,* 2007, **110**, 4123–4128.
39. R. J. Kelly, A. Hill, L. M. Arnold, G. L. Brooksbank, S. J. Richards, M. Cullen, L. D. Mitchell, D. R. Cohen, W. M. Gregory and P. Hillmen, *Blood,* 2011, **117**, 6786–6792.
40. R. P. de Latour, J. Y. Mary, C. Salanoubat, L. Terriou, G. Etienne, M. Mohty, S. Roth, S. de Guibert, S. Maury, J. Y. Cahn, G. Socie and H. French Society of and H. French Association of Young, *Blood,* 2008, **112**, 3099–3106.
41. V. M. Moyo, G. L. Mukhina, E. S. Garrett and R. A. Brodsky, *Br. J. Haematol.,* 2004, **126**, 133–138.
42. A. Emadi and R. A. Brodsky, *Am. J. Hematol.,* 2009, **84**, 699–701.
43. G. Socie, P. Muus and H. Schrezenmeier, *Blood,* 2009, **22**, 4030.

44. S. T. van Bijnen, R. S. van Rijn, S. Koljenovic, P. te Boekhorst, T. de Witte and P. Muus, *Br. J. Haematol.*, 2012, **157**, 762–763.
45. Y. Kanakura, K. Ohyashiki, T. Shichishima, S. Okamoto, K. Ando, H. Ninomiya, T. Kawaguchi, S. Nakao, H. Nakakuma, J. Nishimura, T. Kinoshita, C. L. Bedrosian, M. E. Valentine, G. Khursigara, K. Ozawa and M. Omine, *Int. J. Hematol.*, 2011, **93**, 36–46.
46. J. S. Kim, J. W. Lee, B. K. Kim, J. H. Lee and J. Chung, *Korean J. Hematol.*, 2010, **45**, 269–274.

CHAPTER 15

The Discovery and Development of Ruxolitinib for the Treatment of Myelofibrosis

KRIS VADDI

Incyte Corporation, Experimental Station, Rte 141 and Henry Clay Road, Wilmington, DE 19880, USA
E-mail: KVaddi@incyte.com

15.1 Introduction

15.1.1 Ruxolitinib

The discovery, preclinical characterisation of ruxolitinib (also known as INCB018424) and subsequent clinical research in patients with myelofibrosis (MF), the most serious of the myeloproliferative neoplasms (MPNs), culminated in its approval in November of 2011 by the US Food and Drug Administration (FDA) for intermediate or high-risk MF, including primary MF (PMF), post-polycythaemia vera MF (PPV-MF) and post-essential thrombocythemia MF (PET-MF). Regulatory approvals in Canada and the EU were obtained for ruxolitinib in 2012. Ruxolitinib, marketed under the brand name Jakafi® in the USA and Jakavi® in the rest of the world, is an oral agent that inhibits dysregulated JAK-STAT (Janus kinase-signal transducer and activator of transcription) signalling in MF patients, leading to clinically meaningful, long-lasting reductions in splenomegaly and improvements in most MF symptoms and patient quality of life (QoL) measures.

RSC Drug Discovery Series No. 38
Orphan Drugs and Rare Diseases
Edited by David C Pryde and Michael J Palmer

Published by the Royal Society of Chemistry, www.rsc.org

Emerging data from controlled clinical trials also suggest that ruxolitinib treatment confers a survival benefit in this patient population compared with placebo or best available therapy (BAT).

15.1.2 Origins of Ruxolitinib Discovery

The foundation for the discovery of ruxolitinib was established in 2002, shortly after Incyte Corporation began the journey of transforming from a genomics company to a drug discovery and development company with a major focus on inflammatory diseases and oncology. The challenge for a small group of scientists, comprising 20–30 chemists and an equal number of biologists, was to identify compelling therapeutic targets in the chosen disease indications that offered opportunities to discover first-in-class or well-differentiated molecules with superior pharmacological, pharmaceutical and/or toxicological properties compared with competitor compounds. Incyte scientists began a comprehensive survey of druggable targets in the chosen therapeutic areas, assessed the strength of the pre-clinical and clinical evidence supporting the targets, evaluated the competitive landscape, and prioritised the targets based on a number of internally established criteria. A JAK inhibitor project specifically targeting JAK1 and JAK2 emerged from that effort as a potential opportunity in both neoplastic and inflammatory diseases. Interestingly, the decision to target the JAK pathway predated the discovery of the *JAK2* V617F mutation in MPNs by 2 years, and through pure serendipity, allowed Incyte to get a head start and win the race to develop JAK2 inhibitors by gaining approval of Jakafi® (ruxolitinib) for MF in 2011.

In 2002, the success of imatinib, a breakthrough therapy for chronic myelogenous leukaemia (CML) that targets the bcr-abl kinase, was an important trigger for the pharmaceutical industry in general and Incyte scientists in particular to consider targeting kinases for the discovery of new drugs. Expanding knowledge of the structural biology of kinases and ATP-mimetic pharmacophores offered new opportunities to design novel chemical scaffolds that selectively targeted specific kinases.

With respect to the selection of the JAK pathway inhibitor programme at Incyte, the author recalls a review series highlighting the role of the JAK-STAT pathway in human disease by Christian Schindler in 2002[1] as the first trigger to take a serious look at this mechanism as a potential drug discovery opportunity. A publication from Jacques Ghysdael's laboratory that demonstrated the development of leukaemia in transgenic mice expressing constitutively active JAK2[2] was another critical piece of evidence that convinced the Incyte team to prioritise the JAK programme over other potential therapeutic targets. Further review and analysis of the literature reinforced the belief that targeting JAKs could offer a very interesting and novel opportunity to discover new drugs for cancer in general and haematological malignancies in particular.

15.2 The JAK-STAT Pathway as a Target for Drug Discovery

15.2.1 Function and Components

The JAK-STAT signalling pathway functions to transduce signals from the cell membrane to the nucleus (Figure 15.1). JAKs are intracellular tyrosine kinases that are activated by phosphorylation in response to cytokines or growth factors binding to their respective receptors.[3] Phosphorylated JAKs phosphorylate the cytoplasmic tail of the transmembrane receptor. The phosphorylated receptor then recruits STATs, cytoplasmic transcription factors, to the cell membrane. JAKs phosphorylate the recruited STATs, which subsequently dimerise. Phosphorylated STAT dimers translocate to

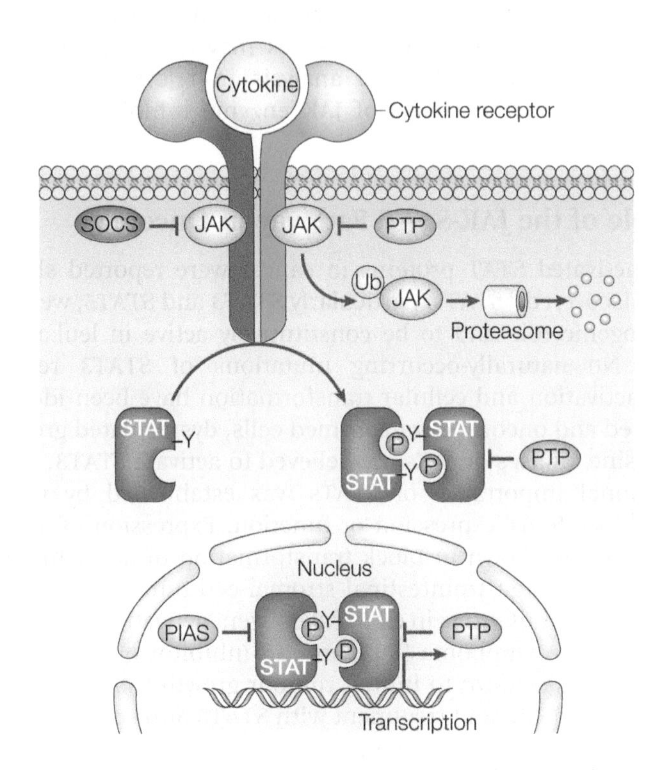

Figure 15.1 JAK-STAT signalling pathway. Cytokine stimulation can activate JAK signalling, leading to downstream dimerisation of STATs and translocation to the nucleus to influence transcription. JAK, Janus associated kinase; PIAS, protein inhibitor of activated STAT; PTP, protein tyrosine phosphatase; SOCS, suppressor of cytokine signalling; STAT, signal transducer and activator of transcription; Ub, ubiquitination.

the nucleus to initiate the transcription of genes involved in the promotion of cell growth, proliferation, differentiation and survival, as well as cytokine and growth factor expression. The pathway is normally activated by growth factor and cytokine receptor stimulation and participates in cytokine signalling and haematopoiesis.[3,4]

Cytokines that signal through heteromeric receptors containing the gp130 subunit, including interleukin IL-6 and IL-11, primarily utilise JAK1 and JAK2.[5] Type II cytokine receptors that bind IL-10, IL-19, IL-20 and IL-22 utilise JAK2 and TYK2 for signalling. Receptors for hormone-like cytokines, such as growth hormone, prolactin and the growth factors erythropoietin (EPO), thrombopoietin (TPO), IL-3 and granulocyte-macrophage colony-stimulating factor (GM-CSF) use JAK2.[6] Receptors for IFN-γ use JAK1 and JAK2, whereas those for IL-2, IL-4, IL-7, IL-9, IL-15 and IL-21 signal through γc chain-containing receptors *via* JAK1 and JAK3.[7] However, usage of different JAK family members by cytokines may also depend on cellular context, suggesting that a detailed analysis in relevant cell systems is necessary to interpret the effects of JAK enzyme inhibition accurately in a given disease state.[8]

15.2.2 Role of the JAK-STAT Pathway in Oncology

Persistently activated STAT proteins in cancer were reported shortly after STATs were discovered.[9] STATs, particularly STAT3 and STAT5, were found to have an oncogenic role and to be constitutively active in leukaemias and lymphomas. No naturally-occurring mutations of STAT3 resulting in constitutive activation and cellular transformation have been identified. In tumour-derived and oncogene-transformed cells, dysregulated growth factor receptor tyrosine kinases or JAKs are believed to activate STAT3.

The functional importance of STATs was established by various approaches to block STAT expression or function. Expression of a dominant negative STAT3 was shown to block transformation of acute myelogenous leukaemia (AML) and gastrointestinal stromal cell tumours (GISTs) associated with activating mutations in c-kit.[10] Persistent activation of STAT3 is also found in Hodgkin's lymphoma (HL), and an inhibitor of JAK2 and STAT3 phosphorylation was shown to inhibit tumour growth.[11] STAT3 is also activated in prostate cancer, and treatment with STAT3 anti-sense RNA induced apoptosis.[12,13] In large granular lymphocyte leukaemia, blockade of STAT3 caused cell death *via* pathways favouring apoptosis.[14,15]

The activity of the JAK-STAT pathway in various cancers may be altered by mutations, oncogenic proteins and excessive cytokine stimulation. Both haematological and solid tumour malignancies have shown dysregulated JAK-STAT pathway activity, making this pathway one of the most commonly altered in cancer. Haematological malignancies with identified JAK-STAT pathway alterations include AML, myelodysplastic syndromes (MDS), CML, multiple myeloma (MM), non-Hodgkin's lymphoma (NHL), HL and chronic lymphocytic leukaemia (CLL). Solid tumour malignancies with evidence of aberrant JAK-STAT pathway activity include breast cancer, prostate cancer, pancreatic cancer, lung cancer, and head and neck cancer.[16]

15.2.3 Targeting the JAK-STAT Pathway in MPNs

The first evidence implicating mutated JAK2 in patients with Philadelphia chromosome (Ph)-negative MPNs came from genetic and functional studies of haematopoietic progenitor cells, which frequently contained a mutation of somatic origin.[17–21] The acquired mutation in *JAK2* implicated in the pathogenesis of Ph-negative MPNs is a point mutation in the negative autoregulatory pseudokinase domain (V617F) that lies adjacent to the tyrosine kinase domain; it results in constitutive activation of the kinase and increased sensitivity to cytokine signalling.[19] It promotes increased cell survival, cell cycle transition and differentiation in progenitor cells. Additional strong evidence supporting a role for mutated *JAK2* in MF comes from a number of independent studies using mouse models. Transgenic mice that constitutively expressed *JAK2* V617F in haematopoietic progenitor cells exhibited leucocytosis, erythroblastosis, marked thrombocytosis and haemoglobin levels consistent with anaemia when examined at 1 month of age.[22] These mice developed marked splenomegaly, and the severity of the phenotype correlated with the level of expression of the mutant transgene; aged mice that expressed high levels of *JAK2* V617F developed MF.

Although the *JAK2* V617F mutation is present in more than 50% of patients with ET and more than 90% of patients with PV, the role of this mutation in the progression of ET or PV to MF is not well understood. In both PV and ET, the risk of developing MF increases with the duration of disease, and the risk is greater in patients who harbour mutated compared with wild-type JAK2.[23] Several experiments in mice performed by independent groups also suggested a role for mutant JAK2 in progression to MF.

Mutated *JAK2* V617F is not the only genetic alteration associated with the development of MPNs, however. In the process of focusing on the JAK pathway following the discovery of the *JAK2* V617F mutation, a number of research groups identified other mechanisms and mutations that contribute to JAK-STAT overactivity. These include a mutation in the myeloproliferative leukaemia virus proto-oncogene (*MPL*; thrombopoietin receptor), mutations in exon 12 of *JAK2*, a mutation in *LNK*, overactive pro-inflammatory cytokine signalling, and epigenetic mechanisms that lead to the silencing of negative regulatory elements known as SOCS (silencers of cytokine signalling).[24] Thus, other mechanisms contribute to overactive JAK signalling in MPN patients who do not harbour the *JAK2* V617F mutation. In addition to JAK2, JAK1 was shown to be constitutively phosphorylated in blood mononuclear cells from patients with primary MF but not in cells from healthy subjects, suggesting hyperactivation of the JAK1 pathway in MF as well.[25] Given that pro-inflammatory cytokines such as IL-6 are highly elevated in MPNs and that JAK1 is the major JAK isoform that mediates the signalling of such cytokines, targeting both JAK1 and JAK2 inhibition is a logical approach for effective treatment of these diseases.

15.2.4 MF

MF occurs *de novo* (PMF) or secondary to PV and ET (PPV-MF and PET-MF, respectively).[26] The incidence of PMF in the USA, based on data collected from 2001 to 2003, was estimated to be 0.22 per 100 000 persons per year.[27] The 10 year risk of progression of PV to PPV-MF and ET to PET-MF was estimated to be <10% and <1%, respectively.[28] The median age at diagnosis of MF is 64 years.[29] Clinically, MF is characterised by bone marrow fibrosis, ineffective haematopoiesis, extramedullary haematopoiesis (EMH), splenomegaly, debilitating progressive constitutional symptoms and shortened survival.[30]

Progressive reactive bone marrow fibrosis in MF accompanies stem cell-derived clonal myeloproliferation and interferes with normal blood cell formation to result in cytopenias, EMH, cachexia and a hypermetabolic state.[31] These processes may lead to thrombo-haemorrhagic complications, infections, splenomegaly and hepatomegaly, and chronic, progressive symptoms. Symptoms include fatigue, night sweats, fever, itching (pruritus), abdominal discomfort, pain under left ribs, early satiety, weight loss and bone/muscle pain,[29,32] and it negatively impacts patient QoL. Splenomegaly is associated with abdominal discomfort and pain and leads to poor nutritional status, cachexia and low cholesterol.[33,34] In a survey that contained 456 MF patients, 84% reported fatigue, and 56%, 50% and 47% reported night sweats, itching and bone pain, respectively.[32]

Although secondary malignancies are rare, in a subset of patients MF will transform to secondary acute myeloid leukaemia (sAML). Survival after leukaemic transformation is poor, with a median of 2.6 months measured in one study.[35] In patients with PMF and a known cause of death, the most common causes were leukaemic transformation (31%), PMF progression without transformation (19%), thrombosis and cardiovascular complications (13%) and infections (11%).[29]

MF has been proposed to be a generalised systemic inflammatory disorder by Dr Hasselbalch.[36] Patients with MF have excessive production of pro-inflammatory, fibrogenic and angiogenic cytokines. Cytokine profiling of 127 patients with PMF and 35 control subjects identified increases in the levels of the following cytokines and related factors: IL-1β, IL-1 receptor antagonist (RA), IL-2R, IL-6, IL-8, IL-10, IL-12, IL-13, IL-15, tumour necrosis factor (TNF)-α, granulocyte colony-stimulating factor, interferon alpha (IFN-α), macrophage inflammatory protein (MIP)-1α, MIP-1β, hepatocyte growth factor, IFN-γ-inducible protein, monokine induced by IFN-γ (MIG), monocyte chemotactic protein (MCP)-1, and vascular endothelial growth factor (VEGF). IFN-γ levels were decreased.[37]

The MF disease course is heterogeneous, and several prognostic scoring systems have been established to predict patient survival based on disease characteristics. The International Prognostic Scoring System (IPSS) uses five adverse prognostic factors present at diagnosis: age > 65 years, presence of constitutional symptoms (weight loss > 10% of the baseline value in the year preceding diagnosis, unexplained fever, or excessive sweats persisting for >1 month), haemoglobin level < 10 g dL^{-1}, leukocyte count > 25 × 10^9 L^{-1}, and circulating blast cells ≥ 1%. Patients with 0 risk factors have low-risk disease; the addition of one risk factor changes the classification to intermediate-1 risk; the presence of two risk factors changes the classification to intermediate-2 risk; and high-risk patients have three or more risk factors. In the population of patients with PMF used to establish this system, median survival ranged from 2.3 years in patients with high-risk disease to 11.3 years in patients with low-risk disease.[29] Other scoring systems that have evolved from the IPSS include the dynamic IPSS (DIPSS), which contains the same risk factors as the IPSS but can be used at any time point during the clinical course of disease, as opposed to only at diagnosis, and assigns a greater weight to disease-related anaemia to account for its influence on survival when acquired over time.[38] The DIPSS-Plus contains the five factors of the IPSS plus three additional factors (unfavourable karyotype, platelet counts < 100 × 10^9 L^{-1}, and red blood cell transfusion dependence) and can be used at any time point during the clinical course of disease.[39]

MF treatment and management approaches have involved the administration of agents to ameliorate disease symptoms and signs. Treatments included hydroxyurea, corticosteroids, thalidomide, lenalidomide, anagrelide, epoetin alfa and danazol.[30] Because these interventions do not alter the natural history of the disease, they provide limited clinical benefits. Additional treatment options for splenomegaly are splenic irradiation and splenectomy, but they are associated with cytopenias, and splenectomy carries risks of perioperative complications and mortality. The only potentially curative treatment for MF is allogeneic stem cell transplantation; however, even with recent advances to improve its safety, the procedure is associated with a high risk of morbidity and mortality.[40]

15.3 Ruxolitinib Drug Discovery

15.3.1 JAK1 and JAK2 Discovery Programme

The discovery programme that targeted JAK1 and JAK2 at Incyte began in 2003. Following careful consideration of the biology of JAK signalling, phenotypes of JAK knockout mice, and the potential roles for different JAK family members in oncogenic, inflammatory and immune function, we decided to focus on JAK1 and JAK2 and deliberately build selectivity against JAK3 in the target compound profile. The decision to avoid JAK3 inhibition was based on the absence of a role for JAK3 in the signalling of key cytokines, such as IL-6 and IL-12/23, which are believed to be important for oncology and inflammatory diseases, and on the desire to minimise the risk for immunosuppression. Given the absence of a specific and dominant role for

TYK2 in any cytokine signalling pathway, we decided to accept some degree of TYK2 cross-reactivity. A rational design process utilising known chemical scaffolds with JAK inhibitory activity, rather than computational chemistry or chemical library screening, was used to discover novel pharmacophores. A series of cell-based assays, including isolated cellular assays using peripheral blood monocytes and T-cells and whole blood assays with IL-6 or TPO stimulation and phosphoSTATs as readouts, was used to identify compounds with good cellular activity. As a result of this effort, several potent and selective JAK1 and JAK2 inhibitors were identified, among which ruxolitinib (INCB018424) was one. Selection of ruxolitinib as a development candidate was based on *in vitro* and *in vivo* pharmacology data, its pharmacokinetic profile and toxicology data, described in part below.

15.3.2 Chemistry

Ruxolitinib phosphate is (*R*)-3-(4-(7*H*-pyrrolo[2,3-*d*]pyrimidin-4-yl)-1*H*-pyrazol-1-yl)-3-cyclopentylpropanenitrile phosphate (Figure 15.2). Its molecular formula is $C_{17}H_{21}N_6O_4P$, and it has a molecular weight of 404.36. It exists as a white to off-white to light pink powder, and it is soluble in aqueous buffers ranging from pH 1 to 8. It is available in dosage strengths of 5, 10, 15, 20 and 25 mg. Ruxolitinib tablets contain ruxolitinib phosphate, along with microcrystalline cellulose, lactose monohydrate, magnesium stearate, colloidal silicon dioxide, sodium starch glycolate, povidone and hydroxypropyl cellulose.[41]

15.3.3 *In Vitro* Data

The potency and selectivity of ruxolitinib for JAK1 and JAK2 were demonstrated in enzyme assays by measuring its 50% inhibitory concentration (IC_{50}) for JAK1, JAK2, JAK3, TYK2, CHK2 and cMET. Testing against 26 additional kinases showed no inhibition by ruxolitinib when used at

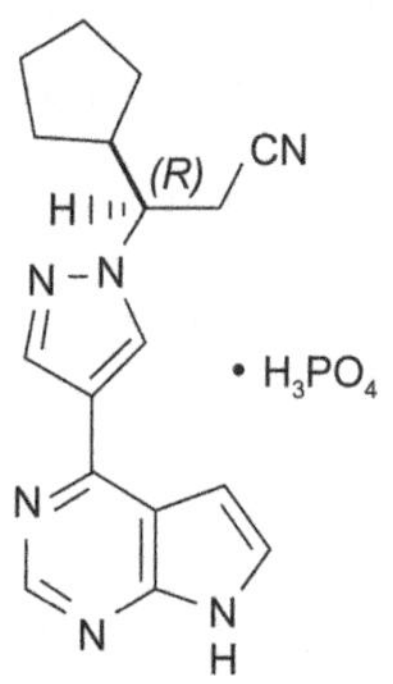

Figure 15.2 Structure of ruxolitinib.

Table 15.1 Enzymatic and functional potency of ruxolitinib.

Enzyme assays	*IC_{50} nM (mean ± SD)*	*N*
JAK 1	3.3 ± 1.2	7
JAK 2	2.8 ± 1.2	8
JAK 3	428 ± 243	5
Tyk2	19 ± 3.2	8
CHK2	>1000[a]	7
cMET	>10 000[a]	1
Whole blood assays		
IL-6 stimulation	282 ± 54	6
TPO stimulation	281 ± 62	4

[a]Highest concentration evaluated. SD, standard deviation; IC_{50}, 50% inhibitory concentration; IL, interleukin; TPO, thrombopoietin receptor.

approximately 100 times the IC_{50} for JAK1 and JAK2. In cytokine-stimulated whole blood assays, pre-incubation with ruxolitinib inhibited IL-6 and thrombopoietin signalling (Table 15.1).

Ruxolitinib also inhibited the phosphorylation of JAK2, STAT5 and ERK1/2 and reduced the viability of cells engineered to express *JAK2* V617F, showing evidence of apoptosis. In MPN patient-derived samples, treatment with ruxolitinib resulted in dose-dependent inhibition of erythroid and myeloid progenitor cell growth and potent inhibition of erythroid colony formation.[25]

15.4 *In Vivo* Data

15.4.1 Pre-Clinical Studies

Pre-clinical characterisation of ruxolitinib was established with a mouse model of a *JAK2* V617F-driven malignancy. The model was constructed by injecting Ba/F3-EpoR-*JAK2* V617F cells into the tail vein of Balb/c mice to induce a progressive splenomegaly over a 3 week period. Oral administration of ruxolitinib after inoculation of the Ba/F3-EpoR-*JAK2* V617F cells prevented splenomegaly, decreased *JAK2* V616F mutant cells in the spleen, normalised uncontrolled JAK-STAT signalling in the spleen, and prolonged survival compared with mice treated with vehicle (Figure 15.3).[25] Ruxolitinib also reduced IL-6 and TNF-α levels in these mice compared with vehicle. No myelosuppressive or immunosuppressive effects were seen in ruxolitinib-treated mice.[25]

15.4.2 Pharmacokinetics and Pharmacodynamics

The single-dose pharmacokinetic parameters of orally administered ruxolitinib were analysed in fasted, healthy, adult subjects. Rapid and almost complete (96%) absorption was observed, and maximum serum concentrations (C_{max}) were reached in 1–2 hours.[42,43] Pharmacokinetic studies also

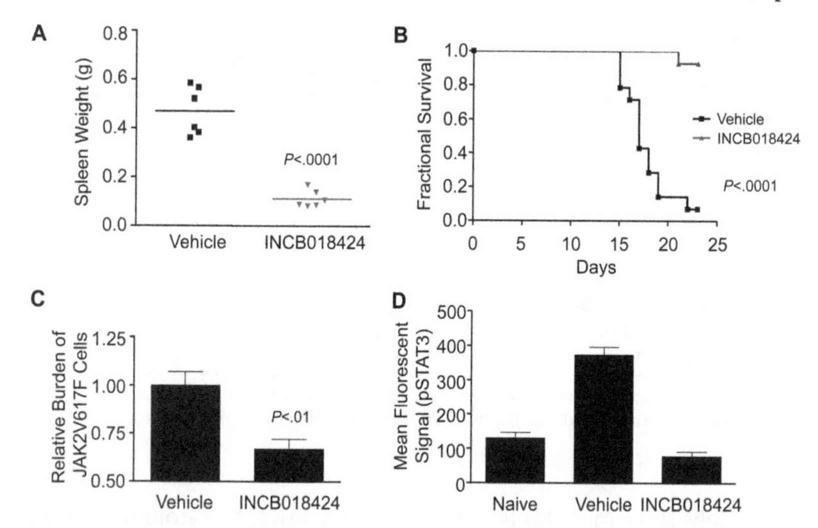

Figure 15.3 Effects of ruxolitinib in a *JAK2* V617F-driven pre-clinical model. Mice were inoculated intravenously with Ba/F3-EpoR-*JAK2* V617F cells and treated with ruxolitinib or vehicle. (A) Effect of ruxolitinib *versus* vehicle on spleen weight after 2 weeks of treatment. (B) Kaplan–Meier analysis of survival with ruxolitinib and vehicle. (C) Levels of *JAK2* V617F-expressing cells after 2 weeks of treatment with ruxolitinib or vehicle. (D) Levels of phosphorylated STAT3 in spleen lysates of naive mice and injected mice 4 hours after dosing with vehicle or ruxolitinib.

demonstrated a low, dose-independent clearance of approximately 20 L per hour, a moderate apparent volume of distribution (range, 75–91 L for doses from 5 to 100 mg), and a high rate of plasma protein binding (97%). The elimination half-life was approximately 3 hours. Dose proportionality of C_{max} and area under the concentration-*versus*-time curve (AUC) were demonstrated over the dose range of 5–200 mg. Although consumption of a high-fat, high-calorie meal at the time of administration of a single 25 mg dose of ruxolitinib slowed its absorption and t_{max}, the AUC was similar to that in fasted subjects, indicating that ruxolitinib can be given with or without food.[42] No appreciable accumulation of ruxolitinib was noted in multiple-dose studies in healthy subjects.

Ruxolitinib is predominantly metabolised by the cytochrome P450 enzyme 3A4 (CYP3A4). Ruxolitinib has two major active metabolites that account for 25% and 11% of the AUC and that have one-fifth and one-half of its activity, respectively. Excretion is primarily renal, as 74% of a single 25 mg radio-labelled dose administered to healthy subjects was recovered in urine, and 22% was recovered in faeces.[43]

The pharmacodynamic effect of ruxolitinib was evaluated by measuring STAT3 phosphorylation, a marker of JAK-STAT activation, in blood samples to determine the extent of JAK1/2 inhibition. An analysis of whole blood samples from healthy subjects who had been treated with ruxolitinib demonstrated dose- and time-dependent inhibition of *in vitro* cytokine-stimulated STAT3 phosphorylation.[42] In MF patients participating in a ruxolitinib clinical trial, dose-dependent inhibition of STAT3 phosphorylation was evident in whole blood samples 2 hours after dosing with ruxolitinib compared with levels of STAT3 phosphorylation measured at baseline. After 28 days of therapy, levels of phosphorylated STAT3 in patients who received ruxolitinib 25 mg BID (20 patients who were *JAK2* V617F positive and four patients who were *JAK2* V617F negative) were reduced to levels similar to those measured in healthy volunteers.[44]

15.4.3 Efficacy and Safety in a Phase I/II Trial

A Phase I/II open-label, non-randomised study of ruxolitinib was conducted in 153 patients with newly diagnosed PMF, PPV-MF and PET-MF at two sites: MD Anderson Cancer Center and Mayo Clinic-Rochester (NCT00509899). Patients were required to have intermediate- or high-risk disease as classified by the Lille system (haemoglobin < 10 g dL^{-1} and/or leukocytes $< 4 \times 10^9$ or $>30 \times 10^9$ L^{-1}); 65% of patients had high-risk disease as classified by the IPSS. In addition, 82% were positive for the *JAK2* V617F mutation. All had Eastern Cooperative Oncology Group (ECOG) scores ≤ 2. Patients with enlarged spleens at study entry were evaluated over a 3 month period to determine the proportion achieving a $\geq 50\%$ reduction in palpable splenomegaly. Various BID and once-daily dosing schedules were assessed, and one cycle of therapy was defined as 4 weeks.[44]

After 3 months of treatment, the highest objective response rates (at least a 50% reduction in palpable splenomegaly) were achieved in the 33 patients who received ruxolitinib 15 mg BID (52%) and the 39 patients who received ruxolitinib 25 mg BID (49%). In a *post hoc* analysis that assessed the pooled data of these responders, response rates were similar between patients with and without the *JAK2* V617F mutation (51% and 45%, respectively) and among patients with PMF (49%), PPV-MF (45%) and PET-MF (62%). Overall, spleen size reductions were rapid and durable, occurring in most patients within the first 1–2 months and lasting for approximately 2 years. A subset of patients ($n = 24$) who received ruxolitinib 15 mg BID underwent abdominal MRI for analysis of spleen and liver volume. The median reduction in spleen volume after 6 months of treatment was 33%. In the six patients of this MRI subset who had hepatomegaly (liver volume $\geq$ 2 times normal) at baseline, after 6 months of treatment, a 14% reduction in liver volume was noted (Figure 15.4).[44]

MF-related symptoms were evaluated with the Myelofibrosis Symptom Assessment Form (MFSAF), which was completed by the patient and required patients to rank their symptoms on a scale of 0 (absent) to 10 (least favourable). After 1 month of treatment with ruxolitinib at doses ranging from 10 mg BID to 25 mg BID, most patients experienced a $\geq 50\%$ improvement from baseline in total or individual symptom scores, including night sweats, itching, bone/muscle pain, and abdominal pain or discomfort. Patients with reductions in spleen size reported reductions in abdominal discomfort. Symptom responses

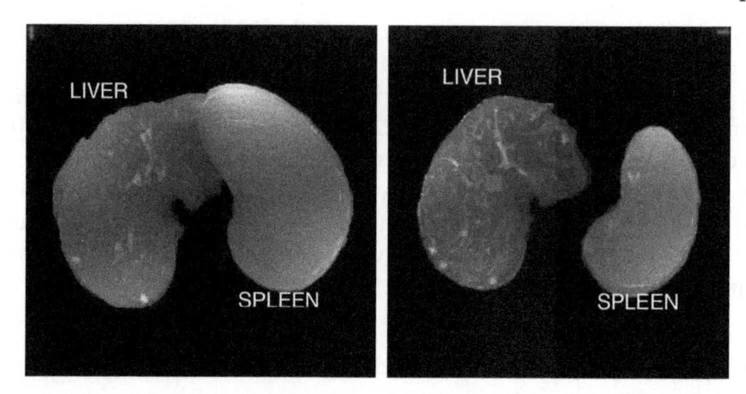

Figure 15.4 Three-dimensional reconstruction of spleen and liver images from an MF patient at baseline and after 6 months of treatment with ruxolitinib in the Phase I/II trial.

were durable during the 6–12 months of follow-up reported. In addition, patients gained weight as early as after two cycles of treatment.[44]

Ruxolitinib treatment also reduced circulating plasma levels of pro-inflammatory cytokines and other biomarkers in MF patients. Levels were assessed at baseline and after one or six cycles of therapy in patients who received ruxolitinib 25 mg BID and compared with those of healthy controls. Heat maps were constructed to show the differences in plasma levels of each of these factors for individual patients at baseline *versus* healthy controls and also for individual patients at day 28 *versus* baseline (Figure 15.5).[43] Decreases of ≥50% in plasma levels of IL-1RA, MIP-1β, TNF-α, IL-6 and C-reactive protein were apparent after six cycles of therapy and correlated with improvements in MF-related symptoms. Levels of erythropoietin and leptin, which were below normal at baseline, increased after ruxolitinib treatment.[44] Decreased cytokine levels were seen in patients with and without the *JAK2* V617F mutation.

Ruxolitinib was well tolerated in this trial. Non-haematological toxicities occurred at an incidence of <10% and were of low grade. The main haematological toxicities were new-onset anaemia in patients who were transfusion-independent at baseline (23%) and dose-limiting grade 3 or 4 thrombocytopenia (20%). Because erythropoietin and thrombopoietin signal exclusively through JAK2, anaemia and thrombocytopenia were expected with ruxolitinib treatment. After an initial decrease in mean haemoglobin levels over the first three to four cycles of therapy, levels stabilised or improved from the nadir with subsequent therapy.[44]

15.4.4 Efficacy and Safety in the Pivotal Phase III Trials

The COntrolled MyeloFibrosis Study with ORal JAK Inhibitor Therapy (COMFORT)-I (NCT00952289) and COMFORT-II (NCT00934544) trials are the pivotal Phase III trials that evaluated the efficacy and safety of ruxolitinib in

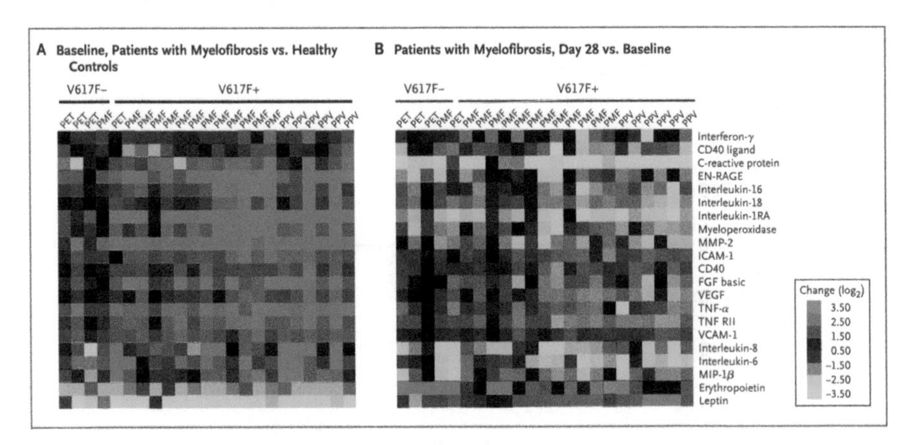

Figure 15.5 Heat maps of selected markers in plasma of MF patients. (A) At baseline *vs.* healthy controls and (B) at day 28 *vs.* baseline. MF patients received ruxolitinib 25 mg BID. Green indicates markers present at lower levels at baseline and markers that decreased with ruxolitinib treatment. Red indicates markers present at higher levels at baseline and markers that increased with ruxolitinib treatment.

patients with intermediate-2 and high-risk MF as classified by the IPSS. As of September 2013, the long-term extension phases of both trials remain ongoing. COMFORT-I is a randomised, double-blind, placebo-controlled trial conducted in the USA, Australia and Canada, and COMFORT-II is a randomised, open-label trial comparing ruxolitinib with BAT that is being conducted in several countries in Europe. In both trials, enrolled patients were ≥18 years of age with PMF, PPV-MF or PET-MF according to the 2008 WHO criteria. Patients were required to have a palpable spleen length ≥5 cm below the left costal margin, platelet count $\geq 100 \times 10^9$ L^{-1}, and ECOG performance status ≤3.[45,46]

In COMFORT-I, the 309 patients studied were refractory to or not eligible for available therapy. They were randomised 1 : 1 to receive ruxolitinib or placebo. The starting dose of ruxolitinib was determined according to the baseline platelet count (15 mg BID for patients with baseline platelet count 100–200 $\times 10^9$ L^{-1}; 20 mg BID for patients with baseline platelet count > 200 $\times 10^9$ L^{-1}).[45] Doses were individually titrated based on safety and efficacy.

The primary end point was the proportion of patients who achieved a ≥35% spleen volume reduction from baseline at week 24 (measured by MRI or CT). A key secondary end point was the proportion of patients with a ≥50% improvement in the total symptom score (TSS) using the modified MFSAF v2.0 diary. The TSS is the sum of individual symptom scores for night sweats, itching, abdominal discomfort, pain under the ribs on the left side, feeling of fullness (early satiety) and muscle/bone pain.

After 24 weeks of treatment, significantly more patients achieved a ≥35% reduction in spleen volume from baseline with ruxolitinib (41.9%) *versus*

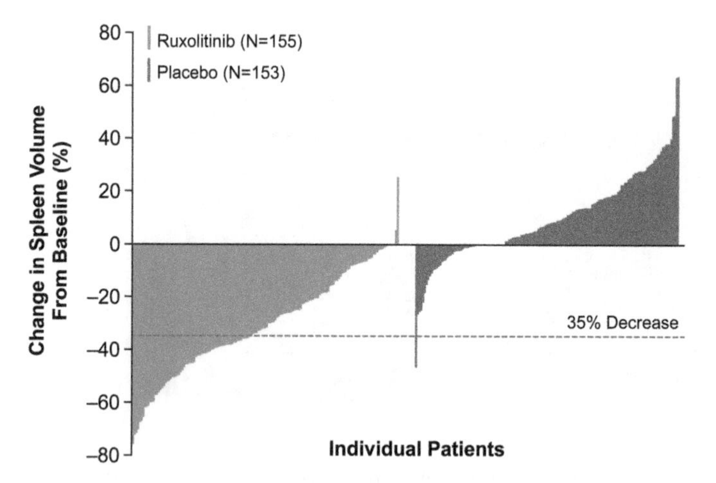

Figure 15.6 Reduction in spleen volume in COMFORT-I. Almost all patients who received ruxolitinib had reductions in spleen volume at week 24; in contrast, most patients who received placebo had increases or no change in spleen volume.

placebo (0.7%; $p < 0.001$). While almost all patients who received ruxolitinib experienced some decrease in spleen volume, the majority of patients in the placebo group experienced increases in spleen volume over the 24 week period (Figure 15.6). Of the ruxolitinib-treated patients with a ≥35% reduction in spleen volume, 67% maintained this reduction for at least 48 weeks.[45]

Accompanying the spleen volume reductions in patients receiving ruxolitinib were favourable changes in MF-related symptoms and QoL measures. A significantly greater proportion of patients who received ruxolitinib achieved a ≥50% reduction in TSS from baseline to week 24 (45.9%) compared with patients who received placebo (5.3%; $p < 0.001$). At week 24, the mean TSS had improved by 46.1% in patients who received ruxolitinib and had worsened by 41.8% in patients who received placebo. Most symptom improvements with ruxolitinib occurred within the first 4 weeks of treatment and were maintained through week 24.

An analysis of pro-inflammatory cytokines and other biomarkers showed that at baseline, patients had elevated plasma levels of TNF-α, IL-6 and C-reactive protein compared with healthy subjects. At week 24, patients who received ruxolitinib treatment showed decreases in these pro-inflammatory cytokines, while patients who received placebo had minimal changes. Plasma levels of leptin and erythropoietin increased by week 24 in ruxolitinib-treated patients and showed no or minimal change in patients who received placebo.[45]

As in the Phase I/II trial, the most common grade 3/4 adverse events with ruxolitinib treatment were anaemia (45.2%) and thrombocytopenia (12.9%). These events generally occurred early in the course of therapy and were managed with red blood cell transfusions and dose adjustments/treatment interruptions, respectively. Non-haematological events generally occurred at similar frequencies in both treatment groups. Ecchymosis, dizziness and headache occurred more frequently in the ruxolitinib group and were mainly grade 1 or 2.

In COMFORT-II, 219 patients were randomised 2 : 1 to receive ruxolitinib or BAT (hydroxyurea, glucocorticoids, other therapy, or no therapy at the investigator's discretion), and the starting dose of ruxolitinib was determined according to baseline platelet count, as in COMFORT-I.[46]

The primary end point was the proportion of patients who achieved a ≥35% spleen volume reduction from baseline at week 48. Secondary end points included the proportion of patients who achieved a ≥35% spleen volume reduction from baseline at week 24, duration of reduction of spleen volume ≥ 35%, and time to reduction in spleen volume ≥ 35%. QoL measures were assessed with the EORTC QLQ-C30 and the Functional Assessment of Cancer Therapy-Lymphoma (FACT-Lym) scale.[46]

At week 48, significantly more patients achieved a ≥35% decrease in spleen volume from baseline with ruxolitinib (28%) compared with BAT (0%; $p < 0.001$). At week 24, these values were 32% and 0%, respectively ($p < 0.001$). The median duration of reduction in spleen volume ≥ 35% was not reached, as 80% of patients had maintained their response at 12 months, and the median time to spleen volume reduction ≥ 35% was 12.3 weeks with ruxolitinib.[46]

Global health status, role functioning and individual symptoms, including fatigue, pain dyspnoea, insomnia and appetite loss, improved in patients receiving ruxolitinib and worsened or stayed the same in patients receiving BAT. On the FACT-Lym, patients who received ruxolitinib demonstrated significant improvements in pain, fever, itching, fatigue, weight loss, loss of appetite, and other patient concerns, as well as in general outcomes and physical condition and functioning, while scores in patients who received BAT worsened.[46]

The long-term adverse event profile with ruxolitinib treatment was consistent with the profile reported in COMFORT-I.[47]

Long-term follow-up of both COMFORT-I and COMFORT-II suggest that ruxolitinib has a survival benefit over placebo and BAT, respectively. In COMFORT-I after a median 2 year follow-up, 27 patients randomised to ruxolitinib and 41 patients randomised to placebo had died (HR, 0.58; 95% CI, 0.36 to 0.95; $p = 0.028$) (Figure 15.7).[47] Similar findings were observed in the 2 year follow-up of COMFORT-II for ruxolitinib compared with BAT (HR, 0.51; 95% CI, 0.26–0.99; $p = 0.041$).[48,49]

Preliminary data presented at the 2013 Annual Meeting of the American Society for Clinical Oncology and European Hematology Association provide encouraging evidence that long-term treatment with ruxolitinib may reverse and/or stabilise bone marrow fibrosis. Bone marrow samples from patients at the MD Anderson Cancer Center site of the Phase I/II trial were assessed for WHO fibrosis grade in an exploratory analysis, and changes from baseline at 24 and 48 months were measured. A group of patients treated with

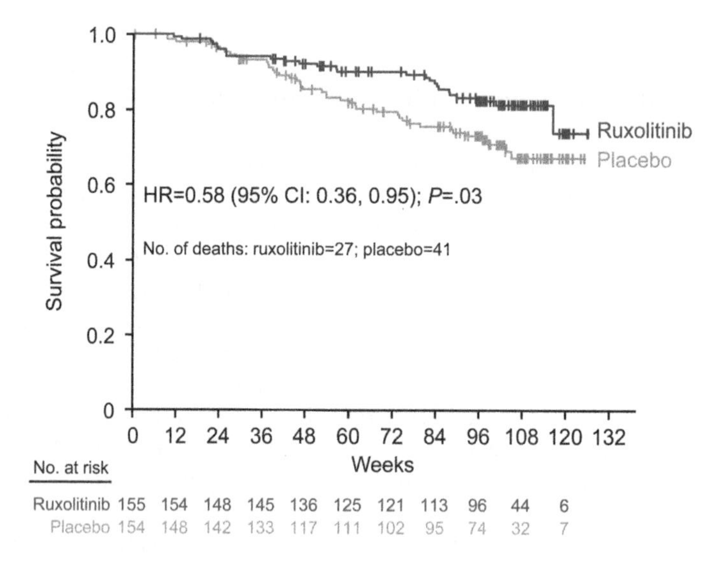

Figure 15.7 Survival analysis in COMFORT-I (2 year follow-up).

hydroxyurea and BAT (including hydroxyurea) were also analysed as controls. A greater proportion of ruxolitinib-treated patients achieved stabilisation or improvement of fibrosis grade at 24 and 48 months compared with patients who received hydroxyurea (at 24 months, 72% with ruxolitinib *versus* 62% with hydroxyurea; at 48 months, 77% with ruxolitinib *versus* 35% with hydroxyurea). Patients who received hydroxyurea had greater worsening of bone marrow fibrosis grade at both time points.[50] Similar findings were observed in the BAT cohort.[51] Although more evidence is needed before firm conclusions can be drawn, the data obtained thus far provide initial support for a potential role of ruxolitinib therapy in altering another important clinical feature of MF.

Ruxolitinib received US FDA approval in November 2011 for the treatment of patients with intermediate or high-risk MF, including PMF, PPV-MF and PET-MF; Health Canada approval in July 2012; and European Commission approval in August 2012 for the treatment of disease-related splenomegaly or symptoms in adults with PMF, PPV-MF or PET-MF. Although the COMFORT trials did not include patients with intermediate-1 risk MF, the US FDA approval includes patients with intermediate-1 risk MF 'since these patients may have symptoms that require treatment'.[52] The recommended starting dose of ruxolitinib is based on platelet count: 20 mg BID for patients with platelet counts > 200×10^9 L^{-1} and 15 mg BID for patients with platelet counts 100–200×10^9 L^{-1}. In June 2013, the US FDA updated the prescribing information for ruxolitinib to include a recommended starting dose of 5 mg BID for patients with baseline platelet counts 50–100×10^9 L^{-1} with dose modifications based on safety and efficacy.[41]

15.5 Future Directions in Ruxolitinib Drug Development in MF

Discovery of the *JAK2* V617F mutation resulted in a great deal of focused effort on JAK-STAT signalling in MPNs and expanded our understanding of the molecular mechanisms underlying the pathogenesis of these diseases. We now know that there are multiple mechanisms by which the JAK pathway is dysregulated in MPNs and that JAK dysregulation is nearly universal in MF patients. However, we also recognised that MF is a complex disease driven by JAK-dependent and -independent mechanisms that require the use of combination approaches to achieve optimal outcomes. Ruxolitinib is currently being investigated in clinical and pre-clinical studies in combination with various agents that target JAK-independent disease mechanisms relevant to MF or that demonstrated some degree of clinical activity in prior clinical trials. These include clinical trials of ruxolitinib with the immunomodulatory drugs (iMiDs), lenalidomide (NCT01375140) and pomalidomide (NCT01644110); a demethylating agent, 5-azacitidine (NCT01787487); a histone deacetylase (HDAC) inhibitor, panobinostat (NCT01693601, NCT01433445); and danazol (NCT01732445). Also being evaluated is the effect of ruxolitinib administration prior to allogeneic haematopoietic stem cell transplantation (NCT01790295). With regard to other MPNs, a Phase I/II trial of ruxolitinib in patients with PV and ET (NCT00726232) and a registrational trial in patients with PV resistant to or intolerant of hydroxyurea (NCT01243944) are ongoing.

15.6 Summary/Conclusion

Ruxolitinib, a selective inhibitor of JAK1 and JAK2, discovered at Incyte Corporation, is the first FDA-approved therapy for MF and the first FDA-approved JAK pathway inhibitor for any indication. The JAK inhibitor discovery programme, originally conceived as a novel approach for the treatment of cancer and inflammation, took a major turn toward MPNs following the discovery of a somatic gain-of-function mutation in *JAK2*, prompting Incyte to advance ruxolitinib into clinical development for MF less than 2 years after the discovery of the mutation. Further research demonstrated that the JAK pathway is more broadly dysregulated in MF than would be predicted by the prevalence of *JAK2* V617F (~50% of patients) and that ruxolitinib showed efficacy in both mutation positive and negative patients. Following a Phase I/II study demonstrating the safety and efficacy of ruxolitinib as measured by rapid and sustained improvements in the major clinical manifestations of MF (*i.e.* splenomegaly and debilitating symptoms), ruxolitinib was rapidly advanced through two registrational studies, COMFORT-1 and COMFORT-II, which demonstrated clear patient benefit with manageable adverse events related primarily to the pharmacology of the drug. These data allowed ruxolitinib to be approved by the US FDA and other regulatory agencies. Long-term follow-up in the pivotal studies has suggested a survival benefit of ruxolitinib therapy *versus* placebo or BAT, in addition to sustained benefits in other efficacy measures. Future clinical trials are addressing improving patient outcomes in MF through the use of various combination therapy approaches, as well as evaluating the efficacy of ruxolitinib in other malignancies.

Acknowledgements

The author would like to thank Stephanie Leinbach, PhD, for editorial assistance.

References

1. C. W. Schindler, *J. Clin. Invest.*, 2002, **109**, 1133.
2. C. Carron, F. Cormier, A. Janin, V. Lacronique, M. Giovannini, M. T. Daniel, O. Bernard and J. Ghysdael, *Blood,* 2000, **95**, 3891.
3. A. Quintas-Cardama, H. Kantarjian, J. Cortes and S. Verstovsek, *Nat. Rev. Drug Discovery,* 2011, **10**, 127.
4. K. Shuai and B. Liu, *Nat. Rev. Immunol.*, 2003, **3**, 900.
5. K. Yamaoka, P. Saharinen, M. Pesu, V. E. Holt, 3rd, O. Silvennoinen and J. J. O'Shea, *Genome Biol.*, 2004, **5**, 253.
6. S. J. Haque and P. Sharma, *Vitam. Horm.*, 2006, **74**, 165.
7. C. Schindler and C. Plumlee, *Semin. Cell Dev. Biol.*, 2008, **19**, 311.
8. K. Ghoreschi, A. Laurence and J. J. O'Shea, *Immunol. Rev.*, 2009, **228**, 273–287.
9. J. F. Bromberg, M. H. Wrzeszczynska, G. Devgan, Y. Zhao, R. G. Pestell, C. Albanese and J. E. Darnell Jr, *Cell,* 1999, **98**, 295–303.
10. Z. Q. Ning, J. Li, M. McGuinness and R. J. Arceci, *Oncogene,* 2001, **20**, 4528–4536.
11. D. Kube, U. Holtick, M. Vockerodt, T. Ahmadi, B. Haier, I. Behrmann, P. C. Heinrich, V. Diehl and H. Tesch, *Blood,* 2001, **98**, 762.
12. B. Gao, X. Shen, G. Kunos, Q. Meng, I. D. Goldberg, E. M. Rosen and S. Fan, *FEBS Lett.*, 2001, **488**, 179.
13. C. L. Campbell, Z. Jiang, D. M. Savarese and T. M. Savarese, *Am. J. Pathol.*, 2001, **158**, 25.
14. P. K. Epling-Burnette, J. H. Liu, R. Catlett-Falcone, J. Turkson, M. Oshiro, R. Kothapalli, Y. Li, J. M. Wang, H. F. Yang-Yen, J. Karras, R. Jove and T. P. Loughran Jr, *J. Clin. Invest.*, 2001, **107**, 351.
15. P. K. Epling-Burnette, B. Zhong, F. Bai, K. Jiang, R. D. Bailey, R. Garcia, R. Jove, J. Y. Djeu, T. P. Loughran Jr and S. Wei, *J. Immunol.*, 2001, **166**, 7486.
16. P. Sansone and J. Bromberg, *J. Clin. Oncol.*, 2012, **30**, 1005.
17. E. J. Baxter, L. M. Scott, P. J. Campbell, C. East, N. Fourouclas, S. Swanton, G. S. Vassiliou, A. J. Bench, E. M. Boyd, N. Curtin, M. A. Scott, W. N. Erber, A. R. Green and the Cancer Genome Project, *Lancet,* 2005, **365**, 1054.

18. C. James, *Hematology/the Education Program of the American Society of Hematology*, American Society of Hematology, Education Program, 2008, p. 69.
19. R. Kralovics, F. Passamonti, A. S. Buser, S. S. Teo, R. Tiedt, J. R. Passweg, A. Tichelli, M. Cazzola and R. C. Skoda, *N. Engl. J. Med.*, 2005, **352**, 1779.
20. R. L. Levine, M. Wadleigh, J. Cools, B. L. Ebert, G. Wernig, B. J. Huntly, T. J. Boggon, I. Wlodarska, J. J. Clark, S. Moore, J. Adelsperger, S. Koo, J. C. Lee, S. Gabriel, T. Mercher, A. D'Andrea, S. Frohling, K. Dohner, P. Marynen, P. Vandenberghe, R. A. Mesa, A. Tefferi, J. D. Griffin, M. J. Eck, W. R. Sellers, M. Meyerson, T. R. Golub, S. J. Lee and D. G. Gilliland, *Cancer Cell*, 2005, **7**, 387.
21. R. Zhao, S. Xing, Z. Li, X. Fu, Q. Li, S. B. Krantz and Z. J. Zhao, *J. Biol. Chem.*, 2005, **280**, 22788.
22. K. Shide, H. K. Shimoda, T. Kumano, K. Karube, T. Kameda, K. Takenaka, S. Oku, H. Abe, K. S. Katayose, Y. Kubuki, K. Kusumoto, S. Hasuike, Y. Tahara, K. Nagata, T. Matsuda, K. Ohshima, M. Harada and K. Shimoda, *Leukemia*, 2008, **22**, 87.
23. F. Passamonti, L. Malabarba, E. Orlandi, C. Barate, A. Canevari, E. Brusamolino, M. Bonfichi, L. Arcaini, S. Caberlon, C. Pascutto and M. Lazzarino, *Haematologica*, 2003, **88**, 13.
24. J. Mascarenhas, N. Roper, P. Chaurasia and R. Hoffman, *Clin. Epigenet.*, 2011, **2**, 197.
25. A. Quintas-Cardama, K. Vaddi, P. Liu, T. Manshouri, J. Li, P. A. Scherle, E. Caulder, X. Wen, Y. Li, P. Waeltz, M. Rupar, T. Burn, Y. Lo, J. Kelley, M. Covington, S. Shepard, J. D. Rodgers, P. Haley, H. Kantarjian, J. S. Fridman and S. Verstovsek, *Blood*, 2010, **115**, 3109.
26. G. Barosi, R. A. Mesa, J. Thiele, F. Cervantes, P. J. Campbell, S. Verstovsek, B. Dupriez, R. L. Levine, F. Passamonti, J. Gotlib, J. T. Reilly, A. M. Vannucchi, C. A. Hanson, L. A. Solberg, A. Orazi, A. Tefferi and International Working Group for Myelofibrosis and Treatment, *Leukemia*, 2008, **22**, 437.
27. D. E. Rollison, N. Howlader, M. T. Smith, S. S. Strom, W. D. Merritt, L. A. Ries, B. K. Edwards and A. F. List, *Blood*, 2008, **112**, 45.
28. A. Tefferi, *Am. J. Hematol.*, 2012, **87**, 285.
29. F. Cervantes, B. Dupriez, A. Pereira, F. Passamonti, J. T. Reilly, E. Morra, A. M. Vannucchi, R. A. Mesa, J. L. Demory, G. Barosi, E. Rumi and A. Tefferi, *Blood*, 2009, **113**, 2895.
30. A. M. Vannucchi, *Hematology/the Education Program of the American Society of Hematology*, American Society of Hematology, Education Program, 2011, vol. 2011, p. 222.
31. A. M. Vannucchi, P. Guglielmelli and A. Tefferi, *Ca-Cancer J. Clin.*, 2009, **59**, 171.
32. R. A. Mesa, J. Niblack, M. Wadleigh, S. Verstovsek, J. Camoriano, S. Barnes, A. D. Tan, P. J. Atherton, J. A. Sloan and A. Tefferi, *Cancer*, 2007, **109**, 68.

33. N. Sulai, B. Mengistu, N. Gangat, C. A. Hanson, R. P. Ketterling, A. Pardanani and A. Tefferi, *Blood,* 2012, **120**, 2851.
34. R. A. Mesa, S. Verstovsek, V. Gupta, J. Mascarenhas, E. Atallah, W. Sun, V. A. Sandor and J. Gotlib, *Blood,* 2012, **120**, 1733.
35. R. A. Mesa, C. Y. Li, R. P. Ketterling, G. S. Schroeder, R. A. Knudson and A. Tefferi, *Blood,* 2005, **105**, 973.
36. H. C. Hasselbalch, *Blood,* 2012, **119**, 3219.
37. A. Tefferi, R. Vaidya, D. Caramazza, C. Finke, T. Lasho and A. Pardanani, *J. Clin. Oncol.,* 2011, **29**, 1356.
38. F. Passamonti, F. Cervantes, A. M. Vannucchi, E. Morra, E. Rumi, A. Pereira, P. Guglielmelli, E. Pungolino, M. Caramella, M. Maffioli, C. Pascutto, M. Lazzarino, M. Cazzola and A. Tefferi, *Blood,* 2010, **115**, 1703.
39. N. Gangat, D. Caramazza, R. Vaidya, G. George, K. Begna, S. Schwager, D. Van Dyke, C. Hanson, W. Wu, A. Pardanani, F. Cervantes, F. Passamonti and A. Tefferi, *J. Clin. Oncol.,* 2011, **29**, 392.
40. V. Gupta, P. Hari and R. Hoffman, *Blood,* 2012, **120**, 1367.
41. *JAKAFI® (ruxolitinib) prescribing information*, Incyte Corporation, Wilmington, DE, June 2013.
42. J. G. Shi, X. Chen, R. F. McGee, R. R. Landman, T. Emm, Y. Lo, P. A. Scherle, N. G. Punwani, W. V. Williams and S. Yeleswaram, *J. Clin. Pharmacol.,* 2011, **51**, 1644.
43. A. D. Shilling, F. M. Nedza, T. Emm, S. Diamond, E. McKeever, N. Punwani, W. Williams, A. Arvanitis, L. G. Galya, M. Li, S. Shepard, J. Rodgers, T. Y. Yue and S. Yeleswaram, *Drug Metab. Dispos.,* 2010, **38**, 2023.
44. S. Verstovsek, H. Kantarjian, R. A. Mesa, A. D. Pardanani, J. Cortes-Franco, D. A. Thomas, Z. Estrov, J. S. Fridman, E. C. Bradley, S. Erickson-Viitanen, K. Vaddi, R. Levy and A. Tefferi, *N. Engl. J. Med.,* 2010, **363**, 1117.
45. S. Verstovsek, R. A. Mesa, J. Gotlib, R. S. Levy, V. Gupta, J. F. DiPersio, J. V. Catalano, M. Deininger, C. Miller, R. T. Silver, M. Talpaz, E. F. Winton, J. H. Harvey Jr, M. O. Arcasoy, E. Hexner, R. M. Lyons, R. Paquette, A. Raza, K. Vaddi, S. Erickson-Viitanen, I. L. Koumenis, W. Sun, V. Sandor and H. M. Kantarjian, *N. Engl. J. Med.,* 2012, **366**, 799.
46. C. Harrison, J. J. Kiladjian, H. K. Al-Ali, H. Gisslinger, R. Waltzman, V. Stalbovskaya, M. McQuitty, D. S. Hunter, R. Levy, L. Knoops, F. Cervantes, A. M. Vannucchi, T. Barbui and G. Barosi, *N. Engl. J. Med.,* 2012, **366**, 787.
47. S. Verstovsek, R. A. Mesa, J. Gotlib, R. S. Levy, V. Gupta, J. F. DiPersio, J. V. Catalano, M. W. N. Deininger, C. B. Miller, R. T. Silver, M. Talpaz, E. F. Winton, J. H. Harvey Jr, M. O. Arcasoy, E. O. Hexner, R. M. Lyons, R. Paquette, A. Raza, K. Vaddi, S. Erickson-Viitanen, W. Sun, V. A. Sandor and H. M. Kantarjian, *Blood,* 2012, **120**, 800.
48. F. Cervantes, J. J. Kiladjian, D. Niederwieser, A. Sirulnik, V. Stalbovskaya, M. McQuitty, D. S. Hunter, R. S. Levy, F. Passamonti, T. Barbui, G. Barosi,

H. Gisslinger, A. M. Vannucchi, L. Knoops and C. N. Harrison, *Blood,* 2012, **120**, 801.
49. J. Mascarenhas and R. Hoffman, *Blood,* 2013, **121**, 4832.
50. A. Kvasnicka, J. Thiele, C. Bueso-Ramos, K. Hou, J. E. Cortes, H. M. Kantarjian and S. Verstovsek, *J. Clin. Oncol.,* 2013, **31**, 7030.
51. H. M. Kvasnicka, J. Theile and S. Verstovsek, *18th Congress of European Hematology Association (EHA)*, Stockholm, Sweden, 2013, p. S591.
52. A. Deisseroth, E. Kaminskas, J. Grillo, W. Chen, H. Saber, H. L. Lu, M. D. Rothmann, S. Brar, J. Wang, C. Garnett, J. Bullock, L. B. Burke, A. Rahman, R. Sridhara, A. Farrell and R. Pazdur, *Clin. Cancer Res.,* 2012, **18**, 3212.

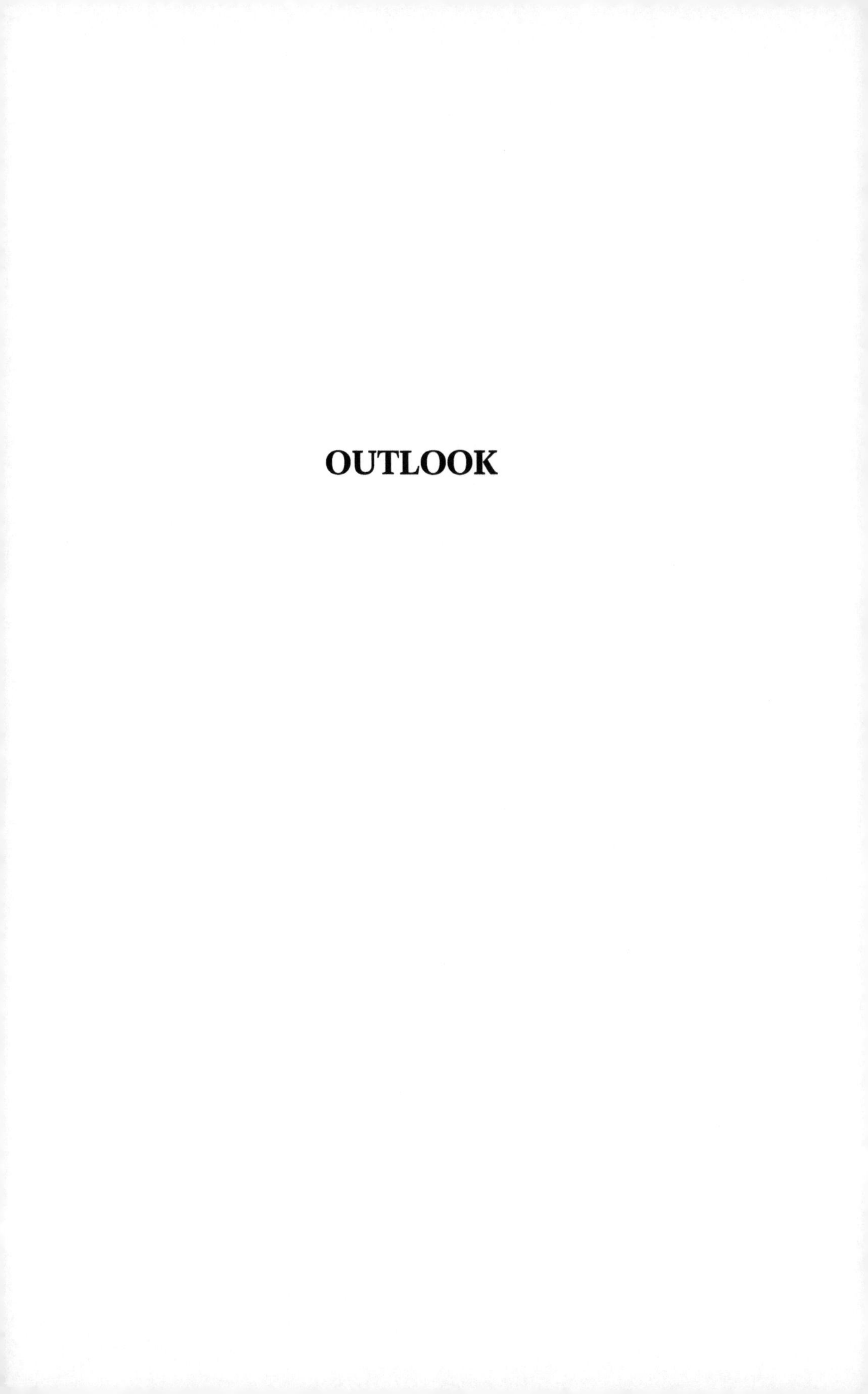

OUTLOOK

CHAPTER 16

Possible Solutions to Accelerate Access to Rare Disease Treatments

MARC DUNOYER

AstraZeneca PLC, 2 Kingdom Street, London W2 6BD, UK
E-mail: marc.dunoyer@astrazeneca.com

16.1 Expediting Rare Disease Research

The introduction of legislative incentives to promote rare disease research, along with changes to regulatory policy, has resulted in increases in the number or orphan drug designations. Despite this, the number of drugs reaching marketing approval across member states of the Organisation for Economic Co-operation and Development remains frustratingly flat. Research into rare diseases faces inherent challenges throughout clinical drug development and regulatory approval. Tailor-made regulatory and access solutions are needed to overcome these problems. Here I suggest areas for consideration that could have an immediate impact in facilitating regulatory approval and access to treatments for rare diseases. Rare diseases are often chronically debilitating, life-threatening or life-limiting. Many have a genetic basis affecting a predominantly paediatric population. With close to 7000 rare diseases identified, these conditions create a sizeable medical and social burden. As science and technologies evolve, in particular through advanced therapy medicinal products (ATMPs), safer and more effective options are emerging to better treat these diseases. However, efficient clinical

RSC Drug Discovery Series No. 38
Orphan Drugs and Rare Diseases
Edited by David C Pryde and Michael J Palmer

Published by the Royal Society of Chemistry, www.rsc.org

development of rare disease treatments is still impeded by an inadequate understanding of how genetics affect clinical manifestation of the disease. This leads to sometimes impractical regulatory expectations, and difficulty demonstrating the public health impact of complex new treatments. Proposals to expedite the approval of drugs for rare diseases, while continuing to meet the (correctly) stringent regulatory environment should help to increase the rate of new treatments reaching the market.

16.2 Improving Disease Understanding and the Research Infrastructure

16.2.1 Heterogeneous Clinical Conditions

The heterogeneity of rare diseases, both in terms of clinical phenotype and underlying pathophysiology, can have a significant impact on the development of effective treatments. For example, congenital ichthyosis (CI), a rare, scaling skin condition, has more than 30 known subtypes, which may involve different gene mutations and biological pathways, often presenting with overlapping clinical phenotypes. It is difficult to envisage an industrial sponsor being able to develop, in isolation, the discovery infrastructure (screening, *in vitro* and *in vivo* model development) that could identify a biological target for each subtype, select a candidate molecule, and progress the molecule to clinical development. Therefore, natural history studies should be encouraged by regulators, and their results published by sponsors to help to characterise underlying disease pathophysiology.

Rare diseases have highly variable presentations (even between siblings) including disease burden, clinical symptoms, age of onset and rate of disease progression. Consequently, it is highly unlikely that all the multiple relevant disease subtypes can be studied prior to the first registration of a new therapeutic agent. Post-marketing studies in heterogeneous populations are therefore important to continue to learn about the wide application and efficacy of a new drug. Patient registries and post-approval studies should also play a more significant role, alongside sponsored controlled clinical trials, to accelerate access to new rare disease treatments.

As the power of genetic diagnoses increases and our understanding of disease pathologies improves, a pharmacogenomics approach can be used to expand clinical results from a specific genetic sub-population to a broader population. An illustration of how this could be used is with oligonucleotide-based medicines, such as in Duchenne's muscular dystrophy. GSK, in partnership with Prosensa, is developing several exon-skipping agents, exquisitely targeted to sub-cohorts of Duchenne patients. Each programme uses the same fundamental technology; a short oligonucleotide that binds to the dystrophin mRNA and introduces a splicing defect to correct the pre-existing mutant frame shift. Pre-selection of patients known to have the targeted genetic defect improves clinical response rates and reduces the size of clinical trials. It would seem appropriate that when developing new treatments in the same disease, but with alternative sequences to correct alternative frame-shift mutations, data from the lead programme should be considered as supporting evidence for safety and efficacy.

16.2.2 Trial Design Complexity

The scarcity of patients with rare diseases causes practical difficulties in turning a discovery finding into a robust clinical development plan. In many cases, it is unfeasible to conduct the type of formal, statistically-powered, randomised, controlled development plans that encompass demonstration of appropriate posology.

Flexible and innovative trial designs (including adaptive study designs) should be encouraged, as this circumvents some of the difficulties of conducting double-blind RCT placebo studies. Wherever possible, historical data and meta-analyses should also be taken into account in support of efficacy claims. Given the associated problems with recruitment in clinical trials, it may be more appropriate to demonstrate proof of concept using a single adaptive trial, as well as a pivotal registration study.

16.3 Predictive Experimental Models

16.3.1 Biomarkers and Clinical Outcomes

Conventional clinical end points can be difficult to assess in the context of rare diseases. The adoption of biomarkers or surrogate markers of clinical meaningfulness could be a viable alternative that enable faster, more efficient clinical trials. It is not practical, in terms of cost or rate of decision making, to rely on disease progression as a clinical end point in these diseases. Sensible application of biomarkers or pharmacodynamic markers can support reasonable dose selection and, in some cases, early registration. For example in Fabry disease, urinary GL-3 levels, which are well correlated with renal function, can be measured to confirm efficacy of novel drugs. Stronger consideration should be given to the use of surrogate markers as primary (or co-primary) end points in pivotal clinical trials of rare diseases where disease progression is slow and definitive proof of efficacy requires prolonged monitoring of patients. Although validation of surrogate end points is challenging in small disease populations, a concerted effort to develop and support their utilisation by industry, academia and regulatory agencies could make this a feasible option. Studies using more traditional clinical end points could then be performed as post-approval commitments.

16.3.2 Modelling

In order to study a meaningful number of patients with a rare disease, clinical assessment usually requires a global development programme. An alternative approach involves computational modelling and pharmacokinetic (PK) simulation, which could be useful to predict outcomes of drug–drug interactions, optimal dosing regimens and evidence of safety and efficacy. Of particular interest in the context of many rare diseases is the prediction of paediatric dosing of an orphan drug. Simulations using non-rare disease populations (*e.g.* healthy volunteers or patients with another illness) can augment the extrapolative ability of the model and provide dosing recommendations for sub-populations not studied. These quantitative prediction systems could provide supporting data in orphan drug applications.

16.4 Regulatory and Access Pathway

The introduction of incentivised legislation for orphan drugs has been highly successful at stimulating patient access in this area, and further steps by regulatory agencies could build upon this momentum. Simplification of drug development requirements for rare diseases, while maintaining rigorous standards of care and an evidence-based approach, could have a big impact on this field. Currently, conducting different development programmes to respond to the differing requirements of separate regulatory agencies can be detrimental to the access to new medicines. Regulators around the world need to harmonise the interpretation and application of technical guidelines and requirements for product registration. For example the European Medicines Agency (EMA) and US Food and Drug Administration (FDA) already share the same application form for orphan drug designation, which is helpful. Expanding on this, regulatory agencies should recognise and utilise the assessment performed by other agencies in order to facilitate and accelerate their own review processes.

Given the high medical need of patients with rare diseases, regulatory agencies should endeavour to grant accelerated approval or conditional approval of rare diseases drugs where possible, and industry should commit to rapid completion of post-marketing commitments. Although accelerated approval represents a greater workload for the agency concerned, the possible benefits to the patients are clear. Initiatives such as pre-approval patient early access schemes should be positively encouraged and supported; indeed innovative systems are already in place in certain countries (Italy 648, France ATU, USA treatment IND protocol). Widely establishing such early access schemes and facilitating cross-border healthcare treatment for diseases, for example where the delivery of breakthrough treatments is only available in very specialised centres, would improve the equity for rare diseases patients to access innovative therapies.

16.5 Conclusion

As research into rare diseases continues to advance, the system in place to bring therapeutics to the market must evolve to best serve this population. Rare diseases often represent uncharted territory; therefore there is a greater need for frequent dialogue between industry, regulators and patient organisations to generate a less risk-averse approach to clinical development and patient access that can be tailored to individual rare diseases. Political backing will also be needed to support the introduction of regulatory solutions leading to faster access to rare diseases treatments. Tackling these key areas is vital in the optimisation of rare disease drug development, and could play an important role in accelerating access to treatments that manage these chronic degenerative, debilitating diseases.

Subject Index